1150

INTRODUCTION TO

Marine Biology

Bayard H. McConnaughey, Ph.D.

Associate Professor of Biology,
Department of Biology,
University of Oregon,
Eugene, Oregon

With 268 illustrations

Contribution from the
Oregon Institute of Marine Biology, Charleston, Oregon;
The Oregon Marine Biological Society;
The Institut Pertanian, Bogor, Indonesia;
and the Lembaga Penilitian Laut, Djakarta, Indonesia

The C. V. Mosby Company

Saint Louis 1970

To

William A. Hilton

long-time professor of zoology at Pomona College, California.
Professor Hilton for many years single-handedly ran a small summer
marine station in the old church at Laguna Beach,
in the days when Laguna Beach was still a tiny, beautiful artists' colony.
Here and at Pomona College he communicated to generations of students
his boundless curiosity about the world of which he was a part, his infectious good
humor, and his respect and enthusiasm for life and for all living things.

*T*he search for truth is in one way hard and in another easy,
for it is evident that no one can master it fully nor miss it wholly.
But each adds a little to our knowledge of nature, and from
all the facts assembled there arises a certain grandeur.

Aristotle

Inscribed in Greek on the National Academy
of Science Building, Washington, D. C.

Preface

Marine biology embraces many diverse disciplines, several of which are currently undergoing revolutionary changes in concepts and methods.

To avoid presenting merely brief unsatisfactory excerpts from such scattered subjects as planktonology, botany, invertebrate and vertebrate zoology, microbiology, submarine geology, oceanography, ornithology, ecology, physiology, chemistry, ethology, and paleontology, it has been necessary to select a few broad unifying concepts or points of view and to build the book around these.

Such a selection and limitation entails the elimination of much that is equally valid. There are other points of view from which the subject could have been presented.

This book is organized around the general theme of the production and consumption of food, the special directions and limitations that marine environments impose, and the ways in which various organisms have become adapted for exploiting marine resources and interacting with each other.

A second theme emerges from this. The ecosystem itself, far from being a static equilibrium in which the effects brought about by various organisms exactly counterbalance each other, is in a state of continuous dynamic change and evolution. Living organisms are responsible for many of these changes. Others are brought about by geological or even by cosmic agencies.

In Part VIII an example of each sort is examined briefly. Considered first is the agency of living organisms in bringing about a major change in the composition of the earth's atmosphere and creating conditions for the evolution of aerobic metabolism. Next considered are the phenomena of sea floor spreading and continental drift and some of the effects these have had in altering the earth's climate and in the evolution of the earth's fauna and flora.

At the present moment of the earth's history, man is exerting effects upon the ecosystem more profound and more rapid than those brought about by any other single species throughout the entire history of life. The fate of the ecosystem as we know it today—of man and of all the animals and plants that populate the earth and the seas—may well be contingent upon our ability to understand and control these changes, a problem best thought of in terms of *self*-control rather than in terms of controlling nature.

Any text of this kind owes its existence to the labors of countless students of science whose careful researches and keen insights have produced the conceptual and factual framework upon which it is based. Some of these scientists and their works are listed in the bibliography. They are responsible for any correspondence between the text and the reality it attempts to elucidate. My role as the writer has been confined largely to the production of whatever errors and mistakes still lurk in these pages.

For the illustrations I am chiefly indebted to the work of two top-notch photographers, Mr. Milo W. Williams and Dr. Richard H. Chesher, and to two fine artists, Kathryn Torvik and Mrs. Lynn Rudy.

Others who have kindly provided photographs are Dr. R. J. M. Riedl, Gnathostomulid; Mrs. Helen Finlayson, intertidal zonation on the rocky coast at Marblehead Neck, Massachusetts; Dr. Kenneth Mesolella, Pleistocene coral reefs on Barbados Island; Dr. D. M. Ross, the swimming sea anemone, *Actinostola* sp.; Wide World Photos, Inc., citizens attempting to clean oil from the beach at Santa Barbara, California.

Mr. James Houk of the Moss Landing Marine Laboratories, Monterey, California, kindly provided the photographs of *Gregarina valettei* and of the dicyemid mesozoan (Fig. 5-2, *B* and *C*). Fig. 16-1 ("animal" gardens near St. Johns, The Virgin Islands) was provided by the U. S. Forest Service through Mr. M. W. Williams.

Fig. 14-1 (mossy chiton), one of the Oregon Institute of Marine Biology photographs, was taken by Dr. Ralph Buchsbaum.

The photographs credited to Oregon Institute of Marine Biology are from sets taken from time to time by students and staff of the Institute at Charleston, Oregon, and from the Biology Department of University of Oregon, Eugene, Oregon. Since these were incorporated into one large set, it is no longer possible to be sure who took which photographs in order to credit the proper individuals.

Bayard H. McConnaughey

Contents

PART I

MARINE ENVIRONMENT
General ecological considerations

CHAPTER 1

The open ocean

The oceans of the world cover approximately 361,000,000 km.², about 71% of the earth's surface, to an average depth of about 4 km. The deepest parts are over 10 km. deep, plunging farther below sea level than our highest mountains rise above it. The seas are the cradle of life. They are so vast and complex, and the assemblages of living organisms in them are so varied, that no book can give a complete view of even a few of the many facets of marine science.

Living things are of primary importance in determining the character of the earth's surface, atmosphere, and waters. Those portions of the earth's crust, together with its waters and atmosphere, which are permeated with living organisms or their products are collectively termed the **biosphere.** The biosphere represents a vast seat of continual physical and chemical change enveloping the earth. The total amount of living matter present at any time is the **biomass.** This term is also used in a slightly different sense, as when one speaks of the biomass produced by a particular area or region, or by a particular species or group of species over a given period of time.

All living things must feed, respire, synthesize new organic molecules, and eliminate unwanted metabolic products. There is a continual flow of materials into and out of every living thing. Furthermore, the organisms themselves are undergoing continual replacement and evolutionary change. The animals and plants of one geological period may be quite different from those of another, and they may exert different effects upon their environment and upon each other.

The manner in which an organism feeds and the quantity and quality of foods available to it determine in large measure not only the structure and mode of life of the organism and its relations with other organisms but also the total biomass of that species which can be produced.

The entire interacting system of organisms, together with the environmental factors with which they interact, is termed

3

the **ecosystem.** With regard to their major roles in the ecosystem, that is, the overall result of their life and activities, organisms can be divided into three categories: **producers, consumers,** and **reducers.**

The producers are those which synthesize new organic matter from inorganic substrates such as carbon dioxide, water, and soluble salts. These are the photosynthetic forms—the plants. All other living things depend directly or indirectly upon the plants for their food and oxygen.

The consumers are organisms that ingest or absorb organic matter directly, assimilate what they need from it, and eliminate the rest. This group includes all animal life. Animals that feed directly on plants, the herbivores, are sometimes termed primary consumers, whereas the carnivores, scavengers, and parasites, which subsist on other animals, are termed secondary consumers. Many animals consume a mixed diet.

The reducers are mostly microorganisms such as bacteria, molds, and actinomycetes, which cannot carry on photosynthesis or ingest particulate organic matter. They liberate exoenzymes into their environment to break down complex organic molecules into smaller units that can be absorbed and assimilated. To the extent that they assimilate and build into their own bodies the organic matter upon which they live, they are, of course, consumers too, but the overall result of their activity is to break down the vast reservoirs of organic matter represented by plant and animal bodies and residues, releasing their constituents in the form of simple ions and gases available once more to the plants. This process is called the **mineralization of organic matter.** It is represented by the various processes of decomposition such as fermentation, decay, and putrefaction.

Through the activities of all these groups of organisms a circulation of the chemicals necessary for life is maintained. This enables life to flourish abundantly for millions of years in spite of the relatively limited amounts of necessary materials available on the earth's surface.

Since the efficiency of conversion of organic matter from one organism to another is rather low, it follows that the total attainable biomass of consumers and reducers together must be less than that of the producers. There is, of necessity, a great deal more plant life than animal life in the world. Likewise the biomass of herbivorous animals is far greater than that of carnivorous animals. This is conveniently oversimplified in the "10% rule," which states that an animal assimilates only 10% or less of the food that it eats—the rest being used for energy production in maintaining its metabolism, or eliminated. In other words, in any food chain proceeding from plants to herbivorous animals to carnivorous animals, there is a 90% reduction in total possible biomass attainable at each step. The farther removed from the plant base of the food chain, and the more specialized or restricted the diet of an animal, the smaller will be the total biomass of that species of animal produced.

Producers, consumers, and reducers comprise a balanced interacting system. None could exist long without the others.

On land, plants are the largest, most numerous, most conspicuous organisms to be found, often growing in such thick masses as to form nearly impenetrable forests and jungles. Since their food is the carbon dioxide in the air around them, and the water and dissolved minerals in the earth, the most efficient manner for them to exist is as sessile organisms with two great absorptive surfaces—the photosynthetic surface, or leaves, which spread out in the air to receive light and carbon dioxide, and the roots and root hairs, which obtain water and dissolved minerals from the soil. Supporting and conducting tissues are required to hold the system up and to bring nutrients to all living cells. Where the soil is damp enough and rich enough in minerals to support an abundant plant growth, their competition is chiefly for light and space. Plants under such conditions tend to grow crowded together, evolving larger and larger forms and overtopping or growing upon each other to reach the light until they reach the limits of size that can be efficiently maintained under the prevailing conditions.

In sharp contrast, when one goes out to sea he might well exclaim: "Where are the plants? Where are the producers in this system?" No obvious plants are to be seen except for the seaweed, marine grasses, and mangroves along the fringes of the continents and islands, or in exceptional areas such as the Sargasso Sea, masses of small floating seaweed. These are so small in amount as to be totally unable to account for the great productivity of the oceans in fish and invertebrate animals, though they are, of course, important in the shoreline ecology.

The answer to this enigma, in fact our awareness that it is an enigma, has come only in the last century. If a finely meshed net capable of retaining microscopic objects is pulled through the upper layers of seawater, a mixture of microscopic plant cells and small animals is obtained. This is the **plankton.** Plankton is a term applied to the communities of floating, drifting organisms carried about primarily by movements of the water rather than by their own swimming activity. Those larger animals that control the direction and speed of their own movements, rather than drifting with the water, constitute the **nekton.** Organisms that live on or in the bottom constitute the

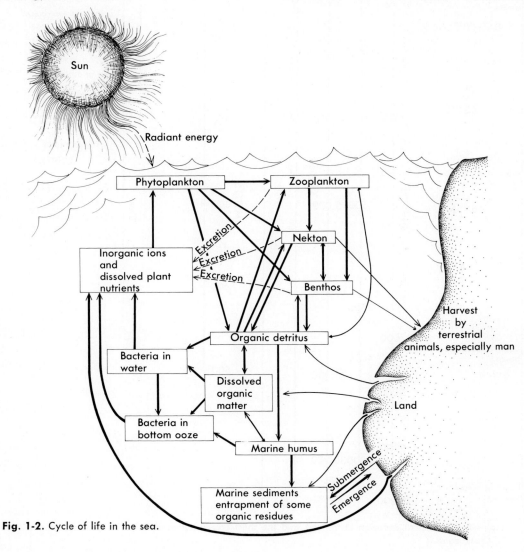

Fig. 1-2. Cycle of life in the sea.

benthos. Phytoplankton is the photosynthetic, or plant, portion of the plankton. Animals constitute the **zooplankton,** the nekton, and most of the benthos. **Demersal** animals are swimming forms which spend much of their time at or near the bottom. They may be considered both as part of the nekton and the benthos. Typical demersal animals include such forms as the flatfishes, octopuses, many shrimps, etc.

Minute animals found in the surface film of the water are termed the **neuston.**

If one filters seawater through a very fine filter, or concentrates it with a centrifuge, he finds also a quantity of even smaller organisms—flagellates, blue-green algal cells, very small diatoms, and bacteria that are too small to be retained by a plankton net. Organisms in this size range have been collectively termed the **nannoplankton.**

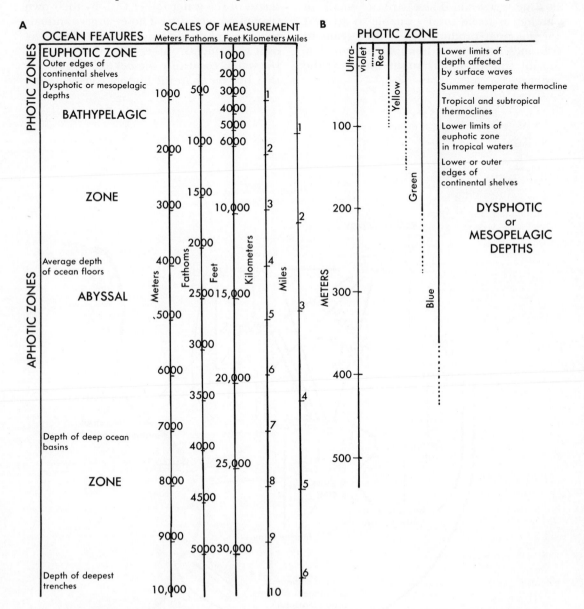

Fig. 1-3. A, Vertical zonation in the sea and the most commonly used units for denoting depth. **B,** Photic zone.

Nannoplankton is, then, simply a size designation that includes parts of both phytoplankton and zooplankton. Strictly speaking, the term nannoplankton is used for organisms ranging from 5 to 60 μ in size. Bacteria and other organisms smaller than 5 μ in length are termed **ultraplankton.**

Small particles of organic matter derived from the disintegration of the bodies, castings, or feces of organisms, together with inorganic particles of similar size, suspended in the seawater or sinking to the bottom, constitute **detritus** or **seston.** Large organic molecules, macromolecules, or aggregations of organic molecules of colloidal size and minute particles of detritus dissolved or suspended in the water comprise the marine **leptopel.** The surfaces of detrital particles in the seawater, as well as all larger surfaces, become coated with dissolved, or colloidal, organic molecules. Some surface-active organic molecules are continually being swept to the sea surface by small bubbles and concentrated, along with other particles and small organisms, in small zones of convergence set up by Langmuir circulation (p. 11). Here they are polymerized into larger aggregates or adsorbed to the surfaces of larger particles, becoming then part of the detritus. Adsorption of leptopel to the surface of larger particles makes it available as food to filter-feeding animals. Fox and associates (1952) have made studies of the amount and distribution of the marine leptopel, and of its significance for the nutrition of marine animals. Numerous animals, known as **detritus feeders,** ingest quantities of this material and digest the organic constituents along with small organisms found in it.

Phytoplankton occurs in vast clouds in the upper 100 meters or so of the waters of all oceans, except portions of the Arctic and Antarctic oceans permanently covered by ice. In the aggregate it comprises a mass of plant life even greater than that found on land.

Because of the necessity of light for photosynthesis, this plant life is restricted to the upper illuminated layers of water—the so-called **euphotic zone.** For each species of plant cell there is a certain depth below which it cannot carry on photosynthesis faster than it metabolizes organic molecules in its body, that is, below a certain depth, **anabolism,** the synthesis of new organic molecules and structure, just balances **catabolism,** the breakdown of these molecules. Under these conditions the species cannot grow or reproduce, although it may continue to live. This depth is known as the **compensation depth.** The precise depth for each species varies according to the turbidity of the water, the amount and angle of incident light, and all factors that affect the rates of photosynthesis and metabolism. Below the compensation depth the respiratory rate will exceed the rate of photosynthesis, and the plant cell will eventually die. Some photosynthetic forms apparently can also live heterotrophically. Many coccolithophores, some dinoflagellates, and others are regularly found at levels well below, as well as above, the compensation depth. Those in the deep waters must be enjoying heterotrophic nutrition.

FLOTATION AND VERTICAL DISTRIBUTION OF PHYTOPLANKTON

Protoplasm is, in general, slightly heavier than water and tends to sink. In a large organism there is far less surface in proportion to volume, and the tendency to sink is accentuated. Minute organisms, with a great deal of surface in proportion to volume, tend to sink far more slowly, especially if they are shaped in such a way as to increase the relative amount of surface and to slow down any sinking that might occur by giving them a spiral or oblique path. Some are equipped with flotation bristles, and many have vacuoles containing less dense fluid or oil droplets. Some of them, for example, the dinoflagellates, can swim actively.

The necessity of remaining in the euphotic zone in order to grow and reproduce is one of the factors that helps to account for the small size and odd shapes of phytoplankton organisms as compared with plants on land.

The absorption of essential nutrients is of equal importance. In addition to helping the cell stay in the euphotic zone, a large surface-to-volume ratio enables the plant cell to absorb the very dilute nutrients from the surrounding water more efficiently, thus compensating for the fact that the nutrients are so dilute. However, when the phytoplankton is abundant, the nutrient supply in the upper stratum of water becomes impoverished. Some sinking then helps the cells to obtain nutrients. The sinking itself increases the depth of the euphotic zone by allowing more light penetration. In summer, maximum concentrations of chlorophyll are found near the bottom of the euphotic zone. At this time of year vertical diffusion is minimal and the quantity of surface nutrients is low. In winter, diffusion rates are higher. Growth, although slower, is greatest at the surface, and there is a linear decrease in chlorophyll with depth.

Increasing the temperature of the surface water decreases its density and increases the tendency of phytoplankton cells to sink. This makes the water more transparent, and the euphotic zone becomes deeper. The interplay of these factors helps account for the wide variation of shapes for chlorophyll curves observed in the ocean. Although many maxima are observed above the compensation depth, others are found just at or below it. Density differences in the water column are insufficient to fully explain the observed distributions.

The correspondence of chlorophyll maxima with an increase of nutrients in deeper strata suggests that a decreased sinking rate occurs as cells enter the darker, nutrient-rich water. Steele and Yentsch (1960) showed in cultures that sinking is increased with cell age but can be slowed down by the use of nutrient-rich water or by darkening the culture. Thus it appears that the general physiological condition of a cell is closely related to its buoyancy, as Gross and Zeuthen (1948) suggested. This is an even more important factor than the changes in water density caused by changes in temperature for explaining differences in seasonal vertical distribution of phytoplankton.

In general both the density changes and the changes in the bouyancy of cells with age, light, and nutrient environment tend to work in the same direction and to reinforce each other in producing the observed seasonal differences in vertical distribution of phytoplankton in the water.

In early spring, when nutrients are abundant in the surface waters, there is rapid growth near the top, where cells tend to accumulate because of good nutrition and a low sinking rate. Later, in summer, the nutrients become depleted near the surface, the sinking rate increases, and cells begin to accumulate near the bottom of the euphotic zone, where better nutrition and darkness again revives them and retards their sinking. Also, at lower light intensities the rate of synthesis of chlorophyll increases.

According to Steemann-Nielsen (1954), maximum chlorophyll content of the euphotic zone is about 300 mg./m.^{2}, about one fourth to one fifth of the highest terrestrial concentrations. In water, the quantity of organic matter produced is also controlled by cell density. When concentrations of cells reach the point where the cells shade each other, growth becomes self-limiting.

At the very surface, the light may often be too intense for many organisms, and the ultraviolet rays might injure them. The greatest bulk of phytoplankton occurs a little below the surface, from about 5 to 70 meters in depth. In northern and southern latitudes the euphotic zone tends to be shallower than in the tropics because of less direct light, increased reflection resulting from the angle of incidence and greater wave action, and the greater turbidity of the water. In turbid coastal waters the euphotic zone may at times be as shallow as 5 meters or even less.

PHYTOPLANKTON BLOOMS

Since phytoplankton organisms are among the most rapidly growing elements of the ocean populations, some of them dividing several times a day under favorable conditions, the populations are very responsive to changes in the environment. This fact is displayed particularly well in the great spring outbursts of phytoplankton in boreal and polar regions. The cold of winter increases the density of the surface water, causing it to sink, which results in a turnover of waters from different depths. Some of the phytoplankton is carried below its compensation depth, which is at a high level in the winter because of the decreased illumination. Because of the decreased amount of phytoplankton in the euphotic zone, the decreased light, and the lower temperature, growth in winter is much slower, and the surface waters become enriched with dissolved plant nutrients from below. Better growing conditions in the spring cause a dramatic outburst of phytoplankton growth. Such outbursts are termed blooms. They may occur at any time and place where local conditions become more favorable for phytoplankton growth. Often there are several blooms between early spring and late fall. In some coastal situations special local conditions commonly produce winter blooms as well.

During a bloom, the quantity of phytoplankton may double daily, producing great clouds of billions of cells, until impoverishment of plant nutrients and the increased grazing by more slowly growing zooplankton populations slow down the rate of increase, bringing it into equilibrium or causing a sharp decline in total phytoplankton of the area and replacing it largely with zooplankton.

When a sudden dramatic bloom occurs in a given area, it tends to be dominated by a rather small number of species that happen to be most favored by the local conditions at the time, or that happen to have a head start numerically when conditions favor a bloom. When a mixed population is subjected to a sudden increase in the carrying capacity of the environment, such as an increase in available nutrients together with better growing conditions in the spring, different species take full advantage of their respective rates of reproduction. These rates vary considerably from one species to another, and the fastest forge ahead. Diversity—in terms of numbers of species represented in roughly the same order of magnitude—decreases, and one or a few species become dominant. Some blooms are so dominated by one form as to

be called **single-species blooms,** but more often several species are present in great numbers. Later, when the carrying capacity of the environment is approached, ecological succession sets in, and diversity increases until a somewhat more stable equilibrium is reached.

Where two great water masses meet and mix, the results may be quite startlingly different, depending on the character of the water masses. If each water mass supplies some complementary nutrient that was depleted in the other one, bloom conditions may prevail at the boundary region, resulting in a marked increase in productivity but a decrease in diversity. If, on the other hand, they do not furnish complementary nutrients, the populations may simply mix at the boundary, resulting in increased diversity.

It is important to emphasize the dynamic and changing character of the plankton populations and the fact that their distributions in the water are not uniform. The ever-changing, drifting aggregations of planktonic organisms may well be compared to clouds in the atmosphere. Like the clouds, they occur at different levels, drift with the movements of their medium, and change in size, shape, depth, configuration, and composition; yet through all this they tend to display certain seasonal and regional regularities.

In any given place, marked changes in the plankton occur from time to time because of the varying seasonal conditions favoring different species in turn, because of interactions within the population itself, such as changes in the level of available nutrients and effects of grazing, and because water movements may bring plankton of a different character to the area and carry away those which were there.

All these factors, plus the difficulties of obtaining truly representative samples from the various depths and of knowing whether the samples obtained are representative, make the study of plankton both difficult and fascinating.

If plankton samples are taken regularly at any given station over a period of years, some of the broad trends and seasonal regularities become evident, and a sort of plankton calendar for the area can be made, showing the principal seasonal changes and the species most likely to be dominant at different times of the year.

The use of the continuous plankton recorder and the continuous, repeating water sampler–thermograph unit, pioneered by A. C. Hardy, that is towed behind commercial vessels on regular runs as well as used with research ships—opened up a new dimension in marine research by making possible the regular plotting, on a wide scale, of marine conditions and the movements of water masses as reflected by the plankton organisms they contain. This new procedure is making possible the prediction of marine conditions on a basis similar to the prediction of weather conditions based on widespread meteorological data. Since the movements of water masses are slower than those of the atmosphere, somewhat longer-range predictions can be made with the same degree of accuracy. This is of great importance to commercial fisheries. The continuous plankton sampler also helps greatly in assessing the general state of a fishery and the size of fish stocks, through its quantitative records of the numbers of fish eggs as well as the quantities and types of fish food in the water. (See Hardy, 1926, 1936, 1939, 1958.)

IMPORTANCE OF CIRCULATION

A vast cloud of phytoplankton may exhaust the dissolved nutrients and bring about impoverishment of the surface layers of water if the nutrients are not replaced. Since plant and animal bodies, feces, and detritus tend to sink below the euphotic zone into deeper water, or even to the ocean floor before mineralization by bacteria has been completed, the deep waters are richer in dissolved plant nutrients than are the surface waters. Below the euphotic zone there is no plant growth to utilize these nutrients.

All the dissolved oxygen in seawater is either dissolved directly at the air-water interface or evolved by photosynthesis in the euphotic zone. Since both respiration of deep-sea animals and bacterial activity diminish the free oxygen, the deeper waters would quickly become stagnant and virtually lifeless, except for anaerobic bacteria, were it not for continual vertical circulation of the waters. Vertical circulation, which brings dissolved plant nutrients to the euphotic zone and oxygen to the depths, is of vital importance in maintaining abundant ocean life. Areas where, for any reason, extensive mixing of surface water with water from deeper strata occurs either periodically or steadily, or where waters from below the euphotic zone are brought to the surface, tend to be areas of high productivity.

Offshore winds that blow the surface waters away from the shore tend to cause **upwelling** of subsurface waters near the shore. Likewise the major ocean currents, when they sweep over the edges of continental shelves parallel to the general curve of the shore, tend to create upwellings along their shoreline sides, as well as eddies and general turbulence over the shelf.

ZONES OF CONVERGENCE AND DIVERGENCE

Where currents flowing at different rates or water masses of different temperatures or salinities meet, large zones of convergence may form at the place where the surface waters from either side meet and are carried below the surface. Some of these zones are sufficiently constant or recurrent to form permanent or semipermanent regional features. Particularly well known are the Antarctic and Antiboreal (Subtropical) convergences of the Southern Hemisphere. A zone of convergence tends to concentrate surface-living forms along its length and is often marked by great numbers of predators attracted to and taking advantage of this concentration. Areas of divergence, where subsurface water is coming to the surface, tend to show fewer surface-living forms and are often marked by the presence, at or near the surface, of organisms usually found at greater depths. Divergences tend to be broad and less well marked. Convergences are often long, narrow, and sharply delimited.

Ocean currents flowing side by side, especially when the water masses of each are of different densities or the currents are of different speed, may form dramatic boundaries, known as **rips.** William Beebe described one observed by the Arcturus in 1926 in the equatorial Pacific, where such a rip extended from horizon to horizon, as a band marked by foam and whitecaps, about 60 feet wide. The Arcturus followed it more than 100 miles. Both currents were warm and flowing in the same direction. The water on the south side of the rip was several degrees cooler than that on the north side, and it was noticeably darker, rougher, and faster flowing.

The water in the rip contained much more life than that on either side of it, a distance of as much as 10 yards to either side making a great difference. The rip was soupy, with smaller plankton and young fish concentrated from both sides by downwelling. Predators of all kinds—sharks, dolphins, sea turtles, large fish and great numbers of sea birds—were gathered along the rip in vast numbers. There were also a number of barnacle-covered logs of palm, coconut, bamboo, and trumpet wood that were riddled with shipworm. The differences in the water masses and the boundary between them extended down to a depth of at least 1000 feet.

Smaller zones of convergence and divergence are set up by a variety of causes, such as changes in temperature, rate of evaporation, and winds. Perhaps most widespread and significant is the so-called **Langmuir circulation,** in which winds blowing steadily over the water surface set up small parallel zones of alternating divergence and convergence. Dissolved and particulate organic matter, together with planktonic organisms, are concentrated in the lines of convergence, facilitating the feeding of small animals. The Langmuir circulation also causes the concentration and polymerization of surface-active dissolved organic substances into larger micelles and facilitates their absorption onto the surfaces of larger particles, thus making some of the dissolved organic matter indirectly available to detritus and filter feeders. (See Fox and co-workers, 1952, on marine leptopel.)

OXYGEN MINIMUM LAYER

Most of the cold, oxygen-rich water that reaches the ocean bottom comes from the sinking of cold waters near the polar regions. This water spreads out north and

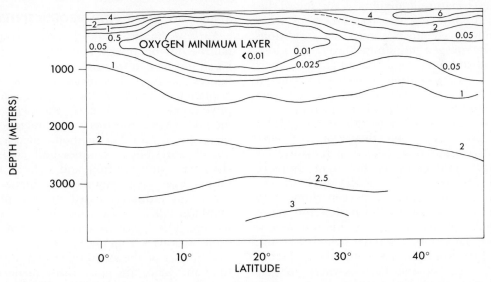

Fig. 1-4. Oxygen minimum layer in the northern Pacific. (Modified from Sverdrup, H. U., M. W. Johnson, and R. H. Fleming. 1946. The oceans: their physics, chemistry, and general biology. Prentice-Hall, Inc., Englewood Cliffs, N. J.)

south over the sea floors, displacing the water masses already there. The result is that for extensive areas in the middle latitudes, oxygen is renewed near the surface and at the sea floor faster than at intermediate depths. This causes the formation of an oxygen minimum layer, which is made up of the water that has been below the surface for the longest time. This layer occurs at about the level of the tropical thermocline, where the water is increasing rapidly in density with depth. A temporary accumulation of organic detritus at this level undergoing bacterial decomposition may accentuate the biological utilization of oxygen, reducing oxygen tension to especially low levels. Oxygen values often reach very low levels in such a layer—sufficiently low to exclude most animals and to form a barrier to their movement through it. Yet some plankton animals in apparently good condition are found even in these strata.

Childress (1968) studied the mysid *Gnathophausia ingens* in the eastern Pacific. In this species very young, small individuals are found above the oxygen minimum layer, most intermediate-sized individuals, 35 to 125 mm. long and weighing 0.3 to 15 gm., are found in the oxygen minimum layer, and most of the very largest, mature individuals are found below it. Studying the respiration of these mysids in chambers simulating conditions in the oxygen minimum layer, Childress found that they were able to regulate their oxygen consumption remarkably well in nearly anaerobic water, with oxygen pressures as low as 0.14 ml. oxygen per liter (most animals begin to fail at around 1.2 ml. per liter). Thus they can live aerobically in waters with oxygen pressures far below those necessary for most animals, and can contribute significantly to the biological depletion of oxygen thought to be responsible for the formation and maintenance of the oxygen minimum layers. Their failure to regulate below 0.14 ml. per liter would also explain their absence from areas of the Pacific where oxygen values in the oxygen minimum layer fall below this value. Both *Gnathophausia* and *Euphausia mucronata* found in the oxygen minimum layer from the same regions are able to consume the oxygen in nearly anaerobic water until the oxygen reaches undetectably low levels. The vertical distribution of the species may perhaps be accounted for by pelagic, low-density eggs that rise into the upper waters before hatching. As the young mysids grow and become capable of better regulation, they seek the protection offered by the deeper, oxygen-low strata of water. Perhaps upon attaining maturity they require more oxygen, become immobilized, and sink until they reach somewhat better oxygenated water below the oxygen minimum layer, where they are revived. Childress found that when subjected to oxygen pressures below 0.14 ml. per liter, these mysids were soon immobilized, but that they would survive for several hours and revive when placed in water with slightly greater oxygen pressure.

In some regions occasional upwellings of water from the oxygen minimum zone into the surface layers may cause mass mortalities of fish and other animals in the water above. When such an event occurs the decomposition of the excess organic matter may continue to deplete the oxygen for a time and slow down the recovery of the area.

STANDING CROP AND PRODUCTIVITY

Early estimates of the standing crop of marine phytoplankton and of productivity, based on net tows, gave misleading low figures because they neglected the nannoplankton. Even for the larger species of phytoplankton retained by the nets, such methods gave erroneous low estimates. Estimates based on other methods, such as the use of water-bottle samples and counting chambers, and dilution and culture techniques have given significantly higher figures than estimates based on net phytoplankton taken at the same time and same station. The factor of increase is not constant but varies from one species to another.

Estimates of the standing crop at a given time and place are more easily made than are estimates of productivity. The standing crop can be estimated from direct counts or measurement of volumes or weights of samples taken from the seawater. Indirect methods such as extracting and measuring the amount of chlorophyll from the plankton concentrated from a known volume of seawater and estimating from this the biomass of the phytoplankton that produced it are also used.

PRIMARY PRODUCTION

To estimate productivity it is necessary to measure directly the amount of new organic matter formed in a given time in a sample of seawater being held under the same conditions that obtain in the water mass for which productivity is being estimated. Alternatively, the productivity can be indirectly estimated from measurements of the rate of oxygen evolution, carbon dioxide incorporation, changes in pH, or some other physiological parameter correlated with the synthesis of new organic matter.

The rate of productivity in a given water mass changes with the diurnal and seasonal cycles, with changes in temperature, illumination, and depth, with the relative amounts of phytoplankton and zooplankton present in the water, and with the species present. There are so many variables and different sorts of factors that affect organic production, and production is so different from one area to another and from one time to another in the same area, that it is difficult to know to what degree the figures obtained can be extrapolated or generalized to give overall estimates of organic productivity in the oceans. It is also difficult to be sure to what extent the techniques used to measure production may have affected the rate of production in the samples. In addition to field and laboratory measurements, which supply most of the information, mathematical models of seasonal and local changes in plant biomass provide a means of predicting changes and of showing where specific research is needed.

METHODS OF MEASURING PRODUCTION

Photosynthesis at sea is measured by the use of bottles containing either a given species, a selected group of species of phytoplankton, or a mixed random sample from the water being studied. The bottles are suspended at different depths in the euphotic zone or held aboard ship under conditions of light, temperature, etc. that simulate as closely as possible those of the water being studied. The **method of Gran** (1931, 1932) uses paired light and dark bottles, the dark bottle serving as a control to give a correction for the simultaneous respira-

Since 1952 the **incorporation of carbon**tion uncomplicated by photosynthesis.
14 has been measured rather than oxygen formation. This is a more sensitive test introduced by Steemann-Nielsen (1952). Solutions of $NaHC^{14}O_3$ are mixed with samples of seawater with phytoplankton. After a suitable interval the phytoplankton is filtered out and the amount of $C^{14}O_2$ taken up is measured by counting with a Geiger counter, usually the windowless type. As the amount of plankton is usually small, it is assumed that there is no self-absorption of the beta particles under these counting conditions. It is, of course, necessary to know the amount of activity initially added, which must be measured under the same counting conditions, i.e., with the same counting geometry and no self-absorption of beta particles. Since adequately thin planchets are nearly impossible to prepare, the zero-thickness activity is extrapolated from absorption curves obtained from thicker planchets. Jitts and Scott (1961), using liquid scintillation counters and new, thinner planchets, showed that this procedure is not accurate.

The possibility that some kinds of phytoplankton may excrete a portion of the C^{14} as soluble organic matter may also prove to be a limitation for this method. Appreciable amounts of C^{14} may at times appear in the form of the photosynthetic intermediate glycolic acid. The quantity of glycolic acid to appear seems unpredictable. Changes in respiration caused by the experimental technique may also cause differences in the results, as estimated by oxygen evolution and by C^{14} incorporations.

The two methods tend to agree fairly well in short-term experiments, but with more prolonged experiments, especially in tropical waters, the oxygen evolution method tends to give significantly higher results. In long-term experiments, depletion of nutrients in the light bottle may affect the results, as may the possible antibacterial effect of the light or of the actively photosynthesizing phytoplankton organisms in the light bottle. Also, population changes in the light bottle and in the dark bottle are of a quite different character in prolonged experiments, so that the correction indicated by the dark bottle may not be valid for the light bottle.

It is also difficult to be sure whether the C^{14} method is measuring net photosynthesis or total photosynthesis. It apparently measures net photosynthesis in long experiments, and something between net and gross photosynthesis in short ones. Another factor of uncertainty lies in the possible importance of heterotrophic assimilation of dissolved carbon compounds. Healthy phytoplankton in the dark fixes about 1% of the amount of C^{14} fixed in the light, but this may be increased to as much as 30% of the amount fixed in light, under conditions of nutrient deficiency.

Because there is a direct relationship in seawater between pH and total dissolved carbon dioxide, it is possible, with **the pH method,** to calculate the amount of carbon assimilated from changes in the pH of a sample during photosynthesis. For example, at 14° C., an increase of 0.15 in pH would be equivalent to the abstractions of 2.24 ml. (STP) of carbon dioxide, or 1.2 mg. of carbon per liter. If we assume that the plant cell converts all of the carbon dioxide taken up into organic matter, this, then, gives us a direct estimate of the amount of new organic matter produced during the period in which the measured rise in pH occurred (Raymont, 1963).

DISCUSSION

In spite of all their limitations, these methods have given a great deal of valuable information. The vertical distribution of photosynthesis in a homogeneously distributed population under bright sunlight in the open ocean is given in Fig. 1-5. This is the sort of curve obtained by suspending a series of bottles, all containing the same

amount and kind of phytoplankton, at different depths for the same amount of time. It must be realized, of course, that in the ocean, the phytoplankton is not homogeneously distributed, and that those species which normally occur near the top tend to have their maximum photosynthetic efficiency at higher levels of illumination than those near the bottom. For species near the bottom of the euphotic zone, **light saturation,** the amount of light that gives maximum photosynthesis, occurs at lower intensities. The depth of the euphotic zone will vary greatly, being a function of the amount of incident light and of water transparency. On clear, bright days in the open ocean, the compensation depth may be approximately 85 meters. In more turbid oceanic waters, it is commonly between 25 and 85 meters, while in turbid coastal waters it may run from 5 to 25 meters. The phytoplankton itself will be somewhat self-regulating. As it becomes more abundant

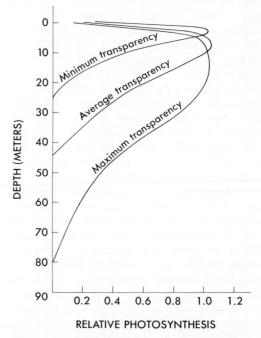

Fig. 1-5. Curves showing the vertical distribution of photosynthesis in homogeneously distributed population under bright sunlight in the open ocean. (Modified from Yentsch, C. S. 1963. *In* H. Barnes [ed.] Oceanography and marine biology. Vol. 1. George Allen & Unwin, Ltd., London.)

during a bloom, the water becomes less transparent, and the depth of the euphotic zone lessens.

The total amount of primary production per unit area of the ocean surface is best calculated by integrating the rates from several depths, relating the total photosynthesis to light saturation (Pmax), to incident light (Io), and to a vertical gradient of intensity defined by the average attenuation coefficient for the euphotic zone (Ze), the base of which is the intensity at extinction (Ie). Talling (1961) calculated the relation between daily photosynthesis, and that occurring at light saturation, and gives the following formula for calculation of primary production per unit area:

$$\Sigma P = \frac{Pmax}{Ze} \left(\ln \frac{Io}{Ie} \right)$$

This does not wholly apply to natural conditions, where the rate of photosynthesis for each species of phytoplankton present also varies with the depth of the particular cells. Steelman (1962) suggests some modifying factors to compensate for light adaptation and efficiency of photosynthesis.

Ryther gives estimates of the gross and net organic production of various systems in grams of dry weight produced per square meter per day. Table 1-1 gives examples of these estimates.

It appears from these estimates that primary production in the ocean on a unit area basis is of the same order of magnitude as that on land, or slightly below it. However, because of the greater extent of the seas, it is estimated that in their entirety, the oceans produce at least two to three times as much as all the land masses together.

A somewhat different approach may per-

Table 1-1. Estimates of gross and net organic production of various systems in grams of dry weight produced per square meter per day*

		Gross	Net
1. Mass outdoor *Chlorella* culture	Maximum		28.0
	Mean		12.4
2. Terrestrial crops in growing season (maxima)			
Sugar			18.4
Rice			9.1
Wheat			4.6
3. Terrestrial habitats			
Spartina (marsh grass)			9.0
Pine forest (best growing years)			6.0
Tall prairie			3.0
Short prairie			0.5
Desert			0.2
4. Marine (maxima for single days)			
Coral reef		24	(9.6)
Turtle grass flat		20.5	(11.3)
Grand Banks (April)		10.8	(6.5)
Continental shelf		6.1	(3.7)
Sargasso Sea (April)		4.0	(2.8)
5. Marine (annual averages)			
Long Island Sound		2.1	(0.9)
Continental shelf		0.74	(0.40)
Sargasso Sea		0.74	(0.35)

*Extracted from Ryther, J. H. 1959. Potential productivity of the sea. Science **130:**602-608.

haps give more reliable estimates of overall productivity, or at least an indication of the upper limits of such productivity. Tables 1-2 to 1-4 give an estimate of the upper limit of ocean productivity based upon what is known of the amount of solar energy available for photosynthesis and the amount of energy required for the fixation of a given amount of carbon. This is then compared with an estimate of annual productivity.

Table 1-2. Solar energy available for photosynthesis in joules/year*

		Joules
1.	Intercepted by earth	5.0×10^{24}
2.	40% reaches surface	2.0×10^{24}
3.	50% of (2) is infrared	1.0×10^{24}
4.	40% of (3) is reflected	6.0×10^{23}
5.	Available in ocean, 75% of (4)	4.0×10^{23}
6.	Assuming 2% efficiency in PS	8.0×10^{21}
7.	Needed per gram of carbon assimilated	5.0×10^{4}
8.	Maximum assimilation per year, tons of carbon	1.6×10^{11}

*From Vishniac, W. 1968. Autotrophy; energy availability in the sea and scope of activity. *In* C. H. Oppenheimer [ed.] Marine biology. Vol. IV. New York Academy of Sciences, New York.

Table 1-3. Energy requirement for photoautotrophic growth*

Stage in biosynthesis	On basis of	Required per gram atomic weight of carbon	
		Cofactors	Energy in joules
$CO_2 \rightarrow$ Carbohydrate	Calvin-Benson pathway	3.0 ATP	1.2×10^{5}
		2.0 NADPH	3.6×10^{5}
Carbohydrate $\rightarrow$ Amino acids	Survey of known pathways	~ 0.5 ATP	0.2×10^{5}
Amino acids $\rightarrow$ Whole cell	10 gm. dry weight/ATP (Bauchop and Elsden)	2.4 ATP	1.0×10^{5}
			6.0×10^{5}†

*From Vishniac, W. 1968. Autotrophy; energy availability in the sea and scope of activity. *In* C. H. Oppenheimer [ed.] Marine biology. Vol. IV. New York Academy of Sciences, New York.
†Minimum energy requirement for photoautotrophic growth is therefore 5×10^4 joules per gram of carbon.

Table 1-4. Ocean productivity*

Carbon assimilation per year	100-200 gm. C/m.² (H. W. Harvey; G. A. Riley)
Area of oceans†	3.5×10^{14} m.²
Annual productivity‡	5.3×10^{10} tons carbon
Maximum productivity	1.6×10^{11} tons carbon

*From Vishniac, W. 1968. Autotrophy; energy availability in the sea and scope of activity. *In* C. H. Oppenheimer [ed.] Marine biology. Vol. IV. New York Academy of Sciences, New York.
†Not counting ice-covered waters.
‡P and N are limiting, so that for this amount of organic matter all available P and N would have to turn over two to ten times each year.

THERMOCLINE

Extensive areas of relatively stabilized waters, such as occur in tropical regions the year round and in temperate regions during the summer, tend to show less overall productivity than do areas of vertical mixing. The warm surface waters become somewhat impoverished in plant nutrients, and because of their lesser density, tend to remain at the surface, separated from the colder, richer waters below by a relatively stable thermocline. A thermocline is a stratum where a fairly rapid transition, or change, occurs between the warmer, less dense, less saline surface water mass and the colder, denser, more saline water masses below. In tropical waters the thermocline may occupy a depth of between 100 to 300 meters and be relatively stable throughout the year. In temperate and boreal waters the thermocline is primarily a seasonal phenomenon occurring only in the spring and summer and tends to be at a higher level.

In the oceans, thermoclines are not as abrupt and well marked as in freshwater bodies, where thermoclines are usually associated with changes of 1° C. or more per meter. In temperate waters and in high latitudes, the surface layers are somewhat warmed during the summer but are kept mixed by winds and by small convection currents set up by evaporation and cooling, so that the surface waters are relatively homogeneous. Below this, the increase in density tends to isolate the upper water from the colder layers below, but the temperature does not change nearly as rapidly with depth as in freshwater bodies.

In parts of the eastern North Atlantic, warm, dense, highly saline water flowing out from the Mediterranean Sea creates a mass of warmer water at intermediate depths so that very little fall in temperature occurs between 100 and 1000 meters.

Since seawater continues to increase in density with cooling, down to its freezing point at −1.9° C., it sinks as it cools in winter. However, its sinking is limited by the permanently cold deeper layers, so that there is not a complete spring and autumn turnover, except in very shallow waters.

The spring and autumn mixing involves only the upper 1000 meters. Only near the polar regions do surface waters become sufficiently cold to sink to the bottom in relatively deep water, causing a complete mixing, with the entire water column becoming very cold.

In temperate waters there is a summer thermocline at about 50 meters, beneath which the temperature continues to decrease more gradually, forming a permanent, much less steep thermoclinal region down to about 1000 meters. As one approaches the tropics, the summer thermocline becomes more marked, deeper in the water, and longer lasting, until in tropical and subtropical waters there is a permanent, well-marked thermocline between 50 and 500 meters in depth.

VERTICAL STRATIFICATION

Since virtually all the plant life of the oceans is confined to the euphotic zone, and all animals depend upon plants either directly or indirectly for food, most animal life of the oceans is also found in or near this zone. Vertically the oceans can be divided into four principle realms by depth and light penetration. (See Fig. 1-3.)

PHOTIC ZONES

At the top is the thin euphotic zone, the upper 100 or so meters, containing all the functioning producers, as well as most of the consumers and a very large portion of the reducers. It is important to bear in mind the diurnal, seasonal, and regional differences and periodicities in the illumination and temperature of the waters in this zone. Organisms of the euphotic zone are often described as **epipelagic.**

Below the euphotic zone are the **dysphotic,** or **mesopelagic,** depths extending down to about 1000 meters. These depths receive too little, and too brief, illumination for effective photosynthesis. Only shorter wavelengths of visible light in the blue end of the spectrum penetrate to these depths. No functional producers are present, but many animals occur here. Some of the animals are regular residents of these

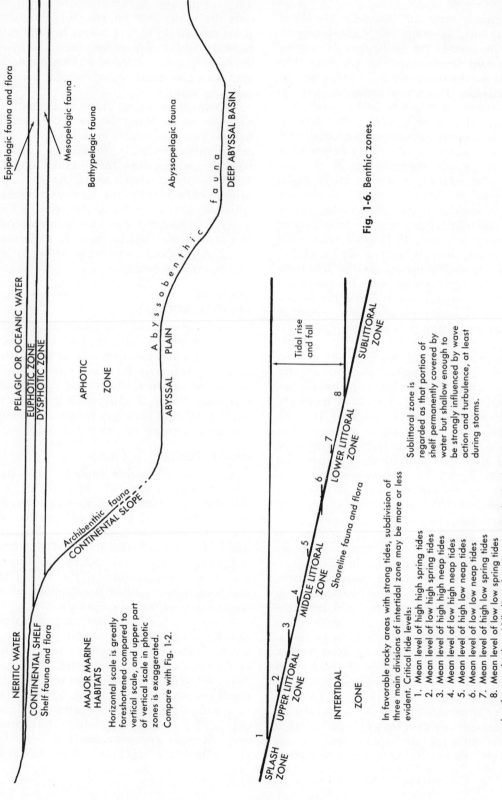

Fig. 1-6. Benthic zones.

PELAGIC OR OCEANIC WATER

Epipelagic fauna and flora

EUPHOTIC ZONE

DYSPHOTIC ZONE

Mesopelagic fauna

Bathypelagic fauna

APHOTIC ZONE

Abyssopelagic fauna

Abyssobenthic fauna

ABYSSAL PLAIN

DEEP ABYSSAL BASIN

Archibenthic fauna

CONTINENTAL SLOPE

NERITIC WATER

CONTINENTAL SHELF
Shelf fauna and flora

MAJOR MARINE HABITATS

Horizontal scale is greatly foreshortened compared to vertical scale, and upper part of vertical scale in photic zones is exaggerated.
Compare with Fig. 1-2.

SPLASH ZONE

UPPER LITTORAL ZONE

MIDDLE LITTORAL ZONE

LOWER LITTORAL ZONE

SUBLITTORAL ZONE

INTERTIDAL ZONE

Shoreline fauna and flora

Tidal rise and fall

In favorable rocky areas with strong tides, subdivision of three main divisions of intertidal zone may be more or less evident. Critical tide levels:

1. Mean level of high high spring tides
2. Mean level of low high spring tides
3. Mean level of high high neap tides
4. Mean level of low high neap tides
5. Mean level of high low neap tides
6. Mean level of low low neap tides
7. Mean level of high low spring tides
8. Mean level of low low spring tides

In other words, the middle littoral zone comprises levels of neap tides, whereas upper and lower littoral zones comprise additional levels covered or uncovered by spring tides.

Sublittoral zone is regarded as that portion of shelf permanently covered by water but shallow enough to be strongly influenced by wave action and turbulence, at least during storms.

strata. Others perform daily or seasonal migrations into and out of them.

APHOTIC ZONES

At their lower limits the mesopelagic depths grade insensibly into the **aphotic zone,** the waters which receive no light at all. This continues to the ocean bottom. The upper portion of the aphotic zone, from around 1000 meters to 3000 or 4000 meters, is termed the **bathypelagic** zone. Below this zone, in the great depths, is the **abyssal,** or **abyssopelagic,** zone.

The terms bathypelagic and abyssal have

Fig. 1-7. Temperatures of marine waters.

not always been used in precisely the same sense by various writers. Bathypelagic is often used rather vaguely in describing organisms from both the mesopelagic and upper aphotic depths, and the term abyssal is used in the same way in referring to the entire aphotic portion of the oceans. The deeper zones are not sharply marked but grade insensibly into each other.

Benthic organisms at corresponding depths are described as **archibenthal** if they are living in the upper bathypelagic and dysphotic depths at the outer edges of the continental shelves and on the continental slopes, and as **abyssobenthal** when they are living on the deep-sea floor.

The water in the abyssal region is cold, usually more saline than in the layers above, and under immense hydrostatic pressure. Water movements are slower. Bathypelagic and abyssal animals depend ultimately on organic matter produced in the euphotic zone. There are no plants.

Organisms limited to relatively narrow ranges of depth may be described as **steno-bathic.** Those tolerant of great differences of depth, provided other conditions necessary for their life are met, are **eurybathic.** Organisms that spend their entire life in the water rather than on the substrate are spoken of as **holopelagic,** and **holoplanktonic** if they are planktonic. Those which are planktonic only part of their life are termed **meroplanktonic.** The term "deep sea" commonly includes mesopelagic and abyssopelagic forms and both archibenthic and abyssobenthic forms. Pelagic animals, of course, face the same problems with respect to flotation, as mentioned in connection with the plankton. Those living only at intermediate depths must be able to prevent excessive rising as well as sinking.

THERMAL RELATIONSHIPS

Temperature is one of the major factors influencing the distribution of marine organisms. Except near the polar regions, surface temperatures are higher and more variable than the temperatures of deeper strata.

Table 1-5. Surface temperatures of open waters in degrees centigrade

	Minimum	Maximum	Mean
Tropical	20	30	24-27
Subtropical	16-18	25-27	20-22
Temperate	9-16	24-26	16-20
Boreal and antiboreal	1-9	10-17	8-14
Arctic and antarctic	−3 to +1	8-10	−1 to +5

Table 1-6. Mean annual temperatures at various depths and latitudes in the eastern Atlantic Ocean in degrees centigrade*

Depth (meters)	60° S.	40° S.	20° S.	0°	20° N.	40° N.	60° N.	80° N.
0	Below 0	15	17	27	20	16	9	2
200	Below 1	10	11	15	15	12	6	2
400	Below 1	10	9	9	12	12	9	1
800	Below 1	4	5	5	8	11	8	0
1000	Below 1	3	4	5	6	9	7	Below 0
2000	Below 0	2	3	3.3	4	4	3	−1

*From Ekman, S. 1953. Zoogeography of the seas. Sidgwick & Jackson, Ltd., London.

Temperature relations are commonly diagramed by drawing **isotherms,** lines connecting points with the same temperature. Various types of data may be used, depending on the purpose of a diagram. For example, one may wish to determine mean annual temperatures, annual maxima, or annual minimal temperatures. In a given region, isotherms divide the water masses of different depths from each other and can be visualized as warped horizontal planes in the water. Isotherms for surface waters have a general east-west trend, indicating progressively colder waters as one goes north or south from the equator. However, they are very much modified by currents, upwellings, convergences, and local weather. Marine waters show a temperature range of −7° to about 42° C. The warmest temperatures are found during the day in shallow tropical lagoons or tide pools. Open surface waters are rarely warmer than 30° C. The coldest temperatures occur in certain abyssal basins of the Antarctic, where the water does not freeze because of its salinity and great pressure.

Table 1-5 gives a rough idea of surface temperatures in the various zones that are commonly recognized. It should be emphasized that in the oceans communities of organisms, especially of pelagic organisms, are associated with water masses displaying certain hydrographic conditions, rather than with particular localities. Movements of water masses may sometimes carry a given community of organisms, or parts of it, for great distances.

Table 1-6 gives some representative temperatures of waters of various depths at different latitudes in the eastern Atlantic. Note that with increasing depth the temperatures tend to be lower and to fluctuate within narrower ranges. Conditions in deep water are commonly more uniform over greater distances than are conditions in the upper layers. For this reason communities of deep-sea organisms often show much wider north-south distribution than do the communities of organisms above them. A rather uniform deep-water community may underlie several different surface communities.

Stenothermal organisms are those which can tolerate a very limited range of temperatures. **Eurythermal** organisms can live in a wide range of temperatures. Many or-

Table 1-7. Examples of spawning season of some animals near Bergen, Norway*

Ecological region	Species	J	F	M	A	M	J	J	A	S	O	N	D
Arctic-boreal	*Strongylocentrotus dröbachiensis*	x	x	x									
	Cucumaria frondosa		x	x	x								
	Dendronotus frondosus	x	x										
Boreal	*Pleuronectes platessa*			x	x	x							
	Mytilus edulis			x	x	x	x						
	Echinus esculentus			x	x	x	x	x					
	Asterias rubens			x	x	x	x						
Mediterranean-boreal	*Psammechinus miliaris*						x	x	x	x	x	x	
	Echinocyamus pusillus						x	x	x	x			
	Echinocardium flavescens						x	x	x	x	x		
	Echinocardium cordatum							x	x	x	x		
	Ciona intestinalis						x	x	x	x	x	x	

*From Sverdrup, H. U., M. W. Johnson, and R. H. Fleming. 1946. The oceans: their physics, chemistry, and general biology. Prentice-Hall, Inc., Englewood Cliffs, N. J.

ganisms are reproductively stenothermal, or stenothermal during young stages even though the adult may be more eurythermal.

The maximum and minimum temperatures of temperate and boreal waters, with wide temperature fluctuations, may be limiting to stenothermal species from warmer or colder waters. However, many eurythermal species usually found in warmer or colder waters can live in these areas, even if they are somewhat stenothermal with respect to reproduction, by spawning during the part of the year that suits them best. Table 1-7 gives some examples of this.

SUBMERGENCE

From Table 1-6 it is evident that for many stenothermal organisms from colder waters, suitable temperatures are found at greater depths as one progresses toward the equator. Cold-water stenothermal species that can tolerate a variety of depths may exhibit the phenomenon of submergence. For example, many species found in shallow areas on the continental shelf in arctic waters are found only at greater depths in the boreal waters of the northern Atlantic and Pacific oceans.

Submergence, together with changes in hydrographic conditions during the ages, accounts for cases of the appearance of the same or related species in temperate, boreal, or polar waters of both hemispheres, but not in the tropics. This phenomenon is called **bipolarity.** Some forms that show bipolar surface distributions may actually exist at greater depths in the tropics, having, then, a continuous distribution. Because of submergence, some cool-water species enjoy a greater geographical distribution than do species limited to warm water.

In the surface waters, however, incursions of pelagic tropical, or warm-water, species into regions normally inhabited by cold-water forms are more frequent and more extensive than are occurrences of cold-water species in normally warm regions. This is largely due to the fact that cold water tends to be more dense and to sink beneath the warmer waters in regions where cold and warm waters meet. Another factor is the fact that most organisms are living closer to the upper limits of their thermal tolerance than to the lower limits, probably because their biochemical processes are more efficient and rapid there. Unaccustomed warming is more quickly deleterious to most organisms than is cooling to the same extent. Cold-water species carried into warm areas die and disintegrate more quickly than do warm-water species carried into cool regions. Warm-water species may often survive a considerable time in cooler-than-normal water and even continue to grow, though they may be unable to reproduce and maintain a population there.

SALINITY

Despite the influx of approximately 3.64 $\times$ 10^4 km.^{3} of fresh water annually from rivers (Sillén, 1967), plus a large amount falling directly on the sea as very unevenly distributed precipitation, and despite the annual evaporation of similar amounts of water from the ocean's surface, also unevenly distributed, the pH, concentrations of major dissolved ions, relative amounts of the principal ions, and osmotic properties of the sea are remarkably constant (Table 1-8). The osmotic properties of seawater are a function of the total amount of dissolved salts. This is commonly calculated in terms of equivalence to sodium chloride, which is the chief component—i.e., it is calculated in terms of the amount of sodium chloride that would give the water the observed osmotic properties. This is usually expressed as parts of sodium chloride per thousand parts of seawater, using the symbol $^0/_{00}$. Thus a salinity of 32.0$^0/_{00}$ would be equivalent to 32 gm. of sodium chloride in 1000 ml. of seawater, or a 3.2% solution of sodium chloride.

Since the ratios of the major ions in seawater are constant, even though the total amount of dissolved salts varies somewhat, the salinity can be calculated from a simple titration of the Cl$^-$ and Br$^-$ ions with silver nitrate of known strength. This gives the **chlorinity** (grams of Cl$^-$ and Br$^-$ per 1000 gm. of seawater, or parts per thousand). From this the salinity (grams of total soluble salts per 1000 gm. of seawater) can be calculated by the following equation:

$$Salinity = 0.03 + 1.803 \text{ (chlorinity)}$$

In the open sea, salinities commonly vary from $32^0/_{00}$ to $37.5^0/_{00}$, the differences reflecting local effects of evaporation, rain, freezing and melting of ice, influx from rivers, and so forth. The pH is quite constant, varying from 7.9 to 8.3. The concentrations of the major components of seawater are shown in Table 1-8.*

The buffering is probably a combined effect of the H_2CO_3, HCO_3^-, $CO_3^=$ system, plus the fine-grained silicates, especially clays, which may do even more of the buffering.

The constancy of conditions in the sea is of great biological significance. It permits the growth under pelagic conditions of organisms that have not yet developed regulatory mechanisms. Terrestrial animals must return to aquatic environments for reproduction or develop complex mechanisms, such as the amnion, that simulate ancestral aquatic conditions for the developing embryos.

The differences in salinity, pH, and osmotic properties of the open seas are not sufficient to constitute major barriers to the distribution of most organisms within the oceans. Salinity, while it is an important barrier between freshwater and terrestrial habitats and the oceans, is not such an important factor within the ocean system itself, except in marginal situations where salinities may be considerably increased or reduced in partially isolated arms of the sea or near the mouths of large

Table 1-8. Concentrations of the major components of seawater in moles per kilogram of seawater at salinity 35⁰/oo*

H_2O	53.557	Cl^-	0.5459
Na^+	0.4680	$SO_4^=$	0.0282
Mg^{++}	0.532	HCO_3^-	0.0023
Ca^{++}	0.0103	Br^-	0.0008
K^+	0.0099	F^-	0.0001
Sr^{++}	0.0001	H_3BO_3	0.0004

*From Sillén, L. G. 1967. The ocean as a chemical system. Science **156:**1189-1196.

Table 1-9. Salinities of various types of water*

Salinity (0/00)	Type of water
0-0.5	Fresh water
0.5-3.0	Oligohaline brackish water
3.0-10	Mesohaline brackish water
10.0-17	Polyhaline brackish water
17-30	Oligohaline seawater
30-34	Mesohaline seawater
34-38	Polyhaline seawater
> 38	Brine

*From Välikangas, I. 1933. Über die Biologie der Ostsee als Brackwassergebiet. Verh. int. Verein. theor. angew. Limnol. **6:**1.

rivers. Table 1-9 summarizes the salinities of various types of water.

The density of seawater increases as salinity is increased and temperature is decreased. In general, deep ocean waters tend to be colder and more saline than surface waters, and to exhibit less variation in these respects. Since colder water will hold more dissolved oxygen than does warmer water, the sinking of cold water from the polar regions is an important factor in maintaining the circulation and oxygenation of the abyssal waters.

Organisms that can tolerate only very narrow ranges of salinity are termed **stenohaline.** Those which can live in a wide range of salinities are called **euryhaline.**

---

*According to Sillén (1967), the most puzzling factor in the composition of both the sea and the atmosphere is the distribution of nitrogen. At the present pH and pE of seawater, one would expect much more of the nitrogen to be in the form of dissolved nitrates in the seawater rather than in the form of N_2 in the air. The known mechanisms for the return of nitrogen from the sea to the air—certain pockets of nitrate reduction—seem wholly insufficient to account for the balance actually observed. No large-scale, near-the-surface process of N_2 return from the sea has yet been discovered.

ENVIRONMENTAL EXTREMES

Unusually high salinities (hypersalinity) are rare in marine environments, occurring only in shallow, isolated arms of the sea where evaporation is high. Simons (1957) studied the ecology of Laguna Madre, Texas, where salinities vary from $27^0/_{00}$ to $78^0/_{00}$. He found that the higher the salinity, above normal ranges, the fewer were the species present, but that the number of individuals of species that were present tended to increase. Fishes present in such environments were of larger-than-average size for their species, but invertebrates such as barnacles and crabs averaged smaller.

In most coastal areas, and especially in bays, river mouths, and estuaries, salinities vary downward from the normal marine values as one ascends. Because of tidal movements of water, salinities may fluctuate considerably at any given station.

Where both temperature and salinity undergo great variation, often reaching extreme values, the number of both species and individuals is usually low.

Where hydrographic conditions are relatively constant, but in extreme ranges, the number of species is usually low, but those species tolerant of the conditions may greatly increase in numbers of individuals, becoming overwhelmingly dominant in that environment. The deep-sea bottom is an exception. The hydrographic conditions there are both constant and extreme, but the scarcity of food reaching these depths imposes a low absolute limit on the numbers of any species that can be present. Thus populations never become as dense as in other extreme biotopes. The diversity, however, is relatively high.

Stable biotopes within normal ranges for marine environments, such as most tropical seas, tend to develop the maximum diversity—that is, the greatest number of species present in significant numbers—but usually do not show the overwhelming aggregates of any one species that commonly dominate more extreme biotopes.

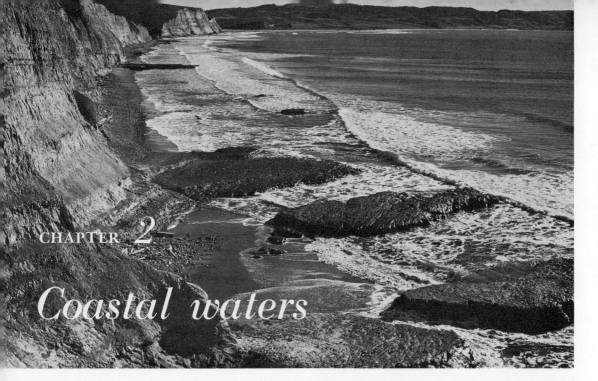

CHAPTER *2*

Coastal waters

CONTINENTAL SHELVES

The ocean basins usually have a narrow margin, or shelf region, surrounding the land masses and extending seaward a variable distance to a depth of about 100 to 200 meters. Being for the most part in the euphotic zone, continental shelves support abundant benthic flora and fauna—the shelf communities—that differ in many respects from the communities found in deeper water.

The water above the continental shelves and its fauna and flora are described as **neritic.** Neritic waters tend to be richer in plant nutrients and more productive than water of corresponding depths in the open sea. One reason for this is that greater mixing that occurs here as a result of turbulence, wave action, upwelling caused by offshore currents or winds, etc., bringing plant nutrients into all strata of the water. Also, additional nutrients are leached from the substrate and washed into the sea by rivers and streams from the adjacent land.

Where the substrate is suitable for attachment, the shallow waters also support a rich growth of seaweed, eelgrass, and, in some areas, other plants.

The increased plant growth is reflected

Fig. 2-1. Heavy winter surf sweeps out the sand from beaches in Drakes Bay, Point Reyes National Seashore, California. Underlying rock platforms and water-worn cobbles are exposed. Driftwood from distant points stacks up against the eroding cliffs. These cliffs are of the Monterey formation. They were formed by ancient seas. Occasionally, bones of prehistoric whales are cut loose by the modern sea. (M. W. Williams photograph.)

in an increase in animal life. Many species are found here that do not ordinarily occur far out at sea or in deeper water. Even the plankton, in addition to being more abundant, is of different character. Some of the oceanic species tend to be rare or absent in neritic plankton, or they occur only sporadically when oceanic water is swept in closer than usual. Many neritic species are limited to the shelf region and are seldom found far out at sea.

In addition to the holoplankton—species that are planktonic throughout their life—neritic waters contain much **meroplankton.** Meroplanktonic animals are those species found in the plankton only part of the time—for example, at night only. They seek shelter in the substrate the rest of the time, or they may be benthic forms that spend part of their life cycle in the plankton. Swarms of larval barnacles, decapods, echinoderms, and other nonplanktonic ani-

25

mals often make up an important part of the neritic plankton at certain times of the year.

The term neritic has often been applied to a species simply because it occurs in greatest numbers in neritic waters, even though the organism may be holopelagic, is not limited to the shelf in any way, and occurs regularly in high oceanic waters as well as in neritic waters. It is probably better to use the term in a more restricted sense for organisms that are bound to the shelf region in some way. Species that spend part of their time, or go through any stages of their life cycle, in or on the shelf substrate, and species that are tied to certain near-to-shore areas by salinity requirements, that feed on detritus washed from the land, or that for any other reason are bound to the shelf waters are more properly termed neritic, even though some of them may at times or at certain stages of their life be found in truly oceanic waters.

TIDES

At its upper margin the continental shelf ends in the **intertidal zone.** This is the relatively narrow strip that is exposed during the lowest low tides and either covered with water or splashed with waves during the highest high tides. The special conditions and stresses caused by alternating exposure and covering with water have resulted in the development of special communities of intertidal animals and plants. The nature of the particular community at any given place is largely determined by the physical character of the substrate and the degree of its exposure to wave action. Differences in temperature are responsible for broader regional differences in fauna and flora.

Intertidal communities are more directly influenced by the local climate than are the subtidal communities. During periods of low tide they may be exposed directly to hot sunlight and drying, to rain, to cold winds and freezing, and to predation from terrestrial animals and man. The tides operate on a lunar rhythm that is out of phase with the diurnal rhythm, resulting in a progression of tides through the lunar month,

the high and low tides coming each day at different times and reaching different levels. That both the lunar tidal rhythms and the diurnal and seasonal rhythms of the earth-sun system exert profound effects upon the lives of organisms is particularly evident in those forms which live part or all their life in the intertidal zone. (See Chapter 8 for a more complete discussion of the intertidal zone.)

There are usually four tides, of different heights, each day, two high tides and two low tides. The tides occurring during that part of each lunar month when the sun, moon, and earth are in line with each other, and when the gravitational effects of the sun and moon reinforce each other, are strongest. Such tides, termed **spring tides,** exhibit the greatest difference in level between the highest high tide of the day and the lowest low tide. The opposite effect occurs when the sun and moon are in such positions as to partially cancel out each other's gravitational effects, resulting in weaker tides exhibiting the least difference between the levels of high and low tide. Such tides are termed **neap tides.** Each lunar month has two series of spring tides alternating with two series of neap tides, with a few days of transitional, or intermediate, tides during each change. The actual time of occurrence and the level reached by a given tide at any particular place are influenced by many complicating factors, such as latitude, local and regional topography, currents, and winds.

The mean, or average, level attained by each of the four daily tides during the year in a given locality has been termed the **critical tide level.** The critical tide levels are the levels of the greatest average change in relative duration of exposure and submergence, and they have been correlated with major changes in the character of the intertidal communities.

Below the level of the lowest low tide, on open coastlines, there is a subtidal zone strongly influenced by wave action. Farther out the deeper waters over the shelf are much more quiet, showing only the circulation caused by currents, upwellings, and so forth.

Fig. 2-2. Exposed seaward-facing rocky slope at Marblehead Neck, Massachusetts, showing zonation of intertidal communities. The lower edge of the barnacle zone shows at the extreme upper left. This is succeeded by a belt of rockweed *(Fucus)*. Below the *Fucus* is an approximately equally broad community dominated by the mussel *Mytilus edulis*. At the waterline, only partly uncovered, is the zone of Irish moss *(Chondrus crispus)*. (Helen P. Finlayson photograph.)

VERTICAL ZONATION

As a result of the differences at various levels, the shelf communities tend to form as bands, parallel to the coast. Those in the intertidal area are the most narrowly compressed and sharply distinguished, particularly where they occur on steep rock faces in areas with strong tides.

The intertidal zone is divided into three major and seven minor subzones (Fig. 1-6) by critical tide levels. The zonation is not everywhere the same. Local conditions may alter relative heights and times of tides in such a way as to eliminate one or more of the critical tide levels as a significant ecological factor. For example, the average levels of the low high spring tides and the high high neap tides may be sufficiently close together to constitute only one critical level. The same sort of thing could happen at the low tide levels. In this case only five zones would be apparent in the intertidal.

Many areas, especially oceanic islands, have so slight a tidal rise and fall that local variations in weather, such as changes in wind direction and velocity, tend to obliterate the small differences in tidal levels, resulting in a narrow intertidal zone without well-marked subzones within it.

For some distance below the level of the lowest low tide of the year, the inshore waters are strongly influenced by turbulence produced by wave action. In the deeper, calmer water beyond the zone of wave turbulence, changes in the benthic biota are less abrupt and less well marked than in the intertidal and shallow subtidal zones.

The subtidal benthic shelf communities tend to be much broader and less sharply defined than the intertidal communities. Their character is largely determined by the character of the substrate, by the amount of illumination reaching the bottom (which in turn is a function of depth, turbidity of the overlying water, and latitude), and by the temperature of the water. The zonation of the continental shelf region is summarized in Fig. 1-6.

There has been considerable difference in usage of the term **littoral**. Some writers use the term to describe the entire shelf, intertidal and subtidal; others restrict the term to the upper edge of the shelf, the intertidal area. In reading a particular work, one must note the manner in which the terms are being used. In this book we shall use the term as indicated in Fig. 1-6, restricting the term littoral to the intertidal zones. The sublittoral areas are divisible into an upper marginal portion, strongly influenced by turbulence and wave action, and the deeper portions with quieter waters, grading into the archibenthal environment at its lower edge.

PART II

MARINE PLANTS

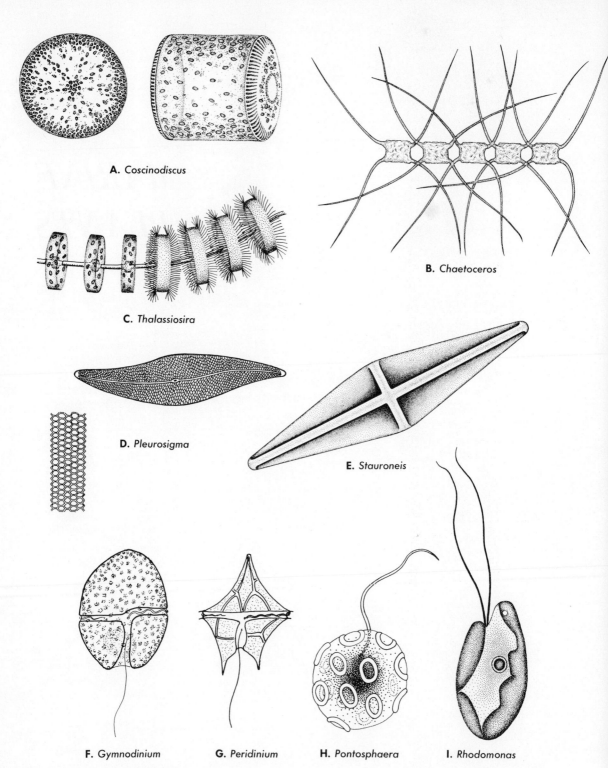

A. *Coscinodiscus*

B. *Chaetoceros*

C. *Thalassiosira*

D. *Pleurosigma*

E. *Stauroneis*

F. *Gymnodinium*

G. *Peridinium*

H. *Pontosphaera*

I. *Rhodomonas*

Fig. 3-1. Representatives of some groups of phytoplankton. **A-C**, Diatoms, centric types; **D-E**, diatoms, bennate types; **F**, naked dinoflagellate; **G**, thecate, or armored, dinoflagellate; **H**, Coccolithophore; **I-K**, others.

J. *Phaeocystis*

K. *Distephanus*

CHAPTER *3*

The phytoplankton

*T*he most obvious and numerous elements of the phytoplankton are diatoms and dinoflagellates. Less abundant generally, but of local importance at some times and in certain regions, are other groups of photosynthetic flagellates and blue-green algae.

DIATOMS

The diatoms are unicellular algae characterized by a skeleton, or capsule, the **frustule,** which is composed of two valves, the **epitheca** and the **hypotheca,** that fit together somewhat in the manner of the two halves of a petri dish. These valves are impregnated with silica, which gives them a glasslike character. A broad, hooplike connecting band is attached at right angles to the incurved edges of each valve. Together, the two connecting bands form the **girdle.** As the diatom grows, the two halves are pushed farther apart and the connecting bands become broader.

Seen in lateral, or girdle, view, diatoms

Most of the figures for this chapter were redrawn from various sources, especially Lebour (1930), Gran (1912), Gross (1937), Marshall (1934), and Grasse (1952). Several have been somewhat simplified.

31

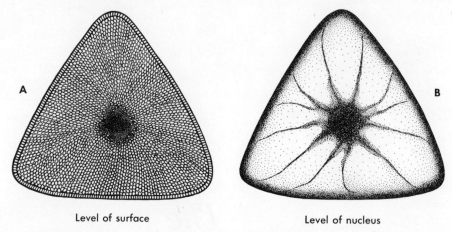

Level of surface   Level of nucleus

Fig. 3-2. *Triceratium*, showing, **A**, level of surface, and, **B**, nucleus and cytoplasmic strands.

commonly look somewhat rectangular. Viewed from above in the top, or valve, view, they exhibit a great variety of forms: circular, triangular, linear, crescentic, wedge-shaped, boat-shaped, and so forth.

The valves are transparent and usually beautifully and symmetrically ornamented with a variety of markings. Often the marks consist of fine lines on the face of the valve, which in many cases, upon very high magnification, can be resolved into rows of separate dots corresponding to minute concavities in the wall. Some genera show complex areolated structures resulting from larger chambers within the wall. Motile naviculoid types have thickenings known as **nodules.** These complex structures, one at each end and one at the center, are generally connected by a long slit called the **raphe,** which is visible in the girdle view.

Most diatoms occur singly, but many species form characteristic chains or aggregates of individuals held together by protoplasmic junctions, mucilage, or interlocked spiny or hairlike projections from the frustules. The groups so formed have shapes that are characteristic for different genera and are often quite complex. Some of them look superficially like filamentous algae.

The cytoplasm forms a relatively thin lining along the inside walls of the valves, surrounding a vacuole filled with cell sap. The nucleus commonly occupies a more-or-less central position, with cytoplasmic strands extending from it to other parts of the cell (Fig. 3-2). The cytoplasm contains chloroplasts in which photosynthesis is carried out. The brownish color characteristic of most diatoms is caused by the pigment **diatomin** in the chloroplasts. Diatomin somewhat resembles the pigment of the brown algae, and masks the chlorophyll. Most diatoms store oil or fatty acids, rather than sugar, as the end product of photosynthesis. Under exceptional conditions a particularly rich growth of planktonic diatoms may produce sufficient oil to form an oily slick at the surface of the sea, up to several miles across. In the tropics somewhat similar slicks are commonly formed parallel to the coastline by the blue-green alga *Trichodesmium.*

Diatoms range in size from less than 10 μ in length (or diameter, for small species) to about 1 mm. in very large ones. Even within the same species there may be considerable variation in size, up to thirtyfold or more in some cases. This is partly the result of the peculiar method of reproduction (Fig. 3-3). At each cell division one daughter cell retains the old epitheca, or larger half of the frustule. The other daughter cell retains the old hypotheca. Each daughter cell secretes a new hypotheca, so that the old hypotheca of the mother cell becomes the epitheca of one of the two daughter cells, which is therefore slightly smaller than the other daughter cell. Thus, as di-

Fig. 3-3. Reproduction in diatoms.

visions proceed, part of the population is continually becoming smaller, resulting in a decrease in the average cell size for the population.

This process clearly cannot continue indefinitely for that part of the population approaching minimal size for the species. Maximum size is commonly restored by formation of so-called **auxospores.** The protoplast discards the old valves altogether, in-creases considerably in size, secretes a membrane with a little silica around itself, and develops new valves of large size. The auxo-spore, then, is not a resting stage but a device for restoration of normal size. Auxo-spore formation may be influenced by external conditions such as light intensity and level of dissolved nutrients, and perhaps by the release of metabolites into the water, as well as by the absolute minimum

C. LACINIOSUS

Relatively typical form

C. DECIPIENS

Summer   Winter

C. DENTICULATUS

Plumose processes

Fig. 3-4. *Chaetoceros,* illustrating variations within and between species.

size attained. There is evidence of formation of haploid sexual gametes prior to auxospore formation, the auxospore being formed by the fertilized diploid zygote.

Some diatoms, especially neritic species, also form resting spores, enabling them to survive severe winter conditions or other unfavorable periods. The formation of resting spores involves a considerable loss of cell sap and the rounding up of the relatively dense protoplast. Such denser resting spores commonly sink into deeper layers of water, or to the bottom in shallow areas. They germinate again and begin reproducing when conditions are favorable once more. Low light intensity, low temperature, and depletion of nutrients in the water tend to favor the formation of resting spores. In cultures, overcrowding also promotes it.

In addition to the planktonic diatoms, all shallow waters where there is available substrate within the euphotic zone have large populations of benthic, attached, and interstitial diatoms. These diatoms constitute an important element of the "algal film," forming diatom slicks on rocks and other submerged surfaces and providing a principal food for many grazing animals, including gastropods and chitons.

Vast numbers of diatoms die because of seasonal changes, local impoverishment in plant nutrients, or grazing by zooplankton, or because they settle below the euphotic zone or are carried below it by vertical movements of the water. The result is an accumulation of great numbers of dead diatoms and their frustules at the sea bottom. In some areas they are a principal component of the marine ooze.

Where ancient marine sediments have been raised during geological time to become part of the land, fossil diatoms—along with the skeletons of Foraminifera, Radiolaria, and other organisms—may be studied to give clues to the conditions prevailing at the time the sediments were laid down. Such studies help in reconstructing and understanding the history of the earth. Where diatoms constitute the bulk of such sediments, they form the so-called diatomaceous earth, which has many industrial and scientific uses.

Diatoms have worldwide distribution but are particularly abundant and constitute a dominant mass of the plankton during the spring and often the autumn, in temperate, boreal, and polar waters. Here they may reach densities exceeding one million per liter of water.

It has been claimed that, in general, truly oceanic species and those living in warmer waters tend to have thinner, lighter frustules than neritic and cold-water species, and that truly planktonic forms have generally lighter, more delicate shells than bottom-living or attached forms. Some very widely distributed species show distinct varieties differing in thickness of the frustules and in average size, in polar and temperate regions.

C. SOCIALIS

Chains embedded in slimy colony

CHAETOCEROS

Spiral arrangement of chains found in several species

Fig. 3-4, cont'd. For legend see opposite page.

Like flowering plants, diatoms appeared rather late in evolution and have come to occupy a dominant role in their environment, exerting far-reaching effects on the evolution of associated organisms and on the physical, chemical, geological, and biological processes of the earth's surface. Among the earliest and best-known fossil deposits of diatoms are those from the Moreno shales in central California, dating from the late Cretaceous period. Some records occur back into the Triassic period.

This period of the earth's history, the latter half of the Mesozoic era, when the great dinosaurs still ruled the earth, foreshadows some of the most striking changes that have ever taken place in the character of life on earth. It was during this period that many of the major groups of animals and plants that characterize life today—the diatoms, flowering plants, mammals, birds, and insects, to name a few of the more outstanding—had their origins. It was during this time that the great reptiles lost their long hold on the earth and disappeared forever, together with many of the archaic plants that characterized the landscapes of their time. It is hard for us to imagine what the world was like before these changes or to visualize the forces that culminated in such a sweeping renovation and alteration of life on earth.

Classification of diatoms

Diatoms are algae, constituting the subclass Bacillariophyceae of the phylum Chrysophyta. Over 5500 species have been described. They are commonly divided into two primary groups, following Schütt (1880)—the Centrales or Centricae, and the Pennales or Pennatae. Recent authorities sometimes divide the group into six or seven primary divisions rather than two. The main features of these classifications can be shown in the combined tabulation below.

Order Centrales or Centricae: Diatom, usually circular, disc-shaped, cylindrical, or triangular. Valves radially symmetrical with radially disposed striae or concentric markings; sometimes with spines, horns, or other protrusions; no raphe; cells nonmotile; cytoplasm with numerous small chromatophores; resting spores and auxospores commonly formed; gametes, where known, anisogamous with motile flagellated male gametes. Mostly marine, planktonic. About 100 genera and 2400 species.

Cells of centric diatoms produce only a single auxospore. Although they were long thought to be asexual, it has been found that for some the process is sexual, and it seems probable that this may be true for all. In auxospore formation the protoplast rounds up to form a single egg cell, pushing the valves apart in the process. In other cells of the same species the protoplast divides repeatedly, forming 8, 16, 32, 64, or 128 motile flagellated antherozoids (called microspores before their role as gametes was understood). An antherozoid swims to and fertilizes an egg. The zygote thus formed becomes an auxospore.

Suborder Biddulphiodeae: Box-shaped; prevalvular axis not much longer than, or shorter than, valvular axis; valves usually oval, sometimes polygonal, circular, or semicircular; unipolar, bipolar, or multipolar—each pole with an angle, horn, or spine, or with both angles and horns. Some representative forms follow:

   Family Biddulphiaceae: *Bacteriastrum, Bellerochia, Biddulphia, Cerataulina, Ditylum, Hemiaulus, Lithodesmium, Terpsinoe*

   Family Chaetoceraceae: *Chaetoceros*

   Family Eucampiaceae: *Climacodium, Eucampia, Streptotheca*

Suborder Discoideae: Disc-shaped or cylindrical, valve circular, flat, or convex, or sometimes hemispherical; zonal diameter shorter than valvular diameter; sculpture radial or concentric, related to a central pole; spines or horns frequent.

   Family Coscinodiscaceae: *Arachnodiscus, Asteromphales, Coscinodiscus, Cyclotella, Planktoniella*

   Family Melosiraceae: *Melosira, Peralia*

   Family Skeletonemaceae: *Skeletonema, Stephanopyxis*

   Family Thalassiosiraceae: *Coscinosira, Lauderia, Schröderella, Thalassiosira*

Suborder Solonoideae or Soloniineae: Valves oval or circular; cells long, cylindrical, with numerous intercalary bands; no internal septa; united in chains by their valves.

   Family Corethronaceae: *Corethron*

   Family Soleniaceae: *Atthenya, Rhizosolenia, Dactyliosolen, Guinardia, Leptocylindrus*

Order Pennales or Pennatae: Elongated, oblong, or feather- or boat-shaped, the valves with two planes of symmetry, usually elongated, with the striae more or less transverse to the apical axis; no spines; a longitudinal groove, the raphe, present in the majority; those possessing a true raphe are motile, with a gliding movement; chromatophores commonly large, platelike, lobed, or somewhat rolled up into a partial cylinder, usually only one or two present; no resting spores formed; gametes, where known, isogamous, not flagellated. About 70 genera and 2900 species, somewhat more than half marine.

This group includes all the soil diatoms, the

DETONULA SCHROEDERI

RHIZOSOLENIA STYLIFORMIS

CHAETOCEROS DERISUS

BIDDULPHIA SINENSIS

BIDDULPHIA MOBILIENSIS

DITYLUM SOL

PLANKTONIELLA SOL

Side view

MELOSIRA BORRERI

Top view

Auxospore formation   *Continued.*

Fig. 3-5. Some representative genera of planktonic diatoms.

THALASSIOSIRA

THALASSIOSIRA GRAVIDA

LAUDERIA ANNULA

Chromatophores in normal position

Chromatophores aggregated at end faces of cells

LAUDERIA BOREALIS

CORETHRON

RHIZOSOLENIA HEBETATA

Fig. 3-5, cont'd. For legend see p. 37.

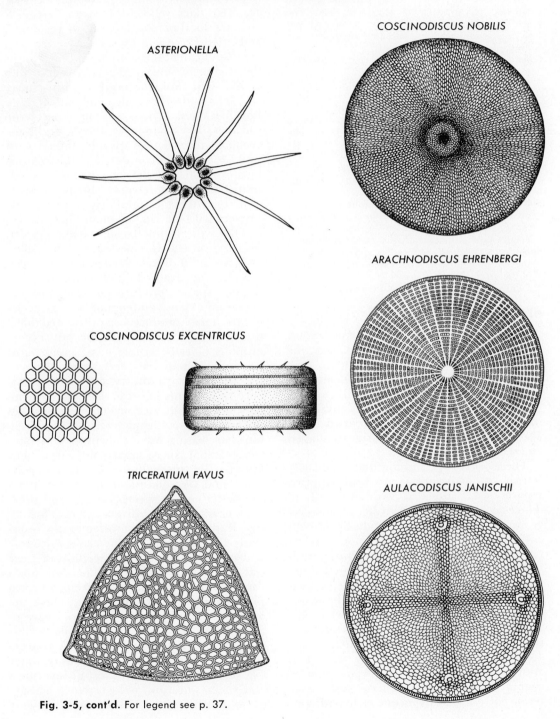

RHIZOSOLENIA ALATA

ASTERIONELLA

COSCINODISCUS NOBILIS

ARACHNODISCUS EHRENBERGI

COSCINODISCUS EXCENTRICUS

TRICERATIUM FAVUS

AULACODISCUS JANISCHII

Fig. 3-5, cont'd. For legend see p. 37.

majority of freshwater species, and a great many marine diatoms, especially the benthic species. In auxospore formation two small cells produced by meiotic division become surrounded by a gelatinous envelope. The valves are pushed apart and the two protoplasts, each acting as a gamete, fuse. The zygote forms an auxospore. In some species each cell first divides into gametes, and two auxospores are formed.

Suborder Araphidineae: No raphe or pseudoraphe; nonmotile.

Family Fragilariaceae: *Asterionella, Fragilaria, Thalassiothrix*

Family Tabellariaceae: *Striatella*

Suborder Biraphidineae: Well-developed raphes on each valve; commonly motile, with gliding motion.

Family Cymbellaceae: *Amphora, Cymbella*

Family Naviculaceae: *Amphiphora, Navicula*

Family Nitzschiaceae: *Bacillaria, Hantzschia, Nitzschia, Phaeodactylum*

Suborder Monoraphidineae: One raphe only; other valve has only a pseudoraphe or a rudimentary raphe.

Family Achnanthiaceae: *Achnanthes, Cocconeis*

Suborder Raphidiodinae: Raphe apparent only at the terminal nodules, not extending the length of the valve.

Family Eunotiaceae: *Eunotia, Peronia*

DINOFLAGELLATA

The dinoflagellates constitute a second well-marked group of unicellular algae, the class Dinophyceae. Like the diatoms, they are an abundant and important element in the plankton of all seas.

The body of a dinoflagellate is enveloped in a cellulose membrane that in many forms also bears hardened plates variously shaped and sculptured.

The most typical and familiar dinoflagellates have a body surface provided with two grooves, each provided with a flagellum. These are the **annulus** or **girdle,** a transverse groove passing around the body like a belt, and the **sulcus,** a longitudinal groove in the posterior part of the body from which a trailing flagellum arises. In the armored dinoflagellates the annulus may be covered by a perforated plate, the **cingulum.** The transverse flagellum is commonly somewhat flattened and ribbonlike. The action of these two flagella gives the dinoflagellates a characteristic swirling motion as they swim. The annulus divides the body into an anterior cone (the **epicone** or **epitheca**) and a posterior portion (the **hypocone** or **hypotheca**).

The nucleus is usually single and rather massive, with evenly distributed chromatin granules. Many planktonic dinoflagellates have numerous small chromatophores that are often dark yellowish brown or greenish and are located peripherally in the body, and often there is a light-sensitive red pigment spot called the **stigma.** In some groups the stigma is developed into a remarkable eyelike structure, the **ocellus,** with a myeloid lens and a dark-pigment ball (family Pouchetiidae). There are no contractile vacuoles, but commonly there is a large vacuole, termed the **pusule,** filled with pinkish fluid and connected to the exterior by a canal. It may be concerned with digestion. Other vacuoles may also be present; sometimes there is one fairly large vacuole surrounded by a ring of small ones.

The typical dinoflagellates are divided into two orders—**Gymnodiniales,** the naked or unarmored dinoflagellates, and the **Peridiniales,** the armored dinoflagellates.

Some species produce powerful poisons. When an extensive bloom of such a species occurs, it may produce a so-called "red tide," discoloring the water with the presence of millions of cells, killing great numbers of fish and other animals, and causing losses to the local fishing industry and great annoyance to residents in areas where large numbers of dead fish wash up onto the beaches or into the bays. People are sometimes poisoned by eating fish or shellfish that have concentrated such organisms with their filter-feeding mechanisms. *Goniaulax polyhedra* and *Gymnodinium breve* are among the better-known species causing red tides.

Marine animals of diverse groups, including Radiolaria, many coelenterates, especially the corals, the giant clam *Tridacna,* and many others, carry, in certain of their cells and tissues, great numbers of small round algalike brownish or greenish brown plant cells—the **zooxanthellae.** These zooxanthellae may be of great importance to their hosts. In corals, for example, a vast amount of animal tissue is often concentrated in a relatively small amount of water —as on an extensive coral reef at low tide— and the respiration and excretory activity of so much animal tissue would quickly foul the water and kill the animals. However, the zooxanthellae use much of the carbon dioxide and nitrogenous wastes before they enter the water, and their photosynthetic activity liberates oxygen into the water and into the tissues of their hosts. Food materials synthesized by the zooxanthellae may also be shared with their hosts. Muscatine (1967) demonstrated that zooxanthellae carrying on photosynthesis in the presence of host tissue excrete up to 40% of the labeled photosynthate as glycerol into the host tissue. In the absence of host tissue, they carried on photosynthesis but produced little or no glycerol or other excreted photosynthate. In some cases part of the zooxanthellae population may be regularly digested either in the host tissues or during times of stress, thus reducing the need to capture as great a quantity of zooplankton food as would otherwise be necessary.

The zooxanthellae at times produce a free-swimming, swarming stage, usually either small cryptomonads or naked dinoflagellates, thus demonstrating the relationships of these symbiotic plant cells to parts of the plankton.

Some dinoflagellates have lost their photosynthetic capacity and exhibit saprozoic, holozoic, or parasitic modes of nutrition and life. Some of these have become so modified that they are difficult to recognize as dinoflagellates.

In addition to the typical dinoflagellates discussed above, there are two groups usually included in the class that differ considerably in structure—the adiniferids or adinidans, and the Cystoflagellata.

The **Adiniferida** or **Adinida** do not possess the girdle and sulcus characteristic of other forms. The cellulose covering is commonly bivalved, the valves being regarded as lateral. The flagella are both at the anterior end, one extending forward, the other curving around more or less horizontally in the manner of the transverse flagellum of other dinoflagellates.

The **Cystoflagellata** are especially interesting. They are large, globular, rather gelatinous, hollow, fluid-filled cells looking almost like minute medusae. They have only the longitudinal groove, or sulcus, and a peculiar tentacle-like organ much thicker than a flagellum and slowly movable. The principle genus is *Noctiluca*—famed for its bioluminescence. *Noctiluca* gives off a flash of light when disturbed. When it is the dominant plankton, the surface waters may be quite dramatic at night, with flashes and trails of light marking the movement of fish, boats, or nets through the water, or with the waves along the shore shining with a strange light as they break.

Classification of dinoflagellates

The classification of the dinoflagellates may be summarized as follows. Only a few of the principal families and genera are mentioned.

Phylum Pyrrophyta
Class Dinophyceae
Subclass Adinida
  Family Prorocentridae: *Exuviella, Prorocentrum*
Subclass Cystoflagellata
  Family Noctilucidae: *Noctiluca*
Subclass Diniferida
  Order Gymnodiniales—the unarmored, or naked, dinoflagellates
    Family Cystodiniidae: *Cystidinium, Glenodinium*
    Family Gymnodiniidae: *Amphidinium, Gymnodinium*
    Family Polydinidae: *Polykrikos*
    Family Pouchetiidae: *Erythropsis, Pouchetia*
    Family Pronoctilucidae: *Oxyrrhis, Pronoctiluca, Protodinifer*
  Order Peridiniales—the armored, or thecate, dinoflagellates
    Family Dinophysidae: *Amphisolenia, Dinophysis*
    Family Peridiniidae: *Ceratium, Peridinium*
    Family Goniaulidae: *Goniaulax*

OTHER PHYTOPLANKTON

Some blue-green algae and some small flagellates belonging to the Chrysophyceae and Chlorophyceae sometimes occur in the plankton in sufficient numbers to be of at least temporary local importance.

Perhaps the best-known members of the blue-green algae (Cyanophyceae) in marine plankton are species of *Trichodesmium*. These slender cells, which float on the water surface, are rich in the pigment phycoerythrin, giving them a reddish color. When present in great numbers, they may produce extensive slicks and give off a characteristic odor of chlorine. *Trichodes-mium* slicks are common in the Java Sea and Indian Ocean, sometimes paralleling the coast for many miles. The Red Sea was so named because of the periodic discoloration of the water by *Trichodesmium,* and the Gulf of California is sometimes called the Vermilion Sea for the same reason.

Species of *Anabaena,* a genus better known from fresh waters, occur in more northern brackish waters. One genus, *Richelia,* is found in diatoms of the genus *Rhizosolenia.*

Flagellates belonging to the class Chrysophyceae or Chrysomonadina are sometimes important. Among the larger forms, *Phaeocystis* forms gelatinous brownish colonies in which the individuals are rounded up in the **palmella phase.** In temperate latitudes, blooms of *Phaeocystis* often color the water brownish and may clog plankton nets. New colonies are developed from free-swimming flagellated zoospores. This genus is impor-tant to fishermen because herring avoid areas where *Phaeocystis* is blooming. Different species have been found to prefer different temperature ranges of the water and hence to characterize separate water masses. *Halosphaera* is a bright green unicellular flagellate sometimes occurring in great numbers.

One of the best-known groups is the family **Coccolithophoridae.** These small uninucleate flagellates usually have two yellowish chromatophores and two flagella, and discs of calcium carbonate, known as coccoliths, are on the surface of the cell. These coccoliths have long been known from marine sediments, where they sometimes make up a significant portion of the "globigerina ooze," or calcareous sediment. The family is divided into two subfamilies on the basis of the structure of the coccoliths. Subfamily **Coccolithophorinae** has perforate (tremalith) coccoliths, while subfamily **Syracosphaerinae** has imperforate (discolith) coccoliths. The majority of species occur in warm seas, but some, such as *Pontosphaera huxleyi* and species of *Syracosphaera,* are abundant in colder waters.

In spite of the fact that the Coccolithophoridae are potentially photosynthetic, many of them are also found at considerable depths well below the euphotic zone, where they must be existing heterotrophically.

Another interesting group are the **Silicoflagellidae** (also called Dictyochidae). These have a siliceous skeleton. The cytoplasm contains small brownish yellow

PONTOSPHAERA HAECKELI

SYRACOSPHAERA BRASILIENSIS

Entire

Aperture

DISCUSPHAERA TUBIFEV

Fig. 3-6. Coccolithophores.

spherical bodies, and the nucleus is near the center of the body. One long flagellum is present. The latticed skeleton commonly consists of two circles of hollow siliceous rods joined by intercalary rods and forming a beautifully latticed skeleton.

Other small flagellates without special skeletal structures, such as *Chromulina,* are also abundant at times.

COCCOSPHAERA ATLANTICA

Single coccolith

PETALOSPHAERA GRANI

ACANTHOICA ACANTHIFERA

HALOPAPPUS VAHSELI

Fig. 3-6, cont'd. For legend see opposite page.

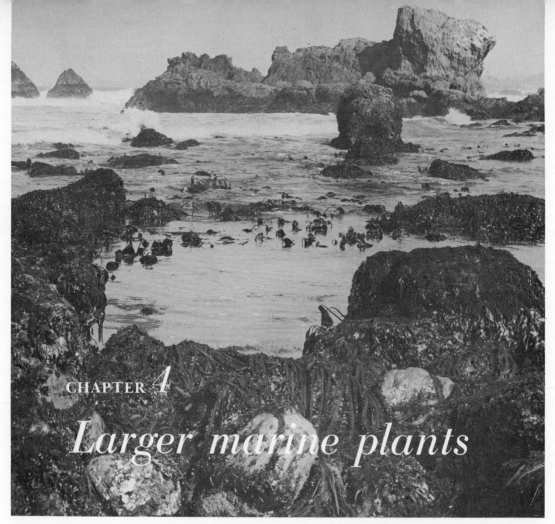

CHAPTER 4

Larger marine plants

Algae

THE SEAWEEDS

Fig. 4-1. Rocky area along California coast showing seaweeds exposed at low tide. Long straplike fringed seaweeds in foreground are *Egregia*. The broad blades on bent-over stems are *Laminaria*. Some eelgrass is visible in the quiet area of water near the rock. (M. W. Williams photograph.)

The great majority of larger marine plants are seaweeds. Seaweeds are for the most part fixed algae growing attached to firm substrates such as rocks, pilings, and shells. Attached seaweeds are of necessity confined to the fringes of continents and islands, or the tops of submarine banks where they can find attachment and enough light for photosynthesis. Larger floating plants are exceptional.

The most conspicuous exception is the subtropical *Sargassum,* or gulfweed. This is a small seaweed growing abundantly at the surface in the Atlantic south of Bermuda. The peculiar patterns of weather and currents produce here a relatively

stable calm region known as the Sargasso Sea. Gulfweed has accumulated in sufficient mass, forming a very special environment, over a long enough period of time to permit the evolution of a community of animals directly or indirectly dependent upon the sargassum weed. Some of these animals exhibit amazing mimicry, imitating both the color and form of the *Sargassum.*

Most seaweeds grow in shallow waters from the intertidal zone down to about 30 or 40 meters. In clear tropical or subtropical waters some of them may occur at ten

44

times this depth. They are usually not abundant where there is nothing firm to which to attach—as on sandy or muddy bottoms. In the intertidal zone on suitable rock surfaces, these algae, together with the associated animals, display the phenomenon of zonation with dramatic sharpness.

Storms or rough water often detach considerable amounts of offshore seaweeds, washing them up onto nearby beaches in windrows, where they serve as food and protection to various scavengers of the beach, such as kelp flies and their larvae, amphipods, and small beetles.

• • •

The larger algae include three great classes or subphyla of thallophytes: the **green algae (Chlorophyceae)**, the **brown algae (Phaeophyceae)** and the **red algae (Rhodophyceae)**.

GREEN ALGAE (CHLOROPHYCEAE OR CHLOROPHYTA)

These algae are clear green. The chlorophyll is not masked by other pigments. This group comprises practically all the more familiar types of freshwater algae and also contains interesting marine types. Morphologically, they range from simple rounded single cells through various sorts of filaments, plates, sheets, and cylinders of cells. There are also curious types without definite cell walls, but rather, with a continuous cytoplasm containing many nuclei, a condition termed **coenocytic.** Some groups of unicellular algae are frequently classed as Protozoa (Phytomastigina). In the marine environment, they are less diversified and less important than the other two classes, the brown and the red algae.

Perhaps the best known and most familiar marine green algae are members of the so-called confervoid algae. They belong to a group, variously termed **Ulotrichales, Confervales,** or **Ulvales,** in which there is formed a multicellular thallus of simple or branched filaments, or a hollow flat or tubular thallus. They reproduce asexually by the production of flagellated zoospores, and sexually with differentiated motile sperm cells, and egg cells. There is an alternation of generations that are reproductively different but that have a similar appearance.

Sea lettuce, *Ulva,* forms a large, flat thallus resembling lettuce leaves and comprised of one or two cell layers. It grows abundantly in protected places in the intertidal and upper subtidal zones. The species are edible and are dried and sold under the names green laver and Iceland sea grass. One species with narrow fronds resembling intestines is called water gut.

Another widespread genus is *Enteromorpha,* which forms large coarse complex branched filaments of tubular structure. It grows abundantly in bays and estuaries and as a waterline fouling plant on boats and pilings.

A very interesting group is the **Codiales** or **Siphonales.** The vegetative body of the alga is without cross walls, so that the cytoplasm is continuous throughout the plant. Such an arrangement is termed **coenocytic** or **syncytial.** The plants may become quite large and have a fleshy, complexly branched thallus, as in *Codium,* which has numerous interwoven, thick, cylindrical filaments with branching, club-shaped apexes.

Some are large hollow bladderlike single or multinucleate cells, attaining sizes up to an inch or more long, as in *Halicystis* or *Valonia* (Siphonocladales). Because of their large size, these cells have attracted the interest of biologists and have been much used in studies of penetration of ions and differential concentration of ions and other substances into the cell sap.

In *Caulerpa,* there is a multinucleate creeping stemlike thallus that has rhizoids on the lower side and gives rise to foliose expansions above.

Classification of green algae

The classification of the green algae is summarized below. Only the more important marine types are included.

Class Chlorophyceae
Order Cladophorales: Cylindrical multinucleate cells joined end-to-end, forming branched or unbranched filaments. The structure could be interpreted as coenocytic but with septa, or as a row of multinucleate cells. This structure shows some similarity to the coenocytic structure of the

Text continued on p. 51.

1	Blue-green algae
2	*Pelvetiopsis*
3	*Halosaccion*
4	*Endocladia*
5	*Fucus*
6	*Gigartina*
7	*Porphyra*
8	*Iridaea*
9	*Alaria*
10	*Cystoseira*
11	*Nereocystis*
12	*Laminaria*
13	*Postelsia*
14	*Odonthalia*
15	*Microcladia*

Fig. 4-2. Intertidal rock face in exposed area near Bastendorf Beach, Oregon, showing zonation of fifteen kinds of larger marine algae present.

Pelvetiopsis limitata

Halosaccion

Continued.

Fucus

Fig. 4-3. Individual algae from Fig. 4-2.

Gigartina

Porphyra

Iridaea

Fig. 4-3, cont'd. For legend see p. 47.

Alaria

Cystoseira osmundacea

Nereocystis luetkeana

Laminaria setchellii

Fig. 4-3, cont'd. For legend see p. 47.

Continued.

Postelsia palmaeformis

Odonthalia floccosa

Microcladia boreale

(detail)

Fig. 4-3, cont'd. For legend see p. 47.

Codiales, which wholly lack cross walls. Sexual reproduction by conjugation of isogametes.
Cladophora
Order Codiales (Siphonales)
Codium, Caulerpa, Halimeda
Order Dasycladales: Marine, warm seas. Erect central axis with whorls of branches along the entire length or only at the top.
Order Siphonocladales: Multicellular thallus attached by a stem of rhizoids. Cells mononucleate and divide in a unique manner. Warm seas.
Halicystis, Valonia
Order Ulvales
Ulva, Enteromorpha

BROWN ALGAE (PHAEOPHYTA OR PHAEOPHYCEAE)

Most of the familiar large seaweeds belong to this group. They are brown or olive-brown in color, the chlorophyll in the chromatophores being masked by fucoxanthin or other carotenoids. Almost all are marine, anchored to the substrate by holdfasts. The primary food reserve is a carbohydrate, laminarin, dissolved in the cell sap. The reproductive cells, both zoospores and gametes, are motile and have two laterally inserted, unequal flagella. All the Phaeophyceae are multicellular and are either filamentous or have a complex, thalluslike portion, the stipe, and a basal holdfast that is often rootlike in appearance. There are about 900 species, all but three, marine.

Brown algae occur in all oceans but are most characteristic of, and often predominate in, the algal flora of temperate-to-cold waters. For the most part they are found near shore in water not more than 20 meters deep.

Classification of Phaeophyta

The class is divided into the following subgroups:

Class Phaeophyceae
Subclass Cyclosporeae: No gametophyte generation. The spores function as gametes instead of developing into gametophytes. The reproductive organs are in round cavities, the **conceptacles,** borne in the swollen tips of the branched thallus.
Order Fucales
Subclass Phaeosporeae: Alternation of gametophyte and sporophyte generation. The gametophytes bear multicellular gametangia, in which each cell contains a gamete. Some have unicellular gametangia with single gametes. The sporophyte

bears unicellular sporangia, with spores in multiples of two (commonly 64 or 128). A few produce multicellular sporangia that look like gametangia.
Grade Isogeneratae: Gametophyte and sporophyte of same appearance.
Orders Cutleriales, Dictyotales (Ectocarpales), Sphacelariales, Tilopteridales
Grade Heterogeneratae: Gametophyte always irregularly branched filaments. Sporophyte may or may not be filamentous but is not similar in appearance to the gametophyte.
Subgrade Haplostichineae: Filamentous sporophytes.
Orders Chordariales, Desmarestiales, Sporochnales
Subgrade Polystichineae: Nonfilamentous sporophytes.
Orders Dictyosiphonales, Laminariales, Punctariales
Order Fucales: See subclass Cyclosporeae, above, for characteristics.
Family Durvilleaceae
Durvillea, Splachnidium
Family Fucaceae
Ascophyllum: Bladder wrack or tang. In eastern North America a dominant member of the intertidal *Balanus, Fucus, Littorina* biome.
Fucus: Rockweed. Olive-green to brown algae. Thallus tough, leathery, flattened, repeatedly forked, with numerous air vesicles. Intertidal in northern temperate zone. Tips of branches are inflated, bearing conceptacles.
Pelvetia: Occupies highest part of fucoid intertidal belt.
Pelvetiopsis
Himanthalia: Sea thong. Temperate to boreal North Atlantic. Both sides.
Sargassum: Gulfweed. Wide distribution in warm waters. Commonest genus of brown algae in warm waters.
Turbinaria
Order Cutleriales: Filaments laterally compacted, forming bladelike or disclike body. Gametophytes with groups of multicellular gametangia. Gametes markedly anisogamous. Sporophytes with unicellular sporangia.
Cutleria
Order Dictyotales: Branched, ribbon, or fanlike growth initiated by solitary or laterally adjoined apical cells. Gametophytes with patches of sex organs called sori. Female gametophyte with unicellular gametangia (oogonia). Male gametophyte with many-celled gametangia (antheridia). About 20 species, 100 genera. Predominant brown algae in tropical seas, often growing in considerable abundance.
Dictyota
Ectocarpus: Slender, filamentous, branching, with abundant gametangia.
Order Sphacelariales: Threadlike thallus, with cells arranged regularly in transverse tiers. Growth of each branch initiated by an apical cell. About

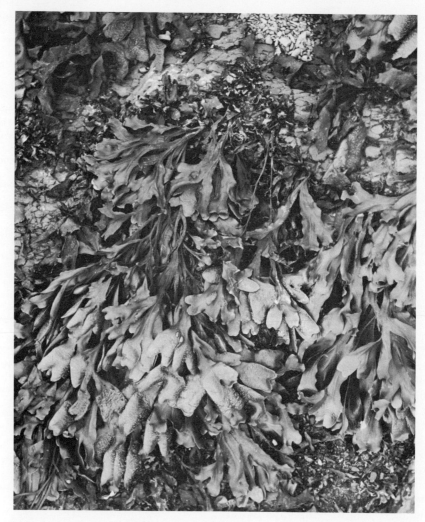

Fig. 4-4. *Fucus* on intertidal rocks at Bolinas, California. (M. W. Williams photograph.)

15 genera, 175 species. Mostly restricted to the Atlantic Ocean.

Sphacelaria

Order Tilopteridales: Freely branching thallus. The lower part resembles *Sphacelaria,* the upper part, *Ectocarpus.*

Order Chordariales: Sporophyte filamentous. Gametophyte resembles irregularly branched gametophytes of *Ectocarpus.*

Leathesia: Bladderweed.

Order Desmarestiales: Sporophyte obscurely filamentous because the branches become covered with irregularly arranged cells. Oogamous gametophytes. Both antheridia and oogonia are unicellular.

Desmarestia

Order Sporochnales: Sporophytes filamentous. Oo-gamous gametophyte. Both antheridia and oogonia are unicellular.

Order Dictyosiphonales: Sporophyte profusely branched and threadlike, with internal differentiation of tissues.

Order Laminariales: This group includes many familiar large seaweeds known as kelps. The sporophyte may become very large, up to 30 meters long or more in some species, with differentiation of tissues internally and of the thallus into holdfast, stipe, blades, and gas floats. Sporangia are unicellular, grouped into sori on the blades.

The gametophytes are microscopic, irregularly branched filaments of rarely more than fifty cells. Male gametophytes have many antheridia, each with one antherozoid. Female gametophytes have

Fig. 4-5. *Ascophyllum* on rock wall at Marblehead Neck, Massachusetts. Zonation shows well here. Above the *Ascophyllum* is a very heavy growth of barnacles. The *Ascophyllum* itself is growing in a patchy manner over and among mussels *(Mytilus edulis)*. Along the upper edge of tbe *Ascophyllum*-mussel zone can be seen numerous carnivorous gastropods (whelks), whereas at the lower edge of this zone and extending well up into it are numerous starfish *(Asterias)*. In the pool at the left can be seen a group of large sea anemones *(Metridium)*. (Helen P. Finlayson photograph.)

one to six oogonia, each with one egg. There are about 30 genera, mostly limited to the North Pacific, but *Laminaria* and a few others are found in all oceans.

Alaria (badderlocks): Olive-brown, in northern seas. Stipe bears numerous short, obovate blades basally and ends as a long, flat blade.

Egregia (ribbon kelp): Thallus grows as a long, beltlike strap with blades and gas floats along the edges. Common along the Pacific coast of North America.

Laminaria (sea tangle): A large genus growing in all oceans. Flat, broad blade or ribbonlike thallus—single blade, with relatively short stipe and strong holdfast. Often grows just below low-tide level, where it is subject to very strong wave action. Sometimes so abundant as to form "laminaria forests," along with *Alaria*.

Macrocystis: One of the giant kelps of the Pacific coast of North and South America. Numerous blades, each with a bladder at its base, borne at regular intervals along the branched stem. Blades about 80 cm. by 40 cm., irregularly corrugated with denticulate margin. Monotypic for *M. pyrifera. Macrocystis* often attains very large size. Reports of specimens 700 feet long have been made. The largest reliable actual measurement along the Pacific coast of the United States was 140 feet. Perennial on rocky bottoms in 7 to 30 meters, especially on open coast, where there is a continuous swell. Forms large kelp beds that may protect nearby shore from excessive wave action.

South of Point Conception, California, *Macrocystis* is practically the only kelp in the kelp beds. North of Point Conception, it tends to become mixed with, or largely replaced, by *Nereocystis.*

Nereocystis (bladder kelp, bullwhip): Second most abundant giant kelp of the Pacific coast of North America, especially north of Point Conception, California. The hollow stem may be many meters long, ending in a single large gas float from which a cluster of long blades float at the surface of the water. May exceed 100 feet in length. *Nereocystis* is an annual. During the growing season in March to June it may grow as much as 10 inches per day. The main species is *N. luetkeana.*

Pelagophycus: Another giant kelp. May attain a length of 100 feet or more.

Postelsia (sea palm): Prominent in lowest inter-tidal zone and top of subtidal zone, on exposed rocks with strong wave action. Stands erect on a thick, cylindrical, flexible stipe, with a cluster of blades arising from the apex. Looks like a small palm tree, about 1 to 2 feet high, growing on the rocks and exposed to the heaviest surf, where relatively few organisms can maintain themselves.

Order Punctariales: Sporophyte bladelike, saccate, without internal differentiation of tissues. Medium size.

Asperococcus, Scytosiphon

RED ALGAE (RHODOPHYTA OR RHODOPHYCEAE)

The red algae contain, in their plastids, in addition to chlorophyll, the red pigment phycoerythrin and sometimes the blue pigment phycocyanin. This group contains the most specialized and complex types of algae. The food reserve formed is an insoluble carbohydrate called floridean starch. Sexual reproduction is unique in that nonflagellated male gametes, spermatia, are passively transported to the female sex organs. There are no flagellated zoospores.

There are about 400 genera and 2500 species. The group is almost exclusively marine, about a dozen genera and less than 100 species being found in fresh water. Red algae are present in all seas, including the polar seas, but are best developed in the tropics, where they often constitute the predominant benthic flora. Most of them grow attached to the substrate in shallow or intertidal areas. In the North Atlantic and North Pacific they rarely occur at depths greater than 35 meters, but in the tropics they have been dredged as deep as 170 meters.

The thallus is commonly a simple or complexly divided blade, or it may have a stemlike portion and separate blades. In some groups, the coralline algae, the body is heavily impregnated with lime. Coralline algae play an important role in the building of coral reefs and atolls.

Classification of red algae

There are two subclasses, the **Bangioideae** and the **Florideae.** The taxonomy of the red algae is still in an unsatisfactory condition, particularly with regard to definitions of larger units such as subclasses and orders.

Subclass Bangioideae: Order **Bangiales:** About 15 genera and 60 species. Cell division occurs anywhere in the body. The fertilized zygote divides directly into **carpospores.**

Bangia fusco-purpurea grows in uppermost levels of the littoral zone. Tolerant to desiccation and warming.

Porphyra (laver). Several species. Grows as thin red or purple blades, tough and leathery, one or two cells thick on rocks in the intertidal zone. The most conspicuous and abundant member of the Bangiales. Of commercial importance for food.

Subclass Florideae: Several orders, about 385 genera, 2400 species. Body filamentous, but the filaments are often laterally compacted to give the macroscopic secondary forms; cell division only in terminal cells of the branches.

The male sex organ, the **spermatangium,** is unicellular and produces one nonflagellated gamete, the **spermatium,** which upon liberation floats to and lodges against a hairlike prolongation of the female unicellular sex organ (**carpogonium**) at the apex of a carpogonial filament.

Upon fertilization, the zygote in some groups grows directly into a special filament, the **gonimoblast filament,** which bears sporangia (**carposporangia**) containing carpospores. In other groups the zygote produces a delicate tube, the **ooblast,** which grows from the carpogonium to another cell of the plant. The zygote nucleus migrates along the ooblast to that cell, and the gonimoblast fibers then grow from this **auxiliary cell.** The fruit of the sexual plant, consisting of the gonimoblast fibers and the carposporangia jointly, constitute the **cystocarp.** The cystocarp has been interpreted as a very small asexual generation growing on the sexual plant.

Each carposporangium liberates one spore. In primitive groups the reduction divisions occur in the zygote stage, and the carpospores then are haploid and develop directly into gametophytes. In more specialized forms the zygote nucleus does not undergo reductional divisions but produces 2n carpospores, which develop into 2n free-living asexual plants of the same size and general structure as the gametophytes. This asexual plant, the **tetrasporophyte,** bears **tetrasporangia** in which the single nucleus undergoes meiosis, producing **tetraspores.**

Upon liberation, the tetraspores develop into gametophytes.

Order Nemalionales: The most primitive of the Florideae. No tetrasporophyte. Gonimoblast filaments develop from the carposporangium. About 35 genera. The taxonomy of this order is in a particularly uncertain condition and in need of a thorough reworking.
Nemalion, Acrochaetium, Gloiopeltis

Order Gelidiales: Gonimoblast filaments grow directly from the carpogonium. No auxiliary cell. Tetrasporophyte present in life cycle. About 6 genera.
Gelidium: Much branched cartilaginous fronds. Cystocarps immersed in swollen branchlets. Includes the agar weed *G. cartilagineum,* an important source of agar-agar.

In the remaining orders the tetrasporophyte is present in the life cycle, and the gonimoblast filaments develop from an auxiliary cell rather than from the carpogonium.

Order Gigartinales: The auxiliary cell is an ordinary vegetative cell of the plant body, not borne on a special filament. There are about 70 genera.
Chondrus: A rather coarsely branched cartilaginous, dark purple seaweed.

On the rocky North Atlantic coasts of Europe and North America *Chondrus crispus,* or Irish moss, forms a broad band of low dense growth in the lowest part of the intertidal zone below the band of *Mytilis* (Fig. 2-2). This growth shelters a multitude of small animals, and the stemlike portions are usually coated with incrusting bryozoans (*Membranipora, Microporella,* and others).
Gigartina: A large, chiefly Pacific genus, with fleshy cartilaginous compressed fronds.
Iridophycus: Forms rather extensive glistening purplish sheets in sheltered low rocks, intertidal areas. Lanceolate to ovate blades.
Hypnea: Often found in subtropical waters. Thick growth just above sublittoral zone below the fringe of coralline algae and zoanthids.
Ahnfeltia
Phyllophora: A major source of iodine.
Agardhiella

Order Ceramiales: The auxiliary cell from which the gonimoblast filaments arise is a special cell at the base of a carpogonial filament. It arises after fertilization. Many species are delicate, lacelike, finely divided forms with fruiting bodies. There are about 160 genera.
Ceramium: Large, widely distributed genus, in both Atlantic and Pacific. Delicate, finely divided.
Dasya: Feathery appearance due to many slender branches from larger axial filaments.

Polysiphonia: Thallus is filamentous. Much-branched, specialized antheridial branches, and complex carpogonium and cystocarp; characteristic tetraspores.
Rhodomela

Order Cryptonemiales: The auxiliary cell is borne on a special filament resembling a carpogonial filament. About 85 genera.
Pikea
Aeodes: Large scattered isolated sheets hanging down from point of attachment on intertidal rocks when uncovered at low tide. Found above *Champia-Vermetus* zone, and extends above, through, and below the Lithothamnium zone.
Schizomenia
Hildenbrandia
This order also includes the coralline algae, family **Corallinaceae,** distinguished by its calcareous habitat. The thallus becomes hardened and brittle from deposition of calcium.
Corallina: Freely branching nodulose growth.
Lithothamnion: Encrusting habit.
Porolithon: Rocklike appearance. Commonly the dominant organism on exposed seaward upper edge of coral reefs.

Green algae of the genus *Halimeda* may help build up a coarse sediment, which may become consolidated by lime deposition or overgrowth of coralline-encrusting red algae, such as *Lithothamnion* and *Porolithon.* This may play a role in the building of noncoral reefs or contribute to the formation of coral reefs. In Funafuti Lagoon, for example, down to depths of 30 meters, 80% to 95% of the sediment is comprised of *Halimeda.* Some of these lime-secreting algae can grow at greater depths than most corals and are thus able to create deposits below the main body of growing corals.

Coralline algae enjoy a worldwide distribution and are not limited to the tropics, as are the corals. In the Indian Ocean numerous "coral reefs" contain no coral but are built largely of coralline algae.

Both encrusting and erect nodulose coralline algae occur commonly on the lower intertidal and subtidal rocky areas in temperate and cool regions. Encrusting species commonly produce a reddish or purplish rose-colored band in the lower intertidal zone, where sea urchins are abundant.

In Scotland some nulliporous white sand beaches are composed largely of ground-up remains of offshore coralline algae, and as far north as Spitsbergen, there are miles of calcareous banks formed by coralline algae.

> **Order Rhodymeniales:** Auxiliary cell is a special cell borne adjacent to the base of a carpogonial filament, developing there prior to fertilization. Ooblastema filament very short. About 25 genera.
> *Champia*
> *Halosaccion*
> *Rhodymenia*
> *Sphaerococcus:* "Worm moss," or "wormweed."

Angiosperms

Very few higher plants have become adapted to marine habitats, although there are a rather large number of estuarine and salt marsh types found around the edges of brackish lagoons and mud flats, and many terrestrial "maritime" plants occur only near the coast.

We shall consider two groups: (1) the eelgrasses and their relatives, which are the only angiosperms to have evolved a wholly marine way of life, and (2) the mangrove complex, which forms a unique and important habitat along many tropical shores and offshore islands.

TRULY MARINE ANGIOSPERMS

By "truly marine" angiosperms we mean those which tolerate the salinity of marine water, grow while wholly submerged, and have the capacity for underwater flowering and pollination. Furthermore, all of them have the ability to anchor themselves firmly enough with their roots to withstand the usual surge of waves and tide in their particular habitats. None are planktonic. All have threadlike pollen, which drifts by and catches on the styluslike stigma.

Five families of plants, all of which be-

Fig. 4-6. Eelgrass in Tomales Bay, California. Open anemone shows that the area is covered by water. Streaming of eelgrass shows direction of tidal flow. (M. W. Williams photograph.)

long to order **Naiadales (Helobieae),** have evolved such species.

Ruppiaceae: *Ruppia*
Zosteraceae: *Zostera, Phyllospadix*
Posidoniaceae: *Posidonia*
Zannichelliaceae: *Zannichellia, Althenia, Syringodium, Amphibolis, Cymodocea, Halodule*
Hydrocharitaceae: *Halophila, Enhalus, Thalassia*

Some authors include the first four families under **Potamogetonaceae.**

Naiadales is a group composed of mostly freshwater aquatic plants. The families that contain marine genera are widely represented in freshwater habitats, and some of their freshwater species had already evolved underwater flowering and pollination. The occurrence of such species as *Potamogeton pectinatus* and *Zannichellia palustris,* which inhabit both fresh and brackish waters, and typically brackish genera such as *Ruppia* and *Althenia,* show how the transition from fresh to salt water may have occurred. Since freshwater Naiadales are themselves a highly specialized group evolved from terrestrial ancestors, it would appear that the marine representatives must be among the most recently evolved groups of angiosperms.

The tropical-subtropical species are quite distinct from those of the temperate and boreal regions, and include the genera *Amphibolis, Cymodocea, Enhalus, Halodule, Halophila, Syringodium,* and *Thalassia.* The species in turn may be grouped as Indo-West Pacific species, ranging in warm waters from the east coast of Africa to the Indo-Malay region and the Pacific islands, and the tropical American species centered in the Caribbean.

Tropical marine angiosperms

Indo-West Pacific group

Enhalus acoroides, Halodule pinifolia, Halodule tridentata, Thalassia hemprichii: Wide distribution from East Africa to the Indo-Malay region and the Pacific islands.

Halophila beccarii: Rather rare, scattered along Malaysian coasts and east to Borneo and the Philippines.

Halophila stipulacea: Red Sea and West Indian Ocean (spread to Mediterranean Sea through canal and now colonizing coasts of Greece).

Syringodium isoetifolium
Halodule uninervis

Tropical American group

Halodule bermudensis: Only in Bermuda Islands.
Halodule beaudettei: Amphi-American.
Halodule ciliata: Caribbean coasts of Panama.
Halophila bacillonus
Halophila englemannii: In West Indies and later on Texas coast.
Syringodium filiforme
Thalassia testudinum: Turtle grass. Florida and West Indies.

Halodule, Halophila, Syringodium, and *Thalassia* have their greatest species density in the Indo-Pacific region and probably originated there. All are absent from tropical West Africa except *Halodule wrightii,* which has an amphi-African distribution, and from the Pacific coast of America, except the amphi-American *Halodule beaudettei,* found in the Atlantic and Pacific coasts of Central America.

It is of interest that some of the Indo-West Pacific species find their nearest related species in the tropical Caribbean region. Such species form pairs of sibling, or twin, species. Examples of twin species are as follows:

Indo-West Pacific species	Tropical American counterpart
Thalassia hemprichii	*Thalassia testudinum*
Syringodium isoetifolium	*Syringodium filiforme*
Halodule uninervis	*Halodule beaudettei*

Temperate-boreal marine angiosperms

The temperate-boreal group is characterized by the genera *Althenia, Phyllospadix, Posidonia,* and *Zostera. Zostera* has the most extensive range, being found on both East and West coasts of North America, and on the coasts or Atlantic Europe, East Asia, Australia, and South Africa. One species of *Posidonia* is found in the Atlanto-Mediterranean region, another in Australia. *Phyllospadix* has an amphi-Pacific discontinuous distribution, being found along the shores of both eastern Asia and western North America, but not along the northern arc between. Since none of these plants are well adapted for very long transport over open ocean, the discontinuous distributions of some of the genera and species have stimulated considerable speculation regarding the modes and routes of transport and the length of time involved.

Zostera, Phyllospadix, Cymodocea, and *Thalassia* are perhaps the best known and the most important genera from an ecological standpoint. They form submarine meadows, contributing a major part of the organic production in such areas, forming a protected habitat for characteristic assemblages of animals, stabilizing the bottom, and serving as a site for increased sediment deposition. *Zostera,* grass wrack, grows especially in protected areas in bays, estuaries, and river mouths, where the water is relatively quiet at least part of the time.

The ecological importance of these beds of eelgrass was dramatically demonstrated in 1931 when the beds of *Zostera marina* along the Atlantic seaboard of America and Europe suffered from a severe epidemic or infestation with a parasitic slime mold, *Labyrinthula,* which wiped out most of the beds. Upon the disappearance of the *Zostera* the numbers of cod, shellfish, scallops, crabs, and other associated animals fell off sharply. The oyster industry of much of northern Europe and the Atlantic coast of America was ruined. Plant and animal life somewhat farther out from river mouths was overwhelmed by raw sewage, which formerly had been filtered out in the *Zostera* beds. Silt from erosion caused by lumbering and farming inland and silt eluted from the substrate of former *Zostera* beds spread farther than before, building new mud flats on formerly gravel bottoms. Smelt and other fish that spawned on these graveled areas could no longer do so and greatly diminished. Formerly richly productive bays and estuaries became wastelands. Brant and other birds that fed on the bivalves and other organisms in the eelgrass community began to die off. Brant decreased to one fifth their former number.

At Great Bay, New Hampshire, the bay became choked with sewage and silt. There were no oysters in Oyster River, no salmon at Salmon Falls. Other fish decreased drastically. The increased effects of pollution prevented eelgrass recovery. Former boatyards and shipping facilities were sludged up and deserted. The effects of the eelgrass catastrophy were still apparent and strongly felt in some areas as much as 20 to 30 years later. In this case, the activities of man, causing excessive pollution and erosion, combined with a natural disaster to produce an overwhelming effect on the local biota.

Smaller eelgrass disasters have occurred in coastal lagoons of eastern Australia when flocks of hundreds of black swans descend upon the *Zostera* beds to eat the stolons. They sometimes denude whole areas of several lagoons. This results in failure of mullet and shrimp fisheries and is harmful to successful spawning and settlement of oysters. Again, this is partly a man-made disaster, the black swans having been deliberately introduced into Australia by man.

Upon removal of the *Zostera,* the redox potential of the sediments increases and much of the silt fraction is washed out by increased water movement. Since *Zostera* roots prefer a reduced environment containing silt and free hydrogen sulfide, these changes seriously impede their return and slow the recovery of these areas.

In late summer *Zostera* flowers, then dies and rots. This rotting of old plants produces rapid growth of microorganisms, especially bacteria, flagellates, and ciliates, which, together with rising temperatures of late summer, stimulates spawning of the Japanese oyster and provides abundant food for its larvae.

THE MANGROVE COMPLEX— TIDAL WOODLANDS

Mangroves do not represent a particular taxonomic group of plants. The term is used for all the woody plants that invade shallow tidal areas, forming "mangrove swamps" or coastal woods extending into zones covered with seawater at high tide. Preeminent among the mangrove plants are members of the family Rhizophoraceae, with their characteristic branching, stilt-like roots holding the rest of the tree above the water. Species of *Rhizophora* are usually the outermost, or fringing, members of mangrove swamps and often constitute the only, or at least the dominant, growth in such areas.

Mangroves are found along shallow protected tropical humid coastlines around the globe, often forming impenetrable thickets. Through their role in stabilizing the substrate, holding sediments, and adding to the detritus, they help build soil and add

Fig. 4-7. Mangrove islands in the Ten Thousand Islands area off the Florida coast. (U. S. Forest Service photograph.)

to the land area of the coast, to form and extend low offshore islands.

Seawater is "physiologically dry." Terrestrial plants using it as a source of water are in danger of excess transpiration. Those which have become adapted to growth in the tidelands tend to have modifications somewhat like those of certain desert plants, such as thickened coriaceous, or waxy, leaves, to reduce transpiration. They are favored by conditions that reduce transpiration, such as humidity and cloudiness. Mangrove forests develop in parallel with tropical rain forests. Cloudiness is apparently an essential climatic condition for both.

Rhizophora is uniquely adapted for life in shallow tidelands. The trees themselves send out adventitious roots from branches above the water, supporting them and extending the growth vegetatively out over the shallow water. Long, pendant seedlings are produced. Upon dropping from the trees, some of the seedlings penetrate the soft substrate, standing upright and taking root where they were dropped. Others are carried away by tides and currents; they may survive for months and be carried for great distances. If they lodge along a shal-

low shore or submerged bar or fringe of an island, they take root and start to grow. In twenty to thirty years the new trees are fully mature and are creating a new mangrove swamp.

One of the greatest mangrove swamps of the world has developed along the southern coast of Florida, from the Florida Keys around the southern tip of the mainland and north along the gulf coast through the Ten Thousand Islands. The thousands of mangrove islands here are for the most part shaped as elongated oval islands, larger and with a lagoon situated at the southeast end. It seems probable that they developed along wave-created sand ridges or ripple marks in the shallow sea floor, the islands of today reflecting the shape and trend of the wave marks of yesterday. Over a period of decades the building of new land, the coalescence of small islands to form larger ones, and the merging of the shoreline with nearby offshore islets can be seen. Whether this entire area will eventually become low-lying land, as now seems to be indicated, or will be reclaimed by the sea, no one can say. A rise in sea level or isostatic changes in the earth's crust could reverse the direction of thousands of years of evolution of the coastline and shallow sea, turning it in new and unexpected directions.

In the Florida–West Indies mangrove swamps the roots of the mangrove are commonly thickly settled with mangrove oysters *(Ostrea frons)* which cling to the roots by curious, fingerlike extensions of the shell. In the mud among the roots and between mangrove thickets are innumerable burrows of fiddler crabs *(Uca pugilator).* Coffee bean shells *(Melampus bidentatus)* are abundant among the roots and along the upper edges of the swamp, where they climb on the stems of salt marsh grasses as the tide rises. Beautiful little shells of the rose tellin *(Tellina lineata)* may lie scattered on the mud, indicating the presence of a colony of these bivalves living nearby just beneath the mud surface.

Prominent among the predators in this community are the large crown conchs, *Melongena corona,* which prey on the oysters. At night on low tides, raccoons wander out into the mangrove swamp to feed on the oysters.

Curiously, as the mangrove plants have evolved as an assemblage of land plants invading the margins of the sea, the swamps they produce have been the site for the evolution of some striking cases in which marine animals have become adapted for life on the margin of the land. Mangrove periwinkles *(Littorina angulifera),* whose relatives live in the sea, have colonized the roots and branches above the tidal level. Land-dwelling crabs, the white land crab *(Cardisoma guanhumi)* and the purple-clawed hermit crab *(Coenobita clypeata),*

Fig. 4-8. A, Mangrove. Same area as shown in Fig. 4-7. Closeup of roots and young plants exposed at low tide. **B,** Young mangrove. (**A,** U.S. Forest Service photograph; **B,** Chesher photograph.)

which return to the sea only to spawn, live along the upper margins, out of the water.

As the mangrove swamp grows it is colonized with a characteristic assemblage of marine animals found nowhere else. It forms nesting and roosting sites for many large birds, which, with their guano, fertilize and help build the swamp.* The interior parts of the swamp become more terrestrial in character and may be colonized by other kinds of plants of the mangrove complex.

In the Indo-Pacific the curious "climbing fish," or mudskippers, *Periophthalmus,* may be seen at low tide, walking and jumping about on the mud or even climbing onto some of the mangrove roots by using their pectoral fins as limbs, and stopping at small puddles to moisten their gills. If frightened, they disappear into burrows in the mud, or if they must cross water, they do not enter it but skip across the top with a series of strong flexures of the body.†

Mangrove swamps of the Indo-West Pacific, from tropical East Africa to the Pacific islands, differ in species composition from those of the tropical Atlantic, found especially along the Caribbean–Gulf of Mexico region and to a lesser extent along the coast of tropical West Africa. Those of the Indo-West Pacific are best developed in parts of India, and in the Malay Archipelago. They are composed of a rather wide assortment of plants, as the following list indicates. Those of the tropical Atlantic are

---

*In some cases small mangrove islands used as nesting sites are "burned out" by excess guano, which kills the trees and eventually destroys the island.
†I once stopped for lunch with my family by a small mangrove swamp on the east end of Java. As we explored the swamp, dozens of curious mudskippers observed us, looking for all the world like some strange cross between frog and fish, with their large, raised eyes bugging out from the top of their heads, and their awkward jumping and walking movements. They seemed to be alert, intelligent creatures, displaying a genuine curiosity about us but staying well out of reach. We discovered that they liked bread crumbs—a new item in their diet. As we walked through the tangled mangrove, scattering bread crumbs, more and more of them followed us, until we felt indeed like the Walrus and the Carpenter.

B

Fig. 4-8, cont'd. For legend see opposite page.

best developed in the region of the Florida Keys and the adjacent gulf coast.

I. Indo-West Pacific
    Family Rhizophoraceae
       Rhizophora mucronata Laup. (the most important and widespread species)
       R. conjugata Linn.
       Ceriops candolleana Arn.
       C. roxburghiana Arn.
       Kandelia rheedii W. & A.
       Bruguiera gymnorrhiza Lamk.
       B. eriopetala W. & A.
       B. caryophylloides Bl.
       B. parviflora W. & A.
    Family Combretaceae
       Lumnitzera racemosa Willd.
       L. coccinea W. & A.
    Family Lythraceae
       Sonneratia apetala Buch-Ham.
       S. acida Linn.
       S. alba Smith.

Family Meliaceae
   Carapa moluccensis Lamk.
   C. obovata Bl.
Family Myrsinaceae
   Aegiceras majus Gärtn.
Family Rubiaceae
   Scyphiphora hydrophyllaceae Gärtn.
Family Verbenaceae
   Avicennia officinalis Linn. and var. *alba* Bl.
Family Acanthaceae
   Acanthus ilicifolius L.
Family Palmae
   Nipa fruticans Thumb.

II. Tropical Atlantic mangrove
    Family Rhizophoraceae
       Rhizophora mangle L.
    Family Combretaceae
       Laguncularia racemosa Gärtn.
    Family Verbenaceae
       Avicennia tomentosa Jacq.
       A. nitida Jacq.

PART III

ANIMAL LIFE
of
THE SEA

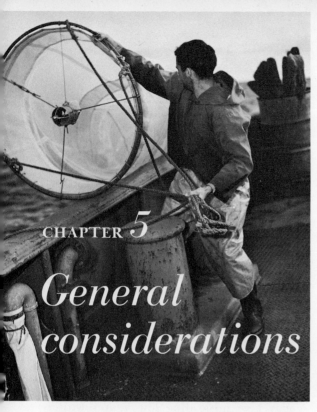

CHAPTER 5

General considerations

Fig. 5-1. One-meter plankton net being launched from deck of M.S. Horizon, a research ship of the Scripps Institution of Oceanography. (M. W. Williams photograph.)

Marine plant life consists largely of simple unicellular forms, microscopic in size. The larger plants, confined to the edges and shallow areas, are far fewer in number of species, and for the most part are relatively simple and primitive, compared to the luxuriant variety and complexity of plant groups dominant on land. Animal life, on the other hand, is rich and diverse in the extreme. All phyla and major classes of animals are represented in the fauna of the oceans. Many are found only there.

PLANKTON FEEDERS

In the ocean, plankton feeders are in a position somewhat analogous to that of the plants on land. That is, their food is all about them in the ambient medium. They need methods of concentrating it. A large majority of the zooplankton organisms, the free-living micro-benthic forms, and a majority of the larger animals are either plankton or detritus feeders. Many of them have ciliary or ciliary mucous mechanisms for creating currents of water and retaining plankton; others have networks of bristles, hairs, or tentacles with which they strain or filter the water. Those which subsist primarily on zooplankton tend to have coarser filtering apparatus than those which eat primarily phytoplankton or minute bits of detritus.

When we consider the benthic animals, where there is substrate to which they can attach or into which they can burrow, we see in the most dramatic way the effect of having a major food supply in the ambient medium. Here the majority of the most abundant and dominant animals, like plants on land, live as sessile organisms. Many, such as the bushy hydroid colonies, gorgonians, sea fans, and sea lilies, are even plantlike in their growth habits. Before biologists had a clear concept of the basic differences between plants and animals,

65

many of marine animals were classified as plants.

In shallow shelf areas and the lower intertidal zones, where plankton is most abundant, we find animals growing crowded together in dense masses, as close together as physically possible, sessile, and often growing on top of each other, much in the manner of land plants. Almost all phyla include forms so modified. A few examples are attached protozoa, sponges, coelenterates, bryozoans, tube worms, bivalve molluscs, barnacles, and tunicates.

USE OF DISSOLVED ORGANIC SUBSTANCES BY MARINE ANIMALS

In 1909 Putter suggested that many of the plankton feeders might also benefit by the uptake of dissolved organic materials in the water. Krogh (1931) reviewed the evidence available at that time and concluded that there was little evidence to support this view. However, it has been shown for some filter feeders, such as mussels (Fox and co-workers, 1952, 1953), that the volume of phytoplankton removed from the water by them is insufficient to account for their growth and metabolism. They also showed that *Mytilus* can remove fish hemoglobin from the water. MacGinitie (1939), using stained proteins, concluded that *Urechis* and *Chaetopterus* can remove particles of the order of 40 Å. Stepheus and Shiuske (1957) demonstrated that a wide variety of ciliary-mucus–feeding invertebrates remove amino acids from the water but that arthropods generally fail to do so. They suggest that whatever the mechanism, it is not direct filtration, as this would imply a pore size too small to allow effective filtration of the volumes of water known to be filtered by such animals. They suggested adsorption of organic molecules onto the mucus as a possibility. Fox and co-workers suggested another mechanism by which such substances might be made available to filter feeders, i.e., by their adsorption onto the surface of colloidal micelles or other particles suspended in the seawater that are large enough to be filtered. Polymerization of surface-active molecules into larger aggregates might also

contribute to this process, especially where such molecules and other particles are concentrated from the water surfaces in small zones of convergence. Values for the amount of dissolved organic material in seawater run from 3 to 6 mg. per liter.

If a 5% efficiency of removal is assumed, an organism would receive 0.15 to 0.3 mg. of organic material per liter of water filtered, which falls within calculated ranges of need. However, it is also questionable whether the composition of the dissolved organic matter is such that most of it would be directly available to the organism even if ingested. It may largely represent compounds refractory to bacterial and probably also to digestive degradation.

The nutritive significance of the uptake is then questionable and probably differs greatly for different animals.

The values for dissolved organic matter obtained by chemical analysis of the water may give misleading low figures with respect to the very classes of compounds most available to and most used by living organisms. Labile compounds may be utilized by bacteria and other organisms as fast as they are produced, so that there may be a rather rapid production and use of quantities of such compounds, even though chemical analysis of the water fails to show their presence in significant amounts. More refractory compounds would tend to accumulate and to be the ones indicated by chemical analysis. The presence and nature of populations of heterotrophic bacteria in the water are probably the best indicators of the presence and probable amounts of such compounds produced in a given water mass.

SUBSTRATE GRAZERS

The surface of rocks or any submerged objects quickly becomes covered with a film of colloidal and molecular organic matter, bacteria, diatoms, other algae, and small attached animals. This "algal film" offers opportunity for another great group of animals that can pick or scrape it off from the rock surfaces efficiently enough to utilize it as food. Such grazers are particularly well represented by gastropod molluscs and chitons, which cling to and slowly creep over the surfaces, scraping the algal film off with a filelike tongue, the radula.

Where the substrate is sand or mud, the organic film in and on it can be utilized by various burrowing or surface-crawling worms, sea cucumbers, balanoglossids, etc., which ingest the substrate, digesting the organic matter contained and ejecting the sand grains as the bulk of their feces.

PREDATORS, SCAVENGERS, AND PARASITES

Anywhere a population of plants and animals exists, a portion of them will evolve as predators, scavengers, and parasites. We find these groups in all major environments. Since scavengers and predators depend upon discrete sources of food (as do also primary consumers on land), rather than upon food throughout their ambient medium, they must be able to move actively or to trap their food. They do not differ so greatly or in such a uniform manner from one sort of environment to another, in either structure or habits, as do the producers and primary consumers. Their modifications everywhere are more diverse and depend chiefly upon the particular kind of food they eat and what they must do to obtain it, together with the physical character of their particular environment.

Parasites are animals that have adopted the habit of living on or in other animals. Commonly, they have become tied in an obligate manner to this association and are unable to live as free-living organisms. Often they have lost the ability to synthesize some vital nutritional factor or factors which they can obtain directly from their host. Thus parasites are commonly deficient in biosynthetic abilities, as compared with free-living relatives. Many groups of parasites have become much modified for parasitic existence, showing reduction or loss of sensory and locomotor functions and organs, and sometimes even of the digestive tract as well, and a great increase in reproductive potential. The latter reflects the difficulty of moving successfully from one host to the next. Commonly there is an exaggerated development of suckers, hooks, or rhizoidal attachment organs. The life cycle is often complex, adapting them to particular successful pathways from host to host and modes of access to their hosts.

Usually a mutual adjustment evolves between parasites and their hosts so that the latter are not overwhelmed by their guests, to the mutual destruction of both. In some instances mutual adjustment has evolved to such a degree that neither can survive alone without the other. Such a relationship is termed **mutualism.** There are all shades of relationship, from the strictly parasitic, in which the parasite lives at the expense of the host, inflicting injury upon it, to the completely mutualistic, such as exists between many cellulose-eating animals and the protozoan and bacterial fauna and flora of their gut. The term **symbiosis** (literally, "living together") is used in its broad, literal sense by some biologists to cover all relationships here included in the term parasitism. Others have used it in the same sense as mutualism. Most parasites fall somewhere between these extremes; they live with their hosts, inflicting little noticeable damage and doing the host no apparent good. Such a relationship is termed **commensalism.** Some biologists prefer to restrict the term parasitism to cases in which the guest organism clearly lives at the expense of its host, and to use the terms commensalism, symbiosis, or mutualism in referring to the others. The distinction is not always sharp, and we will use the term parasite in its broader and more general sense to refer to any organism that lives in or on another (usually larger) animal.

Since most free-living animals harbor one or more kinds of parasites, often in great numbers, there are probably more kinds of parasites, and certainly many more individuals, than there are free-living animals. Because of their small size and hidden habitats, we often fail to realize this. We sometimes speak of three major habitats, the marine, the freshwater and the terrestrial. The advent and proliferation of living things on earth provided a fourth habi-

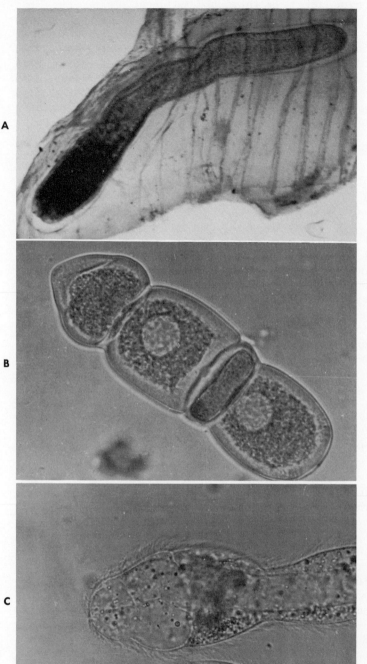

Fig. 5-2. Variety of parasites.

A, *Ouwensia catostyli,* a vermiform parasite of the rhizostome jellyfish *Catostylus,* discovered in 1965 by Miss S. H. Moestafa in jellyfish from near New Guinea. Its affinities are unknown. It may be a larval stage of some familiar but incompletely known group or, perhaps, represents a new class or phylum of worms. The same, or every similar, form was independently rediscovered by P. J. Phillips in 1969 in a related jellyfish, *Stomolophus,* in the Gulf of Mexico—a region thousands of miles away, and completely cut off, from New Guinea. This may indicate great antiquity for the relationship between these worms and the jellyfish hosts—perhaps going back to the Mesozoic era when an uninterrupted tropical marine biota extended by way of the Tethys Sea from the Indo-Pacific to what is now the Gulf and Caribbean region. It is hoped that Mr. Phillips will have an opportunity to carry out further studies to elucidate its life cycle and relationships.

B, *Gregarina valettei* from *Pollicipes polymerus.*

C, Calotte of a dicyemid "mesozoan" from *Octopus.* Dicyemids have long been a controversial, little-understood group. Their life cycle, morphology, and systematic relationships have been given the most diverse interpretations by various zoologists who have studied them.

(**B** and **C,** Photographs by James L. Houk, Moss Landing Marine Laboratories; ×400.)

tat—the parasitic—of such importance that more than half of all the kinds of animals now living occupy it.

LIFE BELOW THE EUPHOTIC ZONE

Below the euphotic zone life is less abundant than in the surface layers and depends for its nourishment entirely upon organic materials drifting down from above, or upon the ability of animals to rise up into the euphotic zone at times to feed. There are no producers.

Conditions here are far more uniform over most of the world than is the case near the surface, and from region to region there is a greater uniformity of animal life. There is darkness, cold, and immense hydrostatic pressure. Currents are slower than at the surface. Seasonal changes are slight or absent. It is not surprising that animals inhabiting such an environment should have evolved what to us appear to be fantastic forms and adaptations.

Pelagic fish from the upper strata of water tend to be silvery or light-colored ventrally and on the sides, resembling the bright reflective air-water interface when seen from below, and darker, bluish, or greenish above, resembling the darker water below when seen from above. The darker color dorsally also serves as obliterative countershading, making the shadow on the ventral portion less conspicuous. Crustacea from upper strata are commonly nondescript gray, whitish, or somewhat transparent. At intermediate depths many of the fishes are much more uniformly silvery and tend to have larger eyes. Here the light intensity is about the same in all directions, and obliterative countershading would have no function. Demersal fish and crustacea at these depths often are partly red, with subcuticular red chromatophores or oil droplets. In very deep water the fish tend to be entirely black (except for bioluminescent organs), and the crustacea are commonly either uniformly red, with the red pigment incorporated into the cuticle, or black. In very deep waters the eyes are usually small and in some cases absent.

Numerous other animals, such as most hydromedusae, ctenophores, siphonophores, salps, and occasional species from other groups, are so transparent that they are difficult to see at whatever level they are found. Deep-sea medusae, however, also tend to be red or purplish.

In the bathypelagic and abyssal zones, food is relatively scarce and hard to find. Many animals have evolved fantastic adaptations for obtaining it and the ability to ingest large amounts when it is found. Some can swallow objects larger than themselves. Often they have a fierce, dragonlike appearance, with large mouths filled with long, sharp teeth, long tactile processes for probing the dark water about them, large eyes (especially animals that live where there is only very dim light), and commonly, photophores. They have evolved as peerers, gropers, and gulpers.

The luminescent organs serve several purposes. They may make species recognition possible, playing a role in keeping schools of the same species together, which is important in mating and reproduction. They may play a role in courtship activity in some species. Some fish use them to attract other food organisms. A few have evolved a dorsal fin ray resembling a fishing pole. It can be extended forward just above and in front of the mouth, dangling a luminescent organ at the end to attract other organisms within easy reach. Luminescence also serves as a defense against predators. A sudden flash of light when a fish is disturbed may momentarily blind or scare a predator, enabling the prey to escape. Some deep-water squid have evolved luminescent ink, which has the same functions in defense and escape in the zones of darkness that the usual black ink of their shallow-living relatives has in better lighted zones.

Species recognition, attraction of food, and defense, then, are some of the functions of bioluminescence.

VERTICAL MIGRATION

Many animals find the twilight zone, where there is a bare minimum of light, a favorable depth during the day. Here it is more difficult for predators to see them, yet they have not gone so deep that they cannot swim up into the richer waters nearer the surface to feed at night. **Diurnal vertical migration** is a striking feature of the life of many marine animals. The usual tendency is to go down during the daylight hours and come up as the light decreases. Some organisms follow an isolume rather closely—that is, they adjust their depth during the day in such a way that they remain throughout the day in water with a particular degree of intensity of illumination. This would, of course, result in their descending to deeper water during the morning and ascending again late in the afternoon, or during cloudy weather. However, this will not serve as a general explanation of diurnal vertical migration, since most plankters do not scatter out at night, as might be expected, and many of them begin their downward migration well before dawn and their upward migration after sunset, at hours when little or no change of illumination is occurring. Nor is the concept that hunger combined with negative phototropism a wholly satisfactory explanation, since in areas where the bulk of the phytoplankton is at some depth in the water, many zooplankters migrate through the most dense strata of phytoplankton to the less dense surface waters.

Many other animals live at different depths at different times of the year or during different stages of their life cycle.

Such periodic changes in depth are often important in other ways than in feeding and avoiding predators. For example, water at different depths may be drifting in different directions. This drifting may play a role in keeping a population of animals in a given region in spite of the movements of the water. In other cases animals may take advantage of the interfaces between different water masses moving in different directions. By concentrating at such a level they can avoid the full effects of drift in either direction.

Because a large proportion of the animals in the euphotic zone descend to lower depths during the day and rise toward the surface at night, and because the phytoplankton remains nearer the surface and waters of different depths move at different rates or even in different directions, there is a continual redistribution of the animals in the phytoplankton. A given cloud, or bloom, of phytoplankton does not necessarily carry the same animals with it from day to day as it drifts across the face of the ocean. Animals finding themselves in phytoplankton-exhausted surface waters one night may find themselves in richer water on a subsequent ascent.

Fig. 5-3. Diurnal vertical distribution of male and female *Calanus finmarchicus*. South of Ireland, August, 1932. (Modified from Farran, 1947.)

For very small animals such as copepods, daily vertical migrations of any extent mean that they must spend most of their time swimming up or down. Hardy and Bainbridge (1954) devised an ingenious plankton wheel in which planktonic animals could be introduced into a large rotating circular glass wheel-shaped tube filled with seawater. The speed of rotation was adjusted to the speed of upward or downward swimming, at various times and under controlled conditions of lighting so that the speed and reactions of the animals could be readily observed and studied. They clocked upward movement of the copepod *Calanus* at 15 meters per hour, *Centropages* at nearly 30 meters per hour, and the very small copepod *Acartia* at 9 meters per hour. The euphausiid *Meganyctiphanes* achieved 135 meters per hour, while the polychaete *Tomopteris* attained over 200 meters per hour. Downward speeds often exceeded the upward speed of the same species by a considerable amount. *Calanus,* for example, can descend 47 meters in an hour and is capable of short bursts of speed at rates exceeding 100 meters per hour. The results of these experiments suggested that the observed daily migrations in the sea were indeed feasible for these small animals.

Enright and Hammer (1967), in carefully controlled, large-scale experimentation with mixed zooplankton just pumped in from the offshore saltwater line at the Scripps Institution of Oceanography, demonstrated that for some species there is an endogenous physiological timing mechanism, or **biological clock,** that causes them to retain the timing pattern of their vertical migration behavior for some time under conditions of constant illumination and temperature. With others this was not the case, and exogenous stimuli were required to elicit the vertical movements. They conclude that "although vertical migration of zooplankton may appear to be a single phenomenon in an ecological context, conceivably evolved on the basis of a common selective pressure, the mechanisms underlying the field behavior are by no means uniform."

The mesopelagic depths also contain many animals that live there as permanent residents, in addition to those which perform daily or seasonal migrations into and out of them.

SURFACE DISTRIBUTIONS

It is commonly observed that the surface distributions of phytoplankton and of zooplankton are often patchy in character, the maxima of zooplankton being found not where the phytoplankton clouds are thickest but near their edges, or in waters between the great phytoplankton clouds. Hardy and Gunther (1935) suggested the **animal exclusion theory** to account for this. According to this theory a dense cloud of phytoplankton may impede the upward movement of many zooplankton species through it, so that they may be prevented from completing their upward migration until the most dense portions of the phytoplankton cloud have drifted past, at which time they rise to higher levels, thus creating a discontinuity between the maxima of phytoplankton and of zooplankton. However, as Bainbridge (1949) found, for *Neomysis vulgaris,* at least some zooplankters will actively swim from diatom-poor waters into diatom-rich waters, especially if they are partially starved. Harvey (1934) proposed that great swarms of zooplankton may consume the phytoplankton to such an extent that they thin it out locally, producing the observed discontinuities. Steemann-Nielsen (1937*b*) noted that differential rates of multiplication might produce similar effects, the fast-growing phytoplankton producing dense blooms where they are locally more abundant in the water at a favorable time. Slower growing zooplankton might graze upon this, and by the time their population became heavy, the bloom of phytoplankton would have passed its peak because of exhaustion of surface nutrients.

It seems probable that all these mechanisms may play a role in producing the discontinuities at various times and places.

Hardy and co-workers (1936) devised a very simple plankton indicator for herring fishermen to aid them in locating areas where zooplankton, especially *Calanus,* abound. It had been observed that herring are usually found where there are clouds of *Calanus,* and tend to avoid areas of dense phytoplankton bloom. The indicator consisted of a cloth disc that was towed behind the fishing boat for a given length of time, and then examined. If it showed a reddish color as a result of having retained many of the red copepods, the area was judged to be favorable for fishing, but if it was green, from retaining mostly phytoplankton, the area was judged to be less favorable. Boats making use of the plankton indicator reported consistently higher catches than those which did not use it.

DEEP-SCATTERING LAYER

When great numbers of animals migrate downward during the day, they form one or more strata at intermediate depths sufficiently dense to reflect an echo from echo-sounding equipment, giving a record of a "false bottom," the so-called deep-scattering layer. As night approaches, the deep-scattering layer rises and becomes more diffuse or even disappears, and then forms again at dawn. The daily increase of animals at such intermediate depths may induce a sort of reverse diurnal migration of certain predators living still deeper. These predators may rise to intermediate levels to feed during the day and return to their murky depths at night.

Large schools of fish at intermediate depths will also produce echoes, often char-

acteristic for particular species. Echo-sounding equipment has become a valuable aid for commercial fishermen in locating and trawling for such schools, in addition to its value as a navigational aid and for mapping the bottom.

Commonly, three fairly well-defined deep-scattering layers are formed—the deepest at about 520 to 580 meters, and sometimes descending to around 700 meters, composed of small myctophid fishes; the second at about 430 meters, mostly of euphausiids; and an upper stratum, at around 230 to 270 meters, of sergestid shrimp. At night the three layers rise and merge into a single diffuse band up to 150 meters thick, but before dawn the three bands begin to sort themselves out again and descend to their characteristic levels, the animals of the lowest strata beginning their descent first. The layers never cross. Other kinds of fish and crustacea, as well as siphonophores and squid, may also be associated with deep-scattering layers.

The animals in the deep-scattering layers are not as profuse and crowded as had been anticipated from early studies of their echograms. Observations by divers in bathyscaphes and small research submarines, and net hauls from deep-scattering layer levels indicate that, more often than not, the animals of these strata are at some distance from each other, and that the deep-scattering layers are not likely to support any spectacular new commercial fishery.

Fig. 6-1. Veliger larva of a gastropod mollusc. (Photographed alive by Ralph Buchsbaum.)

CHAPTER *6*

The zooplankton

Protozoa

*T*here are three groups of protozoa that are almost exclusively marine and that are found in the plankton of all oceans and seas. These are the Foraminifera, the Radiolaria, and the Tintinnidae.

ORDER FORAMINIFERA

The Foraminifera are shell-bearing rhizopods. They are primarily benthic or attached, but there are a number of important pelagic species. The cytoplasm is rather homogeneous and streams out through one or more apertures in the shell, and in perforate forms, through the many small pores as well, forming long, fine, branching, anastomosing, radiating, or net-like pseudopodia, the so-called **reticulopodia** characteristic of this group. In living specimens these pseudopodia are in constant activity, extending, contracting, fusing with others, etc. Minute granules can be seen moving along them, those in the inner core of a pseudopodium going, in general, in an outward direction, and those in the outer sheath of the pseudopodium passing inward. The cytoplasm may contain brown granules that seem to be waste matter, and in some are extruded prior to the formation of a new shell or new chambers of a shell. In some species the cytoplasm is colored. No contractile vacuoles are present.

The test, or shell, varies greatly in form and color. In most species it is less than 1 mm. in diameter or length, although a few attain several millimeters. It may be almost wholly or partly arenaceous, with foreign particles such as sand grains incorporated into it, or calcareous—either perforate (pierced by many small pores in addition to the main aperture) or imperforate and porcelaneous. The young of all species with calcareous tests are perforate, only later depositing a porcelaneous, smooth layer over the outside of the test. The majority of species remain perforate.

Polythalamous species, with tests of more than one chamber, first form a single chamber, the **proloculum,** to which other chambers are added later during growth. In some species the chambers themselves may become subdivided into chamberlets. The septa between chambers are perforated by one or more pores, called stolon canals, through which the protoplasm is continuous to all chambers. The last chamber bears one or more apertures, which are the main openings of the shell.

73

Foraminifera

Fig. 6-2. Foraminifera. **A**, *Globigerina bulloides* d'Orbigny, dorsal view (after H. B. Brady). **B**, *Globigerinoides sacculifera*. **C**, *Globorotalia fimbriata* (after H. B. Brady). **D**, *Hastigerina pelagica* (after H. B. Brady). (Redrawn from Cushman, J. A. 1959. Foraminifera: their classification and economic use. Harvard University Press, Cambridge, Mass.)

Foraminifera, for the most part, feed on diatoms and algae, but pelagic species may also catch other protozoans and microcrustacea. Some of them carry symbiotic zooxanthellae in their cytoplasm.

Most Foraminifera exhibit dimorphism, occurring in two forms that differ from each other in the size of the first chamber, the proloculum, and commonly in other details as well. The **megalospheric** form possesses a large proloculum. The organism has a single nucleus, usually found in the middle chamber and shifting its position to maintain this relationship as new chambers are added. When full growth has been attained, the nucleus divides rapidly many times and isogametes are formed

which leave the parent test and fuse in pairs to form zygotes.

The zygote secretes a **microspheric** proloculum around itself. As growth occurs and new chambers are added, the nucleus continues to divide, forming a multinucleate individual. The nuclei in the various chambers vary in size proportionately with the size of the chambers.

Eventually, after full growth is attained, the cytoplasm around each nucleus becomes individualized and the young mononucleate cells desert the parent shell, secrete a megalospheric proloculum around themselves, and begin the cycle again.

This reproductive sequence may be variously modified in some species, especially by the interposition of more than one asexually produced megalospheric generation before gamete formation, so that megalospheric individuals tend to be more abundant than microspheric ones. For some species, no microspheric generation is known.

In a general sense, in dimorphic species, the microspheric phase may be said to be the more conservative and the more likely to include stages indicative of the phylogeny of the species. The megalospheric form is more likely to be specialized and to skip over such early primitive stages in its development. In fossil Foraminifera, therefore, study of early microspheric and mature megalospheric stages of a species shows something of its ancestry and of its relationship to later descendants, respectively.

Movement is slow in most species, up to 6 cm. per hour, but usually only 1 cm. or less per hour. Some species have been observed to abandon their test, move about for a time as naked protoplasts, and then make a new adult test. In species that incorporate foreign materials into their tests, these are first ingested, and then later concentrated at the surface from within and held in place with a secretion of cementing substance. This explains the growth of one-chambered forms and also the predominance of adult shells in collections.

The earliest, most primitive sort of test is chitinous. Next in order are the arenaceous tests, which are developed by adding foreign material to the exterior of chitinous tests with various sorts of cementing materials. The cement may be chitinous, ferruginous, calcareous, or siliceous. As the cement becomes a more dominant feature, its secretion renders foreign material unnecessary, and a smooth test results. Those with ferruginous cement are usually yellowish or brown, and most are arenaceous. Those with calcareous cement constitute the majority of species. They are all perforate when young, but some species later become covered with a smooth, imperforate layer. The change from chitinous-arenaceous test to purely calcareous test can sometimes be seen in one individual, the early chambers being arenaceous and later

ones being purely calcareous. Among arenaceous types, more primitive species show less selectivity of materials, while advanced types often show remarkable selection and arrangement of foreign materials. Species with arenaceous tests are benthic, and below 2000 meters, they are usually the only ones present.

Some holoplanktonic types, such as species of *Globigerina,* are so abundant that the bottom ooze for many square miles may be composed largely of their calcareous shells, which have sunk to the floor of the seas, accumulating over the centuries as the so-called **globigerina ooze.** Such sediments may, in later periods, become consolidated and made into calcareous rock or chalk, as in the famous white cliffs of Dover. Globigerina ooze, such as that found in large areas in the North Atlantic, commonly contains approximately three-fourths Foraminifera shells and one-fourth coccolithophores, by weight. The most abundant species are the foraminiferan *Globigerina bulloides* and the coccolithophore *Pontosphaera huxleyi.*

Faunal groups can be recognized by co-occurrences of species in particular water masses. There are some 28 species of very abundant, widely distributed pelagic Foraminifera, 22 of them occurring in tropical and subtropical waters, four in boreal or cool temperate waters, and one each characteristic of transitional and of polar waters. Most species spend their early stages in the upper part of the euphotic zone, and then mature at somewhat deeper levels below the euphotic zone, but for the most part in the upper 200 meters. Bé (1968) determined the **porosity** of the shells (a function of the number of pores per unit area, and pore size) for many of the species. He found that in addition to the fact that Foraminifera from cold waters tend to make, on the average, smaller, more compact shells than do those from warm waters, they also have smaller, more closely spaced pores. Porosity (the proportionate area occupied by pores) is lower, however. There is, then, a correlation between the temperature at which the shells are formed and their porosity, and thus, secondarily, between latitude and porosity. In this connection, he also called attention to the work of Wiles (1960), who found that in deep Atlantic cores, the concentration of pores in one species, *Globigerina eggeri* Rhumbler (= *G. dutertrei* d'Orb), varied at different levels in the core, and that changes in pore concentration could be correlated with Pleistocene climatic changes. High pore counts appeared in the interglacial sediments when an assemblage of warmer-water species was present, and low counts appeared in sediments corresponding to glacial epochs. Although Wiles did not determine the shell porosities, Bé points out that if the size of the pores remained fairly constant in this species, the results would be in line with the general differences found between cold- and warm-water species.

According to Waller and Polski (1959), salinities of less than $27^0/_{00}$ are in general limiting to the survival of planktonic Foraminifera. Benthic forms from inshore shelf regions tend to be more euryhaline than do strictly pelagic species, since they commonly encounter more variable salinity.

Most of the planktonic Foraminifera are in the upper 200 meters of water, and none of them are below 2000 meters. At great depths the calcareous shells begin to dissolve. Shells of planktonic species are absent from sediments at depths of 5000 meters or more. Berger (1967) suspended samples of washed shells of mixed species from eastern Pacific core samples at various depths for a month and noted that the rate of dissolution not only increased with depth but was quite differential among the various species, those with lighter or thinner tests disappearing first. Thus the depth at which a sediment is formed not only will determine whether or not tests of pelagic Foraminifera will be preserved in it but also, in those containing Foraminifera, will strongly influence the species composition.

Classification of Foraminifera

Some authorities divide the Foraminifera into more than 300 genera distributed among 45 or more families. For details, spe-

cial monographs of the group should be consulted.

ORDER RADIOLARIA

The Radiolaria are mostly rather large, spherical, pelagic rhizopods from 50 μ to several millimeters in diameter, sometimes forming colonial aggregations up to several centimeters across. The pseudopodia are radiating axopodia or filopodia, and in many there are also radiating skeletal elements as well. The cytoplasm is divided into a central portion surrounded by a capsule and containing the nucleus or nuclei and an outer, or extracapsular, portion. The nucleus, when single, is sometimes very large and may be polymorphic. Some radiolarians have many smaller nuclei instead.

The **central capsule** is a tectinous membrane with pores through which there is a continuity of intracapsular and extracapsular cytoplasm. The pores may be uniformly distributed or arranged in one to three **pore fields.** The extracapsular cytoplasm, the **calymma,** is commonly rather stiff and gelatinous and filled with large vacuoles containing a watery fluid, giving it a frothy appearance. There are no contractile vacuoles. In several groups of Radiolaria the calymma also contains rounded yellowish bodies, symbiotic zooxanthellae.

Other cytoplasmic inclusions, such as fat droplets, pigment granules, and crystals, are sometimes present. The pseudopodia continue as denser rays of cytoplasm through the calymma to the central capsule.

The skeleton, usually of silica, except in the Acantharia, which possess skeletons of strontium or calcium aluminum sulfate, may be radiate, concentric, or of mixed type. Many radiolarians form extremely intricate and beautiful lattice-type skeletons. Radiolarians are holozoic, feeding on diatoms and small zooplankton caught among their pseudopodia. Those which have zooxanthellae may also obtain some nutrients from them.

Like the calcareous skeletons of Foraminifera, radiolarian skeletons accumulate at

Fig. 6-3. Radiolaria. **A,** *Dictyacantha tetrogonopa.* **B,** *Acanthostaurus purpurascens.* (**A** redrawn from Schewiakoff, W. *In* P. P. Grassé [ed.] 1953. Traité de zoologie. Vol. 1. Masson & Cie, Paris.)

the bottom of the seas and are readily fossilized. The oldest known Precambrian fossils are radiolarian skeletons. In some of the deeper places in the oceans subject to immense hydrostatic pressure, the calcareous shells of Foraminifera may be dissolved, but the glassy skeletons of Radiolaria persist and come to characterize the bottom deposit, so that a radiolarian ooze rather than a globigerina ooze may be formed.

Classification of radiolarians

Four suborders are commonly recognized. The first is sometimes separated from the Radiolaria and treated as an independent group of equivalent rank with Radiolaria and Heliozoa.

Suborder Acantharia. The skeleton has long, radiating spines, mostly of strontium sulphate, and often united outside the animal by latticework. The spines radiate from the center of the central capsule and are regularly arranged. Each is surrounded by a circle of contractile elements, the myonemes, as they leave the calymma, constituting a hydrostatic mechanism. Very young individuals are uninucleate, but when fully grown, the animal possesses numerous small nuclei arranged in two or three layers. The inner cytoplasm is often colored because of assimilation of pigments from their prey. Symbiotic zooxanthellae and sometimes aberrant parasitic dinoflagellates are present. The animal is usually spherical and rather large.

Suborder Nassellaria (= Monopylea or Monopylaria). These are oval uninucleate forms. The skeleton is siliceous, characteristically in the form of a tiara or helmet. The extracapsular cytoplasm is nonpigmented, but the inner cytoplasm may be colored. The central capsule has a single large aperture through which the axopodia protrude, grouped into bundles.

Suborder Phaeodaria (= Tripylea). The central capsule commonly has a double membrane. There are three apertures, or pore fields: a basal one, the astropyle, and two lateral parapyles. A mass of dark pigment, the phoeodium, is found near the astropyle. The skeleton is extracapsular and composed of hollow, siliceous elements containing a considerable amount of organic matter—an organic silicate or an amorphous silica. There are no zooxanthellae and no axopodia. The animal is polykaryotic and multiplies by binary fission

A   **B**

Fig. 6-4. Radiolaria. **A,** *Thalassicolla pellucida.* **B,** *Carposphaera nodosa.* (Redrawn from Haeckel. *In* P. P. Grassé [ed.] 1953. Traité de zoologie. Vol. 1. Masson & Cie, Paris.)

after division of the nuclei. Sometimes more than one central capsule may be present in one animal. Formerly all radiolarians with more than one central capsule were grouped into suborder Polycyttaria, but they are now distributed among various other suborders, and this name is descriptive rather than taxonomic.

Suborder Spumellaria (= Peripylea or Peripylaria). The central capsule is pierced all over with pores. The skeleton is absent or composed of concentrically arranged spicules or latticed spheres. Radial spines or spicules may be attached to the shell in some species.

CLASS CILIATA: FAMILY TINTINNIDAE

This is a large family of mostly marine ciliates belonging to the order Spirotricha. Characteristically, tintinnids secrete a chitinous or pseudochitinous **lorica** about themselves, within which they live. The lorica tends to be quite characteristic for a given species, and since it is commonly the only part that can be studied satisfactorily in ordinary plankton samples, it has been extensively used as a basis for the description and classification of species. The lorica is commonly shaped like a vase or bowl, and in many species it also incorporates foreign particles such as sand grains, coccoliths, or fecal detritus. Both sessile and free pelagic species occur. Some are also found in brackish or fresh waters.

Kofoid and Cambell (1929) described over 300 species, which they grouped into 12 families (here regarded as subfamilies) and 51 genera.

The body is somewhat trumpet-shaped and is attached aborally at the narrow end to the base of the lorica. Strong myonemes enable it to contract rapidly back into the lorica. It bears a conspicuous aboral zone of sixteen to twenty-two membranes, sometimes plumose, or feathery, in appearance. There may also be four or more longitudinal rows of shorter cilia on at least the anterior portion of the body. A cytopharynx with neurofibrils is also present.

There are one to four (commonly two) macronuclei and micronuclei.

Sometimes a rounded cell bearing a large nucleus is seen in the cytoplasm. This is a parasitic dinoflagellate, *Duboscquella* of the family Blastodiniidae.

Nongelatinous Metazoa

CRUSTACEA

By far the most numerous and important members of the zooplankton are the Crustacea, which swarm in the seas as insects do on land. Wherever zooplankton is particularly rich in crustaceans, it attracts and supports a great population of fish and other predators.

As with other arthropods, the body of a crustacean is encased in a jointed cuticular **exoskeleton** secreted by the epithelial cells, in arthropods commonly termed the **hypodermis.** This exoskeleton is shed at intervals throughout the crustacean's life, allowing for growth and change in form, and dividing its life into a succession of **instars** marked by **ecdysis,** or molting, at the end of each instar.

Crustacea are characterized by five pairs of head appendages: two pairs of **antennae,** one pair of **mandibles,** and two pairs of **maxillae.** The first and second pairs of antennae are often termed **antennules** and **antennae,** respectively. The first and second pairs of maxillae are commonly called **maxillulae** and **maxillae.** These are followed by a variable number of thoracic and abdominal appendages. The thorax is often termed the **pereion,** and its appendages, the **pereiopods.** The abdomen is termed the **pleon** and its appendages, **pleopods.** The abdomen ends in the **telson,** which bears the anus, but it has no appendages and is not regarded as a somite. The number of thoracic and abdominal somites varies greatly in different groups of Crustacea. The number of apparent segments of the body, as seen from above, is even more variable because of the frequent occurrence of coalescence or fusion of two or more adjacent somites to form one apparent segment.

The Crustacea are divided into five subclasses: **Branchiopoda, Ostracoda, Copepoda, Cirripedia,** and **Malacostraca.** The first four subclasses are often collectively termed the **Entomostraca.** Entomostracans are usually smaller than most malacostracans and are regarded as more primitive. Most of them hatch from the eggs as nauplial larvae. The **nauplius** is a minute oval free-swimming larval form with a single median eye near the anterior end, and, in the first instar at least, only three pairs of appendages, all pediform, representing the two pairs of antennae and the mandible.

COPEPODS

More than 4500 species of copepods have been described, of which about 90% are marine. Approximately 3000 are free-living, and the remainder are parasitic. Copepods comprise the bulk of the zooplankton. In both numbers of individuals and numbers of species they exceed all the rest of the metazoan plankton combined. They are a key group in the economy of the seas. The free-living copepods are usually very small, ranging from 0.2 mm. to about 2 cm. in length.

Copepods have fifteen **somites.** The first five comprise the head, or **cephalon.** These somites bear the two pairs of antennae, the mandibles, and the two pairs of maxillae. The thorax is composed of six somites. These bear the maxillipeds and the five pairs of swimming legs. The genital somite occurs behind the hinge joint, or major articulation. In females it is usually united with the first abdominal somite, forming a compound genital somite as the last thoracic somite (some researchers consider it to be the first abdominal somite). In any case it is part of the urosome and bears either no appendages or very rudimentary ones. The abdomen consists of the last three somites plus the anal segment, or telson. In copepods these do not bear appendages. Behind the abdominal somites is the telson (anal segment of Davis and regarded as a true segment), which bears the caudal rami. In some free-living copepods the fifth pair of legs is much modified, reduced, or absent. Parasitic copepods in some groups, especially the females, show such great modification as to be hardly recognizable as copepods in the adult condition.

The number of apparent segments, as seen from above, is fewer than fifteen because of coalescence, or fusion, of certain somites. The somites of the head plus one or two thoracic somites are fused together into a complex that looks like a single anterior segment. In some groups the last two thoracic somites are fused. In females the genital and first abdominal somites are usually fused together as the "genital segment." In some the last two abdominal somites are fused.

The body of most copepods is divided into a broader oval convex anterior region of several somites, the **prosoma,** in which the segments are firmly united and immovable, and a narrower posterior movable portion, the **urosome.** The articulation between these two regions of the body is termed the **hinge joint,** or **major articulation.** Some authorities define the urosome as the genital somite plus all somites behind it. It seems preferable to define it functionally as that part of the body behind the major articulation. The major articulation does not fall between the same somites in all groups of copepods. This fact, together with the differences in apparent segmentation caused by various combinations of fusion between somites in different groups of copepods, has caused some confusion in terminology and in the interpretation of copepod morphology. When one is reading about copepods, it is important to determine how the author is using the terms. We shall use the terms in the sense outlined in the tabulation below.

prosoma The anterior part of the body in front of the major articulation (= metasome in Davis usage and that of some other writers).

cephalosome The apparent first segment when this comprises the head plus the first thoracic somite.

cephalothorax The apparent first segment when it comprises the head plus two or more thoracic somites.

metasoma The thoracic segments behind the cephalosome or cephalothorax, and in front of the major articulation.

urosome The part of the body behind the major articulation. This includes the abdominal somites and anal segment, or telson, with the caudal furca, and in some copepods, it also includes the posterior thoracic segment.

Those copepods in which the major articulation occurs between the last pedigerous somite and the genital somite, and in which, therefore, the urosome bears no appendages, have been classed as the **Gymnoplea.** They comprise the suborder **Calanoida.** In the other suborders the major articulation occurs anterior to this, so that the urosome includes one or more thoracic somites in addition to the genital somite, and hence may bear one or more pairs of appendages or rudiments thereof. These have been termed the **Podoplea.**

Locomotion is jerky and rapid, and produced by strong, synchronous beating of the swimming legs, during which the long antennules lie back along the sides of the body. When the beating stops the antennules are extended laterally, slowing down sinking. Bursts of active swimming alternate with short periods of gliding or resting. The gliding movements of forms such as *Calanus* and *Diaptomus* are caused by the currents set up by the anterior feeding appendages. Some gliders can reverse the currents and glide backward as well as forward.

Diatoms constitute the principal food of most planktonic copepods. In many, as in *Diaptomus* and *Calanus,* the appendages around the mouth (second antennae, mandibular palps, and first maxillae) are clothed with setae and are rapidly vibrated (between 600 and 2640 times a minute), creating swirls of water that pass along both sides of the body. The resulting eddy is sucked forward into a median filter chamber by the maxillipeds, and out of the chamber laterally by the first maxillae. There is a dense screen of setae on the second maxillae. The food particles caught in these setae are scraped off by the endites of the first maxillae and by special setae of the base of the maxillipeds and passed to the mandibles and mouth.

In others, such as *Acartia,* instead of acting as passive filters, the maxillae are modified to act as scoop nets. Some copepods capture larger organisms in addition to filtering for phytoplankton or detritus, and others are wholly predaceous.

Copepods are the only entomostracans that form **spermatophores.** The male grasps the female and deposits spermatophores on the genital segment of the female at or near the opening of the seminal receptacles, by means of a special adhesive cement.

When the eggs are laid they are surrounded by an **ovisac** produced from secretions of the oviduct, so that they stay attached to the female in the ovisacs, which serve as brood chambers. Each sac contains from a few to about forty eggs, and clutches may be produced frequently. Some species set the ovisacs free in the water rather than retaining them attached to the female. Many calanoids simply shed the eggs singly into the water. The egg hatches into a free-swimming planktonic nauplius, a small oval form with three pairs of appendages and a single median eye. The appendages represent the first antennae, second antennae, and mandible. The two latter in this stage are biramous pediform appendages. No trunk segmentation is evident. Successive molts gradually add more appendages, beginning with those at the anterior end. There are commonly six nauplius instars (five molts), after which the body form changes to one similar to the adult form. Immature stages of this kind are termed copepodids. Usually there are five or more molts, the mature individuals constituting the sixth instar after attainment of the copepodid form.

Classification of copepods

The copepods are divided into seven suborders, of which three, Calanoida, Cyclopoida, and Harpacticoida, each with more than 1000 described species, contain free-living forms. Of these the Calanoida are for the most part planktonic, the Cyclopoida both planktonic and benthic, and the Harpacticoida largely benthic in habit, though some of them are also found in the plankton. These three groups may be distinguished as follows:

The major articulation in the suborder **Calanoida** is between the last thoracic and first abdominal somites. Second antennae are biramous. First antennae are commonly very long, nearly equal to, or exceeding the body length, and are composed of more than 20 segments. Heart is usually present. Last pair of legs of male is asymmetrical and modified for clasping and transfer of spermatophores. In those species in which the female carries the eggs there is only a single ovisac, carried ventrally beneath the abdomen. This suborder contains the majority of important planktonic species of copepods.

In the suborder **Cyclopoida,** the major articulation is between the fifth and sixth thoracic somites. Second antennae are uniramous, and the first antennae are usually clearly shorter than the body and composed of twenty or fewer (usually seven to sixteen) segments. Ovisacs are paired and carried symmetrically, one at each side of the abdomen. Includes numerous benthic and parasitic forms as well as planktonic species.

The metasome-urosome articulation in the suborder **Harpacticoida** is between the fourth and fifth thoracic segments, but the body is often tapered rather gradually, so that there is not as distinct a break between the metasome and urosome, as in other copepods. Abdomen may be nearly as broad as the thorax. Antennules are very short, rarely more than eight or nine segments, and second antennae are biramous. Both second antennae of male are prehensile and used to hold female. Mostly benthic, a few are parasitic and a few planktonic.

The families of calanoid copepods were grouped into two subgroups by Giesbrecht (1892)—the **Amphascandria** and the **Heterarthrandria.** In the Amphascandria the antennules are symmetrical or nearly so in both sexes, those of the male bearing more sense organs (aesthetasks) than those of the female. The last pair of legs of the male are modified for grasping and transfer of spermatophores, while those of the female may be normal, reduced, or absent. The eyes are fused into a typical copepod single median eye, without a cuticular lens. Last, thoracic and first abdominal segments of the male are not markedly asymmetrical.

The Heterarthrandria are characterized by having one of the antennules (usually the right) of the male geniculate and modified as a prehensile organ, and the last thoracic and first abdominal segments asymmetrical. The last legs of the male are very asymmetrical, the right often developed into a remarkable pincerlike structure used in transferring spermatophores. The eyes are often large and separate, and in some are provided with cuticular lenses. The fifth feet in the female are of normal or reduced size but are not absent.

Amphascandria.   Some representative planktonic copepods of the Amphascandria group are discussed below.

GENUS *Calanus* LEACH. This is one of the most abundant, widespread, and important genera of copepods. Most of them are rather large species, some attaining a length of ¼ inch. Most are about the size of a grain of rice. The appendages are much alike in both sexes, though the left fifth leg of the male is larger than the right. The metasome consists of five distinct segments. The exopodites of the second to fourth feet bear two spines on their outer margin. The second segment of the basipodite of the first swimming feet is smooth, without a hooklike armature.

Calanus finmarchicus is probably the most intensively studied of any copepod

Fig. 6-5. Copepod, *Calanus finmarchicus.*

species. The body is about the size of a rice grain, rather transparent, with a varying amount of red color caused by red oil drops. Some individuals, infected with parasitic *Peridinium,* are much redder than usual. The name *C. finmarchicus* was formerly used in a broader sense than now, including the closely related *C. glacialis,* which tends to replace *C. finmarchicus* in the high arctic latitudes, and *C. helgolandicus,* which appears to be a more temperate-water species.

C. finmarchicus, often termed red feed, herring feed, or krill, occurs in such great swarms in the colder waters of the North Atlantic as to give the water a reddish tinge for several miles. It extends through the polar sea to the North Pacific and is a major food for several fish, including such commercially important species as herring and mackerel, as well as for baleen whales such as *Balaenoptera borealis* and *B. physalis.*

C. finmarchicus exhibits a rather wide bathymetrical distribution, dominating both the surface and subsurface communities of planktonic organisms of some regions. It exhibits well-marked diurnal vertical migrations in some areas, as well as the phenomenon of submergence in warmer latitudes. The diurnal migratory behavior differs regionally and seasonally, and in a given region the two sexes and various immature stages may differ considerably in their vertical distribution and movements.

Breeding of *C. finmarchicus* commonly begins at the time of the great early spring blooms of phytoplankton, with the population maxima coming somewhat later. The eggs are planktonic. Early nauplii are so small as to pass through all but the finest plankton nets, and are usually found in the upper strata of water. Older copepodid stages commonly inhabit the deeper strata of water occupied by the species. Where the species occurs submerged in warmer regions, it does not occur in dense swarms or with such marked seasonal fluctuations. It reappears nearer the surface in cold-temperate or antiboreal southern waters, but not in the great swarms characteristic of the North Atlantic. Since it is not present in such great numbers in regions where it is submerged beneath warmer waters, it has been suggested that such living conditions are unfavorable for it. This may not be true, however, since even in these warmer regions, it is just as numerous as some of the species characteristic of and limited to the warm regions. A similar argument can be made with regard to its occurrence in neritic waters. Even though the greatest swarms are commonly found in cold neritic waters, the species is not tied to the shelf in any way and is of regular occurrence as a holopelagic high oceanic species.

If anesthetized, *Calanus* sinks fairly rapidly. Living individuals maintain their level in the water by almost constant upward swimming. Experiments with the Hardy-Bainbridge plankton wheel indicate that *Calanus* can climb 50 feet in an hour, and can descend at rates up to 154 feet per hour. Maintenance of stocks in some areas has been correlated with the vertical movements of the copepods. Off the coast of Norway they descend into deep water during the winter, rise to near the surface in early spring, and reproduce actively through the summer while drifting mainly north and northeast toward arctic waters. Near the end of summer many of them descend into deeper water, 200 to 300 meters deep, drifting back toward their original location to rise again and reproduce the stocks next spring.

Metabolism during the breeding season is rapid. Individuals seem to need about 2% to 4% of their body weight in food each day. Experiments placing them alternately in water containing abundant phytoplankton and in clear water indicate that they require about an hour to fill or empty the gut. They may actively seize diatoms with their antennae as well as obtain them through their filtering mechanism. Experiments of Harvey indicate that they can filter up to 70 ml. or more of water per day, which, in the presence of adequate normal amounts of mixed phytoplankton, would give them the amounts calculated as necessary to maintain their metabolism. However, as they often migrate below most of the phytoplankton for part of the day, their hourly filtration rates while actively feeding must often be greater than this. During the winter, when much of the population spends all of the time in deeper water, starvation may be a real problem, even though the metabolic rates and food requirements are lower. The number falls to a minimum about February or March in the Clyde area. Here the population consists largely of stage V copepodids in December, which begin molting to the adult stage in January, and the small early breeding population begins shedding eggs by late February and early March. Soon a high percentage of nauplii appear. They mature about early April, when a second and larger spawning occurs, followed by a sudden rise in the overall population and augmented again by a third spawning about June and July. A few may have a fourth spawning in late summer, but most are in about the fifth copepodid stage in late autumn, in which stage they overwinter. Males seem to have a distinctly shorter lifespan than do females and tend to feed less extensively. The different generations during the breeding season may show differences in the mean size of the individuals, the most robust individuals appearing in the spring brood, when the population of copepods is relatively small and the phytoplankton is abundant. In colder waters of the far north, the number of generations per year is reduced and the individual lifespan increased, up to two years.

With their rich, oily bodies and tendency to gregariousness, they commonly form a great floating mass that does not mix readily with other water masses of pure oceanic water. The line of separation between a swarm of C. finmarchicus in one water mass, and another water mass of pure Atlantic water, is often like the interface between two fluids of different viscosity which have a slight tendency to mix—the difference in viscosity in this case being at least partly created by the presence of the swarm of copepods. The fact that Calanus is a principal food of herring, and that herring tend to congregate in areas where crustacean zooplankton predominates, was used by Hardy and co-workers (1936) in devising a simple plankton indicator for fishermen. This consisted of a disc of bolting cloth to be towed behind the fishing boat for a given length of time. If it came up reddish, a predominance of calanoid plankton was indicated, the degree of redness roughly indicating the richness of the zooplankton. If it came up green, showing a predominance of phytoplankton, the area was judged less favorable. Fishermen using this indicator obtained consistently better catches than those not using it.

Since the species is both euryhaline and eurythermal to a greater extent than many other common plankton organisms, it is not a very reliable indicator of hydrographic conditions, currents, or water stratification.

Marshall and Orr (1953), who have done a great deal of research on this species, brought together the results of their long and careful studies, together with the other available information on the biology of the species, in a fine book that should be consulted by everyone interested in this copepod or in the biology of copepods generally.

Another species, C. hyperboreus, occurs in the colder parts of the range with C. finmarchicus. It is somewhat more stenothermal, tends to occur only in the colder parts of the range or in slightly deeper water, and does not range as far south. It can readily be distinguished from C. finmarchicus by the more acute angles of the posterior edge of the metasome.

C. tenuicornis is a widely distributed species in slightly warmer waters, transitional between boreal and temperate in character, and overlaps much of the range of C. finmarchicus.

C. simillimus is a species characteristic of the cold waters of the Southern Hemisphere, except the coldest parts of the Antarctic. It occupies much the same ecological position in the Southern Hemisphere that C. hyperboreus does in the Northern Hemisphere, and occurs in cool waters on both sides of the Antarctic Convergence.

GENUS *Undinula* SCOTT, 1909. These are large copepods, very much like *Calanus,* but differing in the segmentation of the antennules (the eighth and ninth segments being partly fused in the females), and in details of the swimming feet, especially the fifth pair. In the fifth pair of swimming feet of the female, both the exopodite and endopodite of both feet have three segments. In the male the endopodite of the left fifth foot is rudimentary (one small segment) or absent, while the exopodite has three segments and ends in a powerful prehensile organ. In the right fifth foot, both rami have three segments. *Undinula darwini* and *U. vulgaris* are among the most frequently taken copepods in tropical waters.

GENUS *Microcalanus.* Very small calanoids, the metasome having only three apparent segments. In North Atlantic waters, as along the Norwegian coast, over the year as a whole it is one of the four numerically most significant zooplankters, though its small size makes it less significant in total volume, or biomass. They tend to reproduce faster than the larger calanoids and may have more generations per year, in the same area. Maxima tend to occur from January to March. *M. pusillus.*

GENUS *Nannocalanus* SARS, 1925. The chief species, *N. minor,* formerly included in genus *Calanus,* is widely distributed in tropical and temperate waters of the major oceans, where it is often one of the commonest copepods present. It is quite eurythermal and euryhaline, and while most frequently occurring near the surface, it exhibits considerable bathymetrical range. A small pointed decurved rostrum is present. The last segment of the metasome is strongly emarginate in dorsal view, but the angles do not flare out, seeming rather to be curved or wrapped rather close around the urosome. Two forms, or varieties, that are often recognized, *major* and *minor,* exhibit slight differences in proportions and other details.

GENUS *Chiridius* GIESBRECHT, 1893. No obvious rostrum, although in ventral view of the "forehead" a very small point can be discerned which may represent one.

Metasome has three apparent segments, the last one produced into spines. *C. obtusifrons* is a high arctic species commonly found in waters of subzero temperatures. *C. poppei* has been taken in the equatorial Pacific, and *C. gracilis* may be a variety of *C. poppei.*

GENUS *Chirundina* GIESBRECHT, 1895. Rostrum is moderately developed and pointed, and the forehead has a distinct crest. In female the last segment of metasome ends in a small knob on each side; has no fifth legs. Male is without knobs on last segment of metasome; has well-developed fifth legs. Taken in tropical waters.

GENUS *Euaetideus.* Fourth and fifth thoracic somites are fused so that metasome has only three apparent segments, the last of which is deeply emarginate and drawn out into posteriorly directed points in dorsal view. Rostrum is well developed. Urosome is quite small compared to prosoma, the spines of the last segment of the metasome extending back about half its length. *E. giesbrechti.*

GENUS *Euchirella* GIESBRECHT, 1888. Rather large calanoids living in deeper strata. Rostrum, if present, is single-spined. Metasome has four (or three) fully developed segments. Male has enlarged, elongate fifth legs, the right leg forming a large grasping organ. Several species. Antennules reach to about the end of the urosome.

GENUS *Gaetanus* GIESBRECHT, 1888. Distinct median spine on front of forehead; rostrum has single point or very small bifurcation at tip. Mostly bathypelagic. In some species the last segment of metasome is strongly excavate in dorsal view and drawn out into sharp points. Antennules are often much longer than endopodite, the latter being much shorter.

GENUS *Calocalanus* GIESBRECHT, 1888. Urosome has two or three segments. Caudal furcae extend laterally or are asymmetrically developed, and bear remarkably long, plumose seta or setae. Antennules usually bear long bristles and plumose sense organs, and the last segment is long. Female is often more conspicuously decorated than the male.

GENUS *Eucalanus* DANA, 1852. Female is

Fig. 6-6. Copepod, *Calocalanus pavo.*

without fifth feet. Frontal part of head is not produced into a head or rostrum. Telson is fused with, not separated from, caudal furcae. Body is often very elongate, the cephalosome occupying about two thirds of the total length and marked off from the greater part of its length by a shallow emargination, so as to appear somewhat like an arrowhead on a broader shaft.

GENUS *Euchaeta* PHILIPPI, 1843. Large, predaceous copepods. Female has four abdominal segments and long inner furcal setae. Male abdomen has five segments. There are no fifth legs in female; those of male have a long styliform process on right side, and the left fifth foot has an exopodite developed into a complex hand for grasping the spermatophores. The endopodite is rudimentary or absent. *E. norvegica* is one of the largest, most beautiful species abundant in the North Atlantic, attaining a body length of about 8 mm. It is rather transparent and has a red color similar to that of *Calanus finmarchicus.* The female carries the dark blue eggs in a single egg sac beneath the abdomen. *Euchaeta norvegica* was separated from this genus by Scott (1909) on the basis of differences in the male fifth feet and the female maxillipeds and made the type of a separate

genus, *Paraeuchaeta. Euchaeta marina* and several other species are abundant in tropical waters.

GENERA *Megacalanus, Bradycalanus, Neocalanus,* ETC. Very large, rather typical calanoids, mostly bathypelagic. Have been segregated into a subfamily, Megacalaninae.

GENUS *Pseudocalanus* BOECK. *P. elongatus* is the second most frequently taken species of copepod associated with swarms of *Calanus finmarchicus* in the North Atlantic. *P. minutus* is a cold-water species abundant around Cape Cod Bay all year, and around other cold, neritic water, but it also extends into oceanic areas.

GENUS *Scolecithrix* BRADY, 1883. Usually fairly large calanoids. Female is without fifth feet. The hind surface of the rami of the third and fourth legs has spines; first legs have one segment; second legs have two segments. Apex of maxillipeds bears a tuft of wormlike sensory filaments. In *S. danae* formalin-preserved samples of the appendages have a delicate mauve tint, enabling them to be recognized at once in a mass of other copepods. Widely distributed; several species.

GENUS *Arietellus* GIESBRECHT, 1889. Head and hind corners of thorax are usually pointed; last thoracic and first abdominal segments are commonly asymmetrical. Fifth feet are characteristic, and the metasome has four segments. Several species.

GENUS *Augaptilus* GIESBRECHT, 1889. Head distinct from first thoracic segment; urosome three-segmented in female, five-segmented in male; rami of all legs three-segmented; grasping antenna of male may be on either the right or left side.

Heterarthrandria. Some representative planktonic copepods of the Heterarthrandria group are discussed below.

GENUS *Centropages* KROYER, 1848. Neritic surface-living species. Last metasomal segment has flared, pointed projections. Antennules have twenty-four segments. All legs except left fifth of male have both endopodites and exopodites of three segments, the male left fifth exopodite having two segments. Male has large pincerlike modified right fifth legs for transferring spermatophores to the female. Widely distributed. *C. hamatus* occurs abundantly in the North Atlantic associated with *Calanus finmarchicus, Tortanus discaudatus,* etc. in cold boreal waters. Metasome has five segments.

GENUS *Eurytemora* GIESBRECHT, 1881. Another chiefly neritic genus, also occurring in brackish and fresh water. Abdomen of male has five segments, of female, three segments. Antennules are as long as prosoma, and have twenty-four joints. Fifth feet are uniramous; endopodites of legs three and four have two segments.

GENUS *Temora* BAIRD, 1850. Caudal furca is long and slender, six times as long as wide, or more; the furcal bristles are short. Metasome has four segments, the posterior corners drawn out into lateral points. Abdomen of body is compact, appearing with cylindrical abdomen and long furcal rami. A widespread genus. The best-known species are *T. discaudata* and *T. longicornis.*

GENUS *Heterorhabdus* GIESBRECHT. Widespread.

GENUS *Metridia* BOECK, 1865. Very widespread genus of essentially herbivorous, diatom-eating copepods. Urosome is usually relatively long and slender; late copepodids of both sexes have four-segmented urosome, but adult female has a three-segmented urosome because of fusion of the first two, and adult male has a five-segmented urosome because of division of the fourth. Fifth legs are uniramous, and have three or four segments in female and five in male. Basal segment of endopodite of second leg is deeply invaginate, with spines along the side of the invagination. Metasome has four segments. *M. longa* is a typical member of the boreal community, frequenting deeper strata at about 40 to 50 meters, and sometimes ascending to the surface. Essentially a cold-water high oceanic or open-water species, with *M. lucens* as a member of the *Calanus finmarchicus–Euchaeta norvegica* association.

GENUS *Pleuromamma* GIESBRECHT, 1898. Distinguished from *Metridia* and other related copepods most readily by the presence of a circular brown-pigmented knoblike luminous organ on one side of the first segment of the metasome. Metasome has four segments. Like *Metridia,* the various species tend to occupy strata below the surface, although they ascend at times. Some are characteristic of warmer waters.

GENUS *Anomalocera* TEMPLETON, 1837. Surface-living, warm- or temperate-water forms. *A. patersoni* of coastal Atlantic waters are blue in color. Both the sides of the cephalosome and the corners of the last segment of the metasome are drawn out into lateral points, giving a very characteristic appearance.

GENUS *Labidocera* LUBBOCK, 1853. Rather small calanoids. Metasome has four segments; abdomen of female has two or three segments, abdomen of male has five; antennules of female have twenty-three segments. Pair of dorsal eyes and a median ventral eye are present, the dorsal eyes having cuticular lenses. Fifth legs of female are biramous, with unequal one-segmented rami; those of male are uniramous and chelate. Posterior corners of last segment of metasome are drawn out into lateral points, and in some species there are lateral points on the cephalosome.

GENUS *Pontella* DANA, 1846. Head with paired cuticular lenses plus additional lenses on anterior and posterior surfaces of rostrum. Median ventral element is also present, forming a projection on ventral surface behind rostrum. Metasome has five segments. Fifth legs of female are biramous and reduced; in male, the left is uniramous and the right is chelate. Endopodites of first legs have three segments; those of second and third legs, two segments. Cephalosome has lateral hooks, or points.

GENUS *Tortanus* GIESBRECHT, 1898. Metasome has five segments. One large dorsal eye; no cuticular lenses. Antennules of female have seventeen segments; fifth legs are uniramous; endopodites of first to fourth legs have two segments. Eggs are planktonic and are extruded individually, not carried in egg sac. *T. discaudatus* is a prominent member of the *Calanus finmarchicus–Tortanus–Centropages* surface-waters assemblage in the North Atlantic.

GENUS *Acartia* DANA. Slender, colorless, nearly transparent body. Metasome has four segments and rather uniform thickness. One large dorsal eye, without cuticular lens. Fifth legs are uniramous; endopodites of ordinary swimming legs have two segments. Several species are among the commonest, most important neritic copepods. Very wide occurrence.

Suborder Cyclopoida. The cyclopoids, like the calanoids, have the body divided into a broad anterior prosoma and a narrow urosome. The major articulation, however, occurs between the fifth and sixth thoracic somites, so that the segment that would normally bear the fifth pair of swimming feet is included in the urosome. In cyclopoids the fifth pair of feet are usually uniramous and vestigial, without an enlarged basal segment. The female carries

Fig. 6-7. Copepod, *Oithona* sp. (Redrawn from Sverdrup, H. U., M. W. Johnson, and R. H. Fleming. 1942. The oceans: their physics, chemistry and general biology. Prentice-Hall, Inc., Englewood Cliffs, N. J.)

the eggs in two ovisacs, one on either side of the urosome. The antennules are shorter than in calanoids, usually being clearly shorter than the body and composed of only six to seventeen segments. Cyclopoids are abundant in fresh water and to a lesser extent in brackish waters, and only a few species have become truly marine. Of these, some are benthic or parasitic, so that the number of planktonic marine cyclopoids is very small compared with that of calanoids.

GENUS *Oithona* BAIRD, 1843. By far the most abundant and widespread genus of marine cyclopoids; occurs in all oceans, and in many regions ranks as one of the commonest and most important genera of copepod present. Several species. The body is elongate-ovate, and much like that of *Cyclops* in general appearance. First antennae in part are rather indistinctly segmented and bear very long sensory bristles; second antennae have two segments. Urosome is moderately long. Bristles of caudal furca are not plumose. Details of mandibular palp and of maxillipeds differ from *Cyclops.* Especially abundant in neritic waters.

Oithona nana is one of the most abundant copepods all year off the California coast, and is also reported from many other parts of the world (English Channel, Monaco, Bay of Algiers, etc.). *Oithona brevicornis* occurs on both Atlantic and Pacific coasts of America. In *Oithona plumifera* the appendages of female have long feathery scarlet setae. Numerous in tropical and warm waters. *Oithona similis* is cosmopolitan and occurs off the California coast and in the North Sea, Irish Sea, Kara Sea, and Barents Sea. Several marine cyclopoids have a pair of eye lenses on the frontal margin, giving them the appearance of possessing built-in opera glasses.

GENUS *Sapphirina.* Second antennae have four segments. Paired eye lenses on frontal margin. Semitransparent ovate or subovate depressed body; short lamelliform furcal joints.

GENUS *Copilia.* Similar to *Sapphirina,* but second antennae have three segments.

GENUS *Oncaea* PHILIPPI, 1843. Body is piriform; antennules have six segments; paired lenses are on frontal margin.

GENUS *Corycaeus* DANA, 1846. Body is elongate and subcylindrical; abdomen has one or two segments; rather large lenses on frontal margin of cephalic segment.

Suborder Harpacticoida. Most members of this suborder are benthic in habit, creeping about, on, or in the substrate, among sand grains, etc. Some appear at times in the plankton. A few are holoplanktonic in habit.

The urosome is often much less sharply distinct from the metasome than in the two previous suborders, the entire posterior portion of the body tapering more or less evenly. The first antennae are short, rarely more than eight or nine segments. In the male, both second antennae are prehensile, used for grasping the female. No heart is present; eye is median, simple. The anterior pair of feet is sometimes modified as second maxillipeds. Some planktonic forms. Eggs are carried in single ovisac beneath abdomen of female.

GENUS *Euterpina* NORMAN, 1903. Widespread in temperate and warm coastal seas. Both the rami of first legs have two segments. *E. acutifrons* is an abundant species.

GENUS *Metis* PHIL. Exopodites of first legs have three segments, the endopodites, two segments. One or both rami are broadened and modified.

GENUS *Macrosetella* SCOTT, 1909. First legs have three-segment exopodites and two-segment endopodites. Neither ramus is broadened or modified. Body slender and cylindrical; prominent beaklike rostrum.

GENUS *Microsetella* BRADY AND ROBERTSON. Both rami of first legs have three segments; urosome as wide as metasome at its articulation.

GENUS *Clytemnestra* DANA, 1852. First legs have exopodites of only one or two segments; endopodites are much longer and have three segments. Both rami of other legs have three segments. Body is rather slender and depressed, with the thoracic segments rather expanded and produced, and pointed posteriorly; triangular in outline when seen from above with blunt-pointed rostrum.

Fig. 6-8. Copepod, *Aegisthus* sp. (From Sverdrup, H. U., M. W. Johnson, and R. H. Fleming. 1942. The oceans: their physics, chemistry and general biology. Prentice-Hall, Inc., Englewood Cliffs, N. J.)

GENUS *Aegisthus* GIESBRECHT, 1891. In first four feet, both rami have three segments; fifth feet are elongate, with two segments. The furcal setae are very long, setigerous spines, each bearing a plumose seta at the jointed distal end. A well-defined, pointed rostrum is present.

THE EUPHAUSIIDS

Second only to the copepods as important elements of the zooplankton are malacostracans of the order Euphausiacea. These are small shrimplike crustaceans 1 to 6 cm. in length, with prominent compound stalkeyes. The eight thoracic somites are covered by a common carapace, which does not, however, form lateral gill chambers. Five abdominal somites bear pleopods, and the sixth bears the uropods, which together with the telson make up the tail fan. The telson bears a subapical spine on each side. All the thoracic limbs are biramous, the second to eighth bearing natatorial setae, and a gill is formed from a branched epipodite. The gills become larger and more branched posteriorly. The anterior pair is not modified into maxillipeds. The last one or two pairs of thoracic legs may be reduced or vestigial. In males the first two pairs of pleopods are modified for copulation. The sperm are transferred in spermatophores, as in copepods. The female does not have oostegites but carries the eggs in a basket formed by setae on the thoracic limbs, or in ovisacs cemented to the ventral surface of the thorax. In some species the two pleopods of the same somite are coupled together by a median process, the appendix interna, or stylambis. Most euphausiids exhibit bioluminescence, the photophores located on the margins of the eyestalks. The body is

nearly transparent in many species, whereas in others it is bright red.

Euphausiids tend to aggregate at times in great swarms or to become stratified at certain levels in the water. In many regions they perform diurnal vertical migrations. Their abundance and tendency to swarm, plus the fact that their bodies are rich in protein, make them a prime food for many predaceous fish, cephalopods, birds, and whalebone whales.

The eggs hatch as nauplii, but the larvae go through more complex metamorphoses than do copepods, including **nauplius, metanauplius, calyptopis** 1-3, **furcilia** 1-6, and finally the adult stage.

Three families are commonly recognized: Euphausiidae, Nematoscelidae, and Bentheuphausiidae. These families, as well as some representative genera, are discussed below.

Family Euphausiidae. Thoracic legs are well developed, all more or less alike, and the second or third pair not markedly elongated nor ending in brushes or claws. Maxillary palps are simple. Eyes are little if at all constricted, and not bilobate.

GENUS *Euphausia* DANA. Last two pairs of thoracic limbs reduced to mere bristlelike processes hidden among the bushy gills and without endopodites; nauplius eye persists throughout life. *E. superba* (Fig. 6-9) is a bright red dominant and characteristic surface-dwelling zooplankter of antarctic waters, the krill upon which baleen whales, many fish, penguins, etc. largely depend for food during the antarctic summer.

Euphausia superba appears in dense shoals in antarctic waters during the summer, feeding on the rich diatom blooms, drifting northward, and spawning for about four months (November to Febru-

Fig. 6-9. Euphausiid shrimp, *Euphausia superba*. The antarctic krill, or whale food. (Redrawn from Bargmann, H. *In* S. Ekman. 1953. Zoogeography of the seas. Sidgwick & Jackson, Ltd., London.)

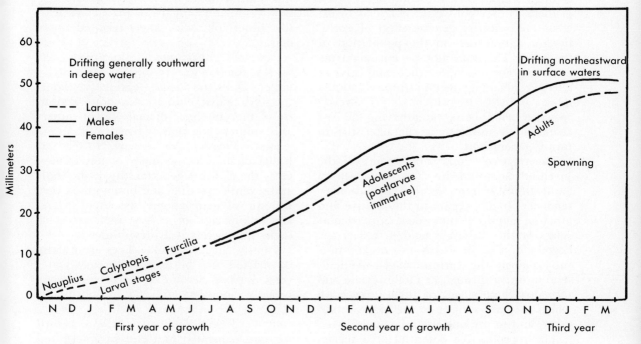

Fig. 6-10. Growth of *Euphausia superba*.

ary). As they approach the Antarctic Convergence, they tend to sink into somewhat deeper water, the eggs sinking farthest. In these deeper strata they are caught in the southerly drift of intermediate-depth waters and gradually returned southward as juvenile stages. Although the calyptopis and furcilia stages show diurnal vertical migrations, the balance is such that the southward drift is continued; the later stages tend to accentuate this by remaining in deeper strata more of the time. Since the drift of the intermediate-depth waters is slower than that of the shallower surface strata, and because of the vertical migrations that occur part of the time, nearly two

seasons are required for the return trip. In September the late juveniles begin to rise to the surface near the Antarctic Continent, molt to the adult stage, and begin the summer's heavy feeding, spawning, and northward drift.

This is an excellent illustration of the adaptation and timing of a life cycle, with different parts of it spent at different depths in the water, enabling a species of planktonic animals to remain in the same general region in spite of continuous drifting of the surface waters out of the region.

Blue whales and other animals that feed largely on krill have an unusually short, efficient food chain, only one step removed from the phytoplankton. Hence they are able to produce a much greater biomass than would otherwise be the case (see Pequegnat, 1958).

Herbivorous animals preyed upon by selective predators commonly evolve highly developed escape mechanisms, such as speed, keen senses, or social defense. It has been claimed that nonselective predators, such as the blue whale, which simply swims through swarms of krill, straining out tons of them as they go, would not tend to exercise any selective genetic effect of evolutionary significance on the population of the prey. This may not be entirely true, however. For example, they may give a selective advantage to rapid reproduction— those individuals which have already spawned before being eaten having the best chance to transmit their characteristics to future generations.

The relatively long submergence of the juvenile stages and the short, single-season adult life illustrate another evolutionary tendency. In any organism the relative and absolute lengths of time spent in particular stages of the life cycle tend to reflect the degrees of risk to which the organism is subjected in the various stages. Animals tend to hurry through or to abbreviate and telescope those stages of the life cycle in which they are subject to the highest mortality, and to prolong the safer stages.

The reproductive potential of a species reflects a summation of all the risks involved in completing the life cycle. Species suffering the highest mortalities before the cycle is completed have the greatest number of potential offspring. If changed conditions of any kind significantly alter the risk a given species runs at any stage of its life from that under which the species evolved, the result may be dramatic increases or decreases in the population of the species in question, which in turn will alter the conditions for all species dependent upon or associated with that species, setting off a chain reaction of adjustments, with what often seem surprising and unpredictable results.

GENUS *Pseudoeuphausia*. *P. latifrons* is a dominant neritic species occurring in tropi-

cal seas. In the Indo-Pacific, it is related to *Nyctiphanes*.

GENUS *Nyctiphanes*. A reflexed leaflet is on base of peduncle of antennules in front of the eyes; peduncle of antennules of male is noticeably stouter than that of female. Seventh and eighth thoracic limbs of female are rudimentary and uniramous; the seventh have two long segments, and the eighth are very rudimentary and lacking setae. Female carries eggs in two egg sacs. Copulatory organs of male have leaf-like inner lobe with finely serrated outer margin.

GENUS *Meganyctiphanes* HOLT AND TATTERSOL. Similar to *Nyctiphanes* but usually larger. Only the eighth legs are vestigial; the sixth and seventh are biramous. Peduncle of first antennae of male is not noticeably stouter than that of female.

GENUS *Thysanopoda* M. EDW. No reflexed leaflet on base of peduncle of antennules. Only the eighth legs are vestigial; the sixth and seventh are alike and biramous. A very widespread genus, some species of which are bathypelagic or at least spend most of their time in relatively deep water.

Family Nematoscelidae. Eyes are usually constricted and bilobate, with dorsal and ventral lobes. Second or third legs are especially elongated and end in a brush or claw.

GENUS *Nematobrachion* CALMAN. Third legs are elongated and end in a terminal group of spines.

GENUS *Nematoscelis* SARS, 1885. Second pair of legs is greatly elongated and slender, ending in bristles on the last or the last two segments. Eyes are large and bilobate. Last five gills are branched. Second and third segments of antennules are longer and more slender in females than in males. Seventh legs are biramous in female, with two-segment endopodite; seventh legs are uniramous in male. Male copulatory organ has three processes on inner lobe, the spinelike process nearly straight and parallel to the others, and the lateral process never hooked.

GENUS *Stylocheiron* SARS, 1885. Third legs are greatly elongated, the penultimate segment broadened and often armed with strong spines, forming with the distal segment a grasping organ. Eyes are large and constricted into two parts, the lower part well behind the upper.

GENUS *Thysanoëssa* BRANDT. Second pair of legs is often much elongated and thicker than the others; last two pairs of legs are reduced, the last pair being uniramous. Rostrum is usually well developed; flagella of antennules are short; two distal segments of peduncle of antennules are more slender in female. Last three pairs of gills have two branches. Numerous species, widely distributed.

Family Bentheuphausiidae. Deep-water forms with imperfect eyes; no legs much longer than the rest; maxillary palps have three segments; no luminiferous organs.

GENUS *Bentheuphausia* SARS.

OTHER PLANKTONIC CRUSTACEA

Several additional groups of Crustacea contain at least some important planktonic members, even though in many of these groups the majority of members may be nonplanktonic.

Order Mysidacea

The mysids, or opossum shrimps, superficially resemble the euphausiids, with which they have sometimes been grouped in the order Schizopoda. Most of them live in shallow waters, burying themselves in the substrate part of the time and appearing only temporarily, usually at night, in the plankton. They are often abundant around river mouths and in brackish water. A few are truly planktonic, and some are bathypelagic or abyssal. As in the Euphausiacea, the thoracic limbs are biramous. They differ, however, in many important respects. The carapace covers most or all of the thorax and forms lateral gill chambers. The first thoracic limbs are modified as maxillipeds and are shorter than the following thoracic limbs. The female has a ventral thoracic brood pouch composed of oostegites (ventral plates representing epipodites from the thoracic limbs are expanded, and flattened and directed medially beneath the thorax), in which the young develop to an advanced stage. The heart is tubular. The telson lacks the subapical spines found on that of euphausiids. Gills, when present, are highly subdivided, each apparently representing one epipodite. Gills are lacking, however, in the family Mysidae, in which respiration takes place through the thin lining of the underside of the carapace. A current of water is drawn under the carapace by action of the epipodites of the maxillipeds. The carapace is often marked by a cervical sulcus, or groove, and the last abdominal somite is usually elongated and marked by a transverse groove. Deep-sea forms sometimes have oddly modified eyestalks, which in some cases extend beyond the eye itself. A well-developed antennal scale is present. Pleopods are always present in males but commonly reduced or absent in females.

Mysids have a worldwide distribution, have about 450 species, and are mostly marine. They are usually 2 to 3 cm. long, although some much smaller forms, only 3 mm. in length, are known, and a few of the largest reach 15 cm. Both the antennules and antennae usually have very long flagella. Representative genera are discussed below.

Anchialina. Wide distribution in tropical shallow waters. Characteristic on Barrier Reef lagoons, Java Sea, etc.

Hemisiriella. Tropical seas; Indo-Pacific.

Eucopia DANA. A typically deep-sea genus, black.

Gnathophausia W. SUHM. Deep-sea form with long, spearlike rostrum.

Austromyces CZERN. A Mediterranean and Caspian genus.

Boreomyss SARS. Arctic abyssal species.

Several brackish-water genera of Sarmantic derivation are characteristic of the Caspian Sea (*Katamysis* Sars, *Lomnomysis* Czern, *Mesomysis* Czern, *Metomysis* Sars, and others).

Mysis LATR.

Siriella DANA. Pelagic.

Order Cladocera

The cladocerans, water fleas and their relatives, are primarily a freshwater group of entomostracans in which the carapace encloses the body but not the head, as a

Fig. 6-11. Cladocera, *Podon* sp.

bivalved but not hinged structure open ventrally. The second antennae are enlarged and used for swimming. A single large compound eye and sometimes an ocellus are present. Some brackish water species and a few truly marine forms are found, especially in neritic plankton. In suborder Calyptomera the carapace encloses the whole body and limbs. In suborder Gymnomera it is somewhat reduced, serving mainly as a dorsal brood sac.

GENUS *Podon* (GYMNO). Family Polyphemidae. Carapace does not enclose legs; four pairs of legs are crowded together. Abdomen is slender, ending in two caudal spines. Head is large, bluntly rounded, and has one eye. Head and thorax are separated by a dorsal depression.

GENUS *Evadne*. Head and thorax are set off by a dorsal depression. Body is oval to piriform, and pointed behind. Abdomen does not protrude.

GENUS *Penilia* (CALYPTO). Carapace encloses body but is broadly open ventrally. Six pairs of legs.

Order Ostracoda

The ostracods are small crustaceans, laterally compressed and wholly enclosed within a bivalved, movable or hinged carapace that gives them, superficially, the appearance of minute bivalve molluscs. They are ubiquitous in both freshwater and marine environments, many of them living in or near the substrate. A number of truly planktonic forms occur. The appendages are reduced in number more than in any other group of Crustacea; usually only seven pairs are present. These are (1) first antennae—uniramous; (2) second antennae —biramous; (3) mandibles, usually with a large biramous palp; (4) first maxillae, serving as jaws; (5) second maxillae, the function of these varying in different groups, serving as jaws, maxillipeds, or legs; (6 and 7) trunk appendages, legs, sometimes absent. The thoracic appendages have lost the functions of swimming, respiration, and feeding, which they serve in most other entomostracans. Both pairs of antennae are used in swimming. Many ostracods carry the eggs in a brood pouch, or space, between the dorsal shield and the body. An unpaired nauplius eye consisting of ventral median and two lateral elements is present. These elements may be united in the midline, or they may be separate. The abdomen is short, ending in a caudal furca. The development of marine forms is usually rather direct, there being no free larval stages or metamorphosis.

Fossil ostracods occur in nearly all geological strata. They abounded during the Silurian period, at which time some attained a length of 9 cm. The largest known living forms are deep-sea species of *Gigantocypris,* which grow to 23 mm. in length.

Suborder Myodocopa Sars. The shell is notched in front, giving play to the antennae. The basal joint of the second antennae is usually wide; the exopodite is its longest ramus, and is composed of many segments.

FAMILY CYPRIDINIDAE. Exclusively marine. The shell is hard, the antennal notch at or below the middle of the dorsoventral extent of the shell. The first antennae are strongly flexed at the articulation between the basal and second joints, and they bear large olfactory hairs. Second maxillae are jawlike and carry a large respiratory plate. Genera include *Cypridina* M. Edw., *Gigantocypris* Mull., *Cylindroleberis* Brady, *Sarsiella* Norman, etc.

FAMILY HALOCYPRIDAE. Exclusively marine. Shell is calcified. Notch above the middle of dorsoventral extent of shell. Numerous unicellular glands are open along margins of shell. Mandibles with well-developed masticatory process. First maxillae

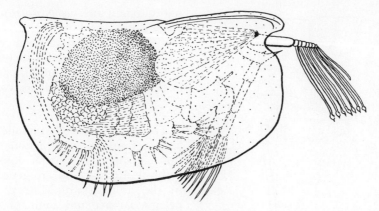

Fig. 6-12. Ostracod, *Conchoecia* sp. (Redrawn from Davis, C. C. 1955. Marine and freshwater plankton. The Michigan State University Press, East Lansing.)

are jawlike. No eyes. Pelagic. Genera include *Conchoecia* Dana, which is circumglobal, occurring in all major oceans, *Halocypris* Dana, etc.

FAMILY POLYCOPIDAE. Minute marine ostracods living at the sea bottom. The shell valves are round or oval, the antennal notch is weak or undeveloped, and the first maxillae are not jawlike.

Suborder Podocopa. The shell is flattened along the ventral border and not notched in front. The second antennae have long, leglike endopodite and short, much-reduced exopodite. Mandibles have short palp that may be biramous. The first maxillae are jaws and carry a large comblike setose plate, the movements of which promote respiration, though the plate does not itself act as a gill.

FAMILY BAIRDIIDAE. Marine forms with shell much arched dorsally and the two halves asymmetrical. The crop forms a well-developed masticatory mill.

FAMILY CYPRIDIDAE. Occurs predominantly in fresh water. Some occur in brackish or neritic marine environments.

FAMILY CYTHERELLIDAE. Regularly arranged, thickened bands in posterior part of body, suggesting segmentation but apparently not corresponding with the limbs. Outer ramus of second antennae larger than in other Podocopa. The divisions of the caudal furca are lamellar and fringed with stout bristles. A representative genus is *Cytherella* Bosq.

FAMILY CYTHERIDAE. Marine, with strongly calcified shell. Second antennae have strongly flexed endopodite ending in strong, clawlike setae; the exopodite is elongate and slender, bent at the tip, and perforated by duct of a large, unicellular "spinning gland," which Müller says spreads a web of fibers over surrounding objects. Divisions of caudal furca are small and weakly developed. No heart. Numerous genera.

Order Cumacea

Small crustaceans living in mud or sand, mostly on or near the bottom. Occasionally appear in plankton in shallow waters or from hauls near the bottom.

Orders Amphipoda and Isopoda

These orders together constitute the group **Arthrostraca,** or **Edriophthalma.** They are mostly benthic forms with no carapace, the bulk of the body being comprised of free thoracic segments. There are seven pairs of thoracic legs, which are uniramous, there being no exopodites. The eyes are sessile. These groups will be considered in more detail in connection with animals of the substrate. A few, however, are typically elements of the macroplankton. Amphipods tend to be laterally flattened and to have the thoracic limbs in two groups, the four anterior pairs directed forward, and the three posterior pairs directed backward and upward. In Isopoda

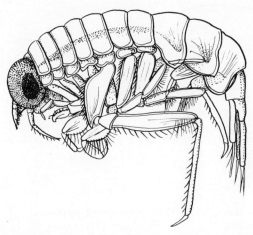

Fig. 6-13. Hyperiid amphipod, *Themisto libellula*. (Redrawn from Sars, 1894.)

all seven pairs of thoracic limbs are more alike, usually shorter, and fitted for crawling or clinging. The body is usually dorso-ventrally flattened. Both are very large groups.

Planktonic amphipods, particularly the family Hyperiidae, are often parasitic on or associated with medusae or salps. Many of the hyperiids have very large heads, comprised for the most part of huge eyes.

The genera *Euthemisto, Hyperia,* and *Parathemisto* are of worldwide occurrence and at times are important members of the zooplankton community.

Other genera of the Amphipoda, including holopelagic species, are *Cuphocaris, Eurythenes, Brachyscelus, Euprimno, Phrosina, Phronima, Themisto, Lanceola, Scina,* and *Vibilia. Themisto libellula* (also known as *Euthemisto libellula*) is important in arctic waters. Its biology has been investigated by Dunbar, 1946. Several genera occur in the antarctic macroplankton widespread in the antarctic waters, including species of *Cyllopus, Eusiris, Vibilia,* and *Primno.*

Only two genera of isopods, *Munnopsis* and *Eurydice,* are commonly planktonic, and even these genera also have benthic as well as pelagic species.

Many amphipods and isopods that spend the day in or on the substrate may enter the plankton of shallow neritic waters at night, types such as *Nototropis* and *Apherusa* often being the most conspicuous elements of such plankton at certain hours. Crustacea with this habit include especially mysids, cumaceans, amphipods and a few isopods, and some euphausiids.

Order Decapoda

The order Decapoda includes most of the larger familiar conspicuous crustaceans such as shrimps, prawns, lobsters, hermit crabs, and crabs, and are generally benthic in the adult state. Most of them have pelagic larvae, and some of those which are pelagic as adults are strong-enough swimmers to be classed as nekton rather than plankton. *Lucifer* is the genus most highly modified for planktonic existence. Some of the deep-sea prawns, such as *Gennadas, Sergestes, Hymenodora, Acanthophyra,* and *Systellaspis,* are weak swimmers and therefore to be included in the macroplankton.

Decapod planktonic larvae will be considered in the section on larval forms in the meroplankton. Gurney (1939, 1942) gives a good treatment and bibliographies of larval decapods.

As has happened in many animals in which the larval stages are very different both structurally and ecologically from the adult stages, some of the larvae were discovered and independently named before their relationships were understood. Some of these invalid generic names, such as zoea and megalops, have continued in use as designations for those stages in the life cycle for which they were originally proposed. The usage is usually broadened to include the same stages in the life cycle of all closely related crustaceans having a similar stage. Others have simply dropped out of sight in synonymy.

A few decapods (*Lucifer* spp.) and some peneid shrimps hatch as nauplii resembling those of most entomostracans, but most decapods pass through the nauplial stage within the egg before hatching and then hatch at a more advanced stage.

Numerous shrimps of the family Sergestidae are planktonic, often highly modified, with long slender bodies, and transparent or nearly so in some species. The rostrum is short or absent. The antennules bear one long and one short flagellum. The short flagellum in males may be bifurcated, forming a short prehensile organ. In *Lucifer,* perhaps the most modified genus for planktonic life, the cephalic region and the eyestalks are much elongated, giving the animal a very strange appearance.

OTHER NONGELATINOUS METAZOA
PHYLUM CHAETOGNATHA

Aside from the Crustacea, the arrow worms or glass worms are perhaps the most widespread and characteristic nongelatinous element of the plankton. The phylum is a small one; only about 50 species have been described. They are, however, in all oceans and often in sufficient abundance to form a significant part of the zooplankton.

Chaetognaths are small, slender, bilaterally symmetrical, transparent, wormlike organisms usually about an inch or less in length. The body is straight and divided into three segments, head, trunk, and tail, each containing a pair of coelomic spaces of enterocoelous origin. The head coelom is nearly obliterated in the adult by the development of a complicated mass of muscles that move the lateral head setae. These strong, curved setae are used in the manner of jaws to seize prey. The head and "jaws" can be retracted into a hoodlike fold of the body wall and protruded to seize prey. The mouth lies subventrally between the jaws. The body has one or two pairs of lateral fins and a terminal horizontal tail fin, which are used as stabilizers in swimming. Propulsion is achieved by alternate contractions of the dorsal and ventral bands of longitudinal muscles, which give the animals a rapid darting and gliding movement. They are actively predaceous, feeding on copepods, small fish larvae, other arrow worms, larval decapods, etc. In turn, they are eaten by fish, medusae, pelagic worms, etc.

The digestive tract is a simple straight tube, or in some, it possesses a pair of diverticula in the anterior region. A large ventral ganglion, connected to the brain by two large connectives, is found in the trunk region. A large ventral ganglion connected to the brain by two large connectives is found in the trunk segment, usually in the anterior half. All chaetognaths are hermaphroditic, the ovaries lying one on each side along the posterior part of the trunk coelom, and the paired testes lying behind them in the tail segment. Mature sperm are stored in lateral seminal vesicles, which are conspicuous in mature arrow worms.

Because different species of arrow worms prefer water masses of different character, they are good indicators of hydrographic conditions or of the origin of waters in which they occur. The geographic and vertical distributions of chaetognaths are of considerable interest.

All chaetognaths are marine, and all but two or three species are strictly holopelagic and planktonic. The one or two presumed benthic species of *Spadella* are sometimes classified as a separate genus, *Bathyspadella.*

Because of their delicate body, which is without hard parts except for the jawlike setae of the head, and because of their pelagic habit, chaetognaths are rarely fossilized. They are, however, apparently a very old group. The only fossil known from mid-Cambrian Burgess shale has been given the generic name *Amiskwia.*

The principal genera can be distinguished as follows:

1. With two pairs of lateral fins—*Sagitta*
   With one pair of lateral fins 2
   Spadella
2. Head cap with a pair of lateral tentacles—
   Head cap without lateral tentacles 3
3. Trunk segment encased in a thick collaret—*Pterosagitta*
   Trunk segment not so encased 4
4. Lateral fin extends to the last third of the tail segment—*Krohnitta*
   Lateral fin not extending posteriorly more than half the length of the tail segment . . 5
5. Tail segment constitutes about one third or more of the total body length. Two rows of teeth on each side of the head—*Heterokrohnia*
   Tail segment constitutes less than one third of the total body length. One row of teeth on each side of the head—*Eukrohnia*

Members of *Eukrohnia* are reported to carry their eggs in a "marsupium" formed from the lateral fins.

Following are notes on some species of the principal genera.

Eukrohnia fowleri. Bathypelagic.

Eukrohnia hamata MÖBIUS. In the Pacific, very abundant in cold waters and subarctic, occasional in transitional waters, and absent from subtropical water masses. In the Atlantic, occurs near the surface in high latitudes, and only in deeper water in middle latitudes. In the antarctic summer it is one of the three dominant species of zooplankton in parts of the Antarctic. Here it shows seasonal rather than diurnal vertical movements, descending to 500 to 1000 meters or more in the winter. A very widespread cool- and cold-water species, showing submergence in warm latitudes.

Eukrohnia subtilis GRASSE. A warm-water species common in subtropical Pacific waters but absent from subarctic waters.

Heterokrohnia mirabilis RITTER-ZAHONY.

Krohnitta pacifica ARDA. High oceanic, cosmopolitan in tropical waters.

Krohnitta subtilis GRASSE. Cosmopolitan in warm and temperate oceans.

Pterosagitta draco (KROHN). Cosmopolitan in warm and temperate oceanic waters.

Sagitta arctica (POSSIBLY A VAR. OF *S. elegans*). In very cold arctic waters. Two distinct age groups usually present, one group about half grown when the other spawns. Spawning probably in autumn; two years to mature.

Sagitta bedoti BERANECK. In tropical neritic regions, Indo-Pacific. Mainly in coastal regions under oceanic influence.

Sagitta bipunctata QUOY AND GAIMARD. Abundant in warm and temperate oceanic waters. Cosmopolitan. Absent from cold waters.

Sagitta decipiens FOWLER. Mesoplanktonic, cosmopolitan. In temperate and warm oceanic waters.

Sagitta elegans VERRILL. Cosmopolitan, abundant, cold and cold-temperate species. An indicator of oceanic water masses in both North Atlantic and North Pacific. In many areas it has three or four broods a year, those spawned at different seasons growing into different-sized adults. Shows diurnal vertical migration during spring and summer. In the English Channel their presence indicates northern and oceanic waters, and has been correlated with best commercial fishing. Apparently the only chaetognath to reproduce successfully in the Gulf of Maine. Over Georges Bank an anticyclonic eddy maintains a year-round population in the region. Associated with medusa *Cosmetira pilosella* and often with *Sagitta serratodentata*. Tends to be absent at times and places where *S. setosa* is abundant, and vice versa (Russell, 1932).

Sagitta enflata GRASSE. Cosmopolitan in tropical and warm waters. Extends north to Plymouth in summer when warm southern Atlantic water masses move into the area. Is carried in Gulf Stream as far north as Gulf of Maine but cannot live there long.

Sagitta exuina. Shows seasonal changes in depth distribution, the upper limit, or occurrence, parallel to or slightly below the 11-degree isotherm. A cool-water species.

Sagitta ferox DONCASTER. Indo-Pacific tropical and equatorial regions. Seems to compete with *S. robusta*. In areas where either one is dominant, the other one is rare or absent.

Sagitta hexaptera D'ORBIGNY. Cosmopolitan in warm and temperate oceanic waters. Often associated with *S. serratodentata*, salps, and doliolids.

Sagitta lyra KROHN. Subtropical and warm temperate waters. Less strictly tropical than *S. enflata*.

Sagitta macrocephala. Bathypelagic.

Some of the species illustrating various types of distributional patterns are tabulated in Table 6-1.

MOLLUSCS AND ANNELIDS

Although the molluscs and annelids are both major phyla of animals and have successfully adapted themselves to freshwater and terrestrial habitats as well as marine, the great majority have remained benthic in habit. Only a few small bizarre groups of gastropods and a few scattered genera of annelids have adopted a fully planktonic mode of existence. Aside from these the only pelagic forms are the squids. They

Table 6-1

Cosmopolitan, oceanic, warm to temperate waters	Cosmopolitan, tropical-subtropical, neritic	Cosmopolitan, cold oceanic or deep in mid latitudes (submergence)	Coastal, temperate to cool, neritic	Bathy-pelagic	Restricted to intermediate	Arctic	Deep, or benthic
Eukrohnia subtilis	Sagitta bedoti	Eukrohnia hamata	Sagitta setosa	Eukrohnia fowleri	Sagitta serratodentata	Sagitta arctica	Spadella cephaloptera
Krohnitta pacifica	Sagitta neglecta	Sagitta arctica					
Krohnitta subtilis	Sagitta oceania	Sagitta elegans					
	Sagitta tropica						

rival or exceed fish in their ability to swim and are to be counted among the nekton rather than the plankton. Most molluscs and annelids do have pelagic larvae, which are commonly abundant especially in neritic plankton.

In the molluscs, the principle modifications involved in the change from benthic shelled ancestral forms to planktonic habits involved reduction or loss of the shell and shell gland, reduction of the dense muscular connective tissue that gives the body of most molluscs its firmness and weight, and the development of special means of flotation.

Two groups, the Pteropoda, a suborder of Opisthobranchia, and the Heteropoda, a suborder of Streptoneura, have become wholly planktonic, as have a few genera belonging to other groups.

In **Pteropoda** the foot is expanded into two large flaps, or fins, that function in both locomotion and flotation. The shell is thin and delicate, or lacking; it is seldom coiled when present. Pteropods are hermaphroditic, usually without gills. Pteropods are epipelagic—usually found in surface waters. They sometimes occur in great swarms, particularly Spiratella (Limacina) and Clione, which are important as whale food in northern waters. The shelled pteropods are largely herbivorous, whereas the naked forms are voracious predators.

Sediment beneath waters in which vast numbers of shelled, or thecosomatous, pteropods live may contain great numbers of the empty shells, which form a significant fraction of the deposit. Such pteropod ooze is especially well developed in areas of the North Atlantic of moderate depth, around 2000 meters. It is not found in the Pacific and Indian oceans, although thecosomatous pteropods are common in the plankton of these oceans as well.

The **Heteropoda** usually have a pair of tentacles and a pair of large, well-developed eyes. As in pteropods, the foot is greatly modified to form an organ of flotation, usually forming from one to three fins. It often also bears a sucker, and in some species, an operculum. The shell is absent in some. Heteropods are confined to warm waters of tropical and subtropical regions.

Aside from these two groups, two other genera of gastropods are regularly encountered in the plankton. Janthina is a thin-shelled streptoneuran that relies on a mass of bubbles for flotation at the surface and whose foot is not greatly modified. Janthina is a beautiful snail with violet- or lilac-colored shell and bright red body. It feeds on the floating siphonophore, Velella, and on medusae. The second, more rarely encountered, is the nudibranch Phyllirhoë, another striking tropical form. These are tabulated below:

Streptoneura:
  Heteropoda
    Principal genera: Atlanta, Carinaria, Pterotrachea
    Others: Janthina

Opisthobranchia:
Pteropoda
    Principal genera: *Cavolina, Clio, Clione, Clionopsis, Creseis, Clumbulia, Herse (Cuvierina), Diacria, Euclio, Halopsyche, Spiratella (Limacina), Styliola*
Nudibranchiata: *Phyllirhoë*

Some of the planktonic Gastropoda can be classified by the following key:

1. No shell present 2
   Shell present 5
   Pseudoshell present; cartilaginous subepithelial pseudo shell formed by connective tissue; larva with calcareous shell and operculum—Cymbuliidae
2. A series of external gills on each side of the body—*Phyllirhoë*
   No external gills along sides of body .. 3
3. Foot compressed laterally, forming a ventral finlike structure—*Pterotrachaea*
   Food expanded anteriorly into a pair of fins 4
4. Body ovate, rounded posteriorly—*Halopsyche (Anopsia)*
   Body elongate—*Clione*
5. Surface-living, using a mass of bubbles as a float; foot not greatly modified; shell purplish or violet, body red—*Janthina*
   Not as above 6
6. Shell covering most of body, spiral; foot operculate 7
   Shell covering only a small part of the body, or not spiral 8
7. Large eyes present—*Atlanta*
   Eyes absent; foot modified to form two large fins—*Spiratella (Limacina)*
8. Shell not spiral, covers most of body .. 9
   Shell spiral, but small, forming a small cap on the end of the soft body—*Carinaria*
9. Shell with lateral spines or angles well back on the shell—*Cavolina*
   Shell without lateral spines or angles, or if present, they are near the anterior end of the shell 10
10. Shell narrow, conical—*Creseis*
   Shell broadened at anterior end if conical; or tubular 11
11. Shell conical; if lateral spines or angles are present, they are at anterior end of shell—*Clio*
   Shell tubular, with parallel or nearly parallel sides
   Most of length is subcylindrical—*Herse (Cuvierina)*

Fossil record

The fossil record is fairly good for thecosomatous forms, indicating a very early origin for the group. The genus *Cornularia,* from Paleozoic and Mesozoic, possessed a larger shell than existing ptero-pods, the shell tapering, four-angled in cross section and having a narrowed aperture. A related genus, *Hyolithes,* is known from the Cambrian. *Tentaculites* are small conical shells abundant in some Paleozoic rocks and thought to be pteropods.

Some important pteropods are listed below.

Pteropods

Gymnosomata

Diacria quadridentata (LESUEUR). In tropical and subtropical waters, South China Sea, Java Sea. Common but not abundant. Rarely more than 100 per 1000 cubic meters of water. Occasional in higher subtropical and transitional waters but absent from subarctic.

Clione antarctica. Fairly important carnivorous member of antarctic macroplankton. Shows less seasonal change in abundance than most of the herbivorous zooplankton species. The fins are relatively short.

Clione limacina. Sea butterfly. Also, *C. papilionacea,* occurring in North Atlantic arctic and cold boreal waters.

Halopsyche gaudichaudi. An oddly shaped, nearly spherical form found in the Indo-Pacific–South China Sea basin in densities of 100 to 400 per 1000 cubic meters.

Thecosomata

Styliola subula. An offshore tropical-subtropical species.

Cavolina gibbosa. An offshore species, common in warm waters, with a peculiar sort of bitropical or amphitropical distribution—not abundant in truly equatorial waters as in warm water on both sides.

Cavolina longirostris. A rather heavy-shelled tropical neritic form often found in high concentrations in areas such as the Gulf of Thailand, near the margins of the gulf but also offshore in the South China Sea.

Cavolina unicinata. Both in the Indo-Pacific and in the eastern warm Atlantic waters, but not in the Mediterranean.

Clio pyramidata. Cosmopolitan in warm and temperate waters. Wide distribution in North and South Atlantic, Mediterranean, around Africa, and in the Indo-Pacific and South China Sea, as well as scattered occurrences in the Central Pacific.

Creseis acicula AND *C. vergula.* Long, slender, conical, sharply pointed shells that look much like detached rostrums of crustacean nauplii. Tropical and subtropical waters. When abundant in a plankton haul, the sharp, slender shells protruding through the meshes in the net make the net painful to handle and are a considerable nuisance during removal of plankton samples from the net.

Spiratella antarctica AND *S. balea.* Cold-water species of antarctic regions.

Spiratella helicina. Cold-water subarctic species in both Atlantic and Pacific. An indicator of arctic or subarctic water masses. Said to occur in the Antarctic.

Spiratella inflata. Widespread warm- and temperate-water oceanic species.

Spiratella retroversa. Oceanic cool waters.

Heteropods

Pelagic gastropods usually swimming at the surface, ventral side up, the foot forming a sort of median fin. Head and foot large compared to the small visceral sac (nucleus). Body and shell usually rather transparent. Best-known genera: *Atlanta, Carinaria,* and *Pterotrachaea.*

ANNELIDA

Very few annelids are holopelagic, although most polychaetes have planktonic larval stages. A number of species enter the plankton for brief periods, as for spawning. Pelagic annelids are never extremely abundant, although they occur with considerable regularity in some regions. They belong to the groups: **Phyllodocidae, Alciopidae, Typhlosolecidae** and **Tomopteridae.** The most common genera can be distinguished as follows:

Calizona. Eyes prominent; parapodia with one appendage.

Greefia. Eyes prominent; parapodia with two appendages.

Pelagobia. Eyes small or lacking; many setae near the aciculae.

Tomopteris. Eyes small or lacking; few or no setae.

NEMERTEA

The nemerteans are another rather large group of benthic worms, mostly with pelagic larvae. The few holopelagic species that have been described are all deep-water forms. Bathypelagic nemerteans are all very greatly modified for a planktonic mode of living and superficially do not look at all like ordinary nemerteans, the body being relatively shorter and flatter, and expanded into a pair of lateroposterior fins and a caudal fin that aid in swimming. Some show sexual dimorphism, with the male bearing a pair of strong tentacles near the anterior end, apparently for use in holding the female during copulation. *Nectonemertes, Pelagonemertes,* and *Planktonemertes* are the most frequently found genera.

UROCHORDATA (ORDER LARVACEA OR COPELATA)

The family **Appendiculariidae** will be the last group discussed under the category of nongelatinous zooplankton. They really fit into both categories (gelatinous and nongelatinous zooplankton), as most of them secrete a temporary gelatinous "house" about themselves which aids in filter feeding and locomotion. They often abandon this house and secrete a new one, and in plankton samples are commonly free from a house, perhaps because of disintegration of such a fragile structure or because the house has been shaken free in handling.

Appendicularians are very much like the larvae of sessile tunicates, from which they differ primarily in the possession of functional gonads and of an anus. The body is divided into two portions, an anterior globular part containing the organs and variously termed the "head," the "trunk," or the "body," and a flat tail attached to the ventral side of the "head" near the middle. The "head" is more sharply differentiated from the tail than in larval tunicates, and the tail has a series of ganglia. They lack the closely investing tunic of larval tunicates. It is probable that the appendicularians are an example of paedogenesis, in which the formation of gonads has occurred in a developmental stage corresponding to the larvae stage of the ancestral forms. This has resulted in the loss of the ancestral sessile habit, freeing the organisms for a pelagic existence.

The "house" is a remarkable adaptation. On the upper surface is a grilled filtering surface the porosity of which allows the smaller plankton organisms (nannoplankton) to pass but excludes larger particles. A current of water is set up by activity of the animal's tail. Water entering through the coarse filter is swept posteriorly, around through a finer internal filter, and then finally into the large ciliated pharynx of the appendicularian. Here food particles are retained, the water leaving through the paired branchial openings, one on each side. When sufficient pressure has built up inside the house through the intake of water, a trapdoor at the posterior end suddenly opens and the water rushes out, propelling the house, with its contained owner, forward by jet propulsion to new feeding grounds. After some hours, if the filters become clogged or the house is damaged, the appendicularian escapes through another trapdoor, or escape hatch, at the front of the house and soon secretes a new house. Appendicularians are the only group of filter-feeding animals in which essential parts of the filtering apparatus are not part of the structure of the animal itself, although it is true that many animals do use a secretion of slime or mucus as an aid in the entrapment and transport of planktonic or detrital particles brought to them by a feeding current.

It is of interest that although coccolithophores had been known to exist for some time because of the coccoliths found in geological deposits, living ones were first discovered by members of the Challenger Expedition in the filters of the houses of appendicularians, coccolithophores being too small to be retained by the plankton nets in use.

Some of the principal genera follow.

Fritillaria. "Head" longer and more slender than in other genera, usually somewhat constricted near the middle at the level of attachment of tail. The anterior part contains the mouth, pharynx, and esophagus; the middle part, the heart and stomach; and the posterior part, the gonads, a spherical ovary, and behind the ovary a single large saccular testis. This is by far the largest genus.

Oikopleura. "Head" stout, short. The branchial openings are well back in the region of the rectum. The tail tapers posteriorly, ending in a single point. The testes usually are paired, one on each side of the median ovary.

Appendicularia. "Head" is short and oval; tail is rather broad and bifid or notched at the end.

Kowalevskia. Tail is broadest in middle, tapering both ways, ending conical or pointed. Branchial openings elongate longitudinally.

ROTIFERA

A few planktonic rotifers are found in marine waters, especially neritic or somewhat brackish waters. Most of these species are derived from euryhaline freshwater forms that have migrated into marginal, somewhat brackish waters.

GENUS *Brachionus.* Has well-developed lorica, usually bearing spines on the anterior border; foot is posterior, with toes of equal size.

GENUS *Keratella* (= *Anuraea*). Anterior edge of lorica has six spines; dorsal part of lorica is reticulate. Gut is complete, with anus. There are no feet. *K. cochlearis* is found in the open sea in such waters as the gulfs of Bothnia and of Finland, which are much less saline than is typical oceanic water.

Gelatinous Metazoa

Gelatinous zooplankton includes those animals which either have gelatinous bodies or are embedded in a gelatinous matrix, and which are either holoplanktonic or have an adult sexual stage that is planktonic. Three groups of animals contribute most of the gelatinous zooplankton: **Cnidaria (Coelenterata), Ctenophora,** and **Thaliacea.** The Larvacea, or appendicularians, considered in the previous section, might also have been included here with some justification.

Where gelatinous zooplankton predominates, the plankton is not as nutritious for larger predators and does not support as good a fishery as does a predominantly crustacean plankton. Larger medusae, because of their bulk and their stinging cells, occasionally constitute a real nuisance to fishermen using nets and are a hazard to swimmers.

CNIDARIA
Medusae

Medusae of two of the three classes of coelenterates are a conspicuous element of the macroplankton. They are saucer- or bowl-shaped organisms. The upper, usually convex, surface is termed the **exumbrella,** and the lower concave surface, the **subumbrella.** The body is comprised chiefly of thick, jellylike **mesogloea,** lined both externally and internally with a thin epithelium. Subepithelial muscle fibers form bands beneath the epithelium responsible for the contractions of the bell by which locomotion is effected. The mouth is at the open end of the **manubrium,** suspended from the center of the subumbrella, and opens into the stomach, or **coelenteron,** which is the only internal cavity. The coelenteron is continued radially as a series of **radial canals** extending through the mesoglea at the edge of the bell, where they usually communicate with a **circular canal** going around the circumference. Most medusae have a series of hollow or solid (filled with a core of endodermal cells) muscular tentacles extending from the margin of the bell and armed with **nematocysts.** In addition, some species also have oral tentacles arising from the manubrium or from the extended lips of the mouth, or the mouth may be drawn out into oral arms. Marginal sense organs—**statocysts, ocelli, rhopalia,** etc.—are also characteristic of most medusae. There is no central nervous system, but the nerve net may be concentrated around the bell edge, forming what amount to nerves along the tentacles and also around the mouth. Like all coelenterates, the medusae are actively carnivorous, capturing prey with the aid of the long, highly contractile, nematocyst-armed tentacles, or, in some, with the aid of the oral arms. Because of the great elasticity of the body, medusae are able to ingest surprisingly large objects. Nematocysts are characteristic of, and unique to, all coelenterates. There are many different types characteristic of various groups.

In the class **Hydrozoa** there is usually an alternation of generations. Gametes produced by the medusa give rise to small ciliated solid-bodied **planula** larvae, which settle to the substrate and give rise to a sessile **polypoid generation.** When mature, the polypoid generation gives rise to **medusae** by asexual budding, and the cycle is completed.

Because of this sessile generation, most hydroid medusae are meroplanktonic and tied to neritic waters. Hydroid medusae are usually rather small, large individuals being only a few centimeters in length or diameter. They are most easily recognized by the presence of a **velum,** a small shelf of tissue growing inward horizontally from the edge of the bell, somewhat restricting the subumbrellar orifice, and probably, in those species in which it is well developed, aiding in swimming. Such medusae are called **craspedote** medusae, in contrast to the **acraspedote**—scyphozoan medusae that lack a velum. The stomach in hydromedusae is simple, not divided into radial chambers by mesenteries and not bearing gastric filaments. The marginal sense organs are statocysts or ocelli, or both. Most coelenterates are active carnivores. The prey is stunned and immobilized, or killed, by nematocysts and swallowed into the large coelenteron, where it is attacked and rapidly broken up by extracellular enzymes. The resulting small particles and dissolved organic material are ingested or absorbed by gastrodermal cells. In medusae, much of this material is distributed into the radial canals, before being taken up by the gastrodermal cells.

Hydrozoa show all gradations, from types such as *Polyorchis* and *Aequorea,* in which the medusa is large and well developed, and is the major stage in the life of the

Fig. 6-14. Hydrozoan medusae. **A,** *Polyorchis.* **B,** *Aequorea.*

flowering and great development of the more successful portion of the life. Many polyps remain active most of the year and may live more than one year, while their medusae are released only seasonally and have a brief life-span. Such medusae are little more than a means for dissemination of the species. Since zygotes and planular larvae may be carried about by currents and tides almost as effectively as small medusae, sessile medusae are just as effective, and may be better protected than free medusae for these species.

Because selective pressures are so different in the two stages of the life cycle, evolutionary changes in polyps and in medusae have not necessarily proceeded at the same rate, or in directions that are closely correlated with each other. Thus it is difficult to correlate a given medusa with a given polyp unless the production of medusae can actually be observed. Furthermore, the methods of collection differ widely, the medusae most commonly being taken in plankton tows and the hydranths collected by low-tide shoreline collecting and dredging so that they are not usually taken in the same collection. The result has been the growth of a double taxonomy for hydroids, one system based on the study of the polypoid generation, and the other based upon study of the medusae. This has caused considerable confusion in the taxonomy of the group, especially where the two stages have been discovered independently, assigned different generic and specific names—sometimes even put in different families—and only much later found to be stages in the life of the same species.

Class Hydrozoa, Order Hydroidea

The polyps of hydroids are grouped into two suborders based upon the presence or absence of a **hydrotheca** (a protective cup of secreted perisarc into which the hydranth can withdraw for protection). These suborders are termed the **Gymnoblastea,** or naked hydroids, which are without such a structure, and the **Calyptoblastea,** which possess a hydrotheca. The medusae to which these hydroids give rise differ in characteristic ways and have been termed

organism, to types such as *Tubularia* and *Plumularia,* in which the polyp or hydranth is highly developed, while the medusa may be reduced to a mere budlike appendage of the hydranth, which never becomes free. This is doubtless the outcome of the relative success of the two stages in the particular lines of hydrozoans. Selective pressures are very different in the two stages. If the habits and environment of a particular sort of hydrozoan favor one stage more than the other, the long-time evolutionary result will tend to be a shortening and reduction of the stage least favored, and a

Anthomedusae and **Leptomedusae,** respectively. In general, the Anthomedusae are taller and always retain a tetraradiate form, and many of them have the gonads on the manubrium. They usually have marginal ocelli. Leptomedusae tend to be flatter and more saucer-shaped, some of them show branching of the radial canals or an increase in their number, which obscures the tetraradiate symmetry, and most of them have statocysts but not ocelli on the bell margin. The gonads are commonly borne on the radial canals rather than on the manubrium.

In addition, there are three more groups of craspedote medusae—the Narcomedusae, the Trachymedusae, and the Pteromedusae, often collectively termed the trachyline medusae—which differ in several respects from both Anthomedusae and Leptomedusae and for which polyp stages do not occur.

In the **Narcomedusae** the tentacles pass through a portion of the bell to emerge from the exumbrellar surface above the bell margin. The margin is scalloped by the tentacle bases. The radial canals are absent or, if present, are in the form of flat radial gastric pouches. The gonads lie in the floor of the stomach. The planular larva develops tentacles, becoming an **actinula,** which either transforms directly into the medusa or buds off medusae. The actinulae are commonly parasitic on the other medusae. The nematocysts are all of one kind, the simplest sort—atrichous isorhiza. **Otoporpae** ("auditory tentacles") with strips of ectodermal tissue termed peronia may be present along the bell margin between the main tentacles.

The **Trachymedusae** resemble ordinary hydroid medusae more closely. The stomach is not lobed. The tentacles arise from the margin of the bell, are solid, and have an endodermal core. There are four, six, or eight radial canals, and in some species there often are blind centripetal canals between the radial canals. Ectodermal gonads lie beneath the radial canals. A band of nematocysts around the edge of the umbrella forms a thickened margin. The nematocysts are of only one kind—heterotri-

chous microbasic euryteles. Development is direct, the eggs giving rise to a swimming planula that becomes an actinula by formation of a mouth and tentacles and transforms directly into a medusa.

The **Pteromedusae** are represented by the genus *Tetraplatia,* a peculiar medusa found in oceanic waters. The body is bipyramidal, with an equatorial groove from which four swimming lappets extend, each bearing two statocysts. These lappets are thought to represent the velum. The upper portion of the body is regarded as equivalent to the bell of other medusae and the lower part as equivalent to the manubrium, with the mouth at the lower pole.

Trachyline medusae, with their direct development, are of considerable phylogenetic interest. Some authorities regard the medusa stage as the primitive one and believe that polypoid stages were derived through prolongation and modification of the actinula larval stages. If the view of Weyl (1968) regarding the Precambrian origins of animal groups in the tropical thermocline are correct, this would appear to be the most likely course.

Class Scyphozoa

Scyphozoans, or true jellyfish, seem to represent a line in which medusa success has been so marked as to lead to the evolution of a group in which the medusae are usually larger, more complex, and longer-lived than those of most hydroids, and the polypoid generation is reduced or wholly suppressed. Scyphozoans retain tetramerous symmetry in both the medusa and the polypoid generations. The stomach is divided into four radial chambers containing **gastric filaments.** The medusa is **acraspedote,** that is, it does not possess a true velum, although in some species the edge of the bell is turned in, constituting a sort of pseudovelum. The marginal sense organs are typically **sense pits** and **rhopalia.** The bulk of the mesogloea is essentially free of cells, although a well-developed layer of circular muscle may grow beneath the epithelium of the subumbrella, and the tentacles and oral arms may have well-developed muscle tissue. There is no stomodaeum such as that found in most hydromedusae. The corners of the mouth are drawn out into oral arms.

There are only some 200 species of true jellyfish, but they are so abundant and widely distributed that they are an ecologically important group. They are found in all oceans and range from surface-swimming forms down to 3000 meters' depth or more.

Scyphozoa are divided into five well-marked orders, four of which are free-swimming and are to be regarded as part of the plankton or nekton. Most of them are much larger than the majority of organisms usually thought of as constituting the plankton, and many of them swim as actively as, or more actively than, many other animals generally regarded as nekton. The largest jellyfish known is *Cyanea arctica*, which sometimes attains a weight of nearly a ton, and may have tentacles 100 feet long.

Order Semaeostomeae. These are the most familiar jellyfish, and the only ones usually seen in temperate and cold waters. In this group the bell is bowl- or saucer-shaped, without any furrow. The tentacles arise from the margin or from the subumbrellar surface near the margin, are hollow, and do not arise from pedalia. The corners of the mouth are prolonged into four long, frilly oral arms. The margin of the bell is scalloped into eight to many lappets. There are eight to sixteen rhopalia borne in marginal notches between special rhopalar lappets. Most semaeostomes have a fixed larval stage known as a scyphistoma, and are hence neritic forms more or less bound to coastal waters. Scyphistomae undergo a peculiar transverse constriction to bud off immature medusae known as ephyrae. *Pelagia,* however, lacks such a stage and is found in high oceanic as well as neritic waters. Scyphistomae of jellyfish have two kinds of nematocysts (atrichous isorhiza and microbasic heterotrichous eurytele), while the medusa has also a third kind (holotrichous isorhiza).

GENUS *Aurelia.* The moon jelly is probably the best-known and most widely studied of all jellyfish. *Aurelia* is abundant in coastal waters over most of the world. In many respects it is, however, an "atypical" jellyfish. The radial canals are much branched, forming a complex system of interradial, adradial, and perradial canals. The tentacles are very numerous and relatively short, and are not used in capturing large prey. *Aurelia* is a flagellar-mucus feeder. Mucus secreted on the exumbrella catches small plankton organisms and particles of detritus, is concentrated on the lappets at the edge of the bell, and is licked off by the long oral arms. Currents produced by strong flagella in the grooves of the oral arms then transport the mucus and its contained food particles to the coelenteron. Scyphozoans feeding on larger prey may use the gastric filaments to help pull it into the coelenteron and further subdue it with additional nematocysts.

GENUS *Cyanea.* Widespread and readily recognized by the fact that the long tentacles arise in eight groups, in V- or W-shaped areas, near the outer edge of the subumbrella rather than from the bell margin. The oral arms are long, wide, and voluminous. When ripe, the four large bunches of gonads may be partly evaginated between the oral lobes and the bunches of tentacles. There are both radial and circular muscles in the subumbrella. *Cyanea* exhibits bioluminescence, although there are no ocelli and the animal does not react noticeably to changes in illumination. There are eight rhopalia, located in marginal indentations. Swimmers brushing against this jellyfish often experience a burning sensation. *C. arctica* is the common large red jellyfish of the Atlantic coast. It sometimes attains a size of more than two meters across, weighing nearly a ton, and is the largest known coelenterate.

GENUS *Chrysaora.* Genus contains common large jellyfish of wide distribution. There are eight rhopalia, twenty-four marginal tentacles, and thirty-two or forty-eight marginal lappets. The stomach is divided into sixteen radial pouches, the eight rhopalar pouches being much narrower than the eight tentacular ones. The exumbrella has numerous minute nematocyst warts. This genus also exhibits bioluminescence. It is exceptional in being hermaphroditic. The scyphistoma stage is well developed, often strobilating into thirteen to fifteen ephyrae. Many species have been named, but the specific characters are vague and unreliable. Some authorities regard the related genera *Dactylometra* and *Kuragea* as merely growth stages of this medusa.

GENUS *Pelagia*. Very widely distributed and prominent, the medusae are large, often brightly colored, and bioluminescent. There is no polyp stage, and the genus has both oceanic and neritic distribution. *Pelagia* has eight tentacles and eight rhopaiia. The margin is scalloped into sixteen marginal lobes, which are themselves partially cleft. The exumbrella is covered with warts of nematocysts. The stomach has sixteen radial pouches, each with two unbranched canals leading to the lappets. Large individuals of *Pelagia* are commonly a meter or more in diameter.

The remaining orders of jellyfish are largely tropical or subtropical in their distributions or are swept farther north in the warm currents along eastern North America and eastern Asia.

Order Cubomedusae. Members of this group are readily recognized by the squarish shape of the bell and the fact that the tentacles are borne on pedalia at the corners of the bell, so that there are either four tentacles or four groups of tentacles. The margins of the bell are turned in, constituting a velarium, or false velum, with septa. There are four rhopalia, occurring along the umbrella margin midway between the bases of the pedalia. At night they are strongly phototropic and are a nuisance as they crowd around fisheries' lights. These jellyfish are strong swimmers and voracious predators, eating mostly fish, and have powerful nematocysts that inflict a very painful sting, giving this group the name sea wasp. They are commonly dull yellowish or livid in color. They seem to stay near the bottom much of the time until fully mature, and then swarm to the surface to shed eggs and sperm. The very serious reaction of some persons to the sting of these jellyfish and of certain other coelenterates may be largely due to hypersensitivity induced by previous exposure to the sting.

The principal genera are as follows.

GENUS *Carybdea*. There are only four tentacles, each arising from a pedalium at the corners of the medusa. Each rhopalium has two large and four or five small ocelli.

GENUS *Tripedalia*. There are three pedalia, each bearing a tentacle, at each corner of the bell.

GENUS *Chiropsalmus*. Each corner of the bell has a single large thick pedalium that branches into several smaller ones, each bearing a tentacle. These jellyfish are much feared by swimmers in many tropical areas.

GENUS *Tamoya*. Large medusae resembling *Carybdea* in having only four pedalia, each with one tentacle. In this genus the bell is taller than wide, and the exumbrella is prominently sculptured in the male. It also differs from *Carybdea* in having four perradial mesenteries joining the stomach to the subumbrella.

Order Coronatae. Mostly small to medium-sized jellyfish, the bell marked with a circular coronal constriction, or groove, that separates the central domelike portion from the marginal scalloped portion. The solid tentacles are borne on pedalia alternating with the lappets. Several are deepwater bathypelagic species, dark purple or reddish brown in color. Cosmopolitan.

GENUS *Atorella*. There are six rhopalia and six tentacles alternating between each pair of lappets. Circular muscles weak. Four or six gonads.

GENUS *Periphylla*. Sixteen pedalia, four of which bear rhopalia and twelve of which bear tentacles. Sixteen marginal lappets grouped into four pairs of rhopalar and four pairs of tentacular lappets. *P. hyacinthina* is cosmopolitan in all oceans. Bell is purple or maroon. Usually from deep water but occasionally occurs near the surface.

GENUS *Atolla*. Sixteen to thirty-two tentacles and rhopalia, the rhopalia bearing pedalia that form a circlet below the tentacle-bearing ones. Eight gonads. Twice as many marginal lappets as rhopalia. Bell is rather shallow, and dark red or purple. Bathypelagic.

GENUS *Linuche (Linerges)*. Eight rhopalia and eight tentacles, and sixteen marginal lappets. Subumbrellar surface has hernia-like protuberances. Wide radial gastric pouches. No circular canal. Central stomach opens by four perradial ostia into a ring sinus that breaks up into sixteen branching radiating pouches in the lappets.

GENUS *Nausithoë*. Eight rhopalia and eight tentacles. No marginal lappets and no hernia-like pouches on subumbrella. Radiating stomach pouches are simple.

Order Rhizostomeae. This is a very distinctive order. It is particularly abundant in shallow tropical and subtropical waters of the Indo-Pacific. The bell is firm and bowl- or saucer-shaped, and it often bears nematocyst warts on the exumbrella. There are no marginal tentacles. The bell margin is scalloped and bears eight or more rhopalia. The upper ends of the oral arms are fused, obliterating the mouth, which runs as branching channels through the oral arms and ends in hundreds of small mouth openings in the lower inner portion of the eight mouth arms. The subgenital pits are, in many of the genera, so large as to form a continuous cavity, or porticus, separating the bell from the oral arms, which are then connected to the bell by four pillars through which run the canallike extensions of the stomach into the mouth arms. The oral arms are armed with filamentous appendages loaded with nematocysts. Subumbrellar musculature is usually strong. In some genera the oral arms terminate in long pendant clublike appendages of unknown function.

GENUS *Catostylus*. Mouth arms without clubs, filaments, or other appendages. There are sixteen radial canals (eight rhopalar and eight adradial). The rhopalar canals extend to the bell margin, and the adradial canals end in the ring canal. The ring canal gives off a network of anastomosing vessels on both sides, some of which connect with the radial canals but not directly with the stomach. The subumbrellar circular muscles are only partially interrupted at the eight chief radii.

GENUS *Crambione*. Similar to *Catostylus,* but the circular muscles of the subumbrella are not interrupted at the radii, and the oral arms bear numerous nematocyst-loaded filaments. There are a pair of ocelli and a large furrowed sense pit at each rhopalium.

GENUS *Rhizostoma*. The mouth arms bear long clublike terminal appendages, about as long as the mouth arms themselves, and in addition, at their upper end, bear frilled scapulets, or shoulder ruffles, with additional mouth openings, near the bell. The eight rhopalar and eight adradial canals, as well as a blind network of canals, all extend to the border. A blind network of canals is present between each pair of radial canals.

GENUS *Rhopilema*. Similar to *Rhizostoma* but with long oral arms, bearing numerous clubs and filaments in addition to the large terminal appendage, and very large scapulets. This is the most northerly occurring rhizostome along the Atlantic coast, occasionally straying into Long Island Sound.

GENUS *Stomolophus*. Bell is deep, covering the upper part of the mouth-arm complex, so that the well-developed scapulets are all up under the bell. There is a persistent central mouth, and the mouth arms do not bear terminal appendages. General appearance is of a thick toadstool on a thick fleshy short stem.

GENUS *Cassiopea*. One of the most remarkable of jellyfish, attaining a rather large size, up to 2 or 3 feet or more in diameter. The bell is flattened, the edge of the flattened area comprised of especially tall columnar epithelium. The animals are sluggish and spend most of their time upside down on the floor of shallow tropical lagoons, often so numerous as to nearly coat the floor. The bell pulsates slowly, and the flattened top portion may act in sucker-like manner to anchor them. The much-branched oral arms, with many vesicular appendages, look for all the world like some rather dense vegetable growth. There is remarkable color variation from one individual to another. By means of mucus and nematocysts, small animals are caught in this bushy mass, swept by flagellar action into the mouths, which open widely to receive them, and carried by flagella in the gastric channels to the stomach. There they are seized by the gastric filaments. Rejected particles are carried by other flagellar tracts back down the brachial canals and discharged through the mouths. Thus *Cassiopea* is a flagellar-mucus feeder somewhat like *Aurelia,* but the details of the mechanism are wholly different. The exposed tissues of the oral arms and subumbrella are also full of zooxanthellae, which are exposed to the sunlight as the jellyfish lies inverted and probably contribute to the nutrition of their host. This makes the analogy of the bushy mass of oral arms with a vegetable growth even more apropos. Numerous species have been named, but the variability of individuals taken in any one lagoon is such that specific characteristics are difficult to delimit. *Cassiopea* may well be regarded as benthic rather than planktonic.

GENUS *Thysanostoma.* The mouth arms are very elongate and bear three rows of frilled mouths from their base to the lower end. There are eight rhopalar canals, a distinct ring canal, and anastomosing networks of canals between the radial canals.

GENUS *Mastigias.* Small rhizostomes with globular or nearly spherical bell, only a few inches in diameter and marked with light spots. The oral arms bear long purple terminal appendages. These small jellyfish are often very abundant at certain times of the year in shallow tropical waters of the Indo-Pacific.

Order Siphonophora. Siphonophora are pelagic colonial hydroids in which several types of modified polypoid and medusoid individuals are united into a colony or higher-grade individual in which the individual components serve as organs. All are connected by common coenosarc and gastrovascular space. Some forms simply float at the surface of the sea, and others swim actively below the surface. Larger floating forms, such as the Portuguese man-of-war, have powerful nematocysts with which they can kill fish and inflict serious injury on man. Siphonophorans are most abundant in tropical waters, to which many are limited, but some occur in all oceans.

The components of siphonophoran colonies are as follows:

1. **Gastrozooids.** Usually possess one hollow tentacle armed with nematocysts for capture of food, a mouth, and a coelenteron.
2. **Palpons.** Simpler than gastrozooids, without a mouth, and serve as accessory digestive organs.
3. **Dactylozooids.** May be long tubular tentacle-like zooids without a mouth, or shorter zooids armed with a single long tentacle. Dactylozooids are armed with powerful nematocysts and are the principal organs of the capture of prey.
4. **Gonozooids.** Short zooids, with or without a mouth, bearing clusters of gonophores.
5. **Gonophores.** Medusoid zooids bearing sex cells on gonodendrons.
6. **Swimming bells** (nectophores or nematocalyce). Used for locomotion.
7. **Bracts.** Thick gelatinous structures of various shapes that aid in swimming and protection of the colony. Bracts have been regarded as altered medusoids, but recent authorities regard them merely as altered tentacles.
8. **Floats** (pneumatophores). Also regarded as an altered medusoid, and now thought to be an aboral invagination.
9. **Stem.** Structure along which other components are strung.

Not all these components are found on any one siphonophoran. Siphonophora in which an elongate, stemlike coenosarc trails behind a nectophore, or swimming bell, have the various individuals arranged along this stem in definite groups, or **cormidia.** Each cormidium consists of a definite number and arrangement of individuals. New cormidia form at the proximal growing end of the coenosarc in the nectophore, and they become larger, more individualized, and more mature distally.

Many siphonophorans are very delicate and extremely transparent in the water. Forms in which many individuals, bracts, etc. are strung out along a long stem are difficult to collect in perfect condition, only disconnected parts usually being found in the usual plankton haul. This has given rise to some confusion, as detached cormidia have often been described as separate species.

A type of nematocyst found only in siphonophorans is the rhopaloneme.

It is thought that Siphonophora may have evolved from trachyline hydroids with a swimming actinuloid stage and an adult medusoid stage—that the merger of these two stages, remaining together and differentiating into the various modified polypoid and medusoid forms found in the colonies, produced the complex forms found today. By using whole individuals as organs, the siphonophorans have achieved an organ grade of construction in spite of their simple diploblastic tissue-grade organization of the individuals. The tendency of colonial animals to specialize nonreproductive members of the colony in order to form organs is one of the strongest lines of support for the theory that metazoan animals originated from colonial Protozoa. A colonial protozoan that developed along these lines, forming a highly complex colony, with various cells performing different functions, would be defined as a metazoan.

Siphonophora are commonly divided into three groups on the basis of the possession of swimming bells and floats: the **Calycophora,** with swimming bells (nectophores), the **Physonectae,** with both floats and swimming bells, and the **Cystonectae,** with floats but no swimming bells.

Hyman (1940) divides the group into two suborders, the **Calycophora,** in which the upper end of the colony possesses one or more swimming bells but not a float, and the **Physophorida,** in which the upper end of the colony is a pneumatophore, or float. Swimming bells may or may not be present in this group.

The following examples will serve to illustrate some of the common types.

Suborder Calycophora. Upper end of colony with one or more swimming bells (nectophores) but no floats. Other zooids commonly arranged along a long stem.

GENUS *Abyla.* Upper end of colony of two unlike bells, replaceable by reserve bells. Upper bell has a rectangular prismatic apical facet and is smaller than lower bell; bears somatocyst and oil cavity. Cormidia are at equal intervals along stem, and each siphon has a bract.

GENUS *Diphyes.* Has two large acuminate swimming bells at upper end of stalk bearing cormidia. An oceanic, very widely distributed genus. Numerous species.

GENUS *Lensia.* Very common cosmopolitan genus.

GENUS *Monophyes.* A single anterior nectocalyx, representing the primary nectocalyx of the larva. No secondary replacement of nectocalyces.

GENUS *Muggiaea.* Single nectocalyx which is a secondary pyramidal one, replacing the primary caplike larval nectocalyx. Widely distributed genus.

Suborder Physophorida. A float, or pneumatophore, is present. Nectophores may or may not be present. The advent of the float above the swimming bells has led to the evolution of forms with shorter stems, crowding the individuals together around a short, massive coenosarc. The cormidia lose their individuality. This line of evolution leads to the Cystonectae, in which nectophores and bracts have disappeared and the various zooids are crowded together at the base of the large float. The Physonectae show intermediate stages of this evolutionary trend, still having nectophores as well as a float and exhibiting varying degrees of shortening and crowding of the stem and its zooids.

SECTION PHYSONECTAE. Both floats and swimming bells are present.

GENUS *Agalma.* Polygastric, with long tubular stem-bearing numerous siphons, palpons, and bracts. Nematocalyces are numerous and biserial. Pneumatophore has radial pouches. Colony is much longer than broad, with an apical float of a simple type in which the inner and outer sacs, the chitinous lining, the funnel, and the simple gas gland are clearly evident. Swimming bells are in two alternating rows closely pressed together. Long stalk-bearing numerous cormidia are composed of gasterozooids, male and female gonozooids, palpons, and large bracts.

GENUS *Forskalia*. Polygastric, with long tubular stem with numerous siphons, palpons, and bracts. Each siphon has a branched tentacle. Nectocalyces are numerous, multiserial, and strobiliform in several spiral rows. Pneumatophore has radial pouches.

GENUS *Physophora*. Biserial nectocalyces, followed by a tight cluster of cormidia with large gasterozooids, palpons, and clusters of small gonophores, below which trail long tentacles bearing cnidosacs on short stalks. No bracts.

GENUS *Porpita*. Brightly colored and pelagic, and occurring in warm seas. Large central-feeding zooid surrounded by smaller gastrozooids and reproductive zooids, with a ring of slender dactylozooids near the margin. The disc has a central float, or pneumatocyst.

GENUS *Stephalia*. Short colony with large, broad float surrounded basally by a corona of nectocalyces, below which many siphons with tentacles are closely crowded, surrounding one larger central siphon. Smaller gastrozooids and gonozooids are clustered between the siphons. The corona of nectocalyces is interrupted by one larger aurophore.

GENUS *Velella*. Complex float, polythalamous, composed of numerous concentric rings and bearing a diagonal vertical crest, or sail. Marginal tentacles are simple, without cnidophores. Gonostyles with mouths. *Velella*, like the Portuguese man-of-war, floats on the surface of warm seas, with the tentacles hanging down into the water. In life they are bright iridescent blue, and sometimes in the tropical Pacific they occur in vast numbers of zooxanthellae, which perhaps supplement the diet of small animals with products of their photosynthesis. *Velella* is a favorite prey of the carnivorous pelagic gastropod *Janthina*. The sails on populations of *Velella* in the Northern and Southern hemispheres are said to be reversed in such a way that the prevailing winds in each hemisphere tend to keep the populations offshore. Nonetheless, great numbers of them occasionally are blown ashore and pile up in windrows on the beaches.

SECTION CYSTONECTAE. A large apical pneumatophore, or float, and the absence of nectocalyces and bracts characterize this group. Pneumatocyst has an apical stigma.

GENUS *Cystalia*. Monogastric, with one large siphon and a tentacle, surrounded by a corona of smaller siphons. Pneumatophore has no radial septa or hypocystic villi.

GENUS *Epibula*. Polygastric, with short, inflated, spirally convoluted stem. Cormidia ordinate in a spiral ring protected by a corona of palpons. Pneumatophore is without pericystic radial pouches but with hypocystic villi.

GENUS *Rhizophysa*. Polygastric, with a long stem, bearing in its median ventral line numerous monogastric cormidia with a single palpon and tentacle. Pneumatophore is large, with radial pericystic pouches and hypocystic villi.

GENUS *Solacia*. Polygastric, with a long stem, bearing in its ventral median line numerous polygastric cormidia. Pneumatophore is large, without radial pericystic pouches but with hypocystic villi.

GENUS *Physalia*. The best-known of all Siphonophora, the Portuguese man-of-war. Polygastric, with a short, inflated stem horizontally expanded along the ventral side of the large horizontal pneumatophore. Cormidia are in a multiple series along the ventral side of the trunk, and are usually dissolved, that is, so crowded together as to have lost their individuality as cormidia. Pneumatophore is large and has a chambered dorsal crest, but is without radial septa or hypocystic villi.

The animals are a beautiful bright blue to azure purple, and the float measures about 1 foot long by 6 inches wide. Mature ones have as many as 1000 polyps (a float, gastrozooids, gonozooids, and dactylozooids). The tentacles trail from 20 to 50 or more feet in the water, and are armed with powerful nematocysts capable of killing fish. When a fish or another prey is taken, it is brought up to the mass of gastrozooids beneath the bell. They spread their mouths out over the prey, forming a sort of communal stomach into which digestive enzymes are poured, so that the victim is rapidly digested. Nematocysts may retain their ability to sting after the death of the animal. Their poison is a protein nerve poison. Fiddler crabs are particularly sensitive to seawater extracts of the tentacle. Curiously enough, there are certain small fish that can swim among the tentacles with impunity. For example, *Nomeus gronovii,* the man-of-war fish, follows *Physalia* about, finding protection from larger predators among the tentacles.

PHYLUM CTENOPHORA

The ctenophores comprise a small, sharply defined phylum of usually delicate transparent pelagic gelatinous animals, often of great beauty. The common name, comb jelly, refers to the unique mechanism of locomotion, which consists of eight meridional rows of strongly flagellated comb plates running from near the aboral pole of the body, over the surface, and toward the oral pole. The body symmetry is biradial around the oral-aboral axis, one plane, the tentacular plane, being defined by the presence of two tentacle pouches, from each of which a long tentacle, bearing short branches armed with colloblasts, or sticky cells, can be protruded. The other plane, the sagittal plane, lies at right angles to the tentacular plane, is situated at the apical pole, and is marked by the polar fields, which are special ciliated sensory areas. The comb-plate rows between which a tentacle pouch lies, that is, those nearest the tentacular plane, are termed subtentacular comb rows; those nearest the sagittal plane are termed subsagittal comb rows. The digestive system consists of a long, heavily ciliated pharynx extending about half way or more from the mouth toward the aboral pole. At the upper end the pharynx is dilated and glandular, serving as a stomach. Beyond this, the ramifications of the digestive tract are thin, transparent canals. The intestine, or aboral canal, continues straight toward the aboral pole, where it divides into four small branches just below the statocyst. These branches pass upward around the statocyst, two of them ending in anal pores. At the junction of the stomach and intestine, two radial canals are given off and pass toward the tentacle pouches. Just before or soon after reaching the tentacle pouches, they give off two oblique, lateral branches, the interradial canals. The interradial canals, in turn, branch dichotomously into adradial canals. Each adradial canal terminates in a meridional canal lying just beneath one of the comb rows. The meridional canals bear the gonads, which are arranged in a very characteristic manner, the ovaries lying along the perradial wall of the canal and the spermaries lying along the interradial wall. The radial canals also give off a pair of paragastric canals that pass down along each side of the pharynx toward the mouth in the tentacular plane.

The above characterization applies primarily to the Cydippida, which are regarded as the most typical, or the least modified, of the ctenophores. Larvae from other groups resemble the cydippids for a brief time and are known as cydippid larvae.

Ctenophores are classified into two classes and five orders. The members of all but one of the orders are planktonic. These classes and orders will be briefly characterized, together with some of the more familiar genera.

Class Tentaculata

Ctenophores with tentacles or oral lobes. In the adults of some groups, the tentacles may be so minute and reduced as to be difficult to find, or even absent.

Order Cydippida. The members of this order retain their general larval characteristics into the adult stage. They are globular or egg-shaped, and have two well-developed tentacles. Their gastrovascular canals are simple, blind, and do not ramify out into a peripheral canal system. The comb rows are usually equal in length or nearly so.

GENUS *Pleurobrachia.* Body is globular or egg-shaped, and circular in cross section. The comb rows extend most of the distance from the apical to the oral poles. The bases of the tentacle pouches diverge, rather than lying close to the pharynx on each side.

GENUS *Hormiphora*. Very similar to *Pleurobrachia* in general appearance, the most easily seen difference being that the bases of the tentacle pouches lie closely adjacent to the pharynx for most of their length.

GENUS *Lampetria*. Comb rows are short, not reaching the oral third of the body. Tentacle pouches are small, lying at the level of the oral ends of the comb rows. Tentacle openings are in the oral half of the body. Stomach is very broad and long, and the mouth can be widely opened and used to creep on, thus foreshadowing the permanent condition in the order Platyctenea.

GENUS *Euplokamis*. Body is taller and subcylindrical, the meridional canals and costae running its whole length. Mouth is not dilatable. Tentacle roots are near the middle level of the body.

GENUS *Moseria*. Rather similar to *Pleurobrachia,* but body may be somewhat compressed in the tentacular plane. The comb rows are wider, and the mouth is surrounded by a broad collar, forming a retractile tube. The costae continue down to the level of this collar.

GENUS *Ganescha*. Of special interest. While showing, in the main, cydippid structure, the genus has some features similar to those of the order Lobata. Dawydoff (1936) regarded this as sufficiently important to warrant the designation of a new order, Ganeshidea, which has intermediate characteristics, for this genus. The pharynx is very large and long, reaching nearly to the apical pole, so that the intestine, or funnel, is very short. The structure forming the definitive mouth of cydippids, Lobata, and Nuda are here pushed up internally, forming a vestibule compressed in the opposite plane to the pharynx. The interradial canals arise directly from the funnel, as in Lobata, and the adradial canals continue as the meridional canals instead of joining them along their course; the lower ends of the meridional canals, instead of being blind, join a circular canal around the margin of the lips. The tentacular pouches are small, close to the body wall, and open toward the oral rather than toward the aboral pole. In very young individuals the tentacular pouches are oriented normally, but as they change to the adult condition, their orientation is reversed.

GENUS *Euchlora*. This genus attracted attention because of the finding of functional nematocysts, giving rise to speculation that they represented a rather direct link with the Cnidaria. However, Komai (1951) has presented evidence that these nematocysts are exogenous in origin and were derived from eating medusae. There have been similar cases in which functional nematocysts found in certain flatworms and nudibranchs had been derived from eating hydroids.

Order Lobata. These commonly rather large ctenophores have a very delicate structure and are difficult to collect and preserve in good condition. The body is commonly helmet-shaped and is compressed in the tentacular plane and expanded in the sagittal plane, into large lobes, with the result that the subsagittal comb rows are decidedly longer than the subtentacular ones. The ends of the subtentacular rows are produced into four heavily ciliated extensions, auricles, or lappets, a pair on each side of the mouth. The ciliated grooves extending from the mouth to the lappets contain additional tentacles. The four interradial canals arise directly from the funnel. There are no tentacle sheaths, and the tentacles are small or absent.

GENUS *Bolinopsis*. Lobes are medium-sized, and the windings of meridional vessels in them are simple. Adradial canals pass directly into the upper ends of the meridional canals; auricles are short. Common along the Atlantic coast of the United States.

GENUS *Eucharis (Leucothea)*. Very large oral lobes, long, slender auricles, long main tentacles, and two deep, narrow pits extending aborally from near the tentacle bases. Body surface has conspicuous papillae tipped an orange or brown. A warmwater genus.

GENUS *Kiyohimea*. Large, very fragile, and colorless. Strongly compressed and with a pair of earlike processes in the transverse plane at the aboral end; auricles are rather stout and not very long. Lappets are of moderate size, with complex meandering canals in them. Tentacles are greatly degenerated.

GENUS *Mnemiopsis*. Very large lobes, the lobes and auricles arising at almost the same level as the funnel; auricles are long and ribbonlike. Deep grooves extend to the level of the statocyst. Bioluminescent. Common along the Atlantic coast in summer and fall.

GENUS *Ocyropsis (Ocyroë)*. Distinguished by the large, muscular oral lobes whose movements serve as the chief swimming mechanism. Abundant in Caribbean area and Gulf of Mexico; also present in the Indo-Pacific.

Order Cestida. Strongly compressed in the tentacular plane and extended in the sagittal plane, becoming ribbonlike, or beltlike. Upper edge bordered by the long subsagittal comb rows; mouth and statocyst are on opposite edges of the middle of the belt; subtentacular meridional canals run lengthwise through the center of the ribbon; lower border has small tentacles. Some become more than a meter in length. Subtentacular comb rows are rudimentary.

GENUS *Cestum*. Venus's-girdle. Tentacles and tentacle sheaths are beside the mouth; tentacles are reduced to mere tufts of filaments; two rows of short tentacles are along whole oral edge of body.

GENUS *Velamin*. Four subtentacular canals do not arch up but proceed directly outward in the equatorial plane. Does not get as large as *Cestum,* and is only about 15 cm. long. Occurs off Atlantic coast.

Order Platyctenea. Aberrant, benthic, creeping ctenophores compressed in the oral-aboral plane; superficially resemble polyclad flatworms. The young have comb rows and two tentacles with tentacle sheaths; these may be lost in the adult in some genera. The creeping surface is homologous to the ciliated pharynx of other ctenophores, opened out to provide a broad, flat, ciliated creeping surface.

Representative genera are listed below.

GENUS *Coeloplana* KOWALEVSKY, 1892. Flat, oval, elongated in tentacular plane.

No comb rows in adult. Four rows of erectile papillae on aboral surface.

GENUS *Ctenoplana* KOROTNEFF, 1886. Comb rows present. No aboral erectile papillae.

GENUS *Gastrodes* KOMAI, 1922. Immature stages parasitic in Thaliacea. Develops into free cydippid larva, which settles to bottom, loses the comb rows, and flattens out as creeping platyctenid.

Class Nuda, Order Beroida

Without tentacles at any stage of development. Mouth and pharynx are very wide and long, occupying most of the body. Meridional canals have numerous side branches often brightly bioluminescent. Body is compressed in tentacular plane.

GENUS *Beroë*. Comb rows do not extend full length of animal. *Beroë cucumis* is cosmopolitan. Oval, with a somewhat constricted mouth. Side branches of meridional canals end in blind twigs. Beautifully bioluminescent; rose tinted. Feeds largely on other ctenophores. *Beroë forskali,* another large, cosmopolitan species. Body is large, up to 10 to 12 cm. long. More or less triangular, with the mouth occupying the broad end. Side branches of meridional canals form an anastomosing network. *Beroë gracilis* is more elongate and cylindrical; about 25 mm. long, and 8 to 10 mm. wide. Costae extend about two thirds of the body length.

GENUS *Neis*. The pharyngeal canals branch at the mouth and form a complete circular canal around the mouth.

SUBPHYLUM UROCHORDATA
Thaliacea

The salps are pelagic transparent tunicates, rather cylindrical, with a large oral opening at one end of the body and a large cloacal opening at the other. The body wall is a thick elastic resilient cellulose tunic, the elasticity of which restores the body shape following muscular contraction. The mantle tissue immediately below the tunic has several conspicuous ring-shaped muscle bands, the contraction of which forces water out through the cloaca, propelling the animal forward. The body cavity is divided by a respiratory partition, which may vary from a complete partition bearing slits, or stigmata, to a single ciliated bar. This partition divides the body into an anterior pharyngeal chamber and a posterior cloacal chamber. An endostyle and dorsal lamina are present in the mantle on the ventral side of the body below the cloacal chamber. In the dorsal wall of the pharyngeal chamber lies a ganglion, which in some species has an eyespot and subneural gland associated with it.

Salps alternate between solitary and aggregated generations, which in the past often received separate generic names. A common practice has been to use both names, the oldest one first, with a hyphen between them.

Salps are ciliary mucus feeders, and they remove phytoplankton and other fine particulate matter from the water. As this is also the food of many larval and some adult plankton-feeding fish, the occurrence of large numbers of small salps may be restrictive to the development of successful stocks of plankton-feeding fish or of fish with plankton-feeding larvae. The salps themselves are a poor food, not much utilized by fish. Salps are characteristic of warm waters. One species occurs in the Antarctic, and there are none in the high northern latitudes.

Two principal groups are recognized, the **Cyclomyaria,** with the large family **Doliolidae,** and the **Hemimyaria,** of which the chief family is the family **Salpidae.**

Cyclomyaria, Family Doliolidae. Body is barrel-shaped, with a series of usually complete muscular rings around it; respiratory partition is platelike, with two rows of stigmata; oral and cloacal openings, at opposite ends of body are lobed. Three types of individuals are produced asexually, and one type hatches from fertilized eggs. A tailed larva is always developed.

GENUS *Doliolum.* Sexual generation has eight muscle bands, twelve oral lobes, and ten cloacal lobes. The first asexual generation has nine muscle bands, and ten oral and twelve cloacal lobes. From the hinder end grows a dorsal protuberance and a ventral stolon. Buds originating on the stolon

migrate to the tubercle, which becomes long and taillike, and bears the buds in five rows. The median row becomes detached and develops into the next asexual generation. The asexual generations are similar to the sexual generation except that they lack gonads and they produce a ventral stolon, the buds of which detach and grow up into the sexual generation. (The buds in the lateral rows on the stolon of the first sexual form are called gastrozooids. They develop into doliolum-like zooids with only rudimentary gonads that soon atrophy, but the gut is well developed, and the zooids actively take nourishment and serve for the nutrition of the growing stolon.)

Hemimyaria, Family Salpidae. In this group the tunic is thicker and the body is often somewhat flattened dorsoventrally. The oral and cloacal openings may be either terminal or subterminal. The respiratory partition is reduced to a single ciliated bar running diagonally from the dorsal wall to the esophagus in the ventral wall. The life cycle includes alternating solitary asexual and aggregated sexual types of individuals. No tailed larva is formed. Gametes produced by the sexual form develop into asexual individuals that produce chains of sexual individuals.

There are four to twenty (usually six to nine) muscle bands, continuous dorsally but usually not ventrally, except for the atrial sphincter.

The aggregate sexual phase may differ greatly in size, shape, and other characters from the solitary phase. Commonly they develop only one ovum per zooid. The muscles are less numerous and less well developed. Each sexual zooid has eight pairs of adhesive papillae on ventrolateral surfaces, by which it is attached to the next zooids in line.

If the zooids become isolated from the breaking of the chain, they continue to exist alone and the adhesive papillae disappear.

GENUS *Salpa.* A typical "nucleus" is present, or if not, the muscle strands are numerous. There is a straight, double chain of aggregate sexual individuals. Digestive tube is usually bent on itself and contained in the nucleus.

GENUS *Cyclosalpa*. No "nucleus" is present. Intestine is not twisted but passes forward along the endostyle in the sexual form and along the dorsal lamina in the asexual form, to open anteriorly into the atrial cavity. Salp chain is often circular, each chain, or wheel, containing about thirteen individuals of the aggregate generations.

Pyrosomida. A suborder of the Ascidiacea, or true tunicates. Most tunicates are benthic throughout their adult life. In the pyrosomids, however, we have a group of colonial forms in which the entire colony behaves as a free-swimming pelagic organism. The colonies are cylindrical, and tapering or flattened, with one end open. The zooids are arranged in one layer in the walls of the gelatinous cylinder, perpendicular to the surface, with their incurrent openings on the outside and their atrial openings on the inner surface. In some species the outer surface of the common tunic may be produced into long, fingerlike processes. As the zooids pass water through their branchial chambers and out the atrial openings for feeding and respiration, the water streaming out from the open end of the colony causes the colony as a whole to swim with the closed end forward. At the fore end of the branchial sac of each zooid are paired luminescent organs. The colonies are noted for the brilliant bioluminescence they display when disturbed. They float horizontally in the sea and have only weak powers of movement. Colonies vary from a few inches to over 4 feet in length.

GENUS *Pyrosoma*. Has characters of the suborder.

Larval forms

Because much of the shelf fauna and some of the pelagic nekton and deep-water benthos have larval forms that are planktonic for a time, and because these larvae are in most cases very different from the adults, from both a morphological and an ecological point of view, and are most commonly taken in plankton tows, we will consider a number of representative planktonic larval types together here, rather than taking them up separately in connection with the corresponding adult groups.

A short description of larval forms in general will be given before the various larvae are described individually. Ecologically they may be thought of as **planktotrophic,** that is, eating other planktonic organisms or detritus, most commonly elements of the phytoplankton, or as **lecithotrophic,** subsisting during part or all of their planktonic existence on stored food from the egg. The latter types usually develop from larger, yolkier eggs. Some of them may subsist for half a month or more, about as long as most planktotrophic forms. Lecithotrophic types grade insensibly into forms in which an increasing portion of their development passes through, in the egg stage, to types that no longer have any planktonic larvae. Thorson (1950) called attention to an interesting correlation between the proportions of species in various groups of benthic animals that have planktonic larvae, and latitude. The highest proportion of species with planktotrophic larvae are found in tropical and subtropical waters, but as one goes toward higher latitudes, or to great depths, the proportion with lecithotrophic larvae or with direct development without a planktonic larvae increases. This phenomenon is thought to be correlated with the highly seasonal character of the outbursts of phytoplankton in the spring in the higher latitudes and with the paucity of suitable planktonic food at great depths. Swarms of planktotrophic larvae might face starvation if reproduction were not timed very nicely at times when there is plenty of phytoplankton. It has been shown repeatedly that in northern latitudes, species with long-lived planktotrophic larvae show much greater fluctuation from year to year in success of settlement than do species that have a very brief period of existence as planktonic larvae, or none at all.

For sessile animals, long-lived planktonic stages give a better chance for successful dispersal to more distant localities with suitable habitat, but there is a greater risk that the larvae may not survive at all. The balance of risk and advantage shifts in favor of planktotrophic larvae in the tropics, where the phytoplankton supply does not fluctuate so markedly, and in favor of direct development in the high latitudes, though in neither case is the shift so drastic as to have led to the elimination of one or the other mode for all the animals. Direct development in the high latitudes seems also to be correlated with a longer average life-span for the sessile stage or stages. It is obviously possible for an animal with a given amount of biomass to produce a great many more planktotrophic larvae than to produce either lecithotrophic larvae or directly developing forms, which must depend upon stored food. Where the chances for their survival are good, this becomes the method of choice.

Having a planktonic larval stage means that such larvae are at the mercy of the water currents of the region. This may have a considerable influence on the ability of such benthic species to live and replace themselves in a given locality. Johnson (1939), reporting on the occurrence of nauplii of the sand crab *Emerita* in offshore waters far from land, found that these larvae are reliable indicators of waters of inshore origin that have been swept out to sea. Such larvae are, of course, doomed, the loss of many larvae being a price paid by the species for the potential of wide dispersal. If the prevailing currents at a particular place are such as to carry away any planktonic larvae liberated, this may be sufficient to exclude benthic species that have such larvae from the area, even though in other respects it may be entirely suitable for them. In some such cases settlement may occur successfully if larvae at a stage ready to settle are also swept in from other areas.

Local rotary drifts and eddies over banks and along shores play an important part in keeping enough planktonic larvae over such areas until they are ready to settle (Fish, 1935), enabling benthic populations to be maintained in places that might otherwise be swept bare by the major currents of the region. The great rotary eddies,

Fig. 6-15. Amphiblastula larva, *Sycandra.*

such as the Sargasso Sea and the eddy in the Gulf of Mexico, probably maintain endemic plankters from which the surrounding currents are continually seeded. Redfield (1941) studied the great eddy in the Gulf of Maine and found that much cold, relatively barren water enters from the north during the winter and spring, but that as it swings counterclockwise around the gulf, the biomass of plankton is greatly increased, so that the water escaping over Georges Bank is relatively rich in plankton, which probably accounts for the very favorable fishing conditions found there.

SPONGE LARVAE

Amphiblastulae of sponges are blastula-like larvae with large, yolk-laden, smooth cells in one hemisphere and small, flagellated cells in the other. They have only a brief planktonic life before settling down to the benthic life of a sponge. The flagellated cells become the choanocytes of the adult sponge. Other cells of the sponge body are derived from the large, smooth cells of the posterior pole. These larvae are seldom very numerous in the plankton, and as they do not feed during this stage, they do not exert much influence on other animals there.

PLANULA LARVAE OF COELENTERATES

The product of sexual reproduction is a short-lived solid ciliated larva resembling an acoel flatworm, but without a mouth or digestive tract. In coelenterates with a sessile polypoid generation, the planula attaches to a suitable place on the substrate and transforms, by way of an actinula-like stage, into the polypoid. In pelagic medusae without a polypoid generation, the planula generally becomes an **actinula** larva that either transforms into, or buds off, medusae. The actinulae of some of the trachyline medusae are parasitic on other species of medusae. Young medusae of scyphozoans budded off by the **scyphistoma,** the **ephyra** might also be considered a larval stage before they have attained the size and proportions of the adult medusa. In the case of the planula, we again have a short-lived nonfeeding form, seldom abundant in the plankton and exerting little effect there.

PLATYHELMINTHES

Small free-living flatworms, particularly of the groups Acoela and Rhabdocoela, are occasionally encountered in neritic plankton, but most of them are, at best, tychopelagic, not typically plankton organisms. Larval stages, **miracidia** and **cercaria,** of digenetic flukes also have a brief planktonic existence, as do the free-swimming ciliated larvae of monogenetic trematodes and the coracidia of pseudophyllidean tapeworms. All of these are nonfeeding larvae that are merely en route from one host to another. They do not constitute an important element in the plankton and are infrequently encountered in plankton hauls.

NEMERTEA

Many benthic nemertean worms, order **Heteronemertea,** have a unique planktonic larval form for dispersal known as the **pilidium larva,** which looks somewhat like a small swimming football helmet with large earlobes. It is derived by gastrulation and the downgrowth of large oral lobes on either side of the blastopore. The invagination produces the foregut and midgut regions, which can be considered endodermal. Ectodermal invaginations produce paired discs that detach and move into the interior, uniting around the gut to form a larva. These discs are the cephalic, cerebral, and trunk discs, together with unpaired dorsal and proboscis discs. When the larva has been fully formed inside the pilidium, it escapes, sinks to the substrate, and grows into an adult worm. The empty pilidium may continue to swim for a time before it undergoes disaggregation or is eaten.

ANNELIDA

Archiannelida and **Polychaeta** usually produce **trochophore** larvae. These larvae are formed through spiral determinate cleavage in which the various blastomeres each give rise to particular parts of the larva. During late cleavage stages a characteristic annelid cross is formed. It is interradial in position and has an apical annelid rosette of four cells. As the embryo grows it develops into a hollow biconical organism with a beltlike band of large cilia in front of the mouth. The band, called the **prototroch,** is equatorial in position. A second band of cilia, situated postorally, is termed the **metatroch,** and there may be one around the anus, the **telotroch.** There is a complete digestive tract, with a mouth near the midlevel just below the prototroch, a curved gut that includes an enlarged stomach, and a short intestine passing to the anus at the lower pole. There is an apical ganglion and a sense organ with a patch of long cilia at the top of the upper hemisphere, a larval nephridium, sometimes a statocyst, and mesodermal bands derived from a pair of posterior teloblasts. In some there is also a pigmented eyespot. Some species, such as *Eupomatus* have only a prototroch. Others, such as *Chaetopterus,* have only the metatroch. Still others have several rings of cilia, and may be described as polytroch (*Ophryotrocha*). Some tube worms have an even more modified trochophore, bearing two bundles of long setae, known as the **metraria** larva (*Myriochele danielssoni*).

Transformation into the adult worm is direct, the lower hemisphere elongating, becoming segmented, and developing segmental setae groups and parapodia. The metatrochophore finally settles to the substrate and develops into an adult individual. The anterior hemisphere becomes the preoral elements of the head.

MOLLUSCA

Many molluscs also have a very similar **trochophore** larva, which differs from that of annelids in details of its development and differs greatly in the manner of its transformation into an adult. The cross of cells formed in late cleavage, known as the molluscan cross, is composed of different blastomeres from those occurring in annelids, and it is radial in position. The anterior hemisphere of the trochophore becomes the head of the mollusc. The ventral portion of the lower hemisphere be-

comes the foot, and the dorsal part becomes the mantle. In gastropods there is a rotation of the visceral hump, termed torsion, which brings the mantle cavity and gills to a forward position over the head and gives the viscera a characteristic twist. For a time the young larvae continue to swim, developing a pair of large, ciliated swimming lobes, collectively termed the velum, from the old prototroch, which grows and gradually shifts to a more dorsal position, forming a large bilobed funnellike velum with prominently ciliated margins. In this stage the larva is known as a **veliger**. Eventually the veliger settles down, the velum is resorbed, and the individual assumes the form char-

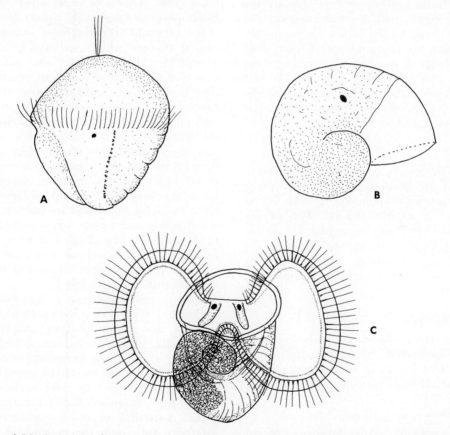

Fig. 6-16. A, Late trochophore larva of a chiton *(Ischnochiton)*. Differentiation has proceeded to the point where the foot, the locations of most of the future shells, and the edge of the mantle can be seen. **B,** Gastropod veliger larva as it usually appears in preserved plankton sample. Soft parts withdrawn into shell. **C,** Veliger larva, *Eulimella nitidissima* (Montagu), with velum expanded. (**B** drawn from Davis, C. C. 1955. Marine and freshwater plankton. The Michigan State University Press, East Lansing.)

acteristic for its species. Gastropods other than Archaeogastropoda tend to pass through the trochophore or its equivalent stage in the egg and to hatch as veligers. In the case of land snails and their derivative freshwater pond snails, the Pulmonata, the veliger stage, too, is passed through in the egg membranes, and the young snail hatches as a fully formed young snail. Bivalves and scaphopods also hatch as a free-swimming form called a trochophore, which transforms into a bilaterally symmetrical veliger of brief duration, during which time the main organs of the adult are laid down.

In the case of Cephalopoda, the eggs are yolky and development is direct. However, in many instances young cephalopods are still very small, still dependent upon the yolk for nourishment when they are liberated from the egg, and spend some time in the plankton as a planktonic larva before the yolk is entirely absorbed. In the case of squids, the young remain pelagic, becoming part of the nekton, while the octopuses are mostly benthic.

BRYOZOA

Adult bryozoans are sessile benthic animals, but the peculiar **cyphonautes** larvae are characteristic members of the meroplankton in neritic waters. The cyphonautae are very flat, triangular larvae encased in a rigid chitinous shell open at the oral end of the animal. It is thought by some to represent a very highly modified trochophore.

BRACHIOPODA

Pelagic larva of brachiopods are· only rarely encountered in the plankton. This phylum, too, occurs in neritic waters. The adults are all fixed benthic forms.

PHORONIDA

These wormlike, tube-building, lophophore-bearing forms have a planktonic larva, the **actinotrocha,** which can be considered to be a very modified trochophore in which the metatroch has developed into a series of tentacles.

ECHINODERMATA

Among the most bizarre and characteristic ciliated meroplanktonic larvae that superficially resemble trochophores are the planktonic echinoderm larvae. They are all derivable from the pluteus type of larva. *Pluteus* was a generic name proposed for certain larvae that were later found to be ophiuroids. Similarly structured larvae in other echinoderms and in certain lower Chordata are referred to this general type, used now as a descriptive name for such larvae.

The coelom in echinoderms and chordate larvae is enterocoelous in origin and undergoes a very complex development in the case of echinoderms, thus differing greatly from the schizocoel developed in the trochophores of annelids and molluscs.

In its simplest early form the **pluteus** can be thought of as a ciliated larva with a complete digestive tract somewhat like that of a trochophore but bent ventrally so that both the oral and anal openings are toward the concave ventral side. A large ciliated band that might be thought of as corresponding to the metatroch passes essentially around the mouth, but is produced into lobes at the anterior and posterior lateral angles. In later stages these lobes are variously divided and extended to form larval arms, and in asteroids there is a meeting and fusion across and in front of the anterior lateral lobes in such a manner as to cut off a circle of the ciliated band in front of the mouth, separating it from the remainder of the band.

The plutei of ophiuroids and echinoids are rather typical and are termed **ophiopluteus** and **echinopluteus,** respectively. They differ chiefly in details of body proportions and in the number, branching, and orientation, of the larval arms. Larvae of holothurians usually have earlike lobes rather than extended larval arms and are called **auricularias,** while the larvae of asteroids, following the cutting off of the separate ciliated band above the mouth, are termed **bipinnarias.** The auricularian larva often changes to form a so-called "pupa," or **doliolarian larva,** before leaving the plankton. The asteroidal larva changes to

Fig. 6-17. Echinoderm and hemichordate larvae. **A,** Early bipinnaria stage. **B,** Brachiolaria stage. **C,** Spatangidae, echinopluteus larva. **D,** Ophiopluteus larva. **E,** Early tornaria larva of *Balanoglossus clavigerus* Delle Chiaje, showing general similarity to echinoderm larva. **F,** Late tornaria, *Balanoglossus clavigerus* Delle Chiaje. (**D** modified from Davis, C. C. 1955. Marine and freshwater plankton. The Michigan State University Press, East Lansing.)

form the **brachiolarian larva,** which develops additional long tentacles, not involving the ciliated band, that later function to provide temporary attachment to the substrate while the larva undergoes metamorphosis.

The larval arms of plutei are not muscular and are used only indirectly for locomotion through the activity of the cilia on the portions of the ciliated bands that traverse them. They seem to be mainly organs of flotation, and are supported by rigid skeletal spicules that would make movement impossible.

Crinoids are mostly deep-sea animals, and they have direct development. Some of the shallower-water tropical forms, such as *Antedon,* produce a short-lived doliolarian type of larva that has only a blind buccal depression representing the mouth, and lives off the yolk of the egg during its few days of swimming existence. It then attaches itself by a small adhesive pit near the apical tuft of cilia and metamorphoses into a stalked sessile form, about 3 mm. long, known as the pentacrinoid stage.

As one passes from the tropics toward the high latitudes, a larger proportion of the echinoderms have yolkier eggs, pass through a greater part of their development in the eggs, and have more abbreviated, less fully developed planktonic larval stages, some of them developing directly, with no planktonic larvae.

CHORDATA
Hemichordata

The wormlike enteropneustans, or balanoglossids, have a ciliated, pluteus type of larva, the **tornaria,** named after the old larval genus *Tornaria,* to which unidentified tornarias are still referred. They are so similar to the bipinnaria larvae of starfish as to have been mistaken for them at times. The similarities in structure and development of the larvae provide the strongest arguments for postulating a relationship between the echinoderms and the chordates. The tornarias even produce such characteristic developmental echinoderm features as a hydrocoel, a pore canal, and a water pore on the left side. In some of the tornarias the ciliated band becomes more complicated than in echinoderm larvae, but it remains essentially a band around the mouth.

Urochorda

The larvae of the tunicates, or sessile urochordata, are very similar to the appendicularians (already described under holoplanktonic forms), except for the lack of gonads, a more developed tunic, the fact that there is no anus, a slightly less differentiated nervous system, and a less sharp differentiation of head from tail. The "tadpole" stage of tunicates is a very briefly planktonic nonfeeding stage that soon settles down; undergoes regression and absorption of the tail and notochord, a great enlargement and increase in the complexity of the pharynx, and an enlargement of the atrium and the enveloping tunic; and becomes a typical sessile tunicate. Tunicates, with their well-developed larval notochord, seem much more closely related to the higher chordates, the cephalochorda and vertebrates, than to the Hemichordata, and their larvae do not show any great resemblance to those of echinoderms.

Cephalochorda

The cephalochorda are benthic in habit but are capable of swimming, and very young ones are occasionally taken in the hypoplankton near the bottom, especially at night.

Vertebrata

Many marine fish shed floating planktonic eggs. Young fry of both these and many species with sessile or demersal eggs also spend some time in the plankton. The eggs and very young fry of many fish are difficult to recognize as to species, but as they grow older the fry become more and more recognizable and begin to behave as part of the nekton. Very few older fish continue to be planktonic in habit, possible exceptions being the huge *Mola mola,* or ocean sunfish, which has such limited powers of swimming that it might be considered planktonic. If the ocean sunfish, the basking shark, and the huge arctic jellyfish *Cyanea arctica* are considered to be part of the plankton, they are by far the largest of

Fig. 6-18. Free-swimming tunicate larva, *Clavelina.*

the plankters. Perhaps the sargassum fish might also be considered planktonic.

ARTHROPODS (NONCILIATED PLANKTONIC LARVAE)

The larval stages of innumerable crustaceans, both holoplanktonic types such as copepods, and benthic types, form a large and characteristic portion of the plankton. Most malacostracans pass through the nauplius stage within the egg membranes and hatch at a later stage. Most of the so-called Entomostraca, as well as a few Malacostraca, hatch as nauplius larvae.

Nauplii are commonly more or less oval, and have a single median eye and three pairs of appendages, at least in their first instar. The appendages represent the first and second antennae and the mandibles, all of which are pediform in this stage. With successive molts, additional appendages are added, beginning with the more anterior ones. Later naupliuslike stages, but with more than three pairs of appendages, are sometimes termed metanauplii. The nauplii of the Cirripedia are distinguished by the fact that the anterolateral corners of the carapace are drawn out into lateral spines.

Fig. 6-19. A, Planktonic larva of *Pleuronectes platessa* (15 mm.). **B,** *Pleuronectes platessa,* just hatched (6 mm.). **C,** Larva of *Mola mola,* newly hatched (1 mm.). (**C** redrawn from Davis, C. C. 1955. Marine and freshwater plankton. The Michigan State University Press, East Lansing.)

Postnaupliar stages are greatly varied in the different crustacean groups. Some of the more commonly encountered ones will be mentioned.

Copepoda

Copepods pass through six instars as **nauplii,** after which they transform into **copepodids,** which look and behave much like adult copepods but are sexually immature and may differ in details of the number and exact form of appendages, and of the apparent body segmentation. The sixth instar, after becoming a copepodid, is the adult stage.

Cirripedia

Swarms of larval barnacles are, at times, a conspicuous element in neritic plankton. The characteristic **nauplii** are readily recognized by the anterolateral spines of the carapace. After a few nauplier instars, they transform into a very different-appearing larva, the **cypris** larva, which resembles a small ostracod, having the carapace developed into a bivalved shell enveloping the body. The cypris larva typically has both compound eyes and a median nauplius eye. The antennules have four segments, and a disc at the end of the second segment on which the duct of the cement gland opens. The cement glands occupy much of the anterior end of the larva. The second antennae have disappeared, and the mandibles and maxillae are rudimentary. Posteriorly, there are six pairs of biramous, setose swimming appendages. A very short abdomen ends in a caudal furca. The mouth lies behind the middle of the ventral region and is directed posteriorly. Two strong apondemes project into the body at the base of the antennae for attachment of the antennary muscles. A transverse adductor muscle closes the valves of the shell.

Fig. 6-20. A, Copepod nauplius larva (*Eurytemora hirundoides*) (fourth instar). Ventral view showing the three pairs of swimming appendages (uniramous antennules, biramous antennae, pediform mandibles) and the single naupliar eye, characteristic of nauplii. **B,** Late nauplius (metanauplius) of a barnacle, *Balanus*. Characteristic lateral extensions of forepart of carapace as "frontal horns" identify nauplii of barnacles. Note presence of both paired compound eyes and the unpaired nauplius eye.

After a very short period of rapidly swimming and crawling about on the substrate, the cypris selects a suitable spot and glues itself firmly to it with its antennules, by means of the secretion from the cement glands. It then undergoes a remarkable metamorphosis to become an adult barnacle.

Malacostraca

Euphausiacea. Most euphausiids simply shed their eggs into the water. A few hold them temporarily attached to the thoracic legs. The egg hatches into a **nauplius.** Following this there is a complicated series of changes. After three or four naupliar instars, the larva changes to a **calyptopis** larva, which has three pairs of appendages behind the mandibles, and in which thoracic segments are clearly differentiated and the telson more fully developed. The calyptopis instar also has a pair of uropods. Soon the calyptopis undergoes another marked change, becoming a **furcilia.** The furcilia has stalked eyes and more thoracic and abdominal appendages, but it still has a natatory function, at which time the young euphausiid is in a postlarval stage and is called a **cyrtopia.** The changes from cyrtopia to adult are so gradual and small as to make it seem proper to consider the cyrtopia a postlarval rather than a larval stage.

Stomatopoda. The eggs of stomatopods hatch at a more advanced stage than the nauplius. Two types of larvae hatch: from *Squilla* and related genera, a **pseudozoea;** from *Lysiosquilla* and other genera, an **antizoea.** They are not comparable to nor derived from the zoea stage of decapod crustaceans.

The carapace of the pseudozoea is produced into a rostrum over the head, and bears two anterolateral spines and two posterolateral spines. Only two thoracic appendages are present. Both are uniramous, and the second is large and raptorial. The abdomen bears five pairs of biramous appendages.

The antizoea has a large carapace covering much of the body but attached only at the head. It bears a large rostrum and a pair of large posterolateral spines, in addition to which there may be other, smaller spines. The thorax and abdomen are not sharply differentiated, but there are five pairs of simple biramous thoracic appendages.

All antizoeae and some pseudozoeae metamorphose into a second larval stage known as the **erichthus.** The rest of the pseudozoeae become **alima** larvae. The

Fig. 6-21. A, Late zoeal larva of a crab *(Inachus).* **B,** Megalops stage of crab *(Portunus).*

erichthus and alima stages are much more alike than are the pseudozoea and antizoea stages.

Decapoda. Most benthic and nektonic marine decapods have planktonic larvae. These larvae are of great variety and are so numerous as to constitute an ecologically important fraction of the zooplankton. Many of them were discovered and given generic and specific names before their relationships were known. Some of these names survive as descriptive names for various stages in the life cycle. Nonuniformity in usage of such names has led to some confusion in earlier literature in the designation of stages of larval decapods. Gurney (1942) has done much to clarify relationships. He divided the life cycle into **naupliar, protozoeal, zoeal,** and **postlarval** stages. In many instances there is no marked change between the protozoean and the zoeal stages, and it seems preferable simply to number these stages without attempting to make a hard-and-fast distinction between them. As a rule, the first zoeal stage does not yet have the thoracic legs de-veloped, and often the spines and rostrum of the carapace are weakly or not developed. Such larvae would be called protozoeae in Gurney's scheme. Following the zoeal stages there is a sudden metamorphosis, and the zoea is transformed into a postlarva in which the appendages have, in general, adult characteristics. Subsequent molts result in a more or less gradual metamorphosis into the adult body form. In the Brachyura the postlarva is termed a **megalops.** It has the general appearance of a crab but still possesses a well-formed abdomen not tightly folded beneath the cephalothorax, as in the adult. The postlarvae of Macrura and Anomura are more varied and still go under a variety of names for various types, such as **mysis stage** (an undesirable term since it implies a relationship that does not exist), **eretmocaris,** with very long, sometimes segmented eyestalks (a convenient designation of certain postlarvae, but it does not represent a particular taxonomic entity), and **phyllosoma,** (larvae of Palinuridae).

CHAPTER 7

The nekton

Fig. 7-1. Mixed herd of herbivorous fishes, tangs and parrot fish, picking over rocky substrate off Guam. These fishes have been shown by Bakus to be important in preventing establishment of beds of larger marine algae in many tropical areas. (Chesher photograph.)

Fish

Fish are by far the most abundant of all vertebrates, in numbers of both species and individuals. It is estimated that there are more than 30,000 different species in the class Pisces. They range in size from a species no more than half an inch long when mature to such giants as the whale shark, which is said to reach 70 feet in length and a weight of several tons. Fish continue to grow throughout their life, although at a much diminished rate after attaining maturity, so that the size of mature individuals of a given species is, at least in many cases, less rigidly fixed than in most other animals.

Being the principal free-swimming aquatic animals, without serious competition from other groups, fish have dominated aquatic environments everywhere. Only the squid (molluscs) and the cetaceans (mammals) have successfully invaded their realm. Competition between species, the balancing of one against another, has led to evolu-

tionary divergence in all directions and their expansion into almost every conceivable ecological niche. The diversity among fish is much greater than that found in any other class of vertebrates, making them a particularly fertile field for the student of evolution. The vastness of the aquatic environment also contributes to this diversity by making possible the survival of various types that have been superseded by different groups in centers of abundance and competition but that may continue to exist in other regions, in isolated or deep waters.

The abundance of fish life is a reflection of the abundance of plankton in localized areas, as well as a reflection of the type of plankton. Since all the phytoplankton and a large portion of the zooplankton are limited to the upper strata of water in or

127

near the euphotic zone, most fish are also found there. Just as there is more plankton in regions where great water masses meet and mix than in monotonous stretches covered by a water mass of relatively uniform character, so too there are more fish and other marine animals in such areas. Neritic waters, which receive materials washed out from the land and which also are enriched by increased turbulence, upwelling, etc., are particularly rich in fish. Such waters provide a diversity of ecological habitats far greater than that found in the open ocean, and they also contain the greatest diversity of fish species.

Coloration

The coloration of fish is of almost infinite variety and beauty and is of great biological importance. The most widespread features are simulation of background color and obliterative countershading, which have survival value to both predator and prey species. Most species that swim near the ocean surface, such as herring, tuna, mackerel, and flying fish, are greenish or bluish above, and whitish, silvery, or light-colored below. Those living in inshore waters are more commonly green, and those in offshore waters, blue. Thus they match the waters in which they live when they are seen from above, while their light undersides blend with the sky when they are seen from below, matching the surface film that shines from below like a mirror.

Demersal fish, which live on the bottom, tend to be uniformly colored, speckled, or mottled, with neutral colors more or less matching the type of bottom on which they are found. Many of them, particularly the flatfish, can rapidly change both color and pattern to match their background. Tide pool fish living among the varicolored plants, coralline algae, and animals of the tide pools are brightly and variously colored; often matching the rocks, coralline algae, or other features of their environment in a striking way.

Obliterative countershading is the rule among fish living in well-lighted waters, the upper part being dark enough to offset the shadow on the lower part, giving the whole

fish a rather flat, indistinct image. Some notable exceptions to this pattern confirm its purpose, such as the Nile catfish *Synodontis,* which habitually swims upside down and has the ventral side darkened, and the remoras, which attach to sharks by means of a dorsal sucking disc—the dorsal surface, which is in the shark's shadow, being light, and the ventral surface, countershaded.

Sharply contrasting bands, stripes, or blotches, although they make a fish that is out of its environment extremely conspicuous, may also serve a concealing and protective function in that they break up the outline of the fish. Such patterns commonly include a dark bar across the eye, making it less conspicuous, and sometimes ocellar, or eyelike, markings on other parts of the body, which may serve to confuse a predator. Some fish mimic particular seaweeds with which they are associated in a most striking manner, involving not only color but orientation and shape of the body and outgrowths from the body surface that resemble the shape of the seaweed.

Fish living most of the time in the mesopelagic dysphotic depths are commonly red when seen in white light, but in their own environment, below the depth of penetration of the red light, they tend to absorb the greenish rays of light reaching them and, there being no red rays to be reflected by their red pigments, are very inconspicuous. Others tend to be colorless, silvery all over, or more or less transparent.

Below the mesopelagic depths, in the deep waters of the bathypelagic and abyssal zones, where there is no light at all, most of the fish are uniformly black, except for luminescent organs.

Some of the pigmentation of fish probably has a protective function in shielding vital organs from excessive light. The concentrations of melanin, commonly centered over the brain, probably have such a function, and the black peritoneal lining of the body cavity of many fish may protect the viscera or certain of the digestive enzymes.

Conspicuous coloration of species with poisonous flesh, foul taste, or formidable defensive stings does not appear to have the same importance among fish as it is supposed to have among insects. Many of these species seem to be just as concealedly colored as their innocuous relatives, and there is not as much mimicry of such species by harmless species as occurs among insects.

Some species that are aggressively territorial and fight off intruders, such as the garibaldi of Southern California, or that engage in strong rivalry for mates show conspicuous warning coloration that may serve both to frighten rivals and to aid in sex recognition. Through selective breeding, aquarists have come up with numerous brilliantly colored "tropical fish" and goldfish strains. In some the male becomes more intensely or deeply pigmented during the spawning season. Species that select mates and that have a complex courtship or breeding behavior tend to show greater difference between the sexes than the majority of fish, in which the sexes look very much alike. The so-called "cleaning fish" are also usually conspicuously colored, advertising their services.

Fins

The fins of fish are thought to have arisen through division and modification of fin folds, such as those found on *Amphioxus,* which has paired lateroventral metapleural folds running along the anterior part of the body and uniting behind the atriopore into a single median fold that runs back around the posterior end of the body and continues forward along the middorsal line as a dorsal fin fold. A division in the paired lateral folds gave rise to two pairs of ventrolateral fins, a forward pectoral pair and a posterior pelvic pair. These fins foreshadow the paired limbs of all higher vertebrates. The unpaired fold has given rise to the caudal fin or tail, a varying number of dorsal fins and finlets, and the ventral or anal fin and finlets behind the anus. In swimming, propulsion is achieved in most fish through strong lateral flexures of the hind part of the body, causing the caudal fin to push strongly against the water. The dorsal and anal fins serve primarily as rudders, and the paired fins as balancing and braking organs. The heavy front end acts as a fulcrum for the tail, which acts as a flexible lever. Sustained swimming is seldom or never faster than about 20 miles per hour, though short bursts may in some cases exceed this rate. In the majority of fish, swimming is considerably slower. Maximum speed for small fish is about ten times body length per second for short bursts. Fast-swimming species show streamlining of the body, which minimizes the water resistance. It has been found that the swimming movements are such that the drag caused by water turbulence behind a swimming fish is less than that on a model of such a fish being propelled or pulled through the water, so that the energy required in swimming is less than one might suppose from a study of such models.

Flotation and maintenance of level

In all except the demersal species of fish, which prefer to live on or near the bottom, the problem of flotation is more acute than in most of the invertebrates and plankton organisms discussed earlier. There is a much smaller surface-to-volume ratio, and the musculature and bones necessary for efficient swimming of a larger organism make the fish proportionately denser. Most teleost fish with heavy skeletons and muscles compensate for this with an air bladder, which, like the lungs of higher animals, is derived from an outgrowth of the gullet. The fact that only the teleosts possess an air bladder lends some support to the hypothesis that both lungs and air bladder presumably arose from a simple pouch of the gullet that originally had a respiratory function. The pouch probably occurred in fishes inhabiting fresh water where low oxygen concentrations are more frequently encountered, and oxygen deficiency was compensated for by gulping air. In teleosts the air bladder lies dorsal to the intestine and serves as a hydrostatic organ. The gas is obtained by secretion from a gas-secreting gland and is a mixture of gases different from that present in the atmosphere or water. A fish can control its depth in the water by adjusting of the volume of gas in the air sac. Control is not a voluntary muscular action but rather a slower process of secretion or absorption of gas into the bladder. The fish then balances its overall density to match that of the water at the level in which it is living, becoming, in effect, a Cartesian diver. Strong-swimming predatory fish, such as mackerel, some tuna, and sharks, which swim at various levels in pursuit of their prey and often rapidly traverse a considerable range of depths, do not have a swim bladder. They sacrifice buoyancy for speed and facility in going up and down. They have to swim to maintain their level. Demersal species with air bladders, brought suddenly to the surface in trawls, are commonly dead on arrival, with the air bladders much distended, sometimes even protruding from the mouth, and vital organs ruptured.

Stretch receptors associated with the air bladder have also been thought to play a role in the perception of changes of hydrostatic pressure, which is particularly useful to fish inhabiting intermediate levels in the water. However, many kinds of fish that do not have air bladders or other compressible gas vesicles also show definite responses to changes in hydrostatic pressure.

Morris and Kittleman (1967) found, during two summers of trapping fish and diving in McMurdo Sound in Antarctica, that there is a well-defined depth stratification of several species of fish, none of which has a gas bladder. Furthermore, there is in these waters no temperature gradient that might serve to keep species at given levels. The entire water column has a uniform temperature of about $-1.9°$ C.

It was known that pressure changes, as well as discrimination of frequencies of sound waves, involves, at least in part, the ear, but the exact mechanism had never been defined. It occurred to Morris and Kittleman that the otoliths might serve both these functions admirably if they were piezoelectric. Examination of otoliths of the antarctic notothemid *Trematomus bernacchii* and the North Pacific demersal flatfish *Parophrys vetulus* revealed that in both cases the otoliths are indeed strongly piezoelectric, oscillating over frequency ranges of 300 to 8,000 cycles per second in *Trematomus,* and from 1 to 15,000 cycles

per second in *Parophrys.* Not all species of fish examined possessed otoliths with piezoelectric properties, but it appears that this may account for discrimination of sound frequency and perception of depth through hydrostatic pressure in some groups of fish. If this proves to be the case, it will be the first instance of a piezoreceptor system reported for animals, and the first really new sensory receptor system reported for some time.

Feeding habits

Perhaps the factor that most profoundly influences the evolutionary diversification and modification of fish is feeding habits. On this basis we can divide fish broadly into plankton feeders, nekton feeders, and benthos feeders. The lines between these categories are not sharp, especially between the last two. Of course, it is also evident that for most fish, the feeding habits must change radically during the growth of the young fry until it attains maturity. The young fry of most fish feed upon smaller elements of the zooplankton or benthic organisms of similar size that they can pick off. As they grow, they begin to take larger organisms and, finally, whatever food is characteristic for the adults of their species.

Plankton-feeding fish and others that eat small-to-minute animals generally have well-developed gill rakers on one or more of the gill arches, on the inner concave side, opposite the gill filaments. As the fish passes water through the mouth and out over the gills for respiration, plankton organisms of a particular size range are filtered out and retained in the throat by the gill rakers until they are brought against the opening of the gullet and swallowed. The opening of the gullet is kept closed most of the time except for the swallowing of food, so that the fish does not swallow too much water. In only a few kinds of fish are the gill rakers numerous enough and fine enough to retain much of the phytoplankton. However, some fish, such as the menhaden, feed directly on diatoms. The majority of plankton-feeding fish feed on zooplankton. The coarseness and number of gill rakers gives some indication of the principal food of such a fish: those feeding on the macroplankton have, in general, fewer and coarser gill rakers than those that feed on the smaller elements of the zooplankton, whereas those feeding on phytoplankton have a great many very fine gill rakers. At the opposite end of the scale, fish that prey upon other fish commonly have only rudimentary rakers or none at all. Fish that feed on macroplankton begin to select and snap up individual macroplankters as well as simply to filter them out of the water. From this point it is a small step to typically predaceous habits, in which all the food items are thus selected and larger and larger organisms are utilized. The herring is a good example of a typical zooplankton feeder, feeding chiefly in clouds of copepods such as *Calanus finmarchicus.* Even at times just prior to spawning, when it is not feeding, the herring tends to remain associated with areas rich in copepods rather than with areas of dense phytoplankton blooming.

Since plankton is by far the major food source in the oceans, plankton-feeding fish can build up by far the greatest biomass of fish. They form the principal food of all the larger fish-eating carnivorous fish, birds, seals, etc. and the basis for our most productive fisheries. Some familiar examples, in addition to those already mentioned, are sardines, pilchards, *Rastrelliger,* and shad.

Most species of very abundant fish that feed on plankton or on a variety of small invertebrates and very small fish, in more or less open waters, show the phenomenon of schooling. By day they tend to swim in relatively compact groups, the fish within the group being all of very nearly the same size and spaced at rather constant intervals from each other. Such schools often act almost as if they were individual organisms, so precisely do the fish swim and turn together. Other species sometimes school shortly before spawning and migrate to the spawning grounds as a school, although they may not remain so organized during ordinary feeding. Some of the schools may attain great size, including thousands of individuals. The individual fish in large schools do not remain in the same relative positions but are continually shifting. The school does not have definite leaders. If it turns suddenly, those fish which are on one side of the school may suddenly be in front, whereas the former front members now find themselves on one side. Fish from the edges of the school tend to work in toward the middle, leaving others on the exposed edges.

Schooling is probably a protection to the individual fish in the school, which would be at the mercy of any larger, faster predator if found alone in the open water. The individual fish, particularly if it is in the central part of a school, may also take evasive and confusing action when approached by a predator, darting rapidly about in a manner that makes it difficult for the predator to select and capture a particular individual. The ability to make a sudden spurt of rapid swimming seems to be more important in evading predators than sustained speed. Most predators are bigger and faster anyway, so that sustained speed would not help much in escaping. Some small fish attain maximum speed within one twentieth of a second in response to an appropriate stimulus. It is true that large schools of fish also attract more predators to the area than would otherwise be there and make it easier for at least some of the predators to obtain their fill, but the balance of the advantage lies with schooling. Otherwise, it could not have become such a widespread hereditary trait in so many diverse fishes as well as in some squid and other animals.

Several kinds of small schooling pelagic fish, such as anchovies, have deciduous scales, which gives them a slightly better chance to escape when seized by a predator.

The appearance of man on the scene, with efficient mechanized equipment for taking advantage of the schooling of fish and taking them in hitherto unheard-of numbers, may well shift the balance of advantage and bring disaster to a number of species, as apparently happened in the case of the sardine fishery off the California coast shortly after World War II. It is possible that such an event will be even harder on the larger predators than on the small schooling prey, both because man also fishes intensively for them and they are far fewer in number to begin with, and because their food supply in a region may be reduced to below the level at which either man or the large predator species could fish efficiently. In such a situation the large predators might be wholly eliminated, but vestiges of the population of plankton-grazing species would survive, from which, now relieved of most of the predation pressure, a population even larger than before might be reconstituted. Most of these species produce tremendous numbers of eggs and mature faster than do large fish, so that population comeback can be relatively fast, given favorable opportunity. A much more ominous possibility facing many species is the destruction of their spawning grounds, particularly when these grounds are in coastal waters and estuaries, through alteration and pollution.

The schooling of large pelagic migratory species such as tuna may primarily serve to keep the fish together for reproduction, rather than serving as a protection from still larger predators. This function would, of course, also be served, in addition to whatever protection schooling may give, in the case of the smaller fish discussed above.

Most of the remaining fish are active predators, feeding upon other fishes or upon invertebrates of comparable size. Those which eat other fish or squid in open waters are, in general, fairly large, fast swimmers and are often well armed with sharp-edged or needlelike teeth. Those of the mesopelagic and bathypelagic depths tend to be smaller than those of the epipelagic waters and to have relatively larger heads and mouths and more exaggerated teeth, giving some of them a fierce, dragonlike appearance. Predatory fish swallow their prey whole, or if attacking something too big to be swallowed in this manner, tear off bite-sized pieces that are swallowed whole. Many of them have a very keen sense of smell and can detect the presence of suitable food in the water far beyond the distance at which they can see it. Some also seem to be attracted to disturbances in the water, such as those created by the swimming of a school of fish or a swimmer.

In shelf areas where there are suitable concealment places among rocks or seaweeds, or on the bottom, numerous species have adopted the habit of lurking in hiding until suitable prey comes within close range and then darting forth to catch it. Fish of this latter group that hide on or near the bottom and habitually lunge upward to catch swimming fish or invertebrates above them usually have large heads and mouths, and the mouth and eyes are directed upward. Bottom-dwelling fish, such as the flatfish, skates, and numerous scorpaenids, usually are sluggish. They have lower metabolic rates, reflected in lower rates of oxygen consumption, and smaller ratios of gill surface to body weight than do more actively swimming types. On the other hand, fish such as the cod, which habitually pick up molluscs and other benthic invertebrates, commonly have the mouth and associated sense organs (nose, barbels, etc.) directed straight ahead, or even on the ventral side of the body, as in rays. Fish that eat molluscs and crabs commonly have rather broad, flat teeth for crushing the shells rather than the sharp, needlelike teeth of fish-eating types. Of special interest are some of the more specialized flatfish, such as the soles, in which the jaws of the eyed side are toothless. Predaceous fish, of course, are the objects of all sport fishing and commercial bait fishing.

Osmotic relationships

Elasmobranchs and marine cyclostomes keep the tissues isotonic with the surrounding water by the resorption of urea through the renal tubules until the concentration in the blood reaches 2000 to 2500 mg. per 100 ml. The tissue osmotic pressure is thus raised above that of seawater, enabling the fish to absorb water through the gill and oral membranes in small amounts adequate for the formation of urine. This system requires that the embryo, prior to developing its own regulatory mechanisms, must be either in a tough-shelled egg or inside the female. Internal fertilization, effected by copulation is therefore necessary and has led to the evolution of claspers in the male. This system apparently arose with the evolution of elasmobranchs and cyclostomes as far back as the late Silurian or early Devonian.

Teleosts, on the other hand, underwent a very different evolutionary history. They appear to have evolved in brackish estuaries or freshwater rivers where the problem was not one of getting enough water into the tissues but rather of excluding or eliminating excess water without at the same time losing valuable salts needed in the tissues and body fluids. The water tended to be absorbed because of the hypertonicity of the tissues relative to fresh water. The glomerular kidney was evolved in response to this need. Each nephron consists of a tuft of capillaries, the glomerulus in the mouth of an excretory tubule, combined with an enzymatic transport system for resorbing most of the salts from the renal tubule before the excess water reaches the exterior as urine. Thus the glomerulus and tube constitute a system for pumping out excess water without undue loss of needed salts. The urine is dilute and copious. When the teleost returns to the sea, the osmotic conditions are reversed, and the tissues tend to lose too much water and gain too much salt. Apparently the return of teleosts to the sea came too late in their evolution for them to change the basic pattern of their protein metabolism or structure of their kidney. The problem was met in a new manner through the evolution of special salt pumps—patches of cells in the gills or mouth that actively excrete excess salt from the blood, leaving the water with which it had been absorbed in the first place free for urine formation. Also, most marine teleosts make much less urine than do freshwater fish. The more urine they make, the more they must drink and the more salt must be excreted. They therefore tend to keep urine formation at the lowest possible level consistent with elimination of their soluble nitrogenous waste products. The glomeruli tend to degenerate, and the urine excreted is much more concentrated than is the case with freshwater fish.

CLASSIFICATION OF PISCES

Authorities differ rather widely in the constitution, ranking, and names applied to some of the higher taxa of fishes. The scheme used in this book does not follow any one source wholly but is a compilation from several, with an effort to adopt what appear to be the simplest, most practical arrangements. Because of their great numbers and diversity in today's oceans, the elasmobranchs and teleost fish will be accorded a much more detailed treatment than the other groups, but even here many interesting groups at various levels have of necessity been omitted. The term "Pisces" is here used in its broadest sense and includes some groups often recognized as independent classes by many specialists, who restrict the name to bony fish. In this discussion the class Pisces is composed of aquatic vertebrates breathing by means of gills throughout life. They do not possess tetrapod-type limbs. The homologues of these, when present, are paired fins. The following outline will show the division of the class Pisces into major groups used here:

Subclass Agnatha
    Order Ostracodermi
    Order Cyclostomata
Subclass Placodermi
    Order Acanthodii
    Order Cladoselachii
    Order Coccosteomorphi
Subclass Chondrichthyes
    Order Elasmobranchii
    Order Holocephali
Subclass Osteichthyes
   Grade Choanichthyes
    Order Crossopterygii
    Order Dipnoi
   Grade Actinopterygii
    Order Polypterini
    Order Chondrostei
    Order Holostei
    Order Teleostei

SUBCLASS AGNATHA (MONORHINA)

Primitive jawless fish, possessing no arches supporting paired fins; nasal aperture is single, median, and on top of the head.

Fig. 7-2. Lamprey, *Petromyzon marinus.*

Fig. 7-3. Hagfish, *Myxine glutinosa.*

Order Ostracodermi (Ostracophora)

Extinct armored fishlike vertebrates that probably arose in the Ordovician, were best represented in the Silurian and Devonian periods. The head was large, with a large bony shield; finlike appendages were at the posterior lateral corners of the head shield; eyes and single median nostril were on top of head; posterior part of body was slender, usually scaly, and ended in a heterocercal tail. Suborders included Anaspida, Antiarcha, Cephalaspidomorphi, and Osteostraci (Aspidocephali), the last named containing the best-preserved forms.

Order Cyclostomata (Marsipobranchii)

Lampreys and hagfish. Mouth is a circular jawless sucking disc armed with rasping teeth. There are no paired appendages, and no scales. Skeleton is cartilaginous; notochord is persistent throughout life; chondrocranium is incomplete, without roof; vertebrae are represented by neural arches. Suborders include Petromyzontes (Hyperoartia), the lampreys; and Myxinoidea (Hyperotreta), the hagfish.

The cyclostomes are of considerable interest as living representatives of a very primitive line of fishes adapted to a semi-parasitic mode of life and showing an eel type of racial senescence. The ammocoetes larvae of lampreys show striking resemblance to the Cephalochordata. Sea lampreys have been of economic importance following their invasion of the Great Lakes through the Welland Canal and the destruction by them of the important white-fish and trout fisheries in these lakes. The United States and Canada have spent about $12 million in control measures over the last twelve years. The economic loss to the fisheries has been much greater.

All remaining fish have sometimes been grouped together into a contrasting group, the **Gnathostomata,** or jawed fish. Recent authorities, however, regard this assemblage as composed of three primary groups, which we shall accord subclass rank, equivalent to that of the **Agnatha.**

SUBCLASS PLACODERMI

Extinct late Silurian to early Permian sharklike fish, some of which had some bone in the internal skeleton as well as bony dermal plates, and scales. Biting jaws and paired fins were present.

Order Acanthodii (Acanthodea)

There were up to seven pairs of paired fins, these and the dorsal and anal fins each with a strong spine along the leading edge. These fish apparently evolved as river-dwelling types and only in the Devonian period began going to sea. The multiple and variable number of paired fins is of particular interest in connection with the fin-fold theory of the origin of paired fins, as are also the very broadly attached fins of the next group, the Cladoselachii.

Order Cladoselachii (Pleuropterygii)

Sharklike but with mouth in more terminal position than in modern sharks. First gill bars were modified into jaws; pectoral and pelvic fins were essentially lateral folds, broad and unconstricted at the base, and supported by nearly parallel rays. Genera include *Cladoselache, Symmorium, Cratoselache.*

Order Coccosteomorphi (Arthrodira)

Head and forepart of body were protected by bony plates, those of the movable head shield articulated with the body shield in such a way that to open the mouth, the lower jaw remained rigid while the loosely hinged skull moved up. What appear to be teeth are only bony projections of the skull and jaw. The homologies of the jaws and fins to those of other fish are uncertain.

SUBCLASS CHONDRICHTHYES

Sharks, rays, and chimaeras. Skeleton is cartilaginous; intestine is spiral-valved. Males have claspers.

Order Elasmobranchii (Euselachii)

Modern sharks and rays. Gill clefts have separate external openings. Skin usually has placoid scales.

Suborder Selachii (Squali, Pleurotremi). The sharks. Pectoral fins are not united to head; gill openings are at sides of more or less cylindrical body. The numerous families are grouped into three assemblages.

Section Galeoidea. A large group comprising the majority of the most familiar sharks, including most of those known to attack people. Most of them are voracious carnivores. But the largest of all, the whale shark and the basking shark, which attain a length of 60 feet or more and weigh several tons, are harmless plankton eaters. Sharks of this group have five gill slits, two dorsal fins, and an anal fin. There are no fin spines. The rostrum is triradiate.

FAMILY CARCHARIIDAE. (ODONTASPIDAE, GALEIDAE, MUSTELIDAE). Tail is much shorter than the body and has a notch below the tip of the upper lobe. Eyes have an inner eyelid. At least one of the gill slits is above the base of the pectoral fins; gill slits do not occupy most of the width of head. Spiracles are small to obsolete. Several species are large, fierce sharks dangerous to man. The following are representative.

Carcharias lamia. Cub shark of the Atlantic, a large shark with a reputation for attacking man.

Galeocerdo arcticus. Tiger shark, a widely distributed, dangerous shark in warm waters. Teeth are large, obliquely set, and serrated on both edges.

Prionace glauca. Great blue shark. Abundant in tropical seas, occasional off West Coast of United States. Grows to 20 feet long; may attack man; has no spiracle.

Galeus galeus. Tope, or oil shark, found in European waters. A rather small shark; spiracle is present; no pit at root of tail; teeth are large and coarsely serrated.

Galeus (or *Galeorhinus) zyopterus.* Soupfin shark, found off California coast. Fins are boiled for a gelatin used by the Chinese in making soup.

Mustelus californicus. Gray smoothhound. Common off Southern California.

Triakis semifasciatum. Leopard shark. About 3 to 5 feet long; gray, with black cross bands across back and round black spots along sides. Teeth have one long median and one or two smaller lateral cusps.

Reniceps (or *Sphyrna) zygaena.* Hammerhead shark. Widely distributed in warm seas; grows to about 15 feet long. The subfamily Sphyrninae comprises the hammerheads, in which the head is flattened and extended laterally, with the eyes at the ends of the extensions.

Odontaspis. Sand shark, shovel-nosed shark. Small sharks living along sandy coasts. Sometimes separated as family Odontaspidae.

FAMILY CARCHARHINIDAE. Requiem shark. GENUS *Carcharhinus.* Small- to medium-sized sharks, much feared because they are aggressive and persistent. Several species.

FAMILY LAMNIDAE. Mackerel shark, porbeagle, etc. Typical members are large fierce pelagic sharks with stout body, wide mouth, large teeth, and wide gill openings. First dorsal fin is large, second is small; tail is lunate; prominent keel is on each side of the caudal peduncle; small anal fin is present.

GENUS *Carcharodon.* Very large, voracious, widely distributed shark with sharp, flattened, triangular finely serrated teeth. Some species attain a length of over 30 feet, and certain fossil species were even larger.

GENUS *Lamna.* Porbeagle, mackerel shark. Somewhat smaller than *Charcarodon,* about 8 to 10 feet long. Pointed nose and crescentic tail. Bluish black above, becoming abruptly white below. Teeth are long, fanglike, and smooth-edged, with a small projection from the base.

GENUS *Isurus.* Mackerel shark. Large, voracious; teeth are slender, with entire edges and no basal cusps.

SUBFAMILY ALOPIINAE. In the thresher sharks, the dorsal lobe of the tail is greatly elongated, making the tail constitute over half the total length. There is no keel on the caudal peduncle. They are said to thrash the tail about violently in schools of small fish, stunning and injuring many, which can then be eaten at leisure. Genus *Alopias.*

SUBFAMILY CETORHINIDAE. Often treated as a separate family, this subfamily comprises the basking sharks, genus *Cetorhinus,* remarkable for their large size, often 40 feet long or more, and the very long gill slits that extend up the whole side of the head. The teeth are minute and the gill rakers are well developed, similar to whalebone. These huge, harmless sharks strain plankton from the water. The liver is very large and about 60% oil. A single shark liver may yield several barrels of oil.

FAMILY ORECTOLOBIDAE. Includes the carpet shark, *Orectolobus,* with a somewhat flattened body and mottled skin, and the whale shark, *Rhineodon,* largest of all sharks, growing up to 60 feet long or even more and weighing several tons. Like the huge basking shark, the whale shark is also

Fig. 7-4. Basking shark, *Cetorhinus maximus.*

Fig. 7-5. Whale shark, *Rhineodon typus.*

a harmless plankton feeder with very small teeth and large whalebonelike gill rakers. *Ginglymostoma,* the nurse shark, a large, pelagic shark of tropical seas, is also sometimes included in this family.

FAMILY SCYLIORHINIDAE (SCYLLIDAE). Cat shark, spotted dogfish, swell shark, etc. Usually rather small sharks, found in fairly deep water. Two dorsal fins, without spines; anal fin is present.

Section Notidanoidea (Diplospondyli). A relatively small group, mostly deep-water sharks with only one dorsal fin and with six or seven gill slits; rostrum is simple.

FAMILY CHLAMYDOSELACHIDAE. Frilled shark. Elongate, eellike body up to 6 feet long or more.

FAMILY HEXANCHIDAE. Six-gilled shark, cow shark. Some are quite large, up to 20 feet long or more; viviparous.

Section Squaloidea. Five or six gill slits; two dorsal fins, each with a stout spine at its leading edge or, if no spine is present, then no anal fin is present either. Rostrum is simple.

FAMILY HETERODONTIDAE (CESTRACIONTIDAE). Bullhead shark. Five gill slits on each side; dorsal fins have strong spines; anal fin is present.

GENUS *Heterodontus.* Port Jackson shark. Small shark with posterior teeth modified into dense pavement for crushing mollusc shells.

GENUS *Pristiophorus.* Saw shark. Small East Asian and Australian sharks with snout like that of a sawfish, but with smaller teeth. Gill openings are lateral (those of the true sawfish are ventral).

FAMILY SQUALIDAE. Dogfish shark, sleeper shark, spiny dogfish, etc. Dorsal fins have stout spine on leading edge of each; no anal fin; five gill slits on each side.

GENUS *Squalus.* Spiny dogfish. Type most widely sold by biological supply houses for comparative anatomy and zoology classes throughout the United States. They are abundant in all seas, often occurring in large schools and hence easily collected in great numbers.

GENUS *Centrophorus.* A deep-water demersal form.

FAMILY SOMNIOSIDAE (DALATIIDAE). Sleeper shark. No spines on the dorsal fins; no anal fin is present; five gill slits on each side. These are rather large sluggish demersal species living on muddy or clay bottoms, coming to surface only to eat. If hooked, they offer no resistance. The head is small, with weak jaws and small teeth. The fins are small and weak. The skin is covered with minute tubercles. These sharks grow to about 25 feet in length. They are probably largely carrion eaters or eat also some live fish. The belief that they bite large chunks out of whales is probably erroneous. Genus *Somniosus.*

FAMILY SQUATINIDAE. Angel shark. This curious shark seems somewhat intermediate in character between typical sharks and rays. The body is flattened, somewhat raylike, with very large broad pectoral and pelvic fins. Head is set off from body by a deep notch, in which the five gill slits open; spiracles are on top of head and in back of the eyes. These are demersal bottom-feeding fish with flattened teeth for crushing molluscs, crabs, sea urchins, and other animals with shells or exoskeletons. Large ones become 6 to 8 feet long. They are found especially in tropical and subtropical seas and in the Mediterranean. Genus *Squatina.*

Section Batoidea (Raiai, Hypotreme). Rays, skates, and their relatives. The pectoral fins are produced forward, joining the sides of the head; gill openings are ventral; hyomandibular cartilage is purely suspensory. The body is commonly much flattened dorsoventrally. Demersal fish, usually lying on, or partly buried in, the bottom much of the time. Typical rays are very broad and flat, and swim by undulations of the wide paired fins, a very different mode of swimming from that of other fishes.

FAMILY PRISTIDAE. GENUS *Pristis.* Sawfish. Large elongate sharklike rays with a flat elongate snout bearing lateral stout toothlike structures on each edge. Sawfish grow as much as 20 feet in length and are usually found in warm, shallow seas, especially of tropical America and Africa, and often around river mouths. Gill slits are ventral.

FAMILY DASYATIDAE (TRYGONIDAE). Whiptailed rays, stingrays. Typical broad ray-

Fig. 7-6. Diamond stingray, *Dasyatis dipterurus.*

Fig. 7-7. Manta ray, *Manta hamiltonia.*

like flattened body; tail is slender, usually bearing one or more spines. The spines, together with a poisonous, slimy secretion on the skin, can inflict intensely painful wounds to the unwary swimmer who happens to step on a stingray in shallow waters. Numerous genera. Representative species include *Dasyatis dipterurus* J. & G., *Pteroplatea marmorata* Cooper, *Urobatis halleri* Cooper, and *Aetobatus californicus* Gill.

FAMILY MOBULIDAE. Manta rays, horned rays. Often very large, with the anterior part of the pectoral fins produced forward into so-called "cephalic fins," or horns; teeth occur only in lower jaw. Found in tropical seas. Viviparous, usually giving birth to only one young at a time. *Manta birostris.*

FAMILY RAJIDAE. Skates. Typical broad flat raylike body; tail is thicker than in the Dasyatidae and unarmed; two dorsal fins, set back on the tail well back of the hind tips of the pelvic fins; snout is longer than interorbital distance, and commonly pointed. Largest genus is *Raja.* Numerous species. Often with short patches of spines, or thornlike prickles, on dorsal surface or along top of tail.

FAMILY RHINOBATIDAE (PLATYRHINIDAE). Guitarfish, fiddlefish. Body is more elongate and sharklike, tail is thicker. Found in warm seas. Examples include *Rhinobatus, Platyrhinoides.*

FAMILY TORPEDINIDAE (NARCOBATIDAE). Electric rays or torpedo fish. Dark-colored rays with broad circular anterior disc. Skin is smooth, without scales or prickles; large electric glands are at the base of the pelvic fins on the dorsal side. The giant electric ray of the North Atlantic, *Torpedo nobiliana,* can deliver pulses of 50 amperes at 50 to 60 volts, sufficient to electrocute a large fish. Smaller electric rays of the genera *Narcine, Narcobatus,* and *Tetranarce* are widespread. The electric organs are derived from modified muscle, which forms masses of large, hexagonal cells on either side of the head at the base of the pectoral fins.

Order Holocephali

Chimaeras or ratfish. A formerly abundant order of the Chondrichthyes, especially in the Mesozoic. Of the four families, only one, family Chimaeridae with three recent genera, *Chimaera, Callorhynchus,* and *Harriotta,* still exists. In contrast to the sharks, the gill clefts open into a chamber with only one external opening covered by a fleshy, or soft, cartilaginous operculum. The skin is soft, without scales. The first dorsal fin bears a long, sharp spine. The head is high, compressed, with a small, narrow mouth, and the teeth reduced to a few plates. The lateral line system is prominent. The body tapers off into a long, narrow heterocercal or whiplike tail. The upper jaw is immovably fixed to the skull, and the lower jaw articulates directly with the skull. The male bears a peculiar stalked knobbed denticulated clasping organ in the middle of the forehead, in addition to having large claspers of the usual elasmobranch type extending back from the cloacal region.

The Holocephali are demersal fish, genus *Harriotta* being found only in very deep waters. In *Callorhynchus* the snout is produced into a pendent tactile organ. The species *C. callorhynchus* is the elephant fish of temperate to cold southern waters. Perhaps the most familiar species are *Chimaera* (= *Hydrolagus*) *colliei* of the western coast of North America and *C. monstrosa,* the king of the herrings, or rabbitfish, of the North Atlantic.

SUBCLASS OSTEICHTHYES

These fish differ from elasmobranchs in having a bony skeleton, or bony dermal plates. The dermal scales are without the dentin, or enamel, of the placoid scale; the intestine is without a spiral valve (except for the sturgeons). A urinary bladder is present; no claspers occur on the pelvic fins; operculum is composed of bony elements. Many species have a single or paired air bladder, a saclike diverticulum from the pharynx, primitively an accessory respiratory as well as hydrostatic organ (in most modern fish possessing an air bladder it is purely hydrostatic in function). These fish appeared in the Devonian and at first dominated fresh water and only later the oceans.

GRADE CHOANICHTHYES (= SARCOPTERYGII)

Lobe-finned fish. The paired fins have a prominent fleshy scale-covered lobe of flesh and bone at the base.

Order Crossopterygii (= Branchioganoidei)

Coelacanths and their relatives. These were the most abundant fish during the Devonian but then declined, so that there were only a few species during the Mesozoic. They were long regarded as extinct until 1938 when a specimen of the coelacanth *Latimeria* was caught off the mouth of the Chalumna River in South Africa. Later others were taken near the Comoro Islands off Madagascar in the Indian Ocean. This was one of the most startling discoveries of a "living fossil." A detailed study of this primitive fish has been made by J. Millot and his students. The skeletal elements of the fin lobes are not homologous to the limb bones of higher vertebrates. The air bladder has bony plates, which may act as a resonator. Unlike most Crossopterygii, it has no internal nares.

Order Dipnoi

The lungfish are characterized by fins with an archipterygial skeletal pattern, the presence of internal nares, an air bladder that remains connected to the pharynx and that may act as a sort of accessory respiratory organ, and a more complex heart than most fish, with a pulmonary circulation. The skeleton is largely cartilaginous, and there is a persistent unconstricted notochord. There is dermal bone and dermal fin rays. The body is covered with imbricate, or overlapping, cycloid scales. The dipnoans were of worldwide distribution during the Jurassic period but are now confined to widely separated localities in Africa, Australia, and South America. Found in fresh water.

GRADE ACTINOPTERYGII

The ray-finned fish. Projecting portion of paired fins are supported by dermal rays, and usually not much of a fleshy lobe appears at base. These fish appeared in the Mid-Devonian. They were rare at first compared to the lobe-finned fishes, but today they are the dominant group in both fresh and marine waters.

Order Polypterini

Polypterus. Freshwater fish, the bichir of the Nile.

Order Chondrostei

Paddlefish and sturgeon. These fish have a long rostral snout with ventral mouth like sharks; heterocercal caudal fin. Fossil ancestral forms had an ossified skeleton, but the living species have only a completely cartilaginous one, which is a secondary degenerate condition. Paddlefish live in fresh water, but sturgeon are found both in fresh water and in marine environments. Sturgeon, family Acipenseridae, are primitive and sharklike in numerous ways. The fin skeleton is sharklike, and there are no functional spiracles; there is a persistent unconstricted notochord covered by a fibrous sheath. The vertebrae are without centra. The mouth is toothless and protrusible. The skin is tough, with five rows of hard, bony dermal plates. Sturgeon are usually rather large fish, becoming several feet long. They are valued as food; the egg mass, or roe, is considered a great delicacy—caviar—and the air bladder has been used in the manufacture of isinglass.

There are sensory barbels ventrally on the snout. Sturgeon are demersal bottom-feeding fish that thrust their protrusible mouth into soft substrate in search of small food animals. Some become very large, one of the largest taken weighing more than 1½ tons.

Order Holostei

Garpikes and bowfins. Freshwater fish with ossified skeleton. Dominant in Jurassic; now only few left. Approach teleosts in structure.

Order Teleostei

This is the dominant group of fish today—an extremely large and diverse assemblage constituting about 90% of all living fish. Teleosts appeared in the Jurassic period, underwent a rapid adaptive radiation, and reached a climax in the Cenozoic period. Teleosts represent the end products of fish specialization and have not given rise to types of vertebrates beyond their own group. The scales have lost most or all of their ganoin and bone and have become relatively small and thin, with a tendency to overlap. Some groups have lost the scales altogether. The internal skeleton is usually well ossified. The vertebrae are amphicoelous, and the tail is usually homocercal. The swim bladder is used as a hydrostatic organ. Commonly there are slender bones lying between the muscles. Some of the more important groups of teleosts are discussed below.

Suborder Isospondyli (Clupeiformes). This suborder includes large parts of Abdominales, Physostomi, and Cycloidei of various classification. It is the most primitive and generalized group of teleosts, as well as one of the most important. In typical members the pelvic fins are abdominal in position, well behind the pectoral. Scales, when present, are cycloid. Air bladder, when present, joins the esophagus by an open duct. All or most of the fin rays are soft, not developed into spines. About 25 families.

Section Clupeoidea. Soft-finned, mostly marine or anadromous fish, with no adipose fin and no photophores.

FAMILY CLUPEIDAE. The major family, which includes numerous well-known commercially important species such as the herring, sardine, shad, and menhaden. Mostly plankton feeders with terminal mouth, narrow, compressed body, forked tail, deciduous scales, teeth small or lacking, and gill rakers long and slender. They lack a lateral line. The dorsal fin is single, in the middle third of the body, well in front of the origin of the anal fin. Head is without scales. Descriptions of representative genera follow.

GENUS *Alosa (Pomolobus)*. Shad. Relatively deep-bodied, anadromous fish. Occurs along the Atlantic seaboard of the United States and carries out spawning runs up the rivers in the spring. Some also occur in Europe. The species *A. pseudoharengus* is the alewife; *A. sapidissima,* the white shad.

GENUS *Brevoortia*. Menhaden, or mossbunker. Along the Atlantic and Gulf coasts of the United States.

GENUS *Clupea*. Herring. Among the most important of all commercial fish. Occur in extensive schools, especially at spawning time. Feed largely on the cope-

Fig. 7-8. Menhaden, *Brevoortia tyrannus.*

Fig. 7-9. Herring, *Clupea harengus.*

pod *Calanus* and occur chiefly in water masses heavily populated by it. Worldwide distribution. *Clupea harengus* is the common herring of the North Atlantic. Often found in huge schools containing many millions of individuals, which greatly facilitates large-scale commercial fishing for them. Many of them are sold as "sardines." They spawn near the coast in shallow water, where the substrate is firm or gravelly and there are stones or other objects to which the eggs can adhere. A distinct race also occurs in the North Pacific, regarded by some as a separate species, *C. pallasii.* Regional populations seem to remain distinct without a great amount of mixing. Sindermann (1957) found that the parasitic and fungal infections of the populations from the Gulf of St. Lawrence and those from along the coast of Maine and southward were quite distinct from each other. Blood group studies on different populations also indicate this. Each female lays about 10,000 eggs in sticky masses.

GENUS *Sardinella.* Sardine, pilchard.

FAMILY CHANIDAE. Genus *Chanos,* the milkfish. Extensively cultivated in tambaks or marine impoundments in the warm parts of the western Pacific, from fry collected in the ocean. Very important in Southeast Asia, from Indonesia to the Philippines.

FAMILY ENGRAULIDAE. The anchovies. Small herringlike fish distinguished by their large mouth, the mouth angle extending to back of the eye, and a pointed snout that extends in front of the mouth. Most species are not over 5 inches long. Extensively fished for pickling or canning, like sardines, etc. Many species, found mostly in warm temperate or shallow tropical waters, often entering brackish water or around river mouths. Huge populations occur in the Peru Current in recently upwelled water where the temperature is 14° to 18° C., shifting both laterally and vertically with movements of water having this range. Typical genus, *Engraulis.* Rich, peculiar flavor. Related families are Alepocephalidae (smooth head) and Dussumieriidae (round herring).

Section Salmonoidea. Usually medium-sized fish, larger than herring, with an adipose fin between the main dorsal fin and the tail. Lateral line is present. Salmon, trout, and their allies. No photophores;

many marine species are anadromous, ascending rivers to spawn. Valued as game and commercial fish.

FAMILY SALMONIDAE. Mouth is large, the hind edge of the maxillary extending to or beyond the eye in the adult. Base of dorsal fin is shorter than length of head; pelvic fins have scaly appendage above the base; dorsal fin has less than twenty-six rays. Eyes are situated laterally. Stomach has more than eleven pyloric ceca.

GENUS *Salmo*. The Atlantic salmon and most freshwater trout. The North Pacific salmon differ in numerous respects, such as a greater number of anal fin rays, branchostegal rays, pyloric ceca, and gill rakers, and are commonly separated as the genus or subgenus *Oncorhynchus*. They spend most of their lives in the sea, two to three years or more. The remarkable spawning migrations of salmon, some of them going hundreds or even a thousand or more miles upstream to the same site at which they lived as young fry, together with the adjustments they must make in going from fresh to salt water and back, have excited the admiration and wonder of students of fish biology. The return to the spawning grounds is particularly difficult to understand, since it involves a long, difficult, dangerous upstream journey and correct choices of many tributaries and branchings of the stream on the way. As they have never traversed these waters since their initial journey downstream to the sea two or three years previously as fingerlings, this would seem to represent an altogether incredible feat of memory. Their ability to migrate has been found to depend upon their sense of smell. The tendency to return to the same place to spawn, together with differences in spawning season between species, doubtless plays a role in keeping various populations and species genetically distinct.

The return of the salmon demonstrates that various tributaries and streams must be more individualized in a chemical sense than one would suppose, since their dilute traces in the larger river systems of which they become a part are sufficient to guide the salmon back.

Salmon and other anadromous fish are threatened by overfishing during their spawning runs, when they are highly vulnerable to commercial fishing methods. An even greater threat is the rapidly increasing, nearly universal pollution of rivers and streams and their alteration by dams and impoundments for flood control and hydroelectric projects.

The complex interrelationships of salmon with their environment are illustrated by the fluctuations in Karluk Lake on Kodiak Island, one of the world's most important spawning grounds for the red salmon. Nelson and Edmondson (1955) showed that the decomposition of 1 million spawned-out salmon contributes the equivalent of 87 tons of commercial phosphate fertilizer and 420 tons of nitrate fertilizer (worth about $28,000) to the lake annually. Salmon runs formerly were on the order of 3 million or more, or equivalent to about 1300 tons of commercial fertilizers. This enabled the lake to support rich phytoplankton production, which in turn supported the vast swarms of zooplankton needed to support the young fry. The great reduction of escapement into the lake because of commercial fisheries resulted in a great decline in the fertilization of the lake and hence indirectly to a further decline in the numbers of salmon beyond the direct effect of the fisheries.

Related families include Coregonidae (whitefish, mostly freshwater), Thymallidae (grayling), Osmeridae (smelts), Galaxiidae, Argentinidae, Bathylagidae (deep-sea smelts), and Opisthoproctidae (barreleyes).

Section Elopoidea (tarpon and bonefish). Large fish, the tarpon often becoming 6 feet or more in length. Good game fish, but the flesh is considered rather poor.

Section Stomiatoidea. Numerous deepwater fish, with rows of photophores on the body. Anglemouths, hatchetfish, stomiatoids, and viperfish.

Section Esocoidea. Pike, blackfish, mudminnow. Mostly freshwater fish.

Section Osteoglossoidea. Bony tongues, mooneyes.

Section Mormyroidea.

Section Giganturoidea.

Suborder Iniomi (Myctophiformes). Resemble suborder Isospondyli in having soft fins, the pelvic fins abdominal in position in most members, scales cycloid, and an adipose fin usually present. The premaxillary bone forms the margin of the upper jaw. Many are bathypelagic species with large mouth and eyes, slender bodies, numerous photophores.

FAMILY MYCTOPHIDAE (SCOPELIDAE). Lanternfish. Perhaps the most common and widespread of the fish at intermediate depths in the oceans, sometimes coming to the surface at night or in stormy weather. On the East Coast, aggregations of myctophids were found to be responsible for an unusual, rather lumpy deep-scattering layer. Related families include Synodontidae (lizardfish), Alepisauridae (lancetfish), and Sudidae (deep-sea fish).

Suborder Lyomeri (Saccopharyngiformes). Gulper eels.

Suborder Ostariophysi (Cypriniformes). The four anterior vertebrae are strongly modified, often ossified together, and support a chain of small bones connecting the air bladder with the ear. This group includes three of the largest families of pre-dominantly freshwater fish, including many popular aquarium fish. Included are such forms as the characins, tetras, suckers, minnows, carps, loaches, electric eels, and catfish. One group of the catfish, the Ariinae (Tachysurinae), are marine and known as sea catfish.

Suborder Apodes (Anguilliformes). The eels. This is a large group of fish showing the syndrome known as "eel type of degeneration or racial senescence." They have snakelike bodies, with no pelvic fins, and some of them have no pectoral fins, either. The shoulder girdle is usually separate from the skull. The scales are rudimentary or absent, the body being smooth, and in some, slimy. In some the dorsal, caudal, and anal fins are continuous around the tail; others have a separate caudal fin.

FAMILY ANGUILLIDAE. GENUS *Anguilla*. Circumtropical except in western Americas. Includes the common European and American eels. They spend most of their adult life in ponds and marshes or other bodies of fresh water. But when they are ready to spawn, they enter the streams, swim down the rivers to the ocean, and make their way to their spawning grounds in the Sargasso

Fig. 7-10. *Cyclothone elongata.* Possibly the world's most common fish. About 3 inches long. Family Myctophidae.

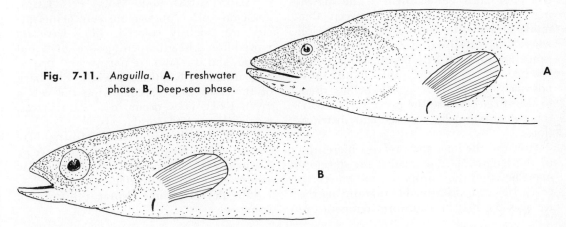

Fig. 7-11. *Anguilla.* **A,** Freshwater phase. **B,** Deep-sea phase.

Sea, hundreds or thousands of miles from their home ponds. As they enter upon this migration they change from a nondescript, dirty-yellowish color to glistening silver. The eyes enlarge greatly. This color change, together with the fact that migrating eels are never found in upper waters between the coastal streams and the Sargasso spawning grounds, indicates that the migrations are done at considerable depths. The eggs are pelagic and hatch into small transparent flattened forms that were given the generic name *Leptocephalus* before their relationships were known, and are now known as "leptocephalus larvae." The larvae come to the surface waters and float as part of the plankton for one or three years. The American eel *(A. rostrata = A. chrysypa)* matures in one year, just as the Gulf Stream has brought the eels alongside the North American coast. The European eel *(A. vulgaris = A. anguilla)* matures a year later, when the Gulf Stream has carried the eel across the Atlantic to the European shores. The young eels, called elvers, migrate up streams and rivers to take up their residence in suitable ponds.

This is the most remarkable instance known of the adaptation of life cycles to the great oceanic currents. Knowledge of the migration was brought to light through the tireless efforts of the Danish scientist Johannes Schmidt (1925), whose researches constitute a classic in fisheries biology. It is probable that a growing stenothermy—the requirement of water temperature at 16° to 17° C. for reproduction—plus negative phototaxis, guides them to the spawning ground, since the area between Bermuda and Puerto Rico is the only region where this particular narrow temperature range prevails at this depth. The newly hatched larvae, on the other hand, show positive phototropism, which brings them to the surface where the great surface currents determine their subsequent distribution.

How did the long and precise differential in the length of time needed for development between the European and American eels arise? It seems probable that during the ice ages the belt of a uniform temperature of 16° to 17° C. at about 500 meters' depth included most of the central Atlantic. The American eel probably spawned in the western edge, and the European eel in the eastern edge of this zone, each having a migration of approximately the same length to make. During geological time the 17° C. temperature zone shrank, especially toward the west, with the result that the European eels had a longer and longer trip to make. Meanwhile, the increasing time needed for the larvae to float back to the European coast placed a selective value on longer larval life—those that metamorphosed too soon perishing at sea. Eventually the shrinking of suitable breeding grounds brought them together in a very restricted region, but the European eel had already evolved a longer larval life.

The eel and the salmon, with their very similar yet opposite biological histories—the one spawning in inland waters far from the sea but living most of its life in the ocean, the other spawning far out in the deep sea but living most of its life in inland waters far from the sea—present fascinating problems in the origin and evolution of behavioral patterns, the guidance mechanisms in animal migrations, and the adaptive changes involved in alternating between fresh and salt water twice during the life cycle, once as immature individuals and once as mature ones ready to spawn and die. They also shed some light on the probable biological and geological history of the world during the evolution of these patterns of life.

FAMILY MURAENIDAE. Moray eels. Large, voracious, often pugnacious eels. Some species are brightly colored. They have no paired fins. The gill openings are small and round, and the jaws are narrow and strong, and armed with knifelike teeth. The back of the head is elevated, giving the eels a mean look. They occur in all warm seas, especially around coral reefs.

Other families include Moringuidae (worm eels), Nemichthyidae (snipe eels), Congridae (conger eels). Ophichthyidae (snake eels), and Echelidae.

Suborder Heteromi (Notacanthiformes). Spiny eels.

Fig. 7-12. A Hawaiian moray eel in the Steinhart Aquarium, San Francisco. (M. W. Williams photograph.)

Suborder Synentognathi (Beloniformes). Slender or elongated fish usually found in surface waters. In flying forms, head is pointed and one or both jaws are produced; fins are soft, supported by rays only, the dorsal fin being set far back on the body, similar to and above the anal fin; pelvic fins are abdominal in position; lateral line is very low on the body; scales are cycloid. In many the tail is emarginate, the lower lobe being the longest, presumably giving maximum push in jumping from the water. Most taste very good.

FAMILY EXOCOETIDAE. Flying fish. In the flying fish, the pectoral fins are very long and firm, and are used in gliding. These fish swim near the surface in all tropical water, some species being cosmopolitan and found in all tropical oceans and others restricted to particular regions such as the Indo-Pacific. They leap into the air, usually to escape predators or avoid the path of a boat, spread their winglike pectorals, and glide for considerable distances. The longest flights are probably not more than a quarter of a mile. In the so-called four-winged flying fish, the pelvic fins are only a little behind the center of the body and are also spread, giving additional lift during gliding. The largest and farthest-flying of the flying fish are of this type, reaching about 18 inches in length. In the two-winged flying fish the pelvic fins are smaller, farther back on the body, and not spread during flight. These fish are seldom longer than 6 or 7 inches. The longest flights are made across the wind in a fairly strong breeze.

FAMILY BELONIDAE (NEEDLEFISH). Billfish, garfish, and snipefish. These are voracious fish with slender bodies, large eyes, and elongated maws forming a beaklike process set with sharp teeth. They are silvery, somewhat greenish above, and swim near the surface in small groups, preying on smaller fish. Large species grow to 4 or 5 feet in length; most are smaller than that. They are abundant along the shores of almost all warm seas. Many have green bones.

FAMILY SCOMBRESOCIDAE. Sauries.

FAMILY HEMIRAMPHIDAE. Halfbeaks. The lower jaw is produced beyond the head as a toothless pointed process. They are small, seldom exceeding a foot in length, and often swim along just below the surface at an angle to the surface. They are small-mouthed, and mainly plankton eaters.

Suborder Anacanthini (Gadiformes). Cod and grenadiers. Tail is isocercal, lacking the slight structural asymmetry found in the homocercal tail of most teleosts. Fins are soft, without spines; there are three separate dorsal and two separate anal fins. The chin bears a barbel in most species. Mostly demersal, swimming near the bottom much of the time, preying on bottom invertebrates and small fish.

FAMILY GADIDAE. Cod. The Atlantic cod, *Gadus morrhua*, is one of the best-known and most-fished commercial species. It is a cold-water boreal species abounding around the off-shore banks on both sides of the Atlantic. Cod frequently reach a weight of 50 pounds and may sometimes reach as much as 200 pounds and a length of about 6 feet. The mouth is large, the upper jaw projecting slightly beyond the lower. The body is covered with small, smooth scales. Cod range along the continental shelves and banks, usually swimming near the bottom in schools, devouring any animals of appropriate size they can find—crabs, molluscs, fish, and squid, sometimes chasing the latter two up into higher water. Several closely related fish, including the haddock (*Melanogrammus aeglefinus*), the pollack (*Pollachius pollachius*), and the tomcod, are also much valued as commercial fish.

The lingcod differ from the typical cod in having a more elongate body and only two dorsal and one anal fin, the posterior dorsal fin being much longer than the anterior. The eyes are large and the chin barbel is well developed. The European ling, *Molva*, reaches a length of about 5 feet and is extensively used for food, es-

pecially in Iceland. In the American ling or hake, *Phycis*, the pectoral fins are reduced to long two-branched filaments arising near the throat. The European hake and American silver hake, *Merluccius* spp. also belong in this group.

FAMILY MACRURIDAE. Grenadiers or rattails. These deep-sea demersal fish are found on the outer portions of the continental shelves and in deeper water. The head is large, often angular, with a small barbel on the chin. Pectoral and pelvic fins occur almost one above the other, with a short anterior dorsal fin above the pectorals; posterior dorsal and anal fins are long and low, sometimes vestigial. Body is long, covered with small scales, tapering into a slender ratlike tail devoid of a caudal fin. They are numerous both in species and individuals.

Suborder Solenichthyes (Gasterosteiformes, Hemibranchii). Several families of fish with reduced branchial apparatus, and usually with a small mouth at the end of an elongated tubular snout, are included here. Some are among the most bizarre-appearing fish. They suck in planktonic organisms. Most of them are rather slender, elongate fish.

FAMILIES AULOSTOMIDAE AND FISTULARIIDAE. Trumpet fish (Aulostomidae) and cornetfish (Fistulariidae) are found in tropical shore waters. In both groups the fins are soft rayed and the dorsal fin is set well back, over the anal fin. Cornetfish are slender but sometimes reach a length of 4 or 5 feet, although they are usually smaller. Trumpet fish grow to 1 or 2 feet. In genus *Aulostomus* there is a row of short spines in the midline of the back, suggestive of the spines of the sticklebacks, which are included by some authorities in this order. It seems probable that the fishes of this order descended from spiny-rayed rather than soft-rayed fish and that the present lack of spines in members of the group is secondary rather than primitive.

FAMILY SYNGNATHIDAE. The best-known and most bizarre fish of this order belong to the family Syngnathidae, which includes the pipefish and the sea horse. These are rather small fish in which the body is pro-

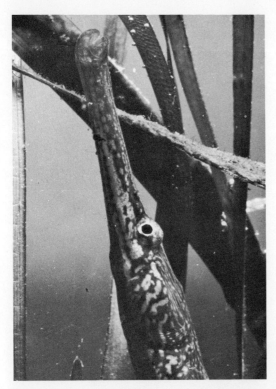

Fig. 7-13. Pipefish. *Syngnathus californiensis* resembles in design, color, and pattern the eelgrass in which it lives. These relatives of the sea horse are common in the eelgrass beds of California bays. (M. W. Williams photograph.)

tected by variously developed bony plates, developed as rings, keels, and ridges, giving the body a stiff, often angular, character. The gills are in tufts on the branchial arches, and the snout is long and tubular, with a very small mouth at the end. Teeth are either absent or minute. The male has a ventral pouch behind the vent where the female deposits eggs, which are then incubated and protected by the male. The female, in this case, initiates courtship and mating. Swimming is commonly a very slow, gliding motion effected by the weak fins. Some can also swim somewhat faster for short distances by eellike movements of the body.

Pipefish (genus *Syngnathus*) are long, slender fish, found commonly among seaweed and eelgrass along the shores of all warm seas, often in protected waters of bays and estuaries. Some species extend into tem-

perate waters. A few may even be found in essentially freshwater at the head of estuaries. *Syngnathus pelagicus* extends across the warm portion of the North Atlantic, associated with *Sargassum*.

Sea horses (genus *Hippocampus*) are very peculiar pipefish with much shortened, thickened, rigid, and laterally flattened bodies. The neck is constricted and the head bent down at right angles to the axis of the body, the whole appearance being much like that of the knight in a chess set, but the sea horse has a long, tubular snout and a tapering, prehensile tail devoid of a caudal fin. Sea horses commonly adopt an erect position propelled by a fanlike motion of the small dorsal fin, which here takes over the propelling functions usually assigned to the caudal fin, albeit rather inefficiently.

Other families include Aulorhynchidae, the tubesnouts; Centriscidae, shrimpfish; and Macrorhamphosidae, snipefish. *Centriscus,* the shrimpfish, is rather translucent and has the peculiar habit of assuming a vertical position in the water, head downward. Small schools, when alarmed, will swim off together rather rapidly, still maintaining this position.

Suborder Salmopercae (Percopsiformes). Trout perch, salmon perch.

Suborder Allotriognathi (Lampridiformes). Oarfish, ribbonfish, crestfish, and opahs. Soft-rayed fish with protrusible jaws differing structurally from those of other fish. They tend to be laterally flattened and often reach a considerable size.

FAMILY LAMPRIDIDAE. Genus *Lampris,* the moonfish. Sometimes up to 6 feet long and weighing over 500 pounds.

FAMILY TRACHYPTERIDAE. Oarfish and ribbonfish. Bandlike body; the largest (genus *Regalecus*) may attain a length of 30 feet.

FAMILY VELIFERIDAE. Crestfish. The dorsal fin in these fish, and in some of the Trachypteridae, forms a flaglike crest.

Suborder Berycomorphi (Berciformes). Squirrelfish, soldierfish, beardfish, and berycids. This is an ancient group known from the Upper Cretaceous and probably close to ancestral types for the Percomorphi.

Squirrelfish or soldierfish (Holocentridae) are abundant in shallow tropical waters around the world. They are moderate-sized fish with two incompletely separated dorsal fins, the anterior one longer and bearing strong spines. The peduncle is narrow, and the tail is firm and well forked. The anal fin bears four spines, of which the first is very small and the third very long, strong, and sharp. The pelvic fins are under or only slightly behind the pectorals and consist of a spine and seven flexible soft rays. The lower hind edge of the opercle also bears a backwardly directed spine. The scales are rather large and ctenoid. Squirrelfish are usually mostly red, with longitudinal white or yellow stripes. The red color is consistent with the hypothesis that squirrelfish are shallow-water invaders from fish commonly inhabiting greater depths. Many fish, as well as other animals, inhabiting somewhat deeper, dimly lit waters, are red in color compared with their shallow-water relatives.

The related families Polymixiidae (beardfish) and Berycidae do inhabit somewhat deeper water.

Suborder Zeomorphi (Zeiformes). Dories and boarfish. Spiny-rayed, thin-bodied, laterally compressed fish living at intermediate depths. The John Dorys are relished as a gourmet item. They grow to be about 3 feet long.

Suborder Microcyprini (Cyprinodontiformes). Includes many small freshwater fish such as the cavefish, killifish, topminnow and livebearers, and four-eyed fish (*Anableps*). Numerous members of Cyprinodontidae also are abundant in coastal salt and brackish waters as well as fresh water, especially in warm regions. Although soft rayed, they are regarded as related to spiny-

Fig. 7-14. Opaleye. The glowing ring around the eye of *Girella nigricans* led to the common name of "opaleye." As a juvenile, this fish is common in the tide pools of southern California. Later, the adults take to offshore waters. (M. W. Williams photograph.)

rayed groups and to have lost the spines secondarily. Most of them are only a few inches long, differing from true minnows in having rounded rather than forked tails. The males of many species are more brightly colored than the females, especially during the breeding season.

Suborder Acanthopterygii (Perciformes). A very large and diverse group that includes several rather distinct assemblages, discussed below.

Section Perciformi. Perchlike fish. Symmetrical fish with spinous anterior and soft-rayed posterior dorsal fins, often more or less confluent and of about equal length; pelvic fins are almost directly under the pectorals, and consist of one spine and five branched rays. Scales are moderate to small and often rough. Tail is rounded, truncate or emarginate, or strongly forked. Several families and numerous important commercial and game fish. Some of the more important families are as follows.

FAMILY PERCIDAE. Perch, darters, walleyes. Mostly freshwater fish.

FAMILY CENTRARCHIDAE. Freshwater sunfish and American bass.

FAMILY KYPHOSIDAE. Chiefly herbivorous. Found in warm seas. Basslike.

FAMILY SERRANIDAE. Sea bass family. A very large, diverse family of mostly marine, carnivorous fish commonly living near the bottom in shallow to moderately deep waters. Some attain considerable size, and most of them are excellent food fish. Included are the striped bass, black sea bass, true groupers, jewfish, hamlet, sergeant fish or robalo, and many others. A few are anadromous or even landlocked. The scales are usually ctenoid. The anal fin never has more than three spines. They are especially abundant in warm seas. The largest ones, certain species of groupers or jewfish, sometimes attain a length of 5 to 7 feet and weigh up to 800 pounds. Many serranids are hermaphroditic, having an ovotestis. Most such species function as males when young, then later as females as they grow larger, but in at least one genus functional sperm and eggs are produced at the same time.

FAMILY LUTIANIDAE. The snappers. Active carnivorous fish with large mouths, strong teeth, and usually deeper, more compressed bodies than serranids. The tail is emarginate. They are fast swimmers and travel in small schools, preying on smaller fish; abundant in warm waters. Species occurring in somewhat deeper water tend to be bright red or rose colored. Shallow-water species are often greenish or variously marked. Commonly about 2 feet long. In the related grunts the inside of the mouth is bright red. Grunts make a peculiar grunting noise when taken from the water. They tend to be somewhat smaller and to have weaker teeth than snappers. Many species occur in the West Indies, in the Caribbean.

FAMILY SPARIDAE. Porgies, sea breams, scup, sheepshead, etc. Related to the snappers and grunts, but some of the teeth along the sides of the jaws are flattened and constitute large, blunt molars. The front teeth are often incisor-like.

FAMILY LEIOGNATHIDAE. Tend to be more active swimmers, with well-forked tail, weakly developed teeth, and unusually protractile mouth that can be extended or retracted.

FAMILY SCIAENIDAE. Croakers, drumfish, squeteague, weakfish or sea trout, kingfish, maigres, etc. A large family, best represented along the tropical shores of the eastern Pacific but generally distributed along sandy shores of all temperate warm seas. Most feed near the bottom, but the weakfish (*Cynoscion*) swims actively higher up. Members of this family usually have a large and complicated air bladder, and some, especially the croakers and drumfish, produce characteristic peculiar sounds with it. Body shape and size and the size of mouth are quite variable in this family. Croakers (*Micropogon*) and kingfish (*Menticirrhus*) tend to have small mouths, the lower jaw somewhat shorter than the upper and often with small barbels on the chin. Croakers have unusually large, smooth otoliths or free ear bones.

FAMILY MULLIDAE. Goatfish. Common in tropical seas. The dorsal fins are of approximately equal size; caudal fin is forked. Snout is elongate and curved, usually projecting slightly beyond lower jaw; a pair of long fleshy barbels are on the chin. The European red "mullet" is one of the better-known species.

FAMILY APOGONIDAE. Cardinal fish. Short bodied, usually less than a foot long. Dorsal fins are separate, of about the same size. Large, oblique mouth; unusually large eyes.

FAMILY CARANGIDAE. Pompano, jacks, crevalles, scad, amberjack, moonfish, cavalla, pilot fish, etc. The spinous dorsal fin is much shorter than the soft-rayed dorsal behind it, and the spines are weakly developed; premaxillaries are protractile, the upper jaw not fused with the snout; the lateral line has a strong arch anteriorly; caudal peduncle is keeled, often with enlarged scutes over the keel; more or less compressed narrow body, tapering to very slender peduncle and widely forked tail. Most carangids are fast swimmers, often

near the surface. Abundant in warm seas, some migrating north in the summer.

FAMILY CORYPHAENIDAE. Dolphins. Among the most truly pelagic of fish. They occur in all warm seas, subsisting largely on flying fish when far from land. The body is streamlined, blue above and bright yellow below, with an even-bordered, undifferentiated dorsal fin running the full length of the back. In older males a high, narrow crest gives the forehead a vertical profile. (These should not be confused with small porpoises, which have also often been called dolphins.) Dolphins grow to be about 6 feet long, and usually swim in small groups, at or near the surface.

FAMILY POMATOMIDAE. Bluefish and allies. The bluefish of the Atlantic and Indian oceans is a highly predaceous, schooling fish somewhat resembling an amberjack but with no keel on the caudal pe-

Fig. 7-15. Viviparous black perch school swims gracefully in the aquarium at the Scripps Institution of Oceanography, La Jolla, Calif. (M. W. Williams photograph.)

duncle and with ctenoid scales that are slightly rough. It is a widely distributed species, valued for food along the Atlantic seaboard.

FAMILY EMBIOTOCIDAE. Surf perch. Pacific distribution, especially off the coasts of North America and Japan. Viviparous, giving birth to a relatively small number of young rather than laying eggs.

FAMILY NOTOTHENIDAE. Found in Southern Hemisphere.

The following three families, among them, contain a majority of the well-known brightly colored coral-reef fish, or coral fish.

FAMILY CHAETODONTIDAE. Butterfly fish. Deep, laterally compressed body; small mouth at the end of short, snoutlike extension of head, for picking off small organisms from crevices in the coral. Dorsal fins are continuous, and their bases and sometimes also the base of the anal fin are covered with fine scales. A very large number of brightly colored, conspicuous small fishes are found around coral reefs in tropical seas.

FAMILY LABRIDAE. Wrasses. Most of the wrasses are also colorful coral fish. They have a small mouth and thickish lips. Jaws are armed with strong, usually pointed teeth, and the throat is armed with blunt grinding teeth. Most of them have rather elongate, compressed bodies, large scales, continuous dorsal fins, and a deep, compressed tail, deeply forked in the adults of some species. Many wrasses swim much of the time with a rowing motion of the pectoral fins, using the tail when extra speed is needed. A few exceptional species are found outside the tropics and become larger than most species. The blackfish, or tautog (*Tautoga onitis*), of the United States Atlantic coast and the burgall, or cunner (*Tautogolabrus adspersus*), which extends as far north as Labrador, are examples. Some wrasses are cleaning symbionts.

FAMILY SCARIDAE. Parrot fish. Rather similar to wrasses but usually thicker bodied, and with the jaw teeth fused to form a hard, parrotlike beak, capable of biting off bits of coral. Usually average a little larger than the wrasses. Parrot fish may be ecologically important in preventing overgrowth of algae and other seaweed on tropical reefs, since they tend to graze off any beginning growth that occurs, often leaving characteristic marks of their teeth on the calcareous surface. If submerged coral rocks, etc. are protected from such grazing, they usually develop an algal cover much richer than any found on exposed areas nearby.

Section Bathyclupeoidea. Family Bathyclupeidae. Deep-sea fish.

Section Acanthuridea. Family Acanthuridae. Surgeonfish, Moorish idol, etc. Usually deep bodied, compressed, with narrow peduncle, emarginate to forked tail. Several have a sharp-edged spine on the caudal peduncle that can be erected forward and outward or depressed into a groove. Largely herbivorous. Fine, inconspicuous scales.

Section Scombroidea. Mackerel, tuna, billfish or swordfish, albacore, sailfish, bonitos, etc. Swift-swimming pelagic fish with elongate fusiform streamlined bodies. The two dorsal fins are followed, in mackerel and tuna, by a series of small finlets. The tail is emarginate, on a slender peduncle. Pelvic fins are thoracic in position. Many are large and highly prized both as game fish and for food.

FAMILY SCOMBRIDAE. Mackerel. Pelagic fish, lacking a swim bladder, which gives them greater facility in going rapidly from one depth to another but means that they must swim to maintain their level in the water. In mackerel (*Scomber*) the spinous dorsal fin is short and set considerably ahead of the soft second dorsal fin. Following the second dorsal fin and the anal fin are a series of small finlets. A pair of small keels are present at the bases of the tail lobes, but none is in the center of the peduncle. Mackerel are medium-sized or small fish usually not exceeding 2 feet in length and weighing 3 or 4 pounds. They swim in close-ranked schools, sometimes of great size, and near the surface. The North Atlantic mackerel (*S. scombrus*) is found throughout the summer, when they go north to spawn in somewhat cooler waters than those in which most of the family are found. In winter the schools disappear. Genus *Scomberomorus* (= *Cybium*) contains somewhat larger fish, with a central keel on the peduncle and the spinous dorsal fin extending along the back to near the front of the second dorsal fin. The Spanish mackerel *(Scomberomorus maculatus)* attains about 10 pounds and is regarded as a choice food fish. The king mackerel *(S. cavalla)* may reach 5 feet in length and weigh up to 100 pounds.

FAMILY THUNNIDAE. Tuna, bonitos, skipjacks, albacore, etc. Closely related to the mackerel, sometimes being classed as a subfamily of the Scombridae. Most are somewhat larger, often deeper-bodied fish. Like the mackerel, some, at least, also lack a swim bladder. They are pelagic, voracious, carnivorous fish, usually found in large schools, and form the basis for some of the world's most important fisheries.

GENUS *Thunnus*. Contains very large typical tuna such as the bluefin tuna, also called horse mackerel or great albacore *(T. thynnus)*. They have a worldwide distribution in all warm seas, often attaining a weight of 1000 pounds or more. The flesh is coarse and oily but is used for food, and the oil is also valued. In this genus the pectoral fins are clearly shorter than the head, and the body does not bear vertical or oblique stripes.

GENUS *Sarda*. Bonitos. Robust, active fish in tropical and subtropical seas. Body is elongate and has minute scales; keel on peduncle is thick, naked; teeth are large, compressed, and strongly curved inward; tongue and vomer are toothless; numerous dark, more or less oblique stripes occur on upper part of body. Bonitos take bait readily, are good sport fish, and make good eating. They congregate in large schools and are important in the economy of some areas.

GENUS *Germo*. Albacore. In the Pacific, associated with transitional waters between the central and subarctic water masses, with surface temperatures of 12° to 18° C., and in summer near the polar front. The pectoral fins of albacore are considerably longer than those of most tuna, reaching back to the level of the anal fin.

Other genera include *Gymnosarda*, the little tunny and ocean bonito; *Neothunnus*, yellowfin tuna, or albacore; *Parathynnus*, bigeye tuna, or ahi; and *Katsuwonis*, skipjack tuna.

FAMILY XIPHIIDAE (INCLUDING ISTIOPHORIDAE). Swordfish and billfish. Large pelagic fish, with the rostrum drawn out into a long, sharp spear above the mouth.

In the sailfish, *(Istiophorus)*, marlins *(Makaira)*, spearfish *(Tetrapturus)*, etc. the dorsal fin is long, the body bears scales, teeth are present, and the pelvic fins are present though reduced to one or a few rays.

The swordfish *(Xiphias)* have a short high first dorsal fin set well forward on the body and a very small second dorsal set far back. There are no pelvic fins, scales, or teeth. The swordfish have very small pelagic eggs, and the young pass through three quite unlike phases while growing up. In one of these phases a long high dorsal fin is present and in another, scales, showing its close relation to the sailfish-marlin group.

A series of deeper-water, mackerel-like

Fig. 7-16. Yellowfin tuna, *Neothunnus macropterus*.

fish leads by small steps to certain very slender deep-water fish in which a very small, although mackerel-like, tail is present, apparently as a vestigial structure, and finally to fish such as the Trichiuridae—slender, bright silver cutlass fish that have a long dorsal fin but no caudal fin at all, the tail end tapering to a slender filament. Like swordfish, these have no pelvic fins.

Section Stromateoidea. Family Stromateidae. Butterfish. Small deep-bodied laterally compressed shore fish superficially resembling the Carangidae but not closely related to them. The young have small pelvic fins, but these fins are very much reduced or absent in the adults. Some pelagic species are usually regarded as related to, if not to be included in, this group. One of the best known is *Nomeus gronovii,* which follows the Portuguese man-of-war *(Physalia)* and finds protection from predators by swimming close up under it when alarmed.

Section Gobioidea. The largest and best-known family of this group is family Gobiidae, the gobies and their close relatives. They are common in shallow waters of warm and temperate oceans everywhere. The skin seems smooth, and no lateral line is in evidence. Scales, when present, are moderate to small in size and smooth; numerous species lack scales. The eyes are commonly placed high on the head, especially in the mudskippers *(Periophthalmus)* of tropical mangrove swamps and estuaries. The mudskipper has a froglike appearance with bulging eyes mounted on the top of its head. At the other extreme is the blind goby *(Typhlogobius californiensis),* in which the eyes are rudimentary and the color pale pinkish, and which lives under stones and in dark crevices in rocky areas along the California coast. Gobies have two dorsal fins, the anterior one with weak spines. The pectoral fins are usually rounded; the pelvic fins are thoracic in position, at the same level as the pectorals or even somewhat in advance of them. In the clingfish (Gobiesociformes) the pelvic fins are modified to form a sucker with which the fish can cling to smooth surfaces. Related families are the Eleotridae (sleepers),

Microdesmidae (wormfish), and Callionymidae (dragonets).

Section Trachinoidea. Weavers, stargazers, and related families. Elongate fish with large heads, the mouth directed upward and the eyes usually rather small and placed well forward on top of the head almost over the mouth. The weavers (Trachinidae) are small to medium-sized, up to about 18 inches in length, with grooved spines in the dorsal fin associated with poison glands. There is also a spine on the opercle. The stargazers (Uranoscopidae) are sluggish benthic fish about a foot long, with an almost vertical mouth and small eyes on top of the head directed upward.

Section Blennioidea. Blennies and related families. A very large and varied group, in which the pelvic fins are reduced to a small number of rays or are absent. When present they are forward in position, arising near the back of the mouth. Blennies are mostly small fish, often found in shallow waters along the shore, together with gobies. Typical blennies (Blenniidae), sometimes called combtooth blennies, lack scales and have a rather large, froglike head with a relatively small, transverse mouth and a row of comblike teeth in each jaw. The dorsal fin is rather high and continuous, with weak spines in the anterior half; caudal fin is rounded. The klipfish (Clinidae) have scales, a somewhat less froglike head, a larger mouth, and sharper spines in the dorsal fin extending along a greater part of its length. Family Pholidae includes the gunnels, most of which live in colder northern waters and have a more eellike body, usually with very small, inconspicuous scales. The dorsal fin is long, low, even, and spinous along its full length. Other fish related to the blennies include the pricklebacks (Stichaeidae) and the eelpouts (Zoarcidae).

Section Ophidioidea. The largest family, Brotulidae, comprises a varied assemblage of about 50 genera mostly found in rather deep water, although the type genus, *Brotula,* includes several species about a foot long, found along tropical shores. Eyes are usually large; bodies are elongate, and covered with fine scales; long even soft-rayed dorsal and anal fins meet in a point around the tail. The pelvic fins are reduced to a pair of filaments arising beneath the gill cleft. The family Carapidae includes numerous curious small fishes often commensal with molluscs or echinoderms. They are slender little fish in which the tail tapers to a threadlike ending with no caudal fin. Some of them enter the cloacal openings of sea cucumbers. The silvery little sand lances (Ammodytidae) may be related to this group, although their affinities are not certain. They abound in the loose, wet sand of many northern beaches.

Suborder Haplodoci (Batrachoidiformes). Family Batrachoididae, the toadfish. Toadfish are mostly shallow-water, rather sluggish, dull-colored fish about a foot long, without scales. The head is large, slightly flattened, with large mouth bearing strong, blunt teeth. The fins are rounded. Toadfish hide by day under rocks or in crevices, shells, etc. The eggs are deposited in masses attached to the surfaces in such hiding places and guarded by one or both parents until hatched.

Suborder Heterosomata (Pleuronectiformes). Flatfish, halibut, flounder, sole, turbot, etc. Demersal fish, with remarkable adaptations to this habit. The flatfish spends much time lying on, or partially buried in, the bottom, on one side. The body is laterally flattened and very deep. The upper side is darkened and variously patterned, whereas the lower side is pale. Most flatfish can vary the tint and even the pattern of the upper side to blend with the substrate, often remarkably well. This color change is under nervous control, and the eyes are necessary for it to take place properly. Although the larva is bilaterally symmetrical, the adult has both eyes on the same side of the head, the upper side. The mouth is also curiously and asymmetrically twisted in many of them. Most species are either right- or left-sided, but some are variable in this respect, some individuals lying on the right side, others on the left. When they swim, by vertical (lateral) undulations of the body, they remain in the same general position as when resting on the substrate. Flatfish are a very suc-

cessful group and have diversified into several families and many genera. Almost all of them are good eating, several being among the most highly prized table fish. Some of the better-known groups of commercial importance are as follows:

Halibut: species with symmetrical pelvic fins and large symmetrical mouth. The Atlantic halibut *(Hippoglossus hippoglossus)* is the largest of the flatfish, sometimes attaining a length of about 9 feet and weighing as much as 700 pounds.

Sole: flatfish with very small eyes, small, crooked mouth, and often lacking pectoral fins.

Flounder: flatfish with a rather small, crooked mouth and nearly or quite symmetrical pelvic fins.

Halibut, sole, and flounder lie with the right side upward.

Turbot: left-sided flatfish with symmetrical mouth and asymmetrical pelvic fins.

Suborder Discocephali (Echeneiformes). Remoras. Peculiar fish in which the top of the head is somewhat flattened and occupied by a large sucking disc that also extends partly onto the back. The sucking disc is oval, with a central longitudinal ridge and a series of transverse ridges. This disc presumably represents a highly modified spinous dorsal fin. Remoras are invariably associated with larger fish, such as sharks and marlins, to which they attach themselves by means of the sucking disc, obtaining a free ride to distant waters plus protection by the host and a share in the host's meals. Most species are rather host-specific with regard to the kinds of fish to which they habitually attach. Remoras are usually grayish in color and lack the countershading found in many free-swimming pelagic fishes. The soft dorsal fin and the anal fin are set well back, the tail is emarginate, and the pelvic fins arise beneath the pectorals. The skin is covered with easily overlooked, minute scales. The relationships of remoras are uncertain; they do not seem closely related to any of the major families of teleosts.

Suborder Loricati (Cottiformes). Mail-cheeked fish. This is a huge group containing many familiar fish, medium-sized to fairly large, such as sculpins, cottids, rockfish, greenlings, scorpionfish, poachers, lumpfish, sea robins, and gurnards. Typically they are demersal fish with a large head and mouth, often bearing spines at the sides of the head. The spinous dorsal fin of many species bears formidable spines, which in the stonefish *(Synanceja),* lionfish *(Pterois),* and a few others have poison glands associated with the anterior spines, enabling them to inflict very painful or even fatal wounds if they are stepped on. A bony process, or stay, of the third suborbital plate extends back to or toward the preopercle, whence the name mail-cheeked fish. In many species this is not obvious without dissection. Although distributed over most of the oceans, this group appears to have its center of origin in, and the greatest number and variety of species endemic to, the North Pacific. Some of the principal families are as follows.

FAMILY COTTIDAE. Includes the sculpins and cottids, commonly rather small fish with a large mouth and with the eyes placed near the top of the head. Cottids commonly rest on the bottom much of the time, move forward with a sudden movement, and then come to rest again. They are voracious eaters, nearly omnivorous. Small, variously mottled and patterned cottids are among the most characteristic fish seen in tide pools along the Pacific coast.

FAMILY SCORPAENIDAE. Common and varied in the North Pacific and also largely replace cottids in the tropics. Mostly larger fish, with a very large, spinous head, and with strong spines in the anterior part of the spinous dorsal fin. Many are red in color and variously mottled. Some of the larger species of *Sebastodes,* the rockfish or "rock cod," and the Atlantic rosefish *(Sebastes marinus)* are of considerable market value.

FAMILY HEXAGRAMMIDAE. Greenlings. Common around kelp beds in the North Pacific coastal waters. Valued food fish, noteworthy for the presence of several well-developed lateral lines. They are usually elongate, with confluent dorsal fins extending the length of the back, and with the spines of the anterior part rather weak. The lingcod *(Ophiodon)* is a related fish that attains a length of over 3 feet and a weight of up to 40 pounds. It is one of the most important food fish from cold waters along the Pacific coast, in spite of the bluish or greenish tint of its flesh. It has only one lateral line.

Suborder Plectognathi (Tetraodontiformes). Porcupine fish, filefish, puffers, triggerfish, mola, etc. This large, mostly tropical or subtropical group is characterized by the lack of pelvic fins, which are either wholly absent or represented by a spine at the end of the long pelvic bone. The mouth is rather small and pincerlike, and scales are lacking or abnormal. They are weak swimmers; the pelagic species tend to have high dorsal and anal fins and swim by moving the fins from side to side. The principal families are as follows.

FAMILY BALISTIDAE. Triggerfish and filefish. Leathery skin and small, nonimbricate scales. In the triggerfish the spinous dorsal fin is well forward in position and consists of one strong and two weaker spines, depressible into a groove. When erected, the strong spine is locked into position by the first weak spine, suggesting the trigger mechanism of a gun. Filefish have more elongate, compressed bodies and a single well-developed anterior spine.

FAMILY DIODONTIDAE. Porcupine fish and puffers. Relatively short, stocky fish, the body surface armed with numerous strong spines. They inflate themselves when alarmed, so that the spines project in all directions.

FAMILY TETRAODONTIDAE. Puffers or swellfish. Related to the porcupine fish but without strong spines, the skin being smooth or with fine prickles. They can inflate themselves with water or air until nearly spherical.

FAMILY OSTRACIONTIDAE (OSTRACIIDAE). Trunkfish, boxfish. The head and body are encased in bony plates, leaving only the tail and fin bases free. The armor is often angled or squarish, somewhat triangular in cross section. Some of them, the cowfish, have hornlike spines projecting forward from above the eyes.

FAMILY MOLIDAE. Mola. The largest of

Fig. 7-17. *Melanocetus johnsoni,* entirely black, with luminescent lure.

the plectognath fishes, attaining a length of up to 10 feet and a weight of over a ton. They have short, deep, compressed bodies with rounded outline, high, stiff dorsal and anal fins, a thick, wrinkled skin, and very small mouth and eyes. The body is abruptly truncated behind, bearing a small fringe-like caudal fin. They are very weak swimmers and are usually seen drifting on one side in the upper strata of water, often far out at sea. The body for the most part has a fatty, cartilaginous character, very different from the flesh of most fish or other large animals. The brain is unusually small and undeveloped for an animal of its size. The young fry are only slightly more than a millimeter long and have a very peculiar spinous appearance.

Suborder Pediculati (Lophiiformes). Anglers, frogfish, etc. Among the most specialized and peculiar marine fish. They are mostly demersal, some of them living in abyssal depths. They have a large, broad head occupying almost half the body length and a very large, transverse mouth, directed upward and armed with numerous long, pointed teeth. The eyes are usually on or near the top of the head, and the body tapers rather rapidly behind the head. The pectoral fins are broad and, in the anglers, supported from short, fleshy arms. The anglers have the anterior dorsal fin ray located well forward on the head, modified into a luminescent lure, or bait, which is dangled over the mouth, luring smaller fish to within easy reach. Anglers and frogfish are sluggish, given to hiding, and either

luring or slowly stalking their prey until it is close enough to take with a quick lunge. Some of the deep-sea species have a remarkable adaptation for reproduction. The males are much smaller than the females, and one (or more) of them attaches itself to the body of the female by its mouth. It remains permanently attached for the rest of its life, organically connected to the female with united vascular systems. The male's digestive organs, etc. are degenerate, only the reproductive organs remaining well developed.

The sargassum fish *(Histrio)* is an exception to the usual demersal habits of the order, but there is a similarity in that it tends to hide in the floating masses of sargassum weed in the Sargasso Sea. It exhibits a wonderful case of mimicry, the body being drawn out into irregular projections so closely resembling the growth habit of sargassum weed in form and color as to be almost indistinguishable in its natural habitat.

Another relatively small, specialized group that deserves mention are the Xenopterygii (Gobiesociformes), or clingfish. These are small fish, usually not more than a few inches long, and characterized by their flattened ventral surface, which bears a large sucking disc between and behind the pelvic fins involved in its structure. This is the most efficient sucking disc formed by any fish except the remoras, in which the sucker is of altogether different location and origin. Clingfish have a smooth, scaleless skin, large head, with the eyes on the top, tapering body, and soft fins, the dorsal and anal fins placed well back on the body. Many are found in shallow coastal or even intertidal waters, clinging to the undersurfaces of stones, which they may match in color. Clingfish from deeper water are often reddish or pinkish.

Cephalopods

The class Cephalopoda of today represents a remnant of an ancient and once dominant group of molluscs that differentiated very early from primitive molluscan stocks, became immensely abundant and di-

Fig. 7-18. Octopus captures crab by surrounding it with tentacles and the connecting web. (M. W. Williams photograph.)

versified during the late Paleozoic and Mesozoic eras, numbering more than 10,000 species, and then dwindled in numbers of species to the approximately 600 living today. Some of the major groups became wholly extinct, whereas others have only a very few living representatives. However, several of the recent cephalopods are very successful, being enormously abundant and widely distributed, and it seems likely that the octopods are at present undergoing active evolutionary diversification.

In contrast with the sedentary benthic habit of most the phylum, cephalopods evolved as active carnivores and scavengers. The earliest cephalopods were benthic, retaining the primitive bilateral symmetry of presumed molluscan ancestors with a posterior mantle cavity but differing in the great growth, dorsally, of the visceral hump, so that the dorsoventral axis became the functional longitudinal body axis. What amount to greatly stretched-out an-

terior and posterior surfaces now serve as the functional dorsal and ventral sides, respectively. The foot became crowded forward and more or less fused with the head, forming an oral crown of long tentacles around the mouth. A portion of the foot became modified as a ventral siphon through which exhalant currents from the mantle cavity are expelled. The tentacles developed suckers and in some cases chitinous hooks as well. The elongated, now "ventral" mantle cavity contained the paired gills, two pairs in the more primitive cephalopods, one pair in most recent species, and water circulation was effected by means of muscular movements of the mantle tissue. Water is drawn in ventrolaterally along the entire open edge of the mantle cavity, and this slitlike orifice is closed and the water expelled through the siphon. The mantle tissue lost the cilia found in the mantle tissue of many molluscs, became increasingly muscular, with sufficient power so that the exhalant jets from the siphon

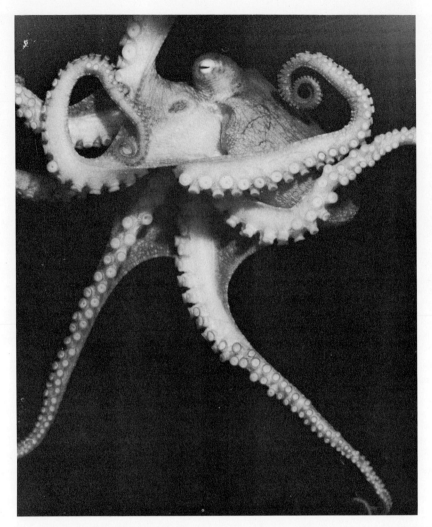

Fig. 7-19. A young specimen of a common southern octopus, *Octopus bimaculoides,* crawling across the aquarium glass. (M. W. Williams photograph.)

were used for locomotion more effectively than pulling the animal over the substrate by means of the tentacles.

Early cephalopods, like most other molluscs, had large external shells, which in many species became variously coiled, and as the animal grew, divided into successive chambers. The cephalopod lived in the last and largest chamber, the posterior chambers being filled with gas to offset their weight, and all chambers were connected to the posterior part of the body of the cephalopod by a membranous tube, the

siphuncle. The four living species of *Nautilus* are the only living representatives of the once-great group of Tetrabranchiata, or four-gilled, shelled cephalopods. They are also primitive in having numerous isopodous tentacles without suckers, imperfectly developed eyes without lenses, no ink sac, four auricles, no branchial hearts, and other features setting them apart from most modern cephalopods.

Most recent cephalopods have very much reduced shells, completely buried in the dorsal mantle tissue, or none at all. The

Fig. 7-20. *Benthoctopus hokkaidensis,* small, purple species from about a 100-meter depth off the Oregon coast.

mantle is much more muscular than in nautiloids, having become developed into a very efficient swimming organ. The nervous system and sense organs are exquisitely developed: the cerebral, pleural, pedal, propedal, and visceral ganglia are fused together in what amounts to a complex brain encased in a cartilaginous "skull." The remarkable cephalopod eye, analogous in every respect to the vertebrate eye—with lens, anterior and posterior chambers, retina, and optic tract—has excited the admiration and wonder of biologists over the years. That such a remarkable structure as the camera eye should have evolved independently twice, in such different evolutionary lines as the molluscs and the vertebrates, is one of the most extraordinary of evolutionary coincidences or cases of convergent evolution. In some ways the cephalopod eye seems even more logically structured than that of the vertebrate. The sensory cells are in front of, instead of behind, the nerves that connect them to the optic tract, and there is no fovea, or blind spot, in the retina for the passage of the optic nerve.

Many cephalopods have chromatophores of two or three different colors under nervous control, enabling them to effect the most remarkable and rapid changes when the animal is at rest, so that when it is looked at closely, a constant play of color change is seen, even though the overall average effect may remain the same. This play of color in the skin may continue for several hours after the death of the cephalopod.

The mouth of cephalopods is equipped with a horny beak somewhat resembling that of a parrot. A radula is also present. Large salivary glands are associated with the mouth, and the bite of some species, at least, is poisonous and sometimes fatal. The general fear of octopuses on the part of the public, however, is not justified, as they are secretive and will avoid people whenever possible. Even when handled, they seldom bite. The writer has caught hundreds of octopuses bare-handed, ranging in size from small, immature specimens weighing less than a gram, to large ones weighing over 80 pounds, without ever having been bitten. Squid seem to show more of a tendency to bite if given the opportunity when captured. However, they are completely helpless out of the water and can be handled safely if one is careful to keep his fingers away from their mouth.

The characteristic defensive mechanism of most cephalopods is a jet of ink from an ink sac lying on, or embedded in, the ventral surface of the liver, and discharging along with a jet of water through the

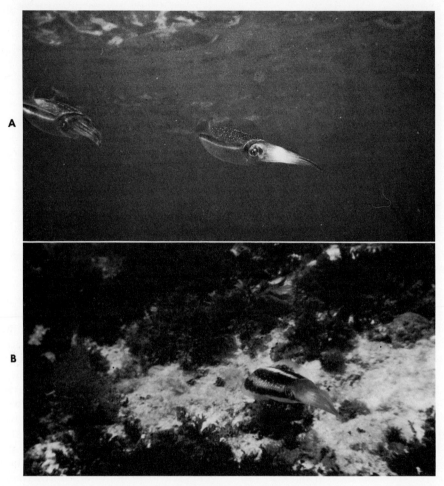

Fig. 7-21. A, Squid over patch reef, Florida Keys. **B,** Same squid as seen 32 seconds later. Note color and pattern change. (Chesher photographs.)

siphon. Most people regard this as primarily setting up a smoke screen behind which the cephalopod can escape. More often the ink probably serves as a dummy in the water to confuse the predator, the small blobs of ink in the water usually being not very different in size and shape from the cephalopod that made them, and often retaining their shape for several seconds. The rich brown ink of cuttlefish, sepia, has long been used by artists. Some deep-water bathypelagic squid have luminescent ink, rather than the black or deep brown ink of their relatives living in well-lighted waters.

Many squid, particularly various oegopsids living in deeper-water strata, have the posterior coelomic space large and filled with a jellylike material that serves as a hydrostatic organ in much the same manner as do the large containers of gasoline of the bathyscaphe. This has an advantage over the swim bladder of fishes, in that it does not change markedly in volume with rapid changes of water level. Such cephalopods often characteristically hang in the water head downward when at rest.

A curious and characteristic modification connected with the mode of locomotion is a pair of flaps, or valves, on the end of the intestine which opens in the mantle cavity back of the siphon. When the mantle contracts suddenly, creating the very considerable pressure in the mantle cavity necessary for jet propulsion, these flaps close the orifice, thus preventing the cephalopod from administering a strong enema to itself whenever it goes somewhere in a hurry.

Some of the squid rival or excel most fish in the speed and grace of their swimming. They can direct the siphon either way, swimming forward or backward. Some of them have even been observed to jump from the water and glide considerable distances in the air, much in the manner of flying fish.

The courtship and mating of cephalopods are also distinctive and varied. Some species exhibit peculiar courtship activities involving rapid changes in color. The sperm of the male come in large spermatophores, which the male transfers to the mantle cavity of the female with one of his arms, which is specially modified for the purpose. In *Argonauta* the end of the male arm, including several suckers holding the spermatophores, is detached and left in the mantle cavity of the female. This was mistaken for a peculiar, many-suckered parasitic worm by Cuvier and given the generic name *Hectocotylus*. Since then the modified male arm has been known as the hectocotylized arm and the modified tip of the arm as the hectocotylus. Most cephalopods do not leave part of the arm in the female mantle cavity but simply use it to transfer the spermatophore. The spermatophore itself is a remarkable structure. One must watch the discharge of one with a stereoscopic microscope to appreciate its complexity and activity. A ripe spermatophore will usually discharge if placed in a little seawater on a slide. In larger octopuses the spermatophores may be over a foot long. They consist of a swollen, saclike portion containing the sperm, a long, tubular, necklike portion with a distal cap and filament. Within the neck is a complex, spirally wound, springlike mechanism, which, when the spermatophore is exposed to seawater, suddenly releases and begins to uncoil, stretching out the tubular portion of the

spermatophore to a much greater length, while the sperm become very active and come boiling up the tube as if under pressure, to escape from the end of the tube or from ruptures along the way. In some cephalopods, for example, *Rossia,* the end of the spermatophore tube, as it elongates, penetrates the tissues of the female, inside the mantle cavity, and then blows up like a small balloon, anchoring the spermatophore firmly in place, so that it looks like a parasitic worm with its head buried in the tissue.

Modern cephalopods are so sharply set apart from other molluscs that they seem out of place in that phylum and are in many respects more like vertebrates. The size range resembles that of vertebrates, the largest cephalopods attaining lengths of more than 40 feet if the two tentacular arms are included, which is much larger and bulkier than any other invertebrates. The body tissue is muscular and has considerable connective tissue resembling that of vertebrates and even containing true cartilage. The shell remnant, or pen, of squid provides a stiffening element in a position similar to the spinal column of vertebrates. Unlike other molluscs they have a closed circulatory system with capillary beds. Their activity, intelligence, and neuromuscular development rivals that of fish, and above all, their eyes give them a peculiarly vertebrate appearance. It is difficult to look an octopus in the eye and believe that you are not seeing something closer to us—something more nearly human than a mollusc.

CLASSIFICATION OF CEPHALOPODA
SUBCLASS TETRABRANCHIATA

The four species of *Nautilus* are the only living remnants of this formerly great group. They are more primitive in many respects than other modern cephalopods. There are four ctenidia, four auricles, four nephridia, no branchial hearts, and no ink sac. The eye does not have a lens, and the arms are numerous, arranged in groups, and without suckers. A large coiled external shell is present, divided into successive chambers, all of which, except the last, are empty and gas-filled, and connected to the body by a centrally located membranous tube, the siphuncle.

Fossil members of the order Nautiloidea number about 2000 species, dating from the Cambrian and reaching their maximum development in the Silurian and Devonian. The shells vary from straight through bent, to loosely or closely coiled. The septa are concave, usually with straight or undulate margins. A scar, the cicatrix, on the initial chamber is thought to show the point of attachment of an embryonic shell. The siphuncle is variable in position, and sometimes quite large.

SUBCLASS AMMONOIDEA

This was an even larger group, existing from the Upper Silurian through the Cretaceous or perhaps into the beginning of the Tertiary. They reached their greatest development during the Mesozoic. Their shells differ from those of the nautiloids in that the margins of the septa are usually more or less intricately plicated and the siphuncle is usually marginal. About 6000 species have been described.

SUBCLASS COLEOIDEA (DIBRANCHIATA)

Cuttlefish, squid, octopuses. In this group, which includes all modern cephalopods except *Nautilus,* the shell is internal or absent. There are two ctenidia, two kidneys, and two branchial hearts. The siphon is tubular. The mouth is surrounded by either eight or ten arms, which bear suckers, hooks, or both, and is equipped with horny mandibles resembling a parrot's beak. An ink sac is present. The eye has a crystalline lens.

Order Decapoda

Squid, cuttlefish, etc. In this large group the arms are of two kinds: eight regular tapering arms, usually connected to each other by the web, and usually bearing suckers their entire length; and two commonly longer arms, the tentacles, which are slender and of rather uniform diameter except for an expansion near the end, which bears suckers, hooks, or both. The body is usually torpedo-shaped, with a pair of posterior

Fig. 7-22. *Chirotanthis veranyi.* A deep-sea squid that hangs head downward in the water, extending the long arms below it. The arms are equipped near the ends with luminescent organs, sharp hooks, and suckers. Prey, attracted to the lights, is snagged by sudden jerks of the tentacles and then brought to the mouth.

lateral fins. Most are pelagic predatory species.

Suborder Belemnoidea (Phragmophora). Extinct. Internal shell has a well-developed apical portion, the rostrum, a dorsal plate-like pro-ostracum, and a partially enclosed phragmocone, divided into chambers.

Suborder Sepioidea. Shell has a reduced rostrum, rudimentary phragmocone, and a modified pro-ostracum. The tentacles can be withdrawn into special pits; coelom is well developed; body is usually short, broad, and saclike. Genera are *Sepia, Idiosepius, Spirula, Sepiola.*

Suborder Teuthoidea. Common squid. Shell or pen is a flattened, chitinous plate buried in the dorsal mantle tissue; body is usually elongate; tentacles are long. Representative genera are *Loligo, Cranchia, Architeuthis, Chiroteuthis, Onycoteuthis.*

Order Vampyromorpha

Vampire squid. Small deep-water octopus-like forms with eight arms united by a web, but also two small retractile tentacles. Genus *Vampyroteuthis.*

Order Octopoda

Octopuses. Eight arms are connected by web, but no tentacles; body is usually globular, without shell or fins. Mostly benthic. Representative genera are *Octopus, Eledone, Eledonella, Vitreledonella, Amphitretus, Cirroteuthis, Argonauta.*

CHAPTER 8

The benthos

The floor of the sea

The ocean bottoms are, for the most part, composed of layers of sediments that have been slowly drifting down from the water above for millions of years. The character of life in the overlying waters, the depth, the distance from land, and the currents all influence the nature of the sediments in any area. Over the ages, changes in climate, temperature, currents, volcanic activity, and isostasy, together with evolutionary changes in the fauna and flora of the seas, are reflected in the sediments and in the fossils preserved in them.

Marine sediments from coastal regions and former shallow seas now raised above sea level are directly available for study and have provided geologists and paleontologists with most of the information we have about the history of the earth. In recent years the study of oceanic sediment cores, together with new sonic, electronic, and magnetic methods for obtaining information about the floor of the sea, has added a new dimension to these studies. Submarine geology has become an exciting field in its own right.

Terrigenous sediments

Near land, the sediments and the rates at which they accumulate are more variable

Fig. 8-1. *Acanthaster planci,* a large, spiny starfish that is destructive to coral reefs in many parts of the tropical Pacific, crawling over a massive meandriform coral head. (From Chesher, R. H. 1969. Destruction of Pacific corals by the sea star *Acanthaster planci.* Science **165**:280-283. Copyright, 1969, by The American Association for the Advancement of Science.)

than in truly oceanic regions. These sediments, largely terrigenous in origin, consist of clay, sand, gravel, volcanic ash, and other materials washed into the adjacent seas from the land and often subsequently transported, sometimes very great distances, by currents, internal waves, ordinary wind, and wave action along the shoreline. They are mixed to varying degrees with materials of oceanic origin. The patterns of sedimentation are strongly influenced by currents and other water movements, which also exert a marked sorting effect. Heavier objects and particles are deposited closer to their point of origin and transported, in general, shorter distances more slowly, whereas lighter, smaller particles may be carried great distances and in various directions, depending on the currents and water densities at various depths and the size, shape, and density of the particles. In some areas, larger, heavy terrigenous objects such as boulders or gravel may be incorporated into the ice from glaciers, which feed icebergs into the water, and be transported with the icebergs until they melt sufficiently to release them.

Oceanic sediments

Truly oceanic sediments are of three principal categories—**calcareous oozes,** in which the shells of foraminiferans or pteropods and coccoliths, commonly predominate; **siliceous oozes,** comprised largely of diatom frustules or radiolarian skeletons; and **red clay,** in which the remains of organisms are less conspicuous.

Calcareous oozes, often termed globigerina ooze in areas where the shells of the pelagic foraminiferan *Globigerina* are abundant in it, characterize especially the warmer and shallower oceanic regions. They cover approximately half of the oceanic part of the sea floor. Below 5000 meters they are largely replaced by red clay or by siliceous oozes because of the solubility of calcium carbonate at great depths.

Red clay is characteristic of the deepest ocean basins and carpets approximately 35% of the ocean floors, whereas siliceous sediments occupy about 14%. Siliceous sediments, largely of diatom shells, occur particularly well developed in the Antarctic Ocean, and also in the far north just south of the Aleutian Islands. Siliceous sediments, largely radiolarian in character, occur in a relatively narrow band across the central Pacific north of the equator, from Central America to west of Hawaii, an area of the equatorial western Pacific, and an area southwest of Indonesia and northwest of Australia.

Andree (1920) classified the sediments into **littoral** (the sediments on the shelf), **hemipelagic,** and **eupelagic.** The eupelagic sediments are those deposited chiefly by pelagic organisms, such as globigerina ooze, together with an admixture of skeletal elements from benthic forms such as shell fragments, sponge spicules, and echinoderm spines. Since the term littoral is now usually restricted to the intertidal region rather than the entire continental shelf, perhaps the term **terrigenous** should be used to cover both Andree's littoral and hemi-pelagic sediments, which contain a higher content of mineral matter of terrestrial origin, and of organic matter, than do the eupelagic sediments.

Sedimentation rates

The accumulation of sediments on the sea floor, particularly the eupelagic sediments, is a very slow process judged by our time standards, averaging perhaps about fifteen thousandths of a millimeter of new sediment per year. Kuenen (1941, 1950), using several independent methods, estimated that they should accumulate to an average depth of 3 km. every 2 billion years. The earth is known to be more than 4 billion years old, which should give an average depth of eupelagic sediments of at least 5 or 6 km.

In actual fact, however, there are only about 0.3 km. of sediments on the floor of the deep seas. This presents something of a paradox that has been resolved only by startling discoveries in marine geology during the last two decades (see Chapter 17).

The sedimentation rate is not uniform over the various regions of the ocean, and it is not the same over a given region throughout the ages. It is fastest in shallow seas and shelf areas and slowest in abyssal depths beneath the open ocean.

Organic content and carrying capacity of sediments

Many animals live at or close to the interface between the bottom sediments and the cold, slowly moving water above. Some burrow or dig themselves into the oozes; others are attached by rootlike or rhizoidal processes; many crawl or creep on or through the ooze, feeding on the organic humus or on other animals they find there; and many demersal fishes and some cephalopods move over the sea floor, feeding on benthic animals. The character of the animal population is largely determined by that of the sea floor. Where there are rocky exposures, cliffs, or other hard substrate, the benthic fauna is very different from that where the bottom is carpeted with soft marine oozes. By far the greater part of the ocean floor is covered with such ooze, although in many areas it may be somewhat compacted, or lithified, rather than soft and fluffy.

The ocean floor is populated with animals at all depths. However, the quantity of animal life decreases markedly with great depth. Some groups disappear altogether, whereas others, not found in shallow waters, form a characteristic part of the deep-sea assemblage. For example, Zenkevitch (1963), reporting the results of a Russian deep-sea expedition in the Northwest Pacific, indicates that between 1000 and 5000 meters' depth there is a sharp decline in the relative importance of sponges. Between 4000 and 6000 meters, starfish, sea anemones, and polychaete worms became much less prominent. Decapod crustaceans were not found deeper than 5000 meters, but holothurians became relatively much more important. Below 8000 meters there were no more starfish, and gorgonians, pennatulids, amphipods, and isopods became rare as did also anemones and molluscs at 9000 meters. Below this depth only a few polychaete and echiuroid worms, pogonophores, and holothurians were taken, the latter constituting 90% of the catch by weight.

At 1000 meters a single haul contained 120 species. At depths between 8000 and 9000 meters three hauls yielded 6, 7, and 17 species, respectively. The weight of living organisms taken per square meter of deep-sea floor was 2 to 5 gm. at depths of 4000 to 5000 meters but decreased to 20 to 30 mg. below 9000 meters.

An important reason for the great decline in biomass with depth is the paucity of available food and the relative uniformity of its character. Organic materials produced in the euphotic zone are very largely cycled in the upper strata of water—eaten by the animals or decomposed by microorganisms long before reaching the bottom. The organic material that finally drifts to the bottom is both small in amount and consists largely of the less digestible residues—the so-called **marine humus,** together with skeletal elements such as foraminiferan and radiolarian shells and diatom frustules.

Eupelagic sediments are markedly poor in organic material. Terrigenous sediments on or near the continental shelves are characterized by a higher content of terrigenous minerals and a much higher content of organic matter, derived in part from the terrestrial fauna and flora, in part from the fauna and flora of the shelf, and in part from the rich life in the neritic waters above the shelf. A large proportion of this material reaches the bottom here, before decomposition is completed, and supports a large population of mud eaters and seston feeders. Much of this sediment is swept out past the edges of the continental shelves so that archibenthal and even some abyssal sediments near the shore are commonly satisfactory from a nutritional standpoint and can support a rather large population of benthic animals. These partly terrigenous sediments swept out beyond the continental shelves constitute the **hemipelagic sediments** of Andree. It is the nutritional content, rather than the mineral composition or the origin of various sediments, that is of greatest biological significance. The line between eupelagic and hemipelagic sediments is, of course, vague. It usually lies some distance beyond the rim of the continental shelves. Typical samples of each, however, differ sharply in organic content and nutritional value.

Distribution of benthic organisms

Although a number of species of marine organisms, especially mid-depth pelagic organisms or benthic forms with long-lived pelagic larval stages, enjoy very wide distributions and are found in all seas, the uniformity of deep-sea fauna has probably been frequently overstressed. Many species are relatively localized in their distributions. Different ocean basins have faunistic elements characteristic of those basins only. This is particularly true of the benthic fauna. Relatively few species are truly cosmopolitan. When the assemblages of species and genera in each region are considered as a whole, the oceans exhibit zoogeographical regional differentiation as distinctive as that which occurs on land. A rise, or sill, between adjacent ocean basins may present a barrier to the distribution of benthic and demersal deep-water animals, comparable to a mountain range on land. If, as is often the case, the waters of the two basins also differ in temperature, salinity, organic content, or other factors, the fauna may be quite sharply differentiated and distinct. Likewise, a great expanse of very deep water may constitute a barrier to the dispersal of shallow-water or shelf fauna. Even where no obvious differences in the physical qualities of the environment occurs, distributions may often be limited and circumscribed by coactive patterns of distribution between ecologically equivalent species and their struggle for living space.

The fact that pelagic larval stages are commonly eliminated from the life cycles of abyssal animals contributes to their isolation and limits their ability to achieve wide distribution rapidly. The scarcity of food and the low temperature in the abyss tend to favor slowly growing animals of relatively low metabolic rate and are probably a cause of the lack of success of planktonic larvae in this environment and their elimination from the life cycles of most of the animals.

The general character of the benthic fauna at any given place is largely controlled by the physical character of the substrates. Substrates may be divided into two quite different series—the rocky, or hard, substrates, and the soft substrates such as sand, mud, and marine oozes. Climatic differences, physical barriers to the spread of particular species, interactions between species, and other factors impress further differences so that the species composition of similar substrate communities from one region to another may be quite different, even though the overall character of the communities and the roles played by ecologically equivalent species may be quite similar. In looking at representative types of benthic communities, we shall consider each from the standpoint of the most important food supplies, the sorts of animals exploiting the food supplies; the special adaptations impressed upon the animals by the physical characteristics of their environment. We will begin with hard substrates in the littoral, or intertidal, zone,

consider the character of shallow-water, intermediate-depth, and deep communities, and then follow a similar scheme for soft substrate communities.

The overriding feature of almost all shallow-water, marine shelf communities is the presence of relatively abundant plankton and seston in the water, constituting the major available supply of food. This permits the development of animal communities dominated by sessile animals equipped with the means of concentrating plankton from the water and by substrate grazers specialized for obtaining their food from the substrate. Such attached or slow-moving animals commonly shed their gametes into the water and achieve dispersal through planktonic larval stages. In communities where there is enough available food in the ambient medium to support large animal populations, other factors, such as space for attachment, oxygen content, wave action, protection from predation, salinity, degree of exposure by tides, and nature of the substrate, become the limiting factors, determining the species that can live there and the biomass of animals produced in a given place. In favorable locations plankton-feeding animals sometimes become very crowded, growing upon each other, even to the extent of smothering the lower members until the mass becomes too unwieldy and is detached from the substrate.

Benthic communities on hard substrates

SHELF COMMUNITIES
THE CORAL REEF

Coral reefs are a dominant feature of shallow tropical marine waters where the mean annual temperature is at least 23.5° C. and where the water does not fall below 20° C. They are so characteristic of the tropics that the northern and southern extremes of coral reef formation may be taken as the boundaries of the tropical zone.

The stony corals, Madreporaria or Scleractinia, are hexamerously symmetrical anthozoan polyps related to sea anemones, from which they differ in lacking a siphonoglyph, in secreting an exoskeleton of calcium carbonate, and in being predominantly colonial through rapid asexual budding.

Typically, each **polyp** sits in a skeletal cup, the **corallite,** the outer portion of which is the **theca.** In the center of the bottom of the cup there is usually a column, or spinelike process, the **columella.** Radiating from the columella to the theca are a series of radially arranged **septa,** which in a general way alternate in position with the mesenteries of the polyp. It is almost as if the polyp were sitting on a tack in the middle of a cup, with radial knife blades passing from the tack to the wall of the cup. Throughout its life the polyp continues to build up the theca, the columella, and the septa. From time to time the polyp withdraws from the bottom of the corallite, secretes a horizontal partition, or **tabula,** as a new bottom for the corallite, so that a coral head in vertical section may show many long, parallel tubes partitioned by tabulae. Former positions of the corallites, may be seen, and new tubules beginning at various levels between the longer ones, show intercalated or budding new polyps, established as the coral grows.

Variations in the pattern of budding of

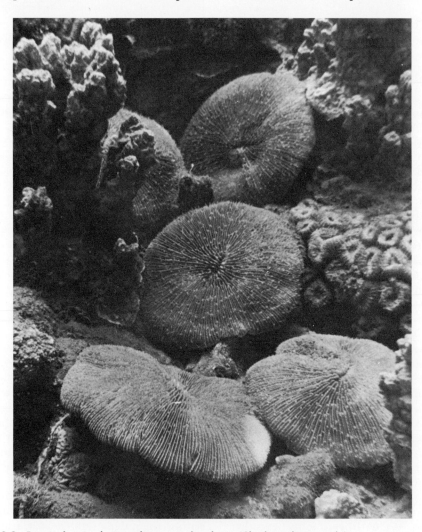

Fig. 8-2. *Fungia fungia,* large solitary coral polyps. (Chesher photograph.)

new mouths, in the degree of separation attained, and in the size of corallites result in a considerable variety of skeletal patterns among corals. In types with extratentacular budding, in which new polyps are formed entirely from tissues outside the ring of tentacles of preexisting polyps or in which there is a complete separation of the new individual, even if it budded from an intratentacular position, skeletons form in which each polyp has an individualized cup, or corallite, and is an easily recognizable unit. In other corals new mouths arise within the ring of tentacles, and the polyp shape alters to accommodate them. This process may be so highly developed as to lead to the formation of one or more polyps, each with a number of mouths arranged in a row. The resulting skeleton is a series of winding ridges alternating with valleys containing septa—the meandriform corals. In the Fungidae we find the development of much larger than usual single polyps, which form a sort of mushroom-shaped skeleton expanding from a short stalk, with the sharp septa radiating from the center of the top of the mushroom. Larger specimens are commonly detached from the stalk portion.

Three suborders commonly recognized are the Perforata, Aporosa, and Fungacea.

In the perforate corals, **Perforata,** the zooids are usually small, often crowded, and usually individualized, and the corallum is porous or reticulate. The septa are relatively few in number, not more than twelve, and sometimes indistinct, and in many of them the columella is absent. Most of the branching corals, such as the staghorn coral, belong to this group. Many important reef-building corals are included.

The imperforate corals, **Aporosa,** possess solid coralla, usually with numerous septa, commonly in multiples of six, without cross bars (synapticula). The disc may have one or more mouths, and there is usually a mesentery in each interseptal locus. Many of the more massive corals and meandriform types, such as the brain corals, belong here.

Fungoid corals, **Fungacea** or **Fungida,** have solid, imperforate skeletons; septa are usually numerous and with cross bars

(synapticula). The discs are not circumscribed and in colonial forms may be confluent. The tentacles are scattered soft lobelike structures, not covered when contracted, and sometimes obsolete.

Despite the striking differences in coral skeletal patterns, the group seems to be a relatively homogeneous one, differing from anemones not only in growth pattern, skeleton formation, and in lacking a syphonoglyph in the pharynx but also in having septal filaments consisting only of a cnidoglandular band and in having rather simple, unspecialized nematocysts of only three kinds.

Spawning of corals commonly displays a marked lunar periodicity. In many species development of the ova proceeds to the planula stage before they are discharged into the water.

Roles of symbiotic algae in corals

All the reef-building corals have abundant symbiotic zooxanthellae (small, rounded algal cells belonging to the groups Cryptomonadida and Dinoflagellida) in their ectodermal cells. In addition to the zooxanthellae there is an even greater amount of skeletal algae forming bands just beneath the polyp zone and in contact with the living tissue through minute pores in the calcareous skeleton. Together, these skeletal algae and the zooxanthellae comprise a mass of tissue about three times as great by dry weight as the animal tissue of the reef. The zooxanthellae and skeletal algae are of vital importance to their hosts, although their role in the alimentation of the host has long been disputed. When grown in culture separately, they do not liberate photosynthates into the water in significant amounts. For this reason it has been thought that their primary importance was in merely providing the coral with oxygen during the daylight hours and perhaps utilizing nitrogenous and other metabolic products of the coral so that these products are not excreted in quantities that would become toxic to the coral.

The amount of animal tissue in an extensive coral reef, such as the Great Barrier Reef of Australia, is so great that during low tide, one might otherwise expect that the corals would be killed through oxygen depletion and the accumulation of metabolic wastes. Indeed, this does occur to some extent on the inner, or protected, side of coral reefs, where many dead corals are found. The corals grow most vigorously at the outer, exposed edges of the reef, and the less favorable growing conditions on the protected side are perhaps responsible in part for the formation of channels and lagoons between the reef and adjacent shorelines. Thus fringing reefs are usually at a little distance from the shoreline.

Recent studies using labeled carbon dioxide have brought to the fore once more the idea that the symbiotic algae may also play an important role in the alimentation of their hosts. Labeled carbon compounds can be found in the host tissue as well as in the zooxanthellae themselves after a period of photosynthesis. Furthermore, calculations made by the Odums at Eniwetok Atoll indicate that the open sea there does not provide enough zooplankton to account for the productivity of the reef. To arrive at a reasonable food budget, or trophic structure, it was necessary to count the algae as major producers for the reef as a whole. The reef is an efficient conserver of nutrients, obtaining only a small part of them from the surrounding ocean and losing only a small portion to the lagoon.

The seeming contradiction between these findings and the failure to find photosynthates in the water following illumination of cultures of zooxanthellae has been cleared up at least in part by Muscatine (1967), who confirmed that if the algae are cultured in seawater separately from the coral, they fix labeled CO_2 but excrete very little soluble radioactive organic photosynthate. However, he found that if they are cultured in the presence of some tissue homogenate from the coral, they secrete as much as 40% of the total carbon fixed, mostly as glycerol, or with long incubation, as glycolic acid. The stimulative property of host tissue demonstrated a previously unsuspected role of the host in the regulation of specific metabolic activities of the symbiont, perhaps in this case through activation of a glycerol permease, enabling the glycerol to be excreted from the zooxanthellae and made directly available to the coral.

Thus it now appears that the algae may supplement the plankton diet of the corals in the clear plankton-thin tropical waters, enabling a greater mass of animal tissue to be maintained than would otherwise be the case and perhaps supplying necessary nutrient components not present in the plankton diet.

Symbiotic ties between the corals and their zooxanthellae and the need of the latter for adequate illumination to carry on photosynthesis help explain why reef-building species of corals are limited to shallow waters.

Other components of coral reefs

Corals are not the only, and in some cases not even the major, components of calcareous "coral reefs." Various coralline algae are also important, as are the calcareous skeletons of the Hydrocorallina, and in some cases even the skeletons of Foraminifera. Shells of molluscs, calcareous tubes of tubeworms, echinoderm spines, and calcareous remnants of bryozoan colonies and other organisms also contribute.

Coral reefs generally have a conspicuous crest of *Lithothamnion* on the windward actively growing edge, which serves to cement the looser coral together and give the reef the strength to withstand the buffeting of wind and waves.

Structure and origin of coral reefs and coral atolls

Perhaps the most characteristic coral formation is the **fringing reef.** Here the growth of corals in shallow waters produces a shelf of coral growth, the top of which is at about the height of the mean low-tide level. If this shelf attains any considerable width, the inner side, protected from wave action and containing quieter water already strained past the outer portion of the reef, as well as perhaps less of the oxygen and plankton and more of the metabolic products, becomes less favorable for coral growth. Here the corals tend to grow more slowly or to die, and eventually a lagoon may be formed between the outer reef and the shore. Darwin was the first to study carefully the formation of coral reefs and to correlate the formation of fringing reefs, isostatic changes in the level of the sea floor, or changes in sea level with the formation of barrier reefs and coral atolls—the isolated rings of coral islands often many miles across and best developed in the tropical western Pacific. He postulated that where there are a series of fringing reefs surrounding an island, either a slow subsidence of the island or a slow rise in sea level, at such a rate that the growth of coral can keep pace with it, will first produce a ring of coral islands surrounding the central island at an increasingly greater distance from it as more and more of the central island is flooded. Finally, only the ring of coral islands will be left. The coral islands comprising an atoll would then be comprised of dead coral remains to a considerable depth, only the top portion in the illuminated zone being alive. Subsequent borings in coral islands have confirmed

Darwin's hypothesis in its general outlines. In some instances several hundred meters consist of the same kinds of coral as that now living at the top of the reef. These must have been near the surface at the time of their growth, proving the hypothesis of slow subsidence in such cases. Darwin's hypothesis first brought the three principal reef forms, the fringing reef, the barrier reef, and the atoll, into relationship with each other.

Accumulation of organic debris and eventual development of vegetation on some of the larger low coral islands have created many habitable islands in the South Pacific.

If barrier reefs and atolls are, in fact, formed by slow subsidence with which the growth of coral has kept pace, we might also expect to find instances where the rate of subsidence exceeded the growth rate of the coral, taking it to greater depths than it could tolerate, and in fact such drowned reefs have been discovered. The best known of these reefs comprise the Chagos Archipelago in the Indian Ocean.

Later students of the problem have sup-

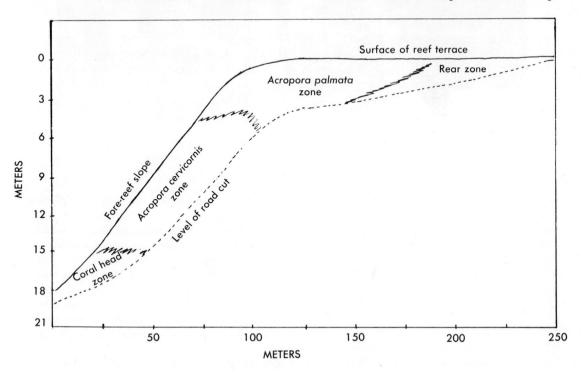

Fig. 8-3. Section of exposed coral reef to show general structure of West Indies coral reef.

Fig. 8-4. Portion of Barbados Island. Aerial view to show present fringing reef, and five successively older Pleistocene reefs that have been raised, forming marine terraces. (Mesolella photograph.)

plemented Darwin's observations and theories. Particularly important is the Glacial Control theory of Daly (1910), who showed, at least for the West Indian coral reefs, that during the Quarternary glacial period an interruption of earlier coral reef development took place as a result of lowering of the sea level through the formation of the great continental ice caps. This caused an emergence of the reefs and, of course, stopped their growth during the periods of emergence. Both the climatic changes of the time and the lowering of the sea level would operate to stop coral growth during these periods.

Another new concept is the possibility that the reef-building organisms themselves may be a contributory factor in causing subsidence. Limestone debris broken from and washed off the coral reefs settles as a sediment in their vicinity, accumulating through the ages in very considerable quantities. It has been calculated that at the present rate of limestone production in a group of coral islands such as the Maldives, sufficient limestone would accumu-

Fig. 8-5. Portion of road cut through a raised coral reef on Barbados Island. *Acropora palmata* zone of the reef. (Mesolella photograph.)

late to overcome the solidity of the earth's crust and cause isostatic subsidence in a period of time ranging from about four hundred thousand to nearly two million years, which is not a particularly long time in geological terms.

Other methods of coral island formation have been suggested, such as growth of a reef on the top of a submerged mountain.

Some areas, such as Barbados, the outermost island of the West Indies, have undergone a series of changes of elevation throughout Pleistocene times, resulting in the formation of a series of fringing and barrier reefs subsequently raised as part of the land, the structure of which may now be readily studied along roadcuts. This, together with the development of scuba diving, has enabled us to obtain a clearer idea of the structure and species succession of typical coral reefs. Coral reef morphology has been studied by Mesolella (1967) on Barbados, where a series of eighteen major reefs formed from time to time, the oldest

ones being found at the highest elevations and most central positions on the island.

Figs. 8-3 to 8-5 show a typical West Indian coral reef comprised of four major zones.

The outer, deepest part of the reef has been termed the **coral head zone** and is comprised largely of species such as *Montastrea annularis,* which form massive, solid-coral heads. Next is the **slope zone,** in the West Indies comprised very largely of *Acropora cervicornis,* a typical branching form. Where the slope approaches the mean low-tide level and flattens to form the main platform of the reef, we have the **crest zone,** characterized by the large expanded flattened heads of *Acropora palmata.* Finally, there is the **rear zone,** or platform back of the crest, which contains a considerable variety of species, including recurrences of numerous species found in the other zones.

The corals in a typical West Indies coral reef can then be tabulated as in Table 8-1.

Table 8-1. Typical components of a West Indies coral reef*

Major elements	Minor elements
Coral head zone	
Montastrea annularis	*Porites astreoides*
Siderastrea siderea	*Agaricia agaricites*
Siderastrea radians	*Favia fragrum*
Diploria strigosa	*Maeandrina maeandrites*
Diploria labyrinthiformis	*Maeandrina brazilensis*
	Colpophyllia natans
Slope zone	*Montastrea cavernosa*
Acropora cervicornis, together with scattered	*Porites porites*
Montastrea annularis, *Siderastrea* spp.,	*Eusmilia fastigiata*
and *Diploria* spp.	*Madracis* sp.
Crest zone	
Acropora palmata	
Rear zone	
Montastrea annularis	*Maeandrina maeandrites*
Acropora cervicornis	*Colpophyllia natans*
Diploria sp.	*Porites astreoides*
Porites porites	*Montastrea cavernosa*
Agarica agaricites	*Occulina diffusa*
Eusmilia fastigiata	*Madracis* sp.
Favia fragrum	Also hydrocoral
	Millipora albicornis

*From Mesolella, K. J. 1967. Zonation of uplifted Pleistocene coral reefs on Barbados, West Indies. Science **156**:638-640. Copyright, 1967, by The American Association for the Advancement of Science.

Several of the same species of corals found in the great coral reefs also occur beyond the tropical zone as individual coral growths, but without forming large reefs, especially along western and southern Africa and in Japan. There are also some deeper-water coral species without zooxanthellae, some of which are found as far north as the Norwegian fjords.

Coral reef associations

Corals are important not only because they comprise gigantic biocenoses of their own, characteristic of a distinct zoogeographical region, the tropical zone, but also because they provide the habitat for a special fauna of surpassing variety, beauty, and zoogeographical importance. No other animal associations can compare in these respects with those of the coral reefs.

Some coelenterates belonging to other groups also contribute to the calcium carbonate structure of the reefs. Among them are, especially, the athecate colonial hydroid *Millepora,* which produces massive branching skeletal structures superficially resembling corals. Certain alcyonarians such as the organ-pipe coral *Tubipora,* and *Heliopora* also come in this category. Some bryozoans also produce calcareous colonial skeletons so closely resembling those of corals that they are almost invariably mistaken for corals by persons who have not studied them carefully. Many of the Paleozoic limestones in America are particularly rich in bryozoan remains. More than 1500 species have been described. Equally rich deposits of bryozoan calcareous rock of Mesozoic age are known in Europe. Bryozoan reefs were formed from the Silurian through the late Tertiary. The contributions of calcareous algae and the many benthic foraminifers found on coral reefs have already been mentioned.

The coral reef provides a firm base for attachment, hiding places, calcareous rock that can be readily penetrated by boring or corrosion, and abundant calcareous sedi-

Fig. 8-6. Soft coral, Alcyonacea, with polyps expanded. (Chesher photograph.)

ment into which animals can burrow. They are inhabited by such a variety and abundance of animals that they constitute rich hunting grounds for a variety of predatory animals especially adapted to taking certain of the inhabitants.

Some of the largest, most beautiful anemones known from shallow waters, such as *Discosoma* and *Actinodendron,* are found here. Fleshy alcyonarians of the group Octocorallia, genera *Lobophytum, Sarcophyton,* and *Sinularia,* and others can sometimes be found carpeting considerable stretches.

Numerous very different sorts of animals have become adapted for burrowing into and living inside the coral heads. Among them are such bivalves as *Lithophaga* and *Lithodomus,* which burrow by means of acid corrosion. The beautiful sinuous many-colored iridescent mantle lobes of the giant clam *Tridacna* can be seen along the surface of stones in which they lie buried. Crabs of the family **Haplocarcinidae** live in galls that they cause to develop on the branches of corals and have become so modified by their peculiar mode of life that it is difficult to determine their systematic relationships. A prawn, *Paratypton,* has adopted a similar mode of life. A peculiar gastropod, *Magilus antiquus,* lives inside the solid *Maeandrina* corals, the last convolution of its shell straightened out during growth and keeping pace with the growth of the coral.

Some of the crustaceans characteristic of coral reefs are the very common widespread *Leptodius exaratus* of the Indo-Pacific, and *Trapezia* spp., also of the family **Xanthidae.** Several pinnotherids are also found here, as well as porcellanids and swimming crabs of the family **Portunidae.**

A variety of echinoderms are also characteristic, especially sea urchins (such as the long-spined *Diadema* spp., *Eucidaris,* and others), many sea cucumbers (such as *Holothuria* and *Stichopus),* which are found in abundance in loose coral sand under coral heads, etc., and the worldwide circumtropical *Holothuria atra* the large, tough black sea cucumber); and some of the most striking and beautiful of the starfish are found in open patches of coral sand between reefs. In shallow lagoons of coral islands one can occasionally find numerous long colorful sea cucumbers of the family **Leptosynaptidae,** sometimes 2 or 3 feet in length and striped like a king snake, with such thin body walls that they stretch out like a long balloon full of water when picked up and may even break from the weight of water and body fluids. Occasionally one also finds comatulid crinoids actively crawling about like highly ornamented giant spiders.

In the West Indies region the small, short-spined sea urchin *Echinometra lacunter* is found in holes and depressions in the coral rock at the level of the low-tide line. This genus is cosmopolitan in warm seas, and the sea urchins are always found only at this particular tide level—never wandering about over the flats.

These same lagoons may at times be almost carpeted by large medusae, *Cassiopea,* lying with the exumbrellar surface flattened against the substrate and the oral arms, with their many appendages, appearing like colorful vegetable growths.

Many striking and beautiful species of molluscs, such as the cone shells (family **Conidae**), cowries **(Cypraeidae),** conch shells **(Cassididae),** strombids **(Strombidae),** and many others, occur in these waters.

Coral reefs provide abundant hiding places and hunting grounds for such predators as octopuses, eels, and blennies.

The so-called coral fish comprise a great

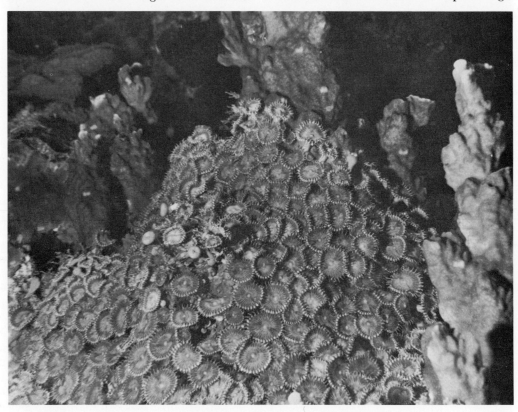

Fig. 8-7. Zoanthids, a type of anthozoan frequently found associated with coral reefs. (Chesher photograph.)

Fig. 8-8. Serpulid worm extending from its tube in a coral reef, radioles and operculum showing. (Chesher photograph.)

Fig. 8-9. Highly colored flatworm, and stellate foraminiferan. Guam. (Chesher photograph.)

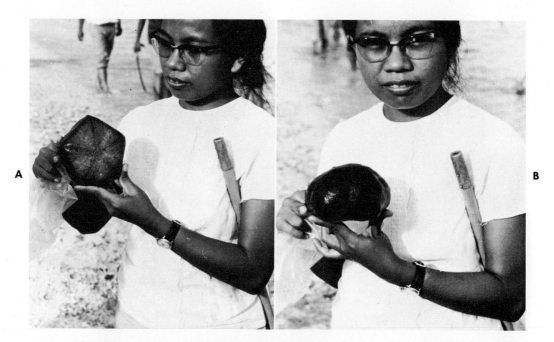

Fig. 8-10. Young Indonesian zoologist with large, black, armless starfish, *Culcita* sp., on the coral islands in the Java Sea off western Java. Ventral view, **A**, and side view, **B**, of the thick, fleshy starfish.

variety of colorful, often peculiarly shaped fish found around coral reefs. Among the most characteristic are the butterfly fish (**Chaetodontidae**), with their vivid coloring, deep, compressed bodies, high dorsal fin, and elongated snout with sharp, bristlelike teeth adapted for pulling small animals from holes in the coral rock. Other prominent members of the coral fish assemblage are the gorgeously colored scarids (*Scarus, Sparisoma,* etc.), numerous members of the family **Labridae**, the Moorish idols (**Zanclidae**), surgeonfish (**Acanthuridae**), trunkfish (**Ostraciontidae**), filefish and triggerfish (**Balistidae**), and the peculiar shrimpfish (**Centriscidae**), which swim about in small schools in a vertical position, head downward.

Cleaning symbionts. Cleaning symbionts constitute a curious and important element of the reef fauna. They are small, colorful fish and certain highly colored shrimp that live by cleaning off parasites and infected tissue from larger fish. Many fish deliberately come to the stations where these cleaning symbionts live in order to

be cleaned and to allow the cleaning symbionts to pick away at parasites or growths on them, even entering the mouth or gill chambers in their search. Conrad Limbaugh (1961) found that when the cleaning symbionts were removed from a reef, the reef was visited by a markedly smaller number of larger fish and that the indigenous fish living in the area of the reef tended to become "ratty" and less healthy in appearance. See Feder (1966).

Certain other species have evolved, closely resembling the true cleaning symbionts, but when fish come to be cleaned by them, they not only take parasites and infected tissue but commonly take a bite of healthy tissue from their unsuspecting customer. Older fish learn to tell the difference and go to the right shrimps. Only the inexperienced young and perhaps a few of the more stupid among the older fish continue to make the same mistake. In the Caribbean region the best-known cleaners are the Pederson shrimp (*Periclimes pedersoni*) and the neon goby (*Elacantinus oceanops*), which sometimes work together

Fig. 8-11. Clown fish, *Amphiprion percula,* a brightly colored coral fish. (M. W. Williams photograph.)

at the same cleaning station. The shrimp usually locates next to the anemone *Bartholomea,* making this a type of triple symbiosis.

Colors of reef fish and other coral reef animals

The striking colors and strange shapes and patterns of many reef fish and other animals associated with coral reefs have long excited the admiration and wonder of visitors. One might suppose that such brilliant coloration would invite disaster where hungry predators are never far away. If this were the case, these bright colors and patterns would long since have been eliminated through natural selection. We must assume that they serve useful functions for their bearers that at least counterbalance any such adverse affects. There is probably no single pat answer to the problem, but rather each case must be examined individually. However, some generalizations can be made which perhaps will shed some light on the question.

For one thing the reefs themselves constitute a colorful background, against which some of these fish may not be as conspicuous as they appear in aquariums and in other waters. Also, a coral reef provides endless small channels, nooks, and crannies in which the coral fish can hide when alarmed, reducing the selective premium on being overlooked by a predator.

Sharply contrasting blotches and bars of color may also serve to break up the outline of the fish so that it is not so easily recognized as such by a predator. In some instances, the color may serve as a warning, advertising distasteful flesh or poisonous spines. However, warning coloration and its mimicry by harmless species do not seem to play as important a role among fish as it does in certain insect groups.

This may be due to the fact that aquatic

vegetation is less diversified than that on land. Numerous groups of terrestrial plants have evolved chemical repellents that protect them from grazing by all except very special insects. These insects have become tolerant to the chemical from a particular protected plant group and, in turn, derive protection from their predators through ingestion of the repellent chemical. Such insects commonly evolve conspicuous warning coloration—and may even serve as models mimicked by other insects not sharing their chemical protection.

The cleaning symbionts seem to use their bright colors and characteristic motions to advertise their presence and attract larger fish to the cleaning stations.

Also, we must remember that we observe these fish through human eyes and brains. What we see does not necessarily appear the same as it does to another fish or squid.

On the positive side, coloration can play important roles in species recognition, in helping keep schools of fish together, in courtship, and perhaps in other ways of which we are not aware.

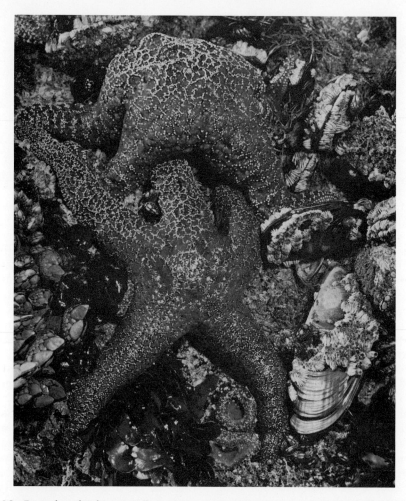

Fig. 8-12. Typical rocky lower midlittoral zone assemblage exposed at low tide along West Coast of United States. Two starfish (*Pisaster ochraceus*), the upper one in humped feeding position and eating a California sea mussel (*Mytilus californianus*), some of which can be seen at the right, encrusted with acorn barnacles (*Balanus glandula*). At left is part of a clump of the stalked barnacle *Mitella polymerus*. (M. W. Williams photograph.)

THE ROCKY INTERTIDAL

Second only to the great coral reefs in the abundance, variety, and beauty of the living communities they support are the rocky coasts, where headlands, reefs, and rock outcroppings of all kinds meet, extend into, and become submerged in the sea. Each such exposure is a unique microcosm, as individualized as a man's fingerprint. The extent, nature, and species composition of the communities occupying it depend on many complexly interacting factors.

In the rocky intertidal there is an alternating exposure and submergence by the tides, to a different extent and at a different time each day, with the lunar progression of the tidal cycles. At low tide there is direct exposure to the sunlight and rain, to winds and desiccation, and to predation from land animals. As the tides roll in there may be exposure to tremendous power of the surf pushed by the incoming tide, especially during the fortnightly spring tides, in exposed situations on the open coast. There are abrupt changes in temperature, salinity, character of active predators, light, available food, etc. The degree of protection and opportunity for settlement vary with the nature and hardness of the rock, the direction and angles at which the rock strata meet the sea, the local tidal range and weather, and the direction and amplitude of currents, waves, and other water movements. Other important factors are whether or not sand or other loose material regularly or ever covers the rocks (and if so, how often and to what extent), the frequency of storms, and the force they attain locally. The geological age of the particular relationship between coast and sea—whether the land in geologically recent times has been rising and emerging from the sea or sinking beneath it, or whether there has been a long period of relatively stable relationship—is also important.

In very recent times the activities of man have become a powerful force for change on a global scale. No place on earth is so remote or protected as to be free from these effects. The intertidal and shallow waters at the margins of the seas, being the most directly exposed, and perhaps the most vulnerable, bear the brunt of this assault upon nature and in all settled regions have been and are being drastically changed. This is even more true of bay and estuarine environments than of the open rocky coasts.

It must be borne in mind in this connection that the intertidal and subtidal communities of animals and plants do not represent just the edge of a vast similar community extending far out to sea and almost infinitely renewable from the sea. Rather, they are very thin communities, often measurable in a few inches or feet vertically, extending parallel to the coastline as a series of very narrow communities. Therefore, they are particularly vulnerable to depredation and to destruction from changes occurring along the coastline.

Supralittoral rocks

Below the margin of typical terrestrial flora, on the seaward-facing rocky slopes, species of rock tripe, the lichen *Umbilicaria,* are the last of the larger terrestrial flora. During the heat of the day they become blackened, thin, and brittle, giving the rock the appearance of scaling of thin layers. When wet with rain or fog they are like thin, pliable sheets of greenish leather. Below the *Umbilicaria,* to the upper extremity of the highest level of the spring tides, the rocks are commonly barren, with only a few insects or other terrestrial animals temporarily resting on them.

The junction between sea and land, at the lower margin of the high areas, where the rocks are dry almost all of the time is usually marked by a darkened zone, which looks more like a stain on the rocks than like a band of living organisms. This **black zone** is formed by a growth of innumerable microscopic plants, most of them too small to be seen with the naked eye, and many encased in gelatinous, or slimy, sheaths. These sheaths protect them from drying during the long exposures to sun and air and make the rock surface of the black zone extremely slippery when wet. This zone of microscopic plants at the extreme upper margin of the intertidal area is perhaps the most universal mark of the transition from land to sea. Above it the environment is essentially terrestrial. Below it the world belongs to the sea. Most of the microscopic plants of the black zone are primitive blue-green algae. A few green algae and minute lichens are often present among them. They are seldom submerged, but receive the splash and spray from the large waves of the fortnightly spring tides.

In some regions small tide pools in the uppermost edge of the intertidal zone, which only occasionally receive water from the sea and are subject to rain and evaporation, when containing calcareous shell fragments that render them alkaline become blood red with the growth of *Sphaerella,* a member of the Volvocales, which are related to species that sometimes produce "red snow" farther inland.

Attached plankton and seston feeders

The dominant animals of the rocky intertidal are usually attached, or sessile, plankton and detritus feeders that draw their nourishment directly from particles suspended in the water. Some, such as many sponges, bryozoans, some coelenterates, tunicates, and others, have developed the encrusting habit, growing as flat growths firmly adherent to, and following the contours of, the surfaces upon which they are growing. Others are raised more from the rock surface but are firmly attached and so shaped as to shed the water readily and offer minimum resistance to the impact of waves. Animals growing in crevices, under ledges, among the holdfasts of marine algae, or in relatively quiet waters, wherever they can obtain enough protection, may assume other shapes and growth habits, such as the many bushy, erect hydroid and bryozoan colonies, globular tunicates and sponges, etc. Still others construct firmly cemented calcareous tubes in which they live or burrow into the substrate, exposing their plankton-gathering apparatus only when the burrow is covered with water. Some of the more conspicuous of these plankton or detritus feeders will be considered group by group.

Barnacles (class Crustacea, subclass Cirripedia). Barnacles comprise one of the most remarkable, widespread, and characteristic groups of animals to be found in the intertidal and shallow marine shelf waters. In rocky coastal areas certain species of acorn barnacles or sessile barnacles, especially genus *Balanus,* are commonly found among the highest of the truly marine animals on the rocks, extending from the upper limits of the highest tides or even into the splash zone down to about the level of the mean high neap tides. They often occur in such numbers as to form a white band marking the uppermost tidal levels, visible from a considerable distance on the darker rock, and in places may grow so crowded together that the shells are distorted—much taller and more cylindrical than normal. Lower on the rocks, these barnacles are replaced by other species, such as the clumps of stalked, or gooseneck, barnacles, at the beginning of the zone of mussels, and often by some much larger species occurring in smaller numbers in the lowest part of the lower littoral zone and subtidally.

It seems remarkable that a sessile plankton-feeding animal should be among those to occur in greatest numbers and most successfully in the highest part of the intertidal zone, where they have the briefest periods of submergence for feeding and respiration, where they must endure the longest periods of exposure to the air, with the dangers of desiccation, overheating or chilling, and exposure to rain, and where they often grow in very exposed situations, enduring the impact of the largest waves during the spring tides or storms. Furthermore, unlike most animals of the intertidal zone, most barnacles are hermaphroditic cross-fertilizing forms with internal fertilization requiring copulation. Thus, if a barnacle of this type happens to be located more than a few millimeters from another individual, it may be unable to reproduce. Some species are capable of self-fertilization. The eggs undergo development within the mantle cavity, and the young are discharged into the water as nauplii.

Within the barnacle zone on the rocks one can often observe that the individuals at the very highest levels are all much smaller than those somewhat lower, near

Fig. 8-13. Acorn barnacles (*Balanus glandula*) and limpets (*Acmaea digitalis*) on rock face in upper littoral zone, California. (M. W. Williams photograph.)

Fig. 8-14. Acorn barnacle *(Balanus tin-tinnabulum)*, feeding. Southern California. (M. W. Williams photograph.)

the lower edge of the zone they occupy. This is probably because the former receive the briefest period of submergence and cannot obtain enough food to support normal growth. Other aspects of their metabolism may also be interfered with by the prolonged periods between high tides, during which they must remain closed up and inactive.

Adult acorn barnacles are enclosed in a calcareous shell secreted by the expanded carapace, sometimes termed the mantle because of its peculiar character and its surrounding of the body. The volcano-shaped shell consists of a basal part cemented to the rock surface, a parapet, or wall, composed of six nonmovable plates (a dorsal carina, two carinolaterals, two rostrolaterals, and a ventral rostrum), and one or two pairs of movable plates (the scutum and the tergum) near the aperture of the parapet, which allow the barnacle to close itself from the outside environment during periods of stress or alarm, but which can be separated to allow protrusion of the six pairs of thoracic limbs, the cirri, for feeding. Various genera of barnacles are characterized, among other ways, by the varying

relationships of the plates of the parapet to each other and in some cases by fusion or elimination of certain of the plates.

A surprising characteristic of some barnacles, discovered by Graham Hoyle in 1963, is that the scuto-tergal adductor muscles and the depressor muscles are composed of giant cells. In the large barnacle *Balanus nubilus* Darwin these are so large that conventional nerve-muscle preparations can be made, using single muscle cells, and microelectrodes can be readily inserted to obtain recordings from individual muscle cells. They are therefore very useful for certain kinds of research in neuromuscular physiology.

In the stalked, or gooseneck, barnacles the body is enclosed in a soft mantle, the capitulum, commonly protected by calcareous plates. The capitulum includes all the body except the preoral portion from which the stalk was developed. The calcareous plates usually include two scuta, posterior in position, two lateral terga, and an anterior keellike carina. Transverse adductor muscles enable the barnacle to draw the edges of the terga and scuta together for protection.

Barnacles undergo a remarkable meta-

Fig. 8-15. Stalked barnacles *(Mitella polymerus)* grow in dense clumps in upper part of the mussel zone. La Jolla, California. (M. W. Williams photograph.)

morphosis during their development. The young are liberated as very characteristic nauplii, with the anterolateral corners of the carapace drawn out into lateral spines, giving them a triangular shape when seen from above or below. Following the naupliar instars they transform into a cypris stage, with the body enclosed in a bivalved carapace. At this time they resemble ostracods somewhat. When the cypris is ready to settle down and metamorphose into the adult stage, it begins to explore the substrate actively. Upon selecting a suitable spot, it attaches to the surface by a small disc on each of the antennules and secretes the basal attachment from the cement glands, which occupy much of the preoral region. A dorsal groove appears and enlarges, nearly separating the portion of the preoral part of the body containing the cement glands and antennules from the rest of the body proper, whereas the bulk of the body turns a half somersault within the carapace, so that the barnacle ends up lying on its back at right angles to the long axis of the carapace in such a position that the six pairs of thoracic limbs point upward. These become the cirri, fringed with setae. When protruded and spread out, the cirri form a net with which the barnacle sweeps the water immediately above the orifice of its shell. Upon retraction the cirri come together, enclosing any food particles encountered and bringing them into the mantle cavity where they are scraped off and transferred to the mouth by the maxillae. Small barnacles that have been exposed out of water for some time become very active upon being immersed, and the feeding reaction is readily observable.

It has been experimentally shown that the barnacles characteristic of the higher parts of the rocks in the intertidal zone do not require the periods of exposure above water. In fact, they grow faster and larger if maintained continually submerged, where they can feed continuously. It is probable that they are limited to the upper fringe of the intertidal zone by competition for space with algae and encrusting organisms that require somewhat longer periods of submersion or that cannot stand such long exposure. In the lower parts of the barnacle zone where the barnacles grow intermingled with small, attached seaweeds such as *Fucus* or *Ascophyllum* it can often be seen that the sweeping motion of the algae over the rock face with the surge and ebb of waves commonly keeps the rock surface swept clean. Predation by carnivorous gastropods and starfish may also play a role in limiting barnacles to the uppermost zones at which plankton feeding is possible.

Barnacles are a prominent element of the "fouling community," organisms that grow on ship bottoms, underwater pilings, pipes, etc.

Bivalve molluscs. Below the barnacle zone, the most abundant and conspicuous filter feeders are various bivalve molluscs. The great majority of bivalves feed by catching small particles of plankton or organic detritus and bacteria on mucus of the gills as the respiratory current passes through them. The mucus, with the contained food particles, is swept to the lower margin of the gills by special tracts of strong cilia and then brought anteriorly along the food groove at the gill margin to the vicinity of the mouth where the palps transfer it to the mouth.

In the rocky intertidal the bivalves can be grouped into three categories on the basis of their relationship to the substrate.

In the first category are the most conspicuous bivalves, such as mussels, which live on the surface of the rock and are held in place by strong byssal threads secreted from a byssal gland near the base of the foot. Mussels commonly occur in great numbers, as closely spaced as possible, forming mats, or beds, of mussels sometimes covering much of the exposed rock surface between the lower margin of the barnacle zone and the upper edge of the zone of permanent seaweeds. Some species,

such as *Mytilus californianus,* the California sea mussel, prefer open, exposed coastline where they meet the full brunt of the waves breaking over them on incoming tides. Others, such as the widely distributed *Mytilus edulis,* often called the bay mussel, prefer somewhat quieter, more protected waters. It is probable that massing together, giving each other additional support, is essential in the case of mussels growing in exposed situations, enabling them better to withstand the wave shock. The mussel bed itself provides protection for many other animals, including worms, gastropods, small crabs and shrimps, and other small bivalves, and often serves as a substrate for settlement of barnacles, bryozoans, tunicates, sponges, and small mussels. Eventually, the mass of animals may become so thick and heavy that big chunks break away during periods of unusually heavy waves, leaving patches of bare rock where the cycle of colonization may begin again.

The principal predators of mussels are starfish, carnivorous gastropods, and man. Of course, if any of the mussels are crushed or damaged, as by heavy objects being cast on them by waves or by people walking on them, the injured individuals are promptly disposed of by scavengers belonging to many groups.

Below the beds of mussels exposed at low tide, beginning at the lower edge of the *Chondrus crispus* (Irish moss) zone where the sweep and backwash of the waves is still heavy, and occasionally in deep crevices or pools in the higher zones are found the larger horse mussels (*Volsella modiolus*), up to 5 inches long, with heavy bulging shells and strong byssal threads attaching them in all directions. They often also attach byssal threads to shell fragments and pebbles brought into their zone by waves. The mat of byssal threads, entangled debris, etc. creates a protected habitat for a variety of small organisms, worms, crustaceans, echinoderms, etc., and for the young of numerous others that would otherwise be unable to live in this rocky zone where the surge of the waves is so strong.

In somewhat deeper water or in places

in warm shallow calm waters, a relative of the mussel, the large, fragile pen shell (*Atrina* spp.), sometimes occurs in great numbers, often where a few centimeters of loose material overlie something firmer to which they can attach. They occur offshore in places from Cape Hatteras south along the East Coast, becoming especially abundant along the Gulf coast of Florida. Millions of them are sometimes cast up on beaches by storms, second only to the arc shells in abundance on the beaches. They are of worldwide occurrence in tropical waters. The byssal threads of pen shells have a beautiful golden sheen and fine texture. In the Mediterranean region they have been used for the production of an especially fine-textured, soft, highly valued "cloth of gold."

The second category of intertidal bivalves consists of such forms as oysters, rock oysters, and some others that cement part or all of one of the valves firmly to the substrate as they grow. In such forms the valve that is cemented to the rock assumes the irregularities in shape of the surface to which it is fastened, and the upper valve becomes somewhat irregular in order to fit snugly.

The third category of bivalves finds protection from the waves and from predators by boring into the substrate, extending only the tips of the siphons to the surface for feeding. Where the rock is relatively soft sandstone, or calcareous, the surface few inches may be riddled with the burrows of pholads and date-nut shells. They have thus become a factor of some geological importance in hastening the disintegration of exposed rock. The boring habit has been assumed independently by different unrelated groups of bivalves. Still others, which do not themselves make burrows, make use of those made by other organisms. Such species may be termed nestlers.

Boring organisms fall into two categories, depending on the method used in making their burrows—the chemical borers, limited to calcareous rocks, and the mechanical borers, in which a portion of the shell is especially modified for scraping out the burrow. The burrows are started soon after transformation of the swimming larval stage and continue to be enlarged throughout the growth of the occupant, with the result that the opening is constricted and the occupant is permanently imprisoned in the burrow.

One of the most characteristic and widely distributed groups of rock-boring clams are the pholads, or piddocks, family **Pholadidae.** In these clams the shell is an elongate oval, without ligament and hinge. There are specialized, filelike rasping areas on the anterior part of the shell. The animal clings to the wall of its burrow with its short, broad foot and rasps with the shell, changing its position from time to time. Piddocks burrow into relatively soft rocks such as consolidated clay, marl, shale, peat exposures, and some sandstones or conglomerates. The boring action can be readily observed by putting young animals in a test tube of the same bore as their burrow, immersed in an aquarium. Piddocks have the mantle cavity closed for most of its length. The siphons are long, and united nearly their whole length, the posterior end of the shell permanently gaping to permit their passage. There are three adductor muscles.

One of the best known is the angel wing shell (*Pholas* or *Barnea costata*) of the East Coast from Massachusetts to the West Indies, which burrows deeply into peat exposures, clay, etc. It grows up to 7 inches long and has a beautiful, fragile, pure-white shell. It is most abundant from Virginia southward.

One genus, *Martesia,* the wood-boring piddock, has taken to boring into wood rather than rock. Unlike the shipworm, the piddock's burrows are superficial, as it excavates only nestling sites and does not continue to bore deeply through the wood.

Perhaps the largest is the rough piddock, *Zirfaea pilsbryi,* which burrows over a foot deep in soft shale along the shores of the Pacific Northwest.

Another characteristic mechanical borer is the rednose, *Hiatella,* a rock-boring clam recognized by its red siphons, also found among crusts of coralline algae and in kelp holdfasts.

The giant clam *Tridacna* is exceptional

among bivalves that grow embedded in hard substrates in that it grows in a supine position, with its large mantle lobes containing symbiotic zooxanthellae exposed, rather than the ends of the siphons. It may be that in this case the clam is too large and massive to support itself solely by plankton feeding and requires the zooxanthellae to supplement its diet.

Among the most remarkable of the mechanical borers are members of the family **Teredinidae**, which includes shipworms. These molluscs have become adapted for burrowing in wood. They grow as long, wormlike organisms, with the shell restricted to two small scraping valves at the anterior end and most of the body being the long, tubular mantle. Instead of enclosing the body with shell, they line their burrows with shell material. Since the opening of the burrow, made by newly transformed larval shipworms, is scarcely larger than a pinhead, the only external sign that a piece of wood is under attack by shipworms is the presence of minute holes in the surface, which are easily overlooked, even though the wood itself may be riddled by the burrows. Shipworms cause immense damage to wooden ship

bottoms and structures such as pilings, etc. in shallow marine waters. In San Francisco Bay in the early part of this century, they caused millions of dollars' worth of damage before anyone was aware of their presence. The problem was recognized only when piers, wharves, and other structures supported by pilings began to cave in.

Tunicates (order Ascidiacea). In the lower intertidal and subtidal zones where there is firm substrate to which to attach, solitary and colonial tunicates are another conspicuous group of filter feeders, with a very different variation in the ciliary-mucus method of gathering food from the water.

The solitary tunicates are globular or elongated saclike animals attached to the substrate at one end and with two openings, an incurrent oral opening and an excurrent atrial opening, usually rather close to each other at the free end. The body is invested with a tough protective cuticular covering, the tunic, made largely of cellulose secreted by the underlying epithelial cells and containing mesenchymal cells that have wandered into it. Most of the body is occupied by a greatly expanded pharynx, or branchial basket, which communicates with the surrounding atrial space by numerous

Fig. 8-16. Solitary tunicate (*Styela montereyensis*), with attached small mussel (*Mytilus californianus*) and hydroid (*Sertularia* sp.).

stigmata, or gill slits. The combined respiratory-feeding current created by the cilia of the pharynx enters the oral opening and passes through the stigmata into the atrial cavity and out the atrial opening. Food particles are retained by the mucus lining the pharynx and passed by special tracts of cilia to the endostyle, a longitudinal, glandular ventral groove, and back into the esophagus. The digestive tract consists of esophagus, stomach, and intestine, which is twisted into a loop so that it comes forward to empty into the atrial cavity, as do also the ducts from the gonads. The circulatory system of tunicates is unique in that the direction of circulation is reversible, the heart pumping the blood alternately in one direction into the pharyngeal vessels and then reversing itself and pumping it into the body and the viscera. Tuni-

cates are hermaphroditic, in most cases shedding the gametes into the water, where they develop into a tailed, tadpolelike stage that possesses a notochord in the tail. They thus display the chordate relationships of tunicates more clearly than do the adults.

Some tunicates can also reproduce asexually, budding off new individuals and forming colonies of more or less closely associated individuals. This highly developed capacity has led to the formation of so-called compound tunicates, in which a great many small individuals or zooids are embedded in a common gelatinous matrix. Different patterns of budding and colony formation and differences in relative amounts of gelatinous material between the zooids in different groups of compound tunicates have led to the development of an amazing variety of color-

Fig. 8-17. Compound tunicate *Distaplia*. (M. W. Williams photograph.)

ful forms, ranging from thin flat encrusting growths closely adherent to the substrate to large fleshy or lumpy colonies such as *Amaroucium* (often misspelled *Amaroecium*).

In some places, such as certain subtidal areas in the Bay of Naples, rock faces have become so coated with compound tunicates and sponges as to form spectacular submarine gardens.

Sponges. Among the most ubiquitous and conspicuous components of lower intertidal and shelf fauna, wherever the substrate presents suitable attachment and the water is reasonably clean, are the sponges. They grow in a great variety of forms ranging from thin encrustations to massive growths and erect branching types. All sponges are sessile animals with a highly porous body permeated with channels and spaces permitting the flow of water through them, a sort of living sieve. The water enters through minute incurrent openings, the **ostia,** and passes through one or more chambers lined with special flagellated collar cells, the **choanocytes,** and leaves, either directly or by way of a series of larger anastomosing channels, through one or more

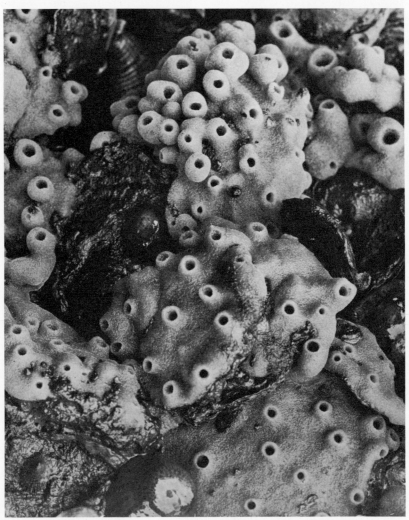

Fig. 8-18. Crumb-of-bread sponge, *Halichondria panicea,* a common West Coast encrusting species with raised, volcano-like oscula. (M. W. Williams photograph.)

larger openings, the **oscula.** The choanocytes maintain the flow of water through the sponge and capture minute food particles as it passes. Oxygen is brought directly to all the cells by the circulating water.

Most functions are carried out at the cellular level, there being no organs and organ systems as we know them in higher animals. The low degree of organization of sponges is shown by the fact that if they are cut into pieces, each piece continues to function as if nothing had happened. Some have even been completely disaggregated by forcing bits of them through bolting cloth, and the cells have been observed to reaggregate into small sponges of the same type as the parent individual. Growth is plantlike, and in many species asexual buds of various types are produced. It is difficult to say just what constitutes an individual in such an animal. Perhaps the question is meaningless at this level. The most obvious functional unit of most sponges would be one osculum, together with the most directly associated flagellated chambers and water channels. In the Huxleyan sense, an individual would be all the growth that proceeds from one fertilized egg cell, whether or not it remains together.

The body of the sponge, or mesenchyme, contains several different types of cells, many of which are amoeboid and some of which secrete the characteristic skeletal elements—spicules or fibers of spongin.

Sexual reproduction in sponges is by gametes developed in the mesenchymal tissue. Male gametes liberated from one sponge may enter another along with the incurrent streams of water and fertilize ripe egg cells there. The fertilized cells commonly undergo part of their development in place and leave the parent sponge as peculiar, nearly spherical larvae, known as **amphiblastula,** containing small flagellated cells in one hemisphere and large, yolky, nonflagellated cells in the other. The amphiblastulae are carried about for a brief time in the plankton. If successful in finding suitable substrate when ready to settle, they affix themselves to it by the flagellated pole and develop directly into small sponges. The flagellated cells give rise to the choanocytes, and the rest of the sponge body is derived from the large nonflagellated cells. The amphiblastula should be regarded as a nonfeeding larval stage, not as a homologue of the embryonic blastula stage of higher animals. Its subsequent development in no way resembles that of an ordinary blastula, and there is little to suggest that the various cells of sponges can be homologized with the ectoderm, endoderm, or mesoderm of the embryos of higher animals.

The classification of sponges is based largely on their skeletal structures, since only they can be studied in ordinary museum specimens. A formidable terminology

Fig. 8-19. Dead-man's-fingers, *Chalina* sp., an erect tubular sponge. (Oregon Institute of Marine Biology photograph.)

has grown up describing the various sorts of spicules and their arrangements. In spite of their primitive degree of organization, the various species of sponges maintain distinctive and recognizable growth habits. Some of them are highly colored—yellow, red, purple, or greenish.

Sponges are preyed upon by relatively few animals. The prickly skeleton of spicules of most of them and probably chemical distastefulness in many seem to give good protection, although some molluscs, notably various nudibranchs and a few other animals, do graze on certain species of sponge. Many small invertebrates are commonly associated with larger sponges, nestling in or under them for protection.

By far the great majority of sponges found in shallow shelf and intertidal waters belong to the huge class **Demospongiae.** This class is characterized by having either a horny fibrous skeletal material of **spongin** or **siliceous spicules** of types other than triaxon (six-rayed), or both. Those that have spicules commonly have two sorts of spicules: **megascleres,** which form the actual supportive skeleton, and **microscleres,** which are small spicules in the mesenchyme not playing a directly supportive role. Flagellated chambers of Demospongiae are usually small and rounded, and the organization of the water channels is of the complex **leuconoid** type.

The commercial sponges, or bath sponges, are large sponges of the order **Keratosa,** in which spicules are lacking. The elastic horny fibrous skeleton, cleaned and freed from other organic remains, comprises the sponge of commerce. In recent years the demand for sponges has decreased because of the development of rubber and synthetic substitutes.

The second class of sponges represented in coastal marine waters, class **Calcarea,** is characterized by having calcareous spicules with one to four rays, not differentiated into megascleres and microscleres. Most of the calcareous sponges are relatively small whitish vase-shaped or tubular sponges of the **asconoid** or **syconoid** type of organization. In the ascon type of sponge the incurrent pores lead directly into the main chamber, or spongocoel, of the vaselike or tubular sponge. This chamber is lined with choanocytes, and bears the osculum at its upper end. The sycon type is somewhat more complex, but can be visualized as an ascon type in which the wall has become complexly folded, so that it consists of alternating incurrent canals and excurrent canals. Choanocytes line the excurrent canals, which open into the spongocoel. The small sponge widely sold under the name *Grantia* is of this type. The ostia in sycon sponges are internal, penetrating the walls separating incurrent and excurrent canals. The outer ends of the incurrent canals constitute new incurrent openings.

The third large class of sponges, the glass sponges or **Hexactinellida,** are usually found on soft substrates in deeper water.

A fourth class of sponges, the **Sclerospongiae,** has recently been proposed (Hartman and Goreau, 1970) to include the coralline sponges. These bear suggestive resemblances to the ancient fossil stromatoporoids (Fig. 8-20).

Hydroids. Hydroids represent the polypoid generation of Hydrozoa. The medusoid generation has been discussed in the section on gelatinous zooplankton. The majority of the hydroids belong to intertidal and shallow shelf fauna. Most are colonial, constituting variously branched bushy or feathery growths in which the body layers and gastrovascular cavities of the polyps or hydranths are continuous throughout the colony, so that the entire colony is nourished by whatever any of the polyps feed on.

Each polyp is provided with a crown of nematocyst-bearing tentacles around the mouth, with which they capture and immobilize zooplankton organisms. A bushy hydroid colony containing dozens of hundreds of such polyps is a formidable peril for any zooplankter swimming near it or carried through it by water movements.

Most hydroids are protected by a noncellular secreted **perisarc,** which covers the stems and stolons (or **hydrorhiza**), giving stiffness, elasticity, and support to the colony. In many it also forms a cup, the **hydrotheca,** which extends up around each

Fig. 8-20. *Astrosclera willeyana* Lister, from Guam. This sponge represents a new class of sponges, the Sclerospongiae, described by Hartman and Goreau (1970). (Chesher photograph.)

polyp and into which the polyp can contract if alarmed, and a protective cover, the **gonangium** or **gonotheca,** which surrounds reproductive polyps.

The **coenosarc,** or living portion of the stems and stolons, is composed of the same two layers of cells, the outer epithelium and gastrothelium, of which the polyps are comprised. These layers have commonly been termed ectoderm and endoderm in coelenterates, but it seems preferable to reserve those terms for their embryological usage and to describe the adult tissues of coelenterates in the same manner employed for similar tissues in other phyla, although recognizing their ectodermal and endodermal origin.

In many hydroids more than one kind of polyp have differentiated. The two most frequent are the **gastrozooids,** or feeding polyps, and the **gonozooids,** or **gonophores,** the reproductive polyps, which bud off medusae asexually. In addition to these,

some species have developed **dactylozooids,** rodlike structures heavily armed with nematocysts and without tentacles or mouth.

The order Hydroidea is divided into two suborders. In suborder **Gymnoblastea** the hydranths and reproductive zooids are devoid of perisarc, that is, there is no hydrotheca or gonotheca, although the stems and hydrorhiza may bear a perisarc. Members of this group are variously termed athecate hydroids or tubularian hydroids. In those that give rise to free-swimming medusae, the medusae are usually tall and bell-like, and bear marginal ocelli but not statocysts, and their gonads are usually borne on the manubrium. Such medusae are termed **Anthomedusae.**

The suborder **Calyptoblastea** contains species in which the perisarc forms hydrothecae and gonothecae. Free medusae of hydroids in this group are commonly flatter and more saucer-shaped, possess marginal statocysts, and bear the gonads on their radial canals. They are termed **Lepto-medusae.**

Alcyonaria. In addition to the hydroids many anthozoans of the group Alcyonaria, also called Octocorallia, form characteristic growths on subtidal rocky areas and coral

reefs. Some of them, especially small colonial "soft corals," may also be found in the low intertidal rocks, under ledges, in crevices, etc.

Alcyonaria are all colonial, with very characteristic polyps, readily recognized by their eight tapering pinnate tentacles with

Fig. 8-21. *Aglaophenia* sp. Ostrich-plume hydroid colony. (Oregon Institute of Marine Biology photograph.)

Fig. 8-22. *Tubularia,* with sessile medusoids between the two rings of tentacles. (Oregon Institute of Marine Biology photograph.)

Fig. 8-23. Soft coral, *Alcyonium* sp., from lower littoral of Oregon. (Oregon Institute of Marine Biology photograph.)

Fig. 8-24. Gorgonian, small portion showing a few side branches of the colony, with polyps extended so that the eight pinnate tentacles are visible. (M. W. Williams photograph.)

pinnules. Anatomically they are also a homogeneous group, the polyps each possessing eight complete septa with the longitudinal muscles located on the sulcal side of each septum (the side of the animal bearing the siphonoglyph) and with the gonads developing on six of the septa (all but the sulcal pair).

The alcyonarian colony is composed of a fleshy **coenenchyma** representing the thickened cellular mesogloea, containing amoebocytes of various kinds and secreted skeletal elements. **Solenia,** or endodermal tubules, continuous with the gastrodermis of the polyps, run through it, forming a continuous gastric network throughout the coenenchyma, connecting all the polyps. All the nematocysts are small, simple atrichous isorhiza (simple tubular types without spines).

The order Gorgonacea in particular, which includes gorgonians (sea whips, sea fans, etc.), includes some of the most colorful and characteristic animals of submarine gardens. The colony possesses an axial horny or calcareous skeleton and grows as bushy or fanlike growths. Sea fans grow in relatively quiet water at depths and locations where they are not subject to excessive surge or turbulence from wave action. They develop with the plane of growth of the colony at right angles to the direction of prevailing water movement so that the water strains through the fan, giving its hundreds of polyps maximal chance to capture small plankton organisms.

The genus *Corallium* of the suborder **Scleraxonia** includes the red corals, or "precious coral" of commerce. Another colorful member of the Alcyonaria, found in warm waters, is the organ-pipe coral, a rather aberrant member of the order **Stolonifera,** which also forms a red calcareous skeleton colored by iron salts incorporated in it and somewhat resembling tier upon tier of organ pipes.

Tube worms. An amazing variety of polychaete worms have adopted the sessile, tubicolous habit, together with elaborate specializations of the body enabling them to gather plankton directly from the water or to obtain particles of organic detritus from the neighborhood of their tube. Among those found on hard substrates, perhaps the most characteristic are the fan worms, families **Serpulidae** and **Sabellidae,** and certain members of the **Cirratulidae, Terebellidae,** and others, which also have many members characteristic of softer substrates. Many terebellids and others actually characteristic of softer substrates are commonly found under rocks on sandy mud or in crevices in rocky areas where detritus collects.

Many serpulids construct calcareous tubes, ranging in size from a few millimeters to several inches in length, strongly cemented to the rock surface. The anterior end of the worm bears two large groups of stiff food-gathering tentacles, the **radioles,** which, when extended, form a beautiful, often highly colored funnel-shaped crown. The cilia on these tentacles create currents that bring food particles to the mouth. If alarmed, serpulids instantly snap back into their tubes, closing the orifice with a specially modified group of tentacles that form an **operculum.** Serpulids possess an anterior collar and a thoracic membrane formed by fusion of dorsal and ventral cirri of some of the thoracic segments and used in molding the tube as calcium carbonate is secreted from glands beneath the collar. Serpulids and other worms of this type, which have a rapid withdrawal reaction in the presence of danger, have giant nerve fibers running through the nerve chain without synaptic interruption at each segmental ganglion.

Among the most familiar and ubiquitous of the serpulids are members of the complex usually called *Spirorbis*. These are very small worms that build flat spiral-coiled tubes encrusting on seaweeds, shells, rocks, pipes, etc. These white tubes, forming little coils only 2 or 3 mm. across, may be so numerous as to practically cover the surfaces on which they have settled and may become a nuisance in saltwater plumbing systems. The larvae tend to settle gregariously and to prefer to settle near adults of their own species. This may be an advantage in that it ensures selection of a location where conditions are suitable for

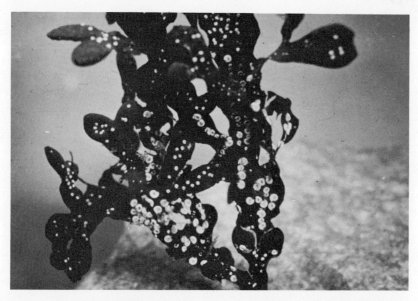

Fig. 8-25. *Spirorbis,* minute serpulid tube worms. (Oregon Institute of Marine Biology photograph.)

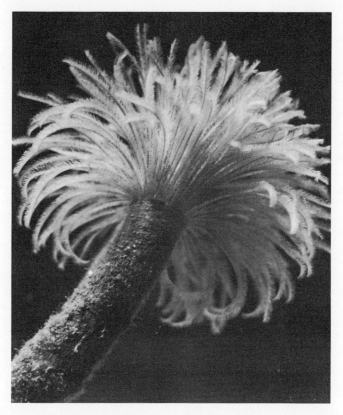

Fig. 8-26. *Eudistylia,* a large sabellid tube worm, sometimes called the feather-duster worm. May attain a length of 2 feet. (M. W. Williams photograph.)

survival as well as for attachment, since the adults have survived there. Some authorities maintain that there are sufficient differences between those species that form sinistrally coiled tubes and those with dextrally coiled tubes to warrant splitting the group into two genera—*Laeospira* and *Dexiospira.*

Among the most spectacular tube worms are various large species of Sabellidae, such as *Eudistylia,* some of which construct parchmentlike tubes more than a foot long, usually in somewhat protected places, such as the undersides of docks and on pilings in harbors, but sometimes clumps of them can also be found on intertidal rocks. These so-called plume worms, or feather-duster worms, have a big cluster of colorful tentacular radioles at the anterior end but no operculum, and the body differs from that of a serpulid, among other ways, in having a longitudinal ciliated "fecal groove," which carries feces deposited in the back of the tube forward and crosses and interrupts the lines of thoracic notosetae in a curious manner. In some sabellids, such as *Schizobranchia,* the tentacular radioles may be branched several times. Many serpulids and sabellids bear eyes either at the base of, or at intervals along, the radioles.

The family Cirratulidae comprises worms in which anterior segments bear long, filamentous outgrowths continuing to arise, in many of them, along the sides for some distance back from the anterior end. The body is not clearly divided into two distinct regions. One genus, *Dodecaceria,* contains rather small, blackish worms with only a few filamentous processes. *Dodecaceria* constructs hard tubes of calcium carbonate. Through asexual reproduction large colonies may be built up, producing thick, firmly attached, rocklike masses comprised of the agglomerated tubes of hundreds or thousands of the worms and sometimes termed "worm rock." *Dodecaceria* is common in the lower littoral zone in rocky areas along the western coast of North America, as are also various larger cirratulids that do not construct calcareous tubes.

The lophophore bearers

BRYOZOA. The Bryozoa, or moss animals (**Ectoprocta**), comprise a group of small colonial animals attached to rocks, shells, seaweeds, etc. Large bryozoan colonies may contain thousands of individual zooids produced asexually by budding. Each zooid is housed in a boxlike or tubular zooecium of hardened, often calcified cuticle secreted by the epidermal cells. The body cavity is extensive and lined by a peritoneum. The lophophore is a circular, crescentic, or double spirally coiled ridge, bearing hollow ciliated tentacles that surround the mouth. The gut is a broad ciliated U-tube, looped in such a way that the anus lies close to the mouth, outside the lophophore. A muscular band, the funiculus, runs from the stomach, near the curve of the looped gut, to the aboral end of the zooecium, enabling the bryozoan to retract the lophophore region into the zooecium when alarmed or during periods of exposure at low tide. Most bryozoans are hermaphroditic, the testes developing from peritoneum lining the funiculus and the ovaries developing from peritoneum lining the lateral walls of the zooecium. Ova and sperm are shed into the coelom, and development of the larvae may proceed there. In those forms in which the zooecium is hard and imperforate, the body wall is reflected back under the anterior edge of the wall of the zooecium, forming a saclike hydrostatic organ, the compensatorium, from which water is expelled upon retraction of the zooid, making room for it in the zooecium.

The form, patterns, and color of bryozoan colonies vary greatly. Many species form thin flat encrusting growths closely following the contours of the substrate and tightly adherent to it. Others form fleshy thickened or branching colonies, and some form erect bushy dendritic colonies. Some bryozoan colonies, such as *Flustrella,* form fleshy growths commonly mistaken for brown algae. Others somewhat resemble certain corals.

Approximately 4000 recent and 15,000 fossil species of bryozoans have been described. Recent marine bryozoans fall into the subclass Gymnolaemata, in which the lophophore is more or less circular. Freshwater species are classed as subclass Phylactolaemata. They possess a horseshoe-shaped oval or double-spiral lophophore and an epistome dorsal to the mouth, give rise to statoblasts, and display other features clearly distinctive of their group.

Recent Gymnolaemata are commonly divided into three orders. In the order **Ctenostomata** the zooecium is not calcified. The colony is gelatinous. The operculum has been described as composed of a fringe of setae forming a thin, membranous operculum that closes the orifice upon withdrawal. Actually, there is a membranous collar that folds in definite creases. Often the creases, but not the membrane between them, are visible, giving rise to the illusion of a fringe of setae. No avicularia or vibracula are formed. A common example of this group is *Alcyonidium*. The order **Cyclostomata** includes forms such as *Crisia*, with a tubelike calcareous zooecium having the wide circular orifice at the end of the tube and no avicularia or operculum. The order **Cheilostomata**, by far the largest order, contains forms with a boxlike, calcareous skeleton with operculum. Usually the zooids are differentiated or polymorphic, such modified zooids as **avicularia, vibracula,** and **oecia** commonly being present. Avicularia are modified zooids commonly resembling a bird's head on a stalk and seem to play a role in defense of the colony and keeping it clean. Vibracula are movable, sticklike structures that probably play a defensive role. Oecia are specially modified zooids, usually with a larger zooecium than the others, containing egg cells. Two suborders are recognized—those with somewhat elastic zooecia and no compensatorium, forming the suborder **Anasca,** of which the erect colonies of *Bugula* are the most familiar; and those with rigid zooecia and compensatorium, forming the suborder **Ascophora.** The first zooid of a colony, which is sometimes distinctive in structure and position in the colony, has been termed the **ancestrula.**

The planktonic larva developed from fertilized eggs is a peculiar flattened triangular **cyphonautes** larva.

Two extinct orders of bryozoans, the **Trepostomata** and **Cryptostomata,** seem to be limited to Paleozoic times, although some trepostomes of questionable identity have been reported from the Triassic. The Cryptostomata appeared in the Ordovician, were most abundant in the Devonian and Mississippian, and died out in the Upper Permian.

The oldest ctenostome is the peculiar threadlike *Marcusodictyon* from the Ordovician of Estonia. Cyclostomata (Stenolaemata) begin in the Ordovician, were the predominant bryozoans in the middle Mesozoic, and have since then been secondary to the Cheilostomata, which seem to have originated during the Jurassic in European seas and rapidly extended their range, being the dominant group of Ectoprocta throughout the world today.

Another group of small animals formerly included in the Bryozoa are the **Endoprocta,** commonly called nodding-heads. They are often found especially on the underside of rocks in the lower intertidal and subtidal and sometimes on mollusc shells and other hard substrates. They are not nearly as common as the ectoproctans. In the endoproctans the anus is inside the circle of tentacles; the tentacles are folded in toward the center, as in hydroids, when the zooid withdraws; and the body cavity is a pseudocoelom rather than a true coelom and is secondarily filled with parenchymatous tissue and obliterated. The zooids of a colony are connected only basally through stolons and do not have zooecia. The stalks of the zooids are muscular, and they exhibit a characteristic nodding movement, which gives them their popular name.

BRACHIOPODA. Brachiopods, or lamp shells, are small bivalved animals, most of them superficially resembling bivalved molluscs. The two valves are dorsal and ventral rather than lateral in position. They usually are unequal and taper toward the posterior end where a stout muscular peduncle attaching the shell to the rock passes out, between the two valves in some, or

Fig. 8-27. Brachiopods dredged near Santa Catalina Island. (M. W. Williams photograph.)

Fig. 8-28. Brachiopod from Coos Bay, Oregon, to show peduncular orifice of the ventral valve. (Oregon Institute of Marine Biology photograph.)

through a hole in the posterior end of the ventral valve in others. Some brachiopods cement the ventral valve to the rock as they grow and have no peduncle. The mantle cavity between the valves is largely occupied by the large double-spiraled lophophore, which is sometimes supported by calcareous extensions of shell from the dorsal valve. Each lophophore bears a set of ciliated tentacles alongside a ciliated groove in which small particles are swept back to the mouth. Larger particles are rejected. The mouth is simple, without special jaws, lips, or palps. It leads into a short esophagus, stomach, and intestine. Saclike digestive glands open into the stomach. Sexes are separate. The shells are composed of calcite rather than the aragonite usually found in mollusc shells.

There are two classes of brachiopods. The **Articulata** (= Testicardines), by far the largest class, possess hinged valves. The peduncle passes out through the ventral valve or in some cases is absent. The lophophore bears an internal calcareous skeletal support, and the digestive tract is blind posteriorly, with no anus. The **Inarticulata** (= Ecardines) have no shell hinge, the shells fitting together more like two saucers. The peduncle passes on between the shells, and an anus is present posteriorly between the shells. The best-known members of the Inarticulata are the Lingulidae, which are rather large brachiopods with a long contractile peduncle. They are found in shallow, muddy bottoms rather than on rocky substrate. They have relatively soft, horny, somewhat rectangular shells fringed with long stiff setae, and very vascular.

Approximately 1400 genera and 25,000 species of brachiopods have been described, of which the living or recent forms constitute about 75 genera and 250 species. Authorities are not yet in general agreement as to the classification within the group into orders, suborders, families, etc., especially within the huge group of articulate brachiopods. Pending better clarification of relationships within the group, Cooper's nonphylogenetic arrangement into **impunctate, pseudopunctate,** and **punctate** Articulata is a convenient and more readily determinable grouping than the various orders and suborders that have been proposed.

In the punctate brachiopods the inner layers of the shell are pierced by many tiny holes into which minute projections of the mantle extend, increasing the mantle surface—the main respiratory organ. Pseudopunctate brachiopods have minute rods of calcite, called **taleolae,** projecting into and often beyond the inner surface. Exfoliation of the shell leaves pits, or pseudopunctae, where these rods have been lost. The impunctate species have neither punctae nor taleolae.

Brachiopods have a long fossil history and have been much used in stratigraphical correlation.

The third group of lophophore bearers, the Phoronida, are more characteristic of soft substrates in estuaries, etc. than of rocky areas, and will be considered in connection with those environments, although some species do construct tubes and live in rocky areas, especially in cracks and crevices.

Algal-film grazers

A second major source of food in the intertidal and subtidal rocky substrate communities is the film of growth and of molecular and colloidal organic matter that coats the surface of all rocks and other objects exposed to marine waters. This is a mixture of diatoms, minute algae, bacteria, early stages in the growth of larger algae, and minute animals, together with absorbed and settled particles and molecules of organic matter—collectively termed the "algal film."

Two groups of molluscs, the chitons and the gastropods, have evolved as algal-film grazers. Both groups are adapted for clinging to the rock surface by means of a broad flat muscular foot and slowly creeping about, scraping off the algal film with a filelike tongue, the **radula.** The tenacity with which some of them can cling is truly remarkable, enabling them to withstand the heavy surf and surge of incoming tides on exposed coasts without being swept

Fig. 8-29. A, *Tonicella lineata,* the lined chiton, a very pretty, small species common along the western coast of North America. **B,** Another small chiton from the Oregon coast, about 1 cm. long or less, probably *Nuttallina.* (Oregon Institute of Marine Biology photographs.)

away. Both groups have developed a calcareous shell for protection from predators and environmental stresses. Basically they have rather similar life cycles, involving a planktonic trochophore stage that later undergoes changes leading to the familiar mollusc form. Both are very ancient groups, going all the way back to Cambrian times. The gastropods have evolved into the largest, most varied group of molluscs, and one of the major animal groups on the earth. They have colonized most marine habitats, have evolved plankton-feeding and carnivorous groups, and have extensively colonized terrestrial and freshwater habitats as well.

The chitons, on the other hand, have been an extremely conservative group from an evolutionary standpoint. From their beginning they seem to have been primarily confined to coastal waters, many of them to intertidal or shallow subtidal waters on rocky substrate. Chitons are an example of a group of animals that appears to have become well adapted to a particular sort of ecological situation early in its evolution and to have found conditions sufficiently constant to enable it to persist through the

Fig. 8-30. Distal end of the radula of *Cryptochiton stelleri*.

ages in this situation with relatively little change.

Perhaps the difference in evolutionary history between the two groups is due in part to a difference that appears during larval life. In the case of the chitons, the primitive bilateral symmetry is retained, the chief features of their evolution apparently having been an elongation and dorsoventral flattening of the body that largely obliterated the large posterior mantle cavity, a growth of the mantle down around the rest of the body as a tough, protective girdle that comes into contact with the substrate on all sides, and the secretion of the shell as eight separate plates partly embedded in the dorsal mantle tissue, giving the body as a whole enough flexibility to fit itself to the contours of the substrate or curl up for protection when pried loose. With the body and head covered and protected by the girdle, most of the sensory functions usually associated with the head either degenerated or were transferred to the edge of the mantle, which bears innumerable minute sensory spines and which in some is adorned with variously formed sensory "hairs." Young chitons may in some species even bear light-sensitive organs, called aesthetes, at the ends of small chan-

nels in the shells on their back. There has also been, in chitons, a secondary multiplication of the ctenidia, which are found in a row on either side of the body in the pallial groove between the foot and the girdle. Water is circulated over them by ciliary action, the chiton lifting the girdle in the anteriolateral areas to admit the water and discharging it posteriorly, often through a permanent notch in the posterior edge of the girdle.

In gastropods, on the other hand, the visceral hump is not flattened but during larval life undergoes a remarkable rotation, termed **torsion**, effected by contraction of the strong right velar muscle of the larva. In some groups torsion is effected more slowly as a growth phenomenon rather than by a sudden contraction of the velar muscle. Torsion rotates the entire visceral hump 180 degrees, bringing the mantle cavity, gills, anus, and excretory and reproductive openings to the front directly over the head and reversing the direction of all organs in the affected part. This produces a twist between the foot and the visceral mass that twists the nervous system into a figure eight (the **streptoneurous** condition) and puts morphologically right structures on the left side and morphologically left structures on the right, in the affected parts. Torsion should not be confused with spiral coiling of the visceral mass, which is another phenomenon, probably an adaptation for better compacting and balancing of the visceral mass. Spiral coiling is brought about by differentials in growth rates on the two sides of the animal.

While enabling the gastropod to protect its head by withdrawing it beneath the shell-protected mantle cavity, torsion also brings several problems, in that the discharge of deoxygenated water from the gill chamber, of feces, and of gametes now occurs directly in front of the mantle opening, over the head of the gastropod, directly in the path of the animal's progression instead of at its posterior end. Also, any debris stirred up by the advancing edge of the foot on the substrate or by feeding activity is likely to enter the mantle cavity and contribute to possible fouling of the gills. Many of the features evolved by various groups of gastropods seem to be in large part adaptations to meet these problems as well as the problem of balancing the high, twisted visceral mass.

Four principal directions of adaptation to the various problems created by torsion have been successfully explored by different groups of gastropods.

Perhaps the most primitive adaptation, found in the pleurotomariids today and in certain very ancient gastropod fossils, is a groove, or notch, in the leading edge of the shell, so that water entering the mantle cavity from the front and sides can escape dorsally over the back rather than directly in the path of the animal. Refinements on this plan consists of closing the leading edge of the notch with subsequent growth of the shell so that it becomes a dorsal hole leading from the mantle cavity. The abalones and various kinds of keyhole limpets, or volcano shells, represent culminations of this line of adaptation.

The most successful, widely found adaptation is a change in direction of the flow of water through the mantle chamber, so that the water comes in from the left side, passes through the mantle chamber, and is discharged on the right. With this basic

change, various secondary modifications have been evolved in various groups, making it more efficient, such as shifting the anus and excretory and reproductive openings to the right, reduction or loss of the right gill with compensating development of the left gill, and development of a portion of the mantle edge as a siphon, which better directs the water flow into the mantle chamber. In burrowing forms or forms that live in soft substrate, such as olive snails, this siphon may become a rather long, wormlike snorkel that can be extended up into the water even though the mollusc is largely or wholly buried. In siphonate gastropods, many of which live in muddy or turbid waters, the gill axis is usually fused to the top of the mantle chamber and has only a single row of filaments (**monopectinate** condition), which is less easily fouled than the large, free bipectinate gills of more primitive gastropods.

Terrestrial gastropods, the Pulmonata, evolved from forms with this left-to-right adaptation of the mantle cavity and its contents. Adaptation to terrestrial life involved loss of gills altogether, wrinkling and vascularization of the walls of the mantle cavity, converting it into a type of lung, and restriction of the opening to prevent dessication. The anal and reproductive openings are usually on the right side and wholly outside the mantle cavity. The life cycle also required modification, with elimination of planktonic larvae.

Finally, in the opisthobranch gastropods are forms that meet the problems of torsion by simply undoing it through subsequent detorsion. In most opisthobranchs the shell is reduced, sometimes enveloped in mantle tissue, or wholly absent. The sea slugs, or nudibranchs, not only undergo detorsion but have altogether lost the shell, mantle cavity, and gills and have developed secondary branchial structures, not homologous with the lost ctenidia, on the dorsal surface of the body. In the dorid type of nudibranchs these structures consist of a circle of retractile "gills" around the anus, which is in a middorsal position near the posterior part of the body. The aeolid type of nudibranchs, on the other hand, have no such anal gills but develop a variety of colorful nonretractile cirri, or processes, from the dorsal surface of the body. In many of them diverticula from the gut may extend up into these processes.

The Gastropoda are the largest, most varied group of molluscs. In addition to algal-film grazers, numerous groups have evolved other feeding habits, ranging all the way from plankton feeding to predatory hunting. The slipper shells (*Crepidula* spp.) have adopted the ciliary-mucous method of plankton feeding, using the gill as a collecting organ much in the manner of bivalves. The Vermetidae form tubular shells firmly fixed to the substrate and looking more like worm tubes than gastropod shells. They extend the gill tips and sweep the water for minute plankton organisms in much the way that a barnacle sweeps the water with its cirri—an adaptation unique among gastropods.

In carnivorous gastropods the radula may be modified as a poison dart, as in cone shells, or as a drill for making holes in other shelled animals such as oysters, mussels, or barnacles, as in the oyster drills, whelks, etc. Others, with less change of structure and habit, have taken to grazing upon sessile animals such as bryozoans, sponges, hydroids, and gorgonians instead of the algal film. The many beautiful nudibranchs are an example of a whole major group that has changed to the carnivorous habit without making a radical change in the method of feeding, but even here we find some surprising individual developments. For example, the small nudibranch *Melibe* has a large, thin frontal membrane dorsal to and partly surrounding the mouth. It uses this membrane in the manner of a butterfly net, spreading it out in front of the head and capturing passing copepods or other small zooplankters.

Perhaps the great evolutionary success and diversity attained by the gastropods, as compared with chitons, are bound up with the problems created by torsion and by their high, unwieldy visceral mass. These continued challenges may have led to selection of diverse lines that met them and improved upon the imperfect basic pattern in a variety of ways. The chitons are, in a way, the victims of their own early success, having become so well adapted for a particular mode of life in a particular sort of habitat, early in the development of the group, that they have never been forced to improve or change. There has been enough similarity in shoreline conditions throughout the ages to enable them to persist as a small, moderately successful group.

Representative intertidal gastropods. In the upper fringe of the intertidal zone, in and extending a little above the barnacle zone, the rocks may be peppered with small periwinkles, snails of the genus *Littorina*. These show an interesting evolutionary series of species adapted to different degrees of exposure, ranging from species that occur only in the lower intertidal, where they are covered with water most of the time, to forms occurring so high on the rocks that only the highest spring tides and the spray from the waves reach them. The periwinkles seem to be midway in an evolutionary course such as that which in earlier days brought forth the present terrestrial snails out of the sea and onto the land. The three common species along the New England coast illustrate different stages of this progression. The smooth periwinkle (*L. obtusata*) is found in the lower intertidal zone, where it is submerged most of the time. It can tolerate only relatively short periods of exposure. The mantle cavity and gill are characteristic of marine gastropods generally. The eggs are laid in small gelatinous masses adherent to fucoid seaweeds, and the young enter the air bladders of these plants for protection. The adults remain among wet seaweeds at low tide. The common periwinkle (*L. littorea*) lives higher on the rocks, where only the high tides cover it, can stand relatively long periods of exposure, and has the wall of the mantle cavity more wrinkled and vascularized and the gill somewhat reduced. The rough periwinkle (*L. saxatilis*) is almost a land snail. The mantle cavity is highly vascularized and functions much like a lung. Continued submersion is fatal to it. It can tolerate a month of exposure to the air before becoming too desiccated.

Fig. 8-31. Periwinkles, *Littorina planaxis,* nestled in crevices or small pockets in rocks between high tides. (M. W. Williams photograph.)

Furthermore, it has freed itself of the reproductive tie to the ocean by becoming viviparous. It is most active during fortnightly high spring tides and during the interval between these tides becomes more sluggish and slightly desiccated. Curiously enough, the rhythm of the tides is somehow deeply impressed in its being. If kept in the laboratory under constant conditions, its activity pattern continues to reflect the rise and fall of the tides over its native rocks for several months. A similar series of species occurs along the Pacific coast and the coasts of Europe.

I have kept several specimens of a related periwinkle *(Tectarius muricatus)* from the West Indies on my desk top as a paperweight for nine months and had them promptly revive and start crawling about when put into seawater again. According to Moore, they have survived in a dry state for at least seventeen months.

The second group of gastropods particularly characteristic of the intertidal zone, and with some species extending into the highest parts of the intertidal, are the limpets, families **Acmaeidae** and **Patellidae.** In limpets the shell is conical rather than spiral, some species having an appearance that has given them the name "Chinaman's-hat." The shell apex is usually closer to the anterior end of the shell, so that the anterior slope is usually somewhat shorter and steeper than the posterior slope, and the apex itself may be inclined anteriorly. In the Acmaeidae there is a large bipectinate gill in the mantle chamber. The Patellidae have lost the gill and developed a number of platelike pallial branchiae, or pseudogills in the mantle groove lateral to the foot. Acmaeids are of worldwide distribution and are the most common limpets in most of the world. Patellids are, for the most part, Old World forms, found along the coasts of Europe and Africa.

Limpets move about and feed when covered with water or while the rocks are still wet. Between tides they tend to cling tightly

Fig. 8-32. Limpets, *Acmaea limatula,* common on intertidal rocks, at a lower level in the intertidal than *A. digitalis.* (Oregon Institute of Marine Biology photograph.)

to the rock, preventing desiccation and predation by terrestrial animals. Many limpets seem to adopt a particular spot on the rock to which they return after feeding excursions. This spot may be worn so that the resident limpet just fits snugly into it. This "homing instinct" has been investigated a number of times ever since Aristotle first reported the habit, but the precise mechanism, the physical basis for the homing ability, and the organs and senses involved are still uncertain.

Constant scraping on rock surfaces subjects the radular teeth to considerable wear. As they are worn down or lost, they are replaced by new portions of the radula pushed up from behind. In both limpets and littorinids, the radula is very long relative to the size of the animal. The continual scraping, year in and year out over the centuries, of countless millions of small radulae also has an appreciable wearing effect on the rock, cutting it away grain by grain. In one tide pool carefully observed for sixteen years, periwinkles lowered the floor about ⅜ inch.

Among the species of limpets living in the lower parts of the intertidal zone, some species have taken on the habit of scraping the cortical cells of seaweeds, rather than the algal film on rocks. Some of them have become so specialized that the foot and shell are specifically shaped to fit the particular plants on which they live. On the Pacific coast of North America the very common large brown alga *Egregia* often has characteristic holes in the beltlike stipe made by *Acmaea insessa,* or it may have numerous depressions, each containing one limpet, where they have not eaten entirely through the stipe. The most highly modified of these plant-eating limpets are probably species such as *Acmaea depicta* and *Acmaea paleacea,* which live on eelgrass and have become shaped to fit the narrow blades of this plant.

In the lower intertidal and subtidal zones are also found other, more primitive limpet-like gastropods, the keyhole limpets, in which both gills are retained and the excurrent water from the mantle cavity is discharged through a hole at the apex of the shell. The resemblance of some of these conical shells to volcanoes has given rise to the popular name "volcano shells" for some of them.

The group of gastropods to which these belong, the **Zygobranchiata,** are the most primitive of living Gastropoda. Others in this group are abalones, which have a series of holes in the shell for the excurrent current, and the deeper-water pleurotomariids, which have a slit in the leading edge of the shell, rather than a hole, for this purpose.

As we go into the lower intertidal and subtidal zones, out onto the shelf, numerous other groups of gastropods are found, in addition to those which have been mentioned.

The algal film itself is better developed and more lush where it is submerged most or all of the time. It supports a considerable fauna of microcrustaceans, nematodes, flatworms, small annelids, protozoans, and other small animals that live in or derive their sustenance from it. Many larger scavengers and predators such as crabs and small fish also pick at it and derive a significant part of their nourishment from elements in the algal film. The very young of many fish, and other animals that later change their diet, may also pick small organisms from algal film as well as from small plankton organisms.

In tropical coral reef areas certain of the coral fish such as filefish and triggerfish have broad, chisellike plates, something like broad, curved incisor teeth, at the front of the jaws, with which they can scrape the surfaces of the calcareous rocks and coral heads, scraping off the algal film and leaving recognizable scars on the rock face. These fish have been shown by Bakus (1964, 1966, 1969) to be an important element in limiting the establishment and growth of larger algae in these areas and to exert a marked effect upon the evolution of the shallow-water benthic invertebrate fauna.

Herbivores of the rocky coast

The oceans have relatively few animals that browse on larger plants as terrestrial herbivores do because of the paucity of such plants in the sea, although, as mentioned, some algal-film grazers, especially certain limpets, have taken to scraping the stipes of larger marine plants instead of the film on rock surfaces. A few

fish browse on certain species of algae. Some of the seaweed-eating sea urchins are perhaps as close as any. Sea urchins eat a variety of organic materials, some are largely carnivorous, some scavenge almost anything they come across, and others eat, for the most part, pieces of the larger seaweeds. Such are the common purple sea urchins of the Pacific coast, *Strongylocentrotus purpuratus,* and the larger red urchin, *S. franciscanus.* Upon opening them, one commonly finds the capacious intestine full of fragments of the larger algae. It has recently been found that these urchins sometimes become so numerous subtidally that they clean out most of the available seaweed fragments that might sustain them, whereupon great numbers of them leave their holes in the rocks and crevices and go foraging. Such groups of foraging sea urchins may cut off growing kelp plants near the holdfast, with the result that the kelp is washed up on shore in windrows and the kelp bed denuded. There may follow a period when, without enough food, the sea urchins either leave the area or die off until the numbers are so reduced that young seaweeds can reestablish themselves, eventually building back the kelp beds and establishing a temporary equilibrium between the available seaweeds and the population of sea urchins. In the area affected by sewage outfalls in and near San Diego Bay, it was found that the sea urchins did not disappear following denudation of the kelp beds but persisted, apparently obtaining some nourishment from dissolved organic materials in the water, perhaps through the tube feet or respiratory structures. In any case the kelp was prevented from reestablishing itself until enough lime was dumped into the area of the former kelp beds to kill off the urchins, whereupon the kelp soon reestablished itself in sufficient quantity for the profitable kelp industry to resume operations there.

Carnivores and scavengers

The great abundance of animal life in the rocky intertidal and subtidal areas, made possible by the supply of food present as plankton and algal film as well as the

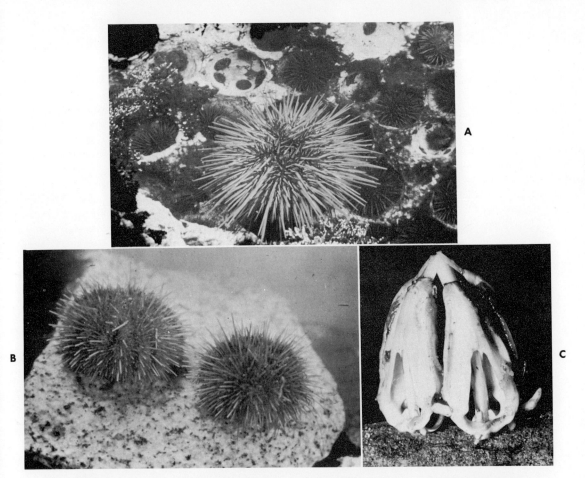

Fig. 8-33. A, The two common intertidal and subtidal West Coast sea urchins: *Strongylocentrotus franciscanus,* the large one in the center, and *Strongylocentrotus purpuratus.* **B,** *Strongylocentrotus dröbachiensis,* the green sea urchin. **C,** Aristotle's lantern, the mouth parts of a sea urchin. Perhaps this is the most complicated mouth of any animal. Five long, chisellike teeth are maneuvered by a complex of over sixty-five calcareous and muscular parts. (Oregon Institute of Marine Biology photographs.)

fact that much of this life is sessile or very slow moving, has left its mark upon the carnivorous and scavenging elements of the community. Here it is possible for a large proportion of the carnivores to browse, much in the manner of herbivores on land. Also, entrapment and ambush techniques become much more feasible methods of capturing prey. A large portion of the more numerous and conspicuous carnivores are either nearly sessile, as are the sea anemones, or slow moving like the starfish, nudibranchs, and whelks, moving about in a leisurely manner and having only to find and select their prey from among the lush growths of sessile or nearly sessile animals. Among the groups that are traditionally more active, numerous species also resort to somewhat similar modes of hunting, waiting at the mouths of burrows or in hiding places for their prey to come close enough to be taken with a swift lunge or thrust.

Among the coelenterates there is a whole gamut of forms, from the bushy hydroid colonies, the polyps of which are so small as to capture only smaller elements of the zooplankton, to the huge sea anemones such as *Discosoma* and *Actinodendron* of the

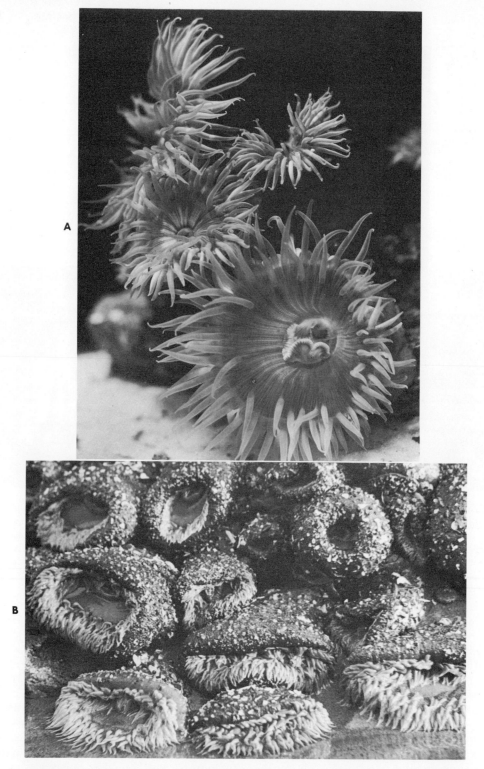

Fig. 8-34. A, The common large, green intertidal sea anemone of the West Coast, *Anthopleura xanthogrammica*. **B,** The even more abundant aggregate sea anemone of the same region, occurring higher in the intertidal, *Anthopleura elegantissima*. (M. W. Williams photographs.)

coral reefs, large species of *Metridium,* and others found offshore or sometimes in bays in cooler waters as well, which are capable of capturing and eating rather large fish and crustaceans.

Along the Pacific coast of the United States and Canada the beautiful large green anemone *Anthopleura xanthogrammica,* colored by symbiotic algae in its tissue, is characteristic of rocky pools and crevices everywhere that the rocks are in exposed localities but afford some protective pools, channels, and ledges. The related *Anthopleura elegantissima* is found somewhat higher in the intertidal and also often in more protected waters, commonly in dense masses formed largely by a peculiar form

of asexual reproduction in which the anemone slowly creeps in opposite directions until it literally pulls itself in half, each half rounding up, healing, and regenerating to form a new individual. This species also has the habit of affixing sand grains, small shell fragments, etc. to the tubercles that line its column. Another common anemone of the rocky shores of this region is the smaller *Epiactis prolifera,* which occurs in an arc capping the North Pacific from Japan to California. It is surprisingly variable in color and pattern and in summer usually has one or more younger anemones of the same species affixed near the base of the column of the parent. The eggs are apparently discharged from the female, and

Fig. 8-35. A mat of the aggregate anemone exposed by low tide. Mats of this sort are produced by a peculiar asexual reproduction. An anemone simply creeps on its pedal disc in opposite directions at the same time, literally pulling itself in half, each portion healing and regenerating to produce a new individual. Sexual reproduction, as in other coelenterates, results in the liberation of free-swimming planula larvae, which seek new locations and start new colonies of the anemone. (M. W. Williams photograph.)

Fig. 8-36. *Metridium senile,* a large anemone with the entire oral disc covered with tentacles. Species of *Metridium* occur along both the Atlantic and Pacific coasts of North America. Along the New England coast, *Metridium* occupies much the same ecological position as does *Anthopleura* on the West Coast. (M. W. Williams photograph.)

some stick to the column mucus. The young remain there, looking like asexually produced buds, until ready to fend for themselves. Species of the genus *Tealia,* with red or scarlet body and sometimes variegated with green patches as in the common West Coast *Telia crassicornis,* are among the more striking species fairly common in the lower intertidal and subtidal. The plumose anemone *Metridium dianthus* occupies a somewhat similar position on the north-eastern coast of the United States to that of the large green anemone of the West Coast. Metridia are commonly pure white, but lavender and rose-colored specimens also occur. When well expanded on their tall columns in a large aquarium, a group of large metridia makes a striking display.

In spite of the delicate flowerlike appearance and exquisite beauty of expanded sea anemones, they are formidable predators capable of paralyzing and devouring whole

Fig. 8-37. A small, colorful anemone, *Epiactis prolifera*, common from Northern California to Japan. The young, when liberated, attach to the column of the mother, where they remain until about one-fifth grown, when they creep off by themselves. (Oregon Institute of Marine Biology photograph.)

fish, crustaceans, or other small animals that brush against their nematocyst-armed tentacles.

Perhaps the animals that more than any others have come to symbolize the seacoast are the starfish and sea urchins, members of the only major phylum of animals wholly confined to marine waters. Although related to the lines of animals from which the chordates arose, the development of a secondary radial symmetry, the unique water vascular system, the dermal calcareous skeleton, and the extensive and complicated coelom set them apart so sharply from all other animals that their affinities were for a long time obscure. Early zoologists classed them together with Coelenterata and Ctenophora in a phylum termed Radiata.

A few species are characteristic of the lower intertidal zones in rocky areas. On the eastern coast of North America two species of *Asterias*—the common starfish, *A. vulgaris,* and the northern starfish, *A. forbesi*—are the most abundant and familiar. Both occur in several color variations. Where they occur together, as in the Massachusetts area, the easiest, most reliable way to distinguish them is by the

color of the madreporite plate—bright orange-red in *A. forbesi,* and light yellow in *A. vulgaris.* On the western coast of North America the same niche is occupied by the colorful, abundant, and widely distributed *Pisaster ochraceus* (Fig. 8-12), which likewise comes in several colors—yellow, orange, reddish, purplish, and chocolate. Sometimes all the colors are found together in the population at a single locality.

These starfish are voracious predators, subsisting on a diet chiefly of bivalve molluscs, especially mussels, but also eating other animals that either cannot get out of their way or fail to do so.

Numerous animals of the intertidal zone exhibit great alarm and make every effort to avoid a starfish that moves near them. They are warned by a chemical exuded into the water by the starfish. Even some animals ordinarily regarded as sessile will move out of their way. For example, in Puget Sound, sea anemones of the genera *Stomphia* and *Actinostola* will detach their pedal discs from the substrate, begin to writhe, and remove themselves from the immediate vicinity by a series of contortions somersaulting them through the

Fig. 8-38. A "swimming anemone" from Puget Sound. Anemones of the genus *Actinostola,* when excited by chemical substances (t) given off by echinoderms or, in this case, by species of *Stomphia,* become very excited, at first withdraw, then begin to expand and writhe, and eventually detach their grip on the substrate and somersault through the water by a series of writhing movements to land some distance from the offending stimulus. (Photographs by D. M. Ross. In Sutton, L. 1967. Science **155**:1419-1420. Copyright, 1967, by the American Association for the Advancement of Science.)

Fig. 8-39. The large, twenty-ray starfish, *Pycnopodia helianthoides,* of the West Coast, turned oral side upward to show the many tube feet. (M. W. Williams photograph.)

Fig. 8-40. The sun star, *Solaster dawsoni,* a West Coast species. (Oregon Institute of Marine Biology photograph.)

water when a starfish is held near them. *Actinostola* also reacts to *Stomphia* if the latter is held close to it.

One of the most spectacular and voracious starfish of the West Coast is the large twenty-ray starfish, *Pycnopodia helianthoides,* found in the lowest parts of the intertidal zone and the upper subtidal areas. This species is so large and soft bodied that it makes a rather sad, droopy specimen when removed from the water and is a very poor dried specimen, but in its natural habitat few animals can rival it in beauty and interest. Its diet is also surprising. The writer has found large, lumpy-looking specimens which, upon examination, proved to have two or even three fully grown sea urchins *(Strongylocentrotus purpuratus)* in the stomach. How such a soft-bodied animal with a delicate thin-walled stomach can accommodate itself to such an unlikely, spiny diet is hard to understand.

Among the more active predators, the most conspicuous groups are the fish, cephalopods, errant polychaete worms, nemertean worms, polyclad flatworms, and various groups of larger crustaceans. These groups are all best developed in the subtidal shelf waters, but characteristic representatives occur also in the intertidal or may invade this zone during high tides.

Various species of tide pool cottids are found in pools and channels well up into the middle and upper parts of the intertidal zone. They are commonly patterned and colored to match the variegated coralline algae and rock surfaces among which they live. Habitually they spend much time resting on the bottom of the pools or along the sides, using the pectoral fins almost as limbs, and swimming and gliding in short, jump-like spurts from one resting place to another. They have large heads and mouths and are voracious eaters. Other fish commonly found in the intertidal include the young of several species found offshore when mature. Eel blennies, with elongate, slippery, almost snakelike bodies, are common in the lower parts of the intertidal under rocks or protecting seaweed as are also clingfish, which cling to the under-sides of rocks with their peculiar modified suckerlike ventral fins. Some species of *Octopus,* such as *Octopus bimaculoides* of the southern and Lower California coast, characteristically occur in the rocky intertidal, but most species are subtidal, and only the fringe of the population occurs in the lower parts of the intertidal.

Most of the crabs, as well as hermit crabs, amphipods, lobsters, spiny lobsters, shrimp, and other crustaceans of the rocky shelf communities, are predators, scavengers, or both. In the deeper, dimly lit or nearly dark portions of the shelf, the fish and crustaceans characteristically take on red hues and patterns.

The really deep, bathypelagic and abyssal, rocky areas are the least well known of all the marine habitats because of the difficulties of sampling and making observations. Most of the level or nearly level regions of the sea floor are soft-bottomed and covered with sediments. For the most part, the exposed, rocky areas either are the steep sides of submarine canyons, mountains, or escarpments or are relatively recently formed lava flows such as are found along some of the midocean ridges. Below the euphotic zone, there are, of course, no seaweeds. The attached growths are entirely animal in character—gorgonians, hydroids, sponges, bryozoans, brachiopods, crinoids, etc. Most of the groups of animals represented in the subtidal rocky shelf areas have representatives extending down into deep regions, but the populations are sparser because of the diminished quantity of planktonic food and the lack of a producing algal film.

The larvae of sessile rock-dwelling animals may drift for hundreds or even thousands of miles in the plankton, ready to settle on any favorable substrate that presents itself. This is dramatically shown by the settlement of relatively isolated rocky outcroppings, or artificial reefs, in regions dominated by otherwise unfavorable substrate.

The eastern coast of the United States from southern New England to the tip of Florida is for the most part sand beaches, and a sandy bottom extends out over the continental shelf. However, off the coast of the Carolinas, here and there, are broken chains of submerged reefs and ledges—outcroppings known as "black rocks" to fishermen because blackfish congregate there. They are sometimes also called coral reefs, although they are not composed of coral but rather of relatively soft, fine grained marl of the Miocene age.

These sunken reefs support forests of *Sargassum* and are in places covered by masses of calcareous tubes of tube worms and tube-building gastropods, or gardens of sponges, tunicates, gorgonians, hydroids, and bryozoans. Where sufficiently free from encrusting coralline algae or tube worms, they are riddled with the burrows of piddocks and date-nut shells.

When storms create deep turbulence reaching down to these submerged reefs, many of these rock-living animals, foreign to the prevailing sandy bottoms of the region, are dislodged and cast ashore, giving mute evidence of the unseen offshore reefs.

Because of the greater difficulty in sampling and observing and because of the greater inherent variety of ecological niches on rocky substrates and their close approximation and interdigitation, the communities of organisms occupying them, except for those in the intertidal zone, long remained much less well defined and understood than those of soft substrates. Only in recent years have adequate techniques for sampling and direct observation become available. In order to understand the communities occupying such substrates, rather detailed data are necessary with collections. These data should include depth, amount, and duration of light reaching the actual spot where the specimen was taken; temperature; whether on seaward, sheltered, or lateral rock face; water movement; associated organisms; and the nature of the rock (granitic, conglomerate, calcareous, etc.). With regard to position, eight categories are commonly recognized: (1) horizontal, with less than a 20-degree slope, (2) moderately inclined, with a 20- to 60-degree slope, (3) sharply inclined, with a 60- to 85-degree slope, (4) vertical, with an 85-to 90-degree slope, (5) slightly overhanging, (6) strongly overhanging, as the roofs of small caves, (7) in crevices, and (8) on the underside of individual rocks. The presence, type, and abundance of seaweeds and their relation to the collected specimens should also be noted.

Benthic communities of soft substrates

Formation and character of soft substrates along shores

Materials eroded from the continental land masses are continually being transported by water, sorted, worn down, partially dissolved, and eventually brought to the sea. Here, some of it forms mudflats, sandbanks, beaches, deltas, and other formations at the edge of the sea. Some of the finer particles may be carried far out to sea, becoming part of the fine, particulate matter that eventually settles to the floor of the abyss. Also, tidal action, waves, and the activities of countless boring and scraping organisms, plus the ordinary erosive factors of wind, rain, frost, etc., continually wear down the rocky areas of the intertidal zone. The world's rivers transport approximately 8 billion tons of sediment to the sea each year. These sediments are deposited most rapidly at the mouths of rivers because of the slowing down of the river as it meets the sea and because of the precipitating action of the sea salts.

In addition to the visible deltas built up at their mouths, large rivers may deposit layers of underwater mud in the form of large fans extending far out to sea. In Fig. 17-1 is shown, for example, such sea-floor fans of alluvial deposits brought to the northern edge of the Indian Ocean on either side of India by the Ganges and Brahmaputra river systems on the east, and the Indus and Narbada rivers on the west. These fans extend out into the Indian Ocean the full length of the Indian subcontinent on the west and for twice this distance on the east.

Long-term changes in sea level or isostatic changes in the level of bordering lands also cause extensive marginal regions

Fig. 8-41. Oyster farm, Tomales Bay, California. Eastern oysters spread out on the flats to fatten. After several months they are gathered in baskets and marketed. (M. W. Williams photograph.)

either to rise, during which time deposits of terrestrial origin may come to overlie older marine deposits in a seaward direction (a phenomenon termed **offlap**), or to sink relative to the sea level, in which case new marine deposits may come to overlie earlier terrestrial ones in the landward direction (a phenomenon termed **onlap**).

At present North America is largely emergent, but ancient marine terraces built up during the last 100 million years, since the Lower Cretaceous, show the extent of former shallow seas on present continental areas. The terraces of the Atlantic coast reach inland to the "fall line" in the eastern Appalachians. Those from the Gulf of Mexico are far more extensive, stretching all the way up the Mississippi Valley to Cairo, Illinois, which was the high-water mark of the shallow Cretaceous seas occupying much of the central portion of the present North American continent. The present continental shelves represent submerged terraces, emergent during periods

of lowest sea level in the past. These sediment platforms formed as stratified wedges, with the thicker edge on the seaward side. At the present shoreline of Louisiana and Texas, the sedimentary platform is about 40,000 feet thick. The bottom of the present Gulf of Mexico is largely muddy because the great muddy Mississippi and Rio Grande rivers empty into the Gulf. The rivers of the Atlantic and Pacific seaboards are much clearer and freer of silt. Much of the silt they have is trapped in large estuaries (Chesapeake Bay) before getting to the ocean. Swifter bottom currents along the open sea margins and stronger tidal action keep the finest particles fairly well swept out to sea, so that the near shore deposits tend to be sandy or gravelly and the shelf is relatively clean. Since fine particles are swept farther out to sea than coarser, heavier ones, offshore deposits are usually finer than those close to shore. During onlap, if the invasion of land areas by sea is sufficiently great, an overlapping of shale on sandstone may occur. During offlap coarser, new deposits may overlap the finer, older mud laid earlier in deeper water.

The marine terrace along the Atlantic coast is only about one fourth as thick as that along the Gulf coast and is cut across by numerous steep submarine canyons, some of which may represent former river canyons. Others were probably formed by turbidity, currents, or slides.

The sediments along the edges of the ocean are far from static, particularly in the intertidal and wave-affected portions of the subtidal zones. Dry sands on the highest parts of a beach may be transported by wind. Every wave on a sandy beach moves a great deal of sand. In winter, large waves of spring tides may carry most of the sand away, depositing it as a bar a short distance offshore and lowering the level of the beach, sometimes exposing underlying rock. The smaller waves of summer tend to wash the sand back up on the beach, restoring its former level.

Most sandy beaches are composed largely of quartz and feldspar, the most abundant minerals and the hardest particles left from weathering of rocks on the mountains. In some areas coralline algae or calcareous remains of coral reef organisms may predominate, or darker minerals from volcanoes may make the sand black.

Since the prevailing direction of the waves is seldom precisely parallel to the beach, sand is steadily moved in the direction of prevailing wave action as an **alongshore current** (commonly termed **longshore current**). If this alongshore current comes to the head of a submarine canyon, a great deal of sand may be cascaded down the canyon. If the source of new sand, such as the large rivers emptying into the ocean, is altered so that the new sand no longer is available, alongshore transport may strip the region for progressively greater distances, leaving rocky or gravelly shoreline where there were formerly extensive sandy beaches.

On the shelf and continental slope beyond the zone of wave turbulence the transporting forces are weaker and slower, but the sediments to be moved are commonly finer. Currents and internal waves also bring about the transport of deeper-water sediments in these areas.

When an alongshore current of sand comes to an embayment, the waves lose some of their force as they are refracted into the embayment. They drop part of their load of sand, so that a bar tends to build up across the mouth of the bay from the side from which the sand is coming. If this is not dredged and deposited on the far side of the bay so that alongshore transport can continue to carry it along the shore, it may eventually close off the embayment. If there is sufficient current into the embayment to maintain an opening, a long bar may form, narrowing the entrance and extending along the side of the channel.

Since smaller, finer particles are more readily transported both by wind and by water, there is a continual sorting by size and density during transport, and the particles deposited at any one place are often quite uniform. On the beach, ordinary wave action tends to push larger, heavier objects to the higher levels reached by the waves, since incoming breaking waves have more force than receding waves.

Embayments are of many types, depending on the recent geological history of the region; the size, water volume, and sediment load carried by rivers; the seasonal fluctuation in amount of water carried; the amplitude of the tidal fluctuations; the frequency, strength, and direction of storms; the amount of precipitation in the region; the direction or trend of rock strata relative to the shoreline; and other factors. If the sea level in recent geological times has risen relative to the land, many embayments may be in the form of large drowned river valleys, with numerous protected secondary embayments, and perhaps miles of relatively stable estuarine conditions, with a salinity gradient ranging from nearly that of open seawater at the mouth to that of fresh water at the extreme upper ends. Chesapeake Bay is a good example of this type of embayment. If, on the other hand, recent geological history has been one of a

Fig. 8-42. Salt marsh association frequently found in upper parts of estuaries. Pickleweed and marsh rosemary. Holes made by Oregon mud crab *(Hemigrapsus oregonensis).* (M. W. Williams photograph.)

rising of the land relative to the sea level, rivers will tend to meet the sea more abruptly, with narrower and shorter estuaries, in which changes in salinity and volume of water with tidal cycles are more marked.

In any case, when the relations between land and sea level are relatively constant for a long-enough period of time, rivers tend to build estuaries and deltas at their mouths. As tides push into river mouths, the water is slowed down and the water level raised. Much of the silt carried may be deposited, tending to make the mouth of the river shallower and broader, which accentuates its tendency to deposit its load before reaching the ocean proper. The salinity may also be a factor causing agglomeration and settlement of particles. A sandbar is commonly built near the mouth, both by sand brought down by the river and by sand brought from the beaches of one side by alongshore transport. If the

flow of water is sufficient throughout the year, channels to the sea will be opened and filled with deposits, and new channels will be made, the process resulting in a fanning out at the river mouth, with the formation of deltas extending into the sea and the deposit of finer muds in the deeper water beyond the deltas. With the slowing of flow near the mouth, more sediments are deposited back of the sandbar, and eventually mud flats or areas of mixed sand and mud are formed, through which the river passes in winding, ever-changing channels and over which the water rises during the high tides and may recede at low tide. Where alongshore transport from the beach is considerable, the sandbar may be built continuously, deflecting the mouth of the river to one side and forming a long estuary parallel to the coastline and separated from the ocean by a sandspit. Examples are the great southward-trending spits extending from the north sides of Delaware Bay,

Table 8-2. Particle sizes of sediment

Size (mm.)	Name*	Phi scale†
256 and larger	Boulder	−8
64 to 256	Cobble	−6, −7
32 to 64	Pebble	−5
16 to 32	Coarse gravel	−4
8 to 16	Medium gravel	−3
4 to 8	Fine gravel	−2
2 to 4	Granule	−1
1 to 2	Very coarse sand	0
0.5 to 1	Coarse sand	1
0.25 to 0.5	Medium sand	2
0.125 to 0.25	Fine sand	3
0.0625 to 0.125	Very fine sand	4
0.0039 to 0.0625	Silt	5, 6, 7, 8
0.0039 or less	Clay	9 or more

*In part from Wentworth, C. K. 1926. Methods of mechanical analysis of sediments. Univ. Iowa Stud. Natur. Hist. 11(11):3-52.
†Log_2 of diameter in millimeters. (Krumbein, W. C. 1936. Application of logarithmic moments to size-frequency distributions of sediments. J. Sed. Petrology **6**:35-47.)

Chesapeake Bay, and Albemarle Sound. Perhaps the best examples of the development of such spits separating long estuarine areas from the main body of the oceanic water are those in the western end of the Gulf of Mexico, forming the outer side of Laguna Madre.

Where the water flow is not sufficient to maintain a channel at all times, such an estuarine area may be entirely cut off from the sea, either seasonally or semipermanently. Such an isolated estuary will become a freshwater lake eventually if the input of fresh water exceeds evaporation. If evaporation exceeds input, it may become even saltier than ocean water. If the inflow has significant seasonal variations, new channels may be cut during the wet season, restoring the connection with the sea, and the estuary may be closed off by sandbars at the mouth during the dry season.

Particle size of deposits

The particle size of a deposit exercises marked biological effects through differences in water retention and movement, organic content, compactness, and the ease with which burrows can be excavated and maintained. Table 8-2 gives a résumé of particle sizes, together with names applied to sediments with particles in certain size ranges and the Phi scale, based on a logarithmic relationship, which has been found convenient in quantitative work with sediments.

In any sediment, the smaller the average particle size, the greater will be the exposed surface, the capillarity and water-holding capacity, and, given the same organic content, the bacterial flora.

Fine sand may give a **capillary lift** ten times as high as coarse sand. This is of major importance to organisms in the intertidal zone, where retention of water during the low tide period may be vital. The **porosity**, or proportion of space between grains, would be 26% regardless of grain size, if the grains were all perfect spheres. Actually, small grain size increases porosity, so that there is not only more surface to exert holding power for water but also more space for water in fine-grain sediment than in coarse-grain sediment.

Where shallow marine and estuarine sediments and beach sands are not stabilized by the growth of plants, they are continually being shifted about by the movements of water, covered during high tides, exposed during low tides, or if they are below the low tide level, swept back and forth by tidal movements, currents, and internal waves. They do not offer the firm anchoring sites of the rocky areas. Survival here depends largely on the ability to burrow into the wet substrate and the possession of adaptations enabling the animal to breathe, feed, and reproduce while lying below the surface, where it is out of the reach of the surf and largely protected from desiccation and predators. A mudflat or sandy beach may appear barren and lifeless to the casual observer, even though it is teaming with countless living things ranging from microscopic bacteria and protozoa to large worms, ghost shrimps, and clams. The finer sediments, deposited where the water is quieter and often containing considerable organic matter, have a slower circulation of water. They are soon rendered anaerobic below the surface by the action of bacteria. Such sediments are blackened because of the accumulation of hydrogen sulfide. Animals living in such sediments must have some kind of direct communication with the surface for respiration, and in the intertidal zone they must be able to survive until the next tide brings oxygenated water over the flats.

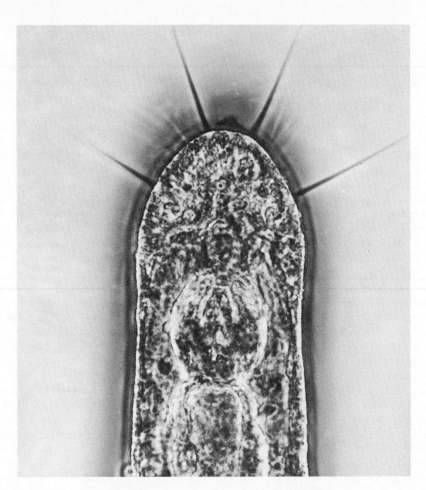

Fig. 8-43. Anterior end of *Guzthostomolz jenneri,* a gnathostomulid. (Photograph courtesy Dr. R. J. M. Riedl.)

Fig. 8-44. A group of the recently discovered phylum Gnathostomulida, a dominant element of the endopsammon in fine sands that are anaerobic below the surface layer. **A,** Jaws and basal plates. **B,** Series of species showing increasing differentiation of the head and sensory cilia. Only the sensory cilia are drawn. There has been a suggestion that perhaps at least some of the enigmatic microfossils known as conodonts may be gnathostomulid jaws or basal plates. (From Riedl, R. J. M. 1969. Science **163:**445-452. Copyright, 1969, by the American Association for the Advancement of Science.)

The endopsammon

Wet or submerged sands usually contain a rather varied microfauna, the endopsammon, living between or on the sand grains. The composition of this microfauna will vary according to the particle size of the sediment, its location relative to tidal or subtidal depth, the salinity and temperature of the water, and other factors. In coarse, aerated sediments typical members include many microcrustaceans (harpacticoid copepods and ostracods), small turbellarian flatworms (Acoela, Macrostomida, Otoplanida, and others), protozoans (especially ciliates and foraminifers), small nematodes, gastrotrichs, rotifers, and tardigrades. In fine sand deposited in sounds and estuaries, the deeper sand below the surface centimeter or so is usually anaerobic and black because of the production of hydrogen sulfide by growths of iron bacteria. Many groups of typically endopsammal species disappear in such sediments. These fine-grained anaerobic muddy sands are commonly dominated by gnathostomulids, a group of lower worms almost wholly overlooked until very recently because they are small, delicate, and recognizable only when alive. They are strongly **haptic,** that is, they stick tightly to the sand grains. They are very resistant to bad water conditions and emerge from deteriorating, stinking samples of substrate only after most other endopsammon has left. The samples would ordinarily be discarded before they come out.

OPEN SANDY BEACHES AND SANDY SUBTIDAL COMMUNITIES

Sandy beaches facing directly into the ocean without the protection of headlands or offshore islands are areas of constant change and motion. Every wave lifts and moves quantities of the sand. The breakers pound the shoreline ceaselessly. The tides move up and down the beach in daily, monthly, and annual rhythms and progressions. Exposed sands are subjected to direct sunlight, drying, wind transport, rain, and freezing. Floating objects and masses of seaweed and subtidal animals torn from their anchorages offshore by the turbulence of storms are cast ashore and pushed to the upper part of the beach by the waves of the highest tides. Each level of the beach is submerged and exposed for different amounts of time and at different times of day or night each successive day. High tides bring with them certain predators from the fringe of the sea, and low tides bring another set of hungry predators from the land.

Only very specialized animals can live successfully and maintain themselves in this turbulent world, where even the sand itself displays something of the fluidity of water. Most of the animals found here are amazingly proficient burrows and are relatively small in size.

Plankton feeders

Mole crabs. One of the most successful animals of the surf zone is the mole crab, or sand crab. These are not "true crabs" but members of the curious group Anomura, which is intermediate in many ways between the macrural, or large-abdomened, decapod crustaceans such as shrimp and lobsters, and the true crabs, or Brachyura, in which the abdomen is small and carried tucked out of sight beneath the cephalothorax. Other familiar examples of Anomura include such diverse forms as hermit crabs, porcellanid crabs, ghost shrimp, galatheids, and lithodids, ranging from shrimplike forms to crablike types. The crablike members of the group can be readily distinguished from true crabs by their long antennae and by the fact that the last pair of walking legs is usually greatly reduced and often held up against or beneath the posterior edge of the carapace, so that the crab appears to be missing one pair of legs.

The mole crabs, however, resemble neither shrimps, hermit crabs, nor true crabs. They have become so modified for existence in the wave-swept sands that one would hardly recognize them as decapod crustaceans at all without careful examination. The body is a smooth, oval shape and is largely covered by the hard carapace. The appendages can all be tucked in, and they fit neatly beneath and at the sides of the body, so that the general streamlined form is maintained. They are flattened and shaped as efficient swimming and digging organs.

Fig. 8-45. Sandy beach, Drake's Bay, California. Piles of large kelp torn from offshore kelp beds during rough weather are cast up on the beach. The beach itself has been somewhat eroded, exposing rocks that are frequently covered with sand, as indicated by the lack of attached animals. (M. W. Williams photograph.)

Like most other animals dominant in marine environments, the mole crabs are plankton feeders. Great numbers of them burrow backward into the soft sands facing the sea where every wave sweeps across the wet beach and recedes as a thin sheet of water. As a wave recedes they extend their plumed antennae, capturing minute planktonic organisms carried in the surf and then draw the antennae through the mouth parts, scraping off the food they have captured. As the tide moves up and down the beach, they move with it in stages, suddenly emerging from the sand and allowing themselves to be carried either up or down, depending on the direction of the tide, and promptly burying themselves at a more favorable level to continue their feeding. Any that are left behind by a falling tide bury themselves a few inches into the wet sand and await the return of the water.

In the summer the females carry bright orange egg masses firmly attached beneath the body. As the time for hatching approaches, they cease feeding and remain lower in the water. The young hatch as zoeae—minute, grotesquely spined planktonic creatures that are sometimes carried great distances along the coast by currents. Sometimes they are swept out to sea and lost. Off California they have been taken as far as 200 miles or more from shore. By the end of the summer they have undergone their larval molts and are ready to assume a benthic existence. The second-stage larvae seek out sandy bottom near the shore, where they undergo another molt bringing transformation to the adult form and feeding behavior.

In the northern parts of their range, where the sand becomes very cold and ice may be forming on the beaches, the mole crabs move out beneath slightly deeper water. In warmer areas they may remain active in the intertidal zone throughout the year. In spring, mating occurs. The males

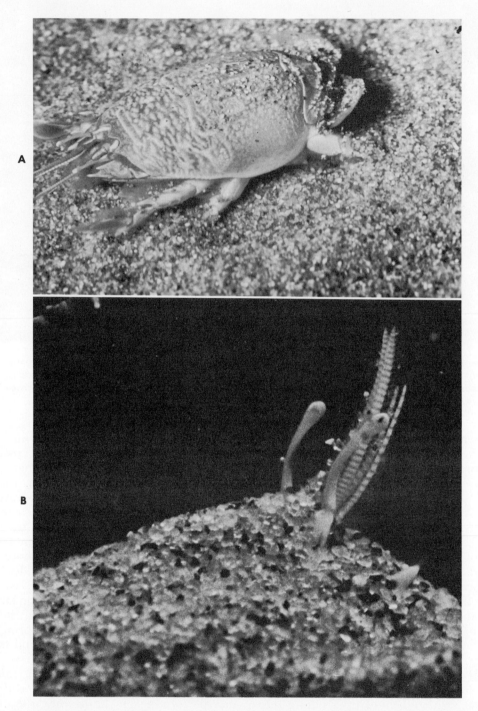

Fig. 8-46. A, The mole crab, or sand crab, *Emerita analoga,* from the Oregon coast, dug out of the sand to show general appearance. **B,** Same, buried in sand beneath water, with only the eyes and antennae protruding. (**A,** Oregon Institute of Marine Biology photograph; **B,** M. W. Williams photograph.)

soon die, and the females carry their eggs through much of the summer and then die as winter approaches.

The common mole crab along Atlantic shores is *Emerita talpoida* and that of the Pacific coast, *Emerita analoga*. A somewhat larger, more grotesque species, the spiny sand crab *(Blepharipoda occidentalis)*, occurs lower in the intertidal and upper subtidal sands.

Bivalve molluscs. Some bivalve molluscs have also become adapted to life in the open sandy beaches between the tide lines or just below the low tide level. These, too, are plankton feeders.

On many open Atlantic beaches countless thousands of small coquina clams *(Donax variabilis)* are found in the surf zone. As waves wash them out of the sand, they promptly burrow back beneath the surface with their muscular foot, to approximately the depth of one shell length, and then extend their siphons to the surface of the sand to bring the life-giving current that bears their food and oxygen and carries away their waste products. Coquinas are not permanent residents of any particular beach. They may be present by the thousands at one time and absent at another. Perhaps, like the sand they live in, communities of coquinas are moved along the shore with the alongshore current of sand.

On many beaches other clams, too, find a living. Most of these clams are somewhat larger and burrow more deeply into the sand where there is a greater degree of stability. Preeminent among rapidly digging, sand-dwelling clams are the razor clams, much sought after by gourmets on the lowest of spring tides.

A large, heavy-shelled clam, the pismo clam, occurs on California beaches at the lowest intertidal and subtidal levels. They

Fig. 8-47. Pismo clam, *Tivela stultorum*, on a California beach. (M. W. Williams photograph.)

lie barely buried in the sand, facing the sea, with only the umbo of the shell reaching the level of the sand surface. The intensity of the rock-dwelling animals' competition for firm substrate to which to attach can be seen in the fact that the shell apexes of the pismo clams, hardly reaching the sand surface, commonly become sites of attachment for tufts of hydroid colonies. Clam diggers often locate the clams by these small feathery hydroids that stick up out of the sand.

Detritus feeders

On many beaches where the sand is of the proper porosity—medium to fine sands, with sufficient organic matter in it—peculiar polychaetes of the family **Opheliidae,** *Ophelia, Thoracophelia,* or others, known as bloodworms because of their bright red color and their red hemoglobin-containing blood, live in dense mats or colonies just below the surface of the sand, and the colony may extend the full length of a beach. These worms, having adopted a permanent burrowing, substrate-eating habit, have lost the parapodia characteristic of most polychaetes. Only the bundles of setae mark their positions. Thus they superficially resemble oligochaete earthworms in form and habit. There is even a clitellum-like glandular region marking the posterior border of the "thoracic" region. Its function is unknown; it is certainly not similar to that of the clitellum of oligochaetes, since these worms do not form egg coccoons. There are also small paired fingerlike respiratory "gills" on most of the trunk segments. Bloodworms continually pass wet sand through their gut. Experiments in which sand stained with methylene blue is used alternately with unstained sand indicate that they can replace the sand in their gut in as short a time as fifteen minutes, and that a thriving colony of *Thoracophelia* may cycle an amount of sand equivalent to the entire beach in which they live every five years.

Like other intertidal animals bloodworms are good burrowers. Their method of burrowing is unique. Unlike the coelom of most annelids, the coelom of the blood-worm is continuous throughout most of the body length, each intersegmental septum being represented only by a strand of tissue running from the dorsal body wall to the gut, throughout most of the body. There is, however, a double septum separating the coelomic cavity of the head from that of the rest of the body. This septum is reflected back on the dorsal side as a large muscular saclike "injector organ" communicating with the head coelom by a pair of small pores in the anterior face of the septum. When bloodworms are burrowing through the wet sand, they do so by first withdrawing the coelomic fluid from the head so that it enters the injector organ. They then constrict the circular muscles of the head and anterior part of the body, pushing the head forward into the sand. The muscles of the wall of the injector organ contract, forcing the fluid into the head and blowing it up like a small bulb by which the anterior end of the worm is anchored. The circular muscles are then relaxed and the body of the worm is pulled forward by contraction of the longitudinal muscles of the body wall. In soft wet, sand blood-worms can burrow with surprising rapidity.

Another peculiarity of bloodworms, which explains a characteristic sign of their presence on a beach, is their specialization for anal respiration. The last few segments of the body are enlarged, and bear setae directed obliquely upward. The dorsal wall of the rectum bears two tracts of strong cilia, which beat in an anterior direction. At the time the falling tide uncovers a bed of bloodworms, the worms are very active, burrowing about close beneath the surface of the wet sand. As a worm comes to just below the surface and turns down again, the enlarged anal segments with their bristles dislodge a few grains of sand that fall into the burrow behind the worm, producing a minute hole in the sand. Thus the area occupied by the bloodworm colony becomes marked with thousands of pinpoint-size holes, which in the aggregate may darken the area and which differ from holes excavated by other burrowing forms such as beach hoppers in being smaller and having no sand piled at the entrance.

Perhaps this serves a ventilating function for the colony, allowing more air into the burrows than would otherwise come through the close-packed sand. At the moment the hole is formed and the posterior portion of the worm passes beyond, one can sometimes observe an expansion of the rectum and a rippling effect produced by the two ciliary tracts, as if the worm was taking a breath. In any event, the holes do not indicate that the worms have come out on the surface for a time during the night, as was formerly thought.

In its basic features the life cycle is typically that of a polychaete. The sexes are separate, and the eggs develop into typical planktonic trochophore larvae that spend about four or five days in the plankton before developing into metatrochophores and seeking the bottom.

Bloodworm colonies are quite sharply delimited at both their upper and lower margins, probably by moisture requirements. At the upper margin of the colony, between tides, the worms may be somewhat deeper in the sand. Apparently they cannot thrive where the sand is almost completely saturated all of the time. They also burrow more deeply as the tide comes in or if seawater is poured over the area of part of the colony. They are thus restricted to the belt where the sands close to the surface remain very moist but not saturated during most of the low tides.

Scavengers

Sand hoppers, or beach fleas. A third abundant source of food on the beaches is the flotsam left by the waves and tides. Animals from the surface waters that have drifted too close to shore and been thrown up on the beach, seaweeds and sessile animals torn from nearby rocky areas by the turbulence of spring tides or storms, and plant and animal debris washed down from the land by rivers provide food for many organisms. The most ubiquitous and successful exploiters of this food source are amphipod crustaceans of the family **Orchestiidae** or **Talitridae**, genera such as *Orchestia, Orchestoidea,* and *Talitrus.*

Like certain species of periwinkles from the rocky areas, these amphipods seem to be at a midpoint in an evolutionary emergence from sea to land. They no longer swim, respire, or reproduce in the ocean waters as do the vast majority of amphipods. In fact, they are poor swimmers, and drown if submerged too long. Yet they seem to require the salty dampness of the sea and are bound to its margin. Their life and activities are tuned to both the diurnal and the tidal cycles. By day and during high tides they remain out of sight in burrows excavated several centimeters into the sand of the upper beach or under piles of damp seaweed left on the upper beach by previous high tides. But when there is a falling tide during the hours of darkness, they emerge from their burrows by the thousands, roaming out over the intertidal sands as the sands are exposed and searching for and consuming all manner of organic remnants, thus cleaning the beach. Shortly before dawn, or at any time when the tide has turned and is again beginning to flood the intertidal sands, they begin to move up the beach near the high-water line. There each sand flea furiously excavates a new burrow into which it will retreat from daylight and from high water. Before finally retiring, it plugs the burrow with sand, and withdraws to the chamber it has made at the lower end. There it will await, and somehow sense, the next occurrence of a falling tide during the hours of darkness.

Ghost crabs. Another group of nocturnal scavengers of the beaches, particularly in warm and tropical regions, are the ghost crabs, members of the family **Ocypodidae.** Ghost crabs are "true crabs" living still higher on the beach than the sand hoppers. They build their burrows near the top of the beach or even at the beginning of dunes at the back of the beach. They are still, however, bound to the sea for both respiration and reproduction. In the branchial chambers they carry a little seawater, keeping the gills wet. From time to time this water must be renewed. At such times the ghost crab approaches the waterline, waiting with one side turned toward the ocean until one wave, larger than the rest, washes over it, renewing the water in the gill chambers, and then retreats again to the upper beach. The females also return to the water to spawn. Other than these two instances and occasional dashes into the water for protection from a predator on the beach, adult ghost crabs live and act like terrestrial animals.

At night they emerge from their burrows in great numbers to forage on the beach near the waterline, sometimes taking some morsels back to their burrows. They can run faster than any other crustacean and almost as fast as a man. As their pale forms move rapidly over the night sands they look strangely insubstantial, like wisps of some small spirit being swept across the beach by a wind.

Like most other crabs, the early larval stages of the ghost crab are peculiar spined planktonic forms called zoeae. After the zoeal stages there is a transformation to a more crablike megalops. In the case of ghost crabs, whose megalops must make a beach landing in the turbulence of the surf, the megalops is adapted much like the mole crab, which spends all its adult life in the surf. The body is oval and protected by a hard, smooth carapace. The limbs and abdomen are tucked in closely, neatly fitting, so that the megalops can tumble in the surf without injury. Having made a safe landing, the megalops digs a small burrow in which it undergoes another molt, transforming to the adult shape. The young crabs make their burrows in the wet sands covered by high tide. As they grow larger the burrows are made farther up the beach until in the fully adult stage they are at the uppermost edge of the beach.

Where the winters are cold, as along the shores of the eastern United States, ghost crabs appear to hibernate in deep burrows through the winter, reappearing on the beach in the spring.

Mole crabs, coquina clams, bloodworms, ghost crabs, and sand hoppers, each in their own different way, have become adapted in structure, physiology, life cycle, and behavior patterns for life in the zone of open beaches. These are areas of pounding waves, ever-shifting sands, fluctuating tides, and exposure to wind, rain, sunlight, heat, and cold. The environmental stresses are so great, so diverse, and so variable that most organisms are prevented from establishing themselves there. Yet there are sources of abundant food—the plankton, the organic film and microscopic animals and plants to be found on and among the sand grains, and the flotsam continually thrown up on the beaches from the sea. Those few animals which have become adapted and are able to succeed in this difficult environment and to exploit these food sources find themselves virtually without serious competition and can dominate their environment in overwhelming numbers.

Thienemann (1920) stated that it is a fundamental principle of biocenotics that the more specialized or extreme the habitat (biotope), the poorer in species but the richer in individuals will be the biocenosis.

Terrestrial forms on sandy beaches

Scavengers. Animals from the sea are not the only ones to have become adapted to live on or regularly visit the beaches. On many beaches, where windrows of seaweeds are frequently thrown up, swarms of anthomyiid flies—the kelp flies, genus *Fucilia,* and others—are as ubiquitous as the sand hoppers. Their maggots live in the decomposing seaweeds, and the small brown sculptured barrel-shaped pupae can often be found there. Small elongate rove bettles with truncated elytra, family **Staphylinidae,** are also regularly found there as well as numbers of other small beetles of the families **Nitidulidae, Tenebrionidae, Scarabaeidae,** and others.

Predators. Since much of the life of the beaches lives burrowed into the sand most of the time, several of the most notable predators are adapted for probing the sands. Such are the long-billed shore birds —sandpipers, godwits, curlews, and others often seen in numbers probing the sands for bloodworms and beach hoppers. The small sanderlings race up and down the beach in tight little groups just at the margin of the waves, probing the sand furiously for a few moments, at the lowest points they can reach as each wave recedes, for young *Crago,* small shrimps, amphipods or isopods, and small mole crabs often found there and then race back as the next wave drives them up the beach.

The sea gulls are ubiquitous predator-scavengers patrolling all coastal areas, rocky, sandy, muddy, protected, and open waters for small fish and for any animal or animal remains they can pick up at the water surface or on the coast. Many of them have developed special habits, such as dropping shelled animals from a height onto rocks or hard sand to break them open. On the East Coast the ring-billed gulls tread the soft sand at the waves' edge to bring up the small coquina clams. When low tides occur during daylight hours, the gulls are always busy, searching the intertidal areas for any animal unfortunate enough to have been left stranded or exposed.

A smaller number of predators are adapted to prey on small animals flying over the sand. Cliff swallows nesting on the coastal bluffs are often seen sweeping low over the beach, capturing kelp flies or other insects flying over the sands. Swift-running, big-jawed tiger beetles (**Cicindelidae**) are common at times on the open sand as well as along the margins of estuaries and other bodies of water.

At night still other animals forage on the beaches. *Peromyscus* (the deer mouse) and other rodents living among the bunch grasses or the woods above the beach venture out onto the upper beach in search of insects or morsels of food left in the flotsam. Skunks and raccoons are also frequent visitors in many areas.

Periodic visitors from the sea

A few animals that are truly marine, driven perhaps by predation pressure on the eggs or very young farther offshore, find protection for these stages by depositing them at a higher level of the subtidal or in the intertidal. There has been some speculation that the emergence of the early ancestors of the amphibians onto the land was more or less forced in this manner. In any event, several species of fish, a few invertebrates, and the sea turtles regularly leave the sea long enough to bury their eggs

in the sand. In the case of the sea turtles, which descended from land-inhabiting reptiles, the reproductive tie to the land probably represents, as does their air breathing, an incompleteness in their adaptation to marine life.

The evolutionary changes involved in changing from fully aquatic to fully terrestrial modes of respiration and reproduction or vice versa seem to be the most difficult to make. Thus a number of marine forms, such as ghost crabs, land-dwelling hermit crabs, some mudskippers and periwinkles, and others that seem now partially emergent from the sea must still provide for gill respiration, return to the water to spawn, and in many cases spend part of their life cycle in the plankton. On the other hand, marine mammals and reptiles all still breathe air and, except for the cetaceans and sea snakes of the subfamily **Hydrophinae,** go onto the land to have their young or lay their eggs. Even the cetaceans in many cases seek out relatively shallow lagoons in which to bear their young.

Horseshoe crabs. Along the eastern coast of North America from Maine to Mexico, and along the eastern coast of Asia and parts of the East Indies, live the peculiar primitive "living fossil" *Limulus,* the closest living relative of the trilobites, which dominated ancient Paleozoic seas. *Limulus* usually lives well offshore, but in the breeding season it comes into shallow waters, sometimes in very great numbers. The female is about 20 inches in length, and the male is considerably smaller. Early in the summer they pair. The female comes ashore with one or more of the small males, each hanging to the tail of the one ahead. She excavates a hole in the sand in which the eggs are deposited, fertilized by the males, and left for the waves to cover. About midsummer the young hatch and make their way into the water, to disappear until they return fully grown the following year. At hatching time a beach may be swarming with thousands of the young. The young, lacking the long caudal spine of their parents, look even more like trilobites than do the adults.

Grunion. Of all the fish that spawn high in the intertidal, the most famous and in many ways most remarkable is the grunion, a small atherinid fish, *Leuresthes tenuis.* From Lower California to San Francisco, from the end of March to about August, every second week beginning the second night after a full moon and just after the spring tides have reached their highest, the moonlit beaches may be alive with silvery fish about 6 to 8 inches long. Each wave brings thousands of fish washing high up onto the beach. The females quickly burrow into the wet sand, tail first, until only the head down to about the pectoral fin is above the sand. While she deposits eggs, the male curls around her on the surface of the sand and deposits milt, which runs down into the sand and fertilizes the eggs. Then with the succeeding wave they slither back into the sea while others are brought onto the beach.

Since the eggs are laid just after the turn of the tide, succeeding waves are weaker as the tide recedes and only push more sand over the eggs rather than washing them out. Also, since the eggs are laid after the highest spring tide, on the descending series of tides, the eggs will be left in the sand for two weeks before the next series of rising spring tides washes them out. The eggs are actually ready to hatch in about ten or eleven days after being deposited in the sand, but they do not hatch until they are agitated in the seawater, which occurs at the next series of spring tides. At this time the young are washed out of the sand and carried into the surf, and a new spawning run of the adults begins. Because of the progression of the tides through the lunar month, alternate series of spring tides occur later at night when few people are on the beach to observe the run, giving the appearance of a four-week cycle.

Reproductive behavior of this sort entails the most exquisite adjustment and timing of physiological processes, instincts, and behavior patterns to the diurnal and lunar cycles and illustrates the profound effect that such subtle influences as the lunar cycle may have on organisms. If the eggs are not laid at precisely the right part of the tidal cycle, they will be washed from the sand prematurely and will perish. Only the high spring tides coming at night can be utilized. If they came in during daylight, the fish would be wiped out by seagulls and other predatory birds as they came ashore. How the fish, living offshore, sense the proper moment in the tidal cycle to make their run and how the developmental rate of the eggs became so precisely attuned to the lunar cycle is a source of wonder. It seems probable that a long period of selection was involved, once the habit of depositing the eggs in the beach sand had begun. Any that were not deposited at the right times would perish, so that the progeny would all be derived from those fish which happened to be properly in tune with the lunar cycle, and all deviants would probably be rigidly eliminated. Continued selection and elimination of deviants keeps the pattern precise for the population as a whole.

Throughout the spawning season the female grunion carries three classes of eggs in the ovary, except when recently spawned out. They are very immature ova, intermediate ova, and bright orange ripe ova, which as the spawning date comes, are shed into the lumen of the ovary ready to be spawned. After spawning, the intermediate-stage ova ripen just in time for the next series of spring tides, and a new batch of immature ova develops.

Studies of growth and of concentric striations at the outer edge of the scales indicate that grunions live about three years, usually spawning two or occasionally three seasons.

Sea turtles. Sea turtles, family **Cheloniidae,** are distributed throughout the tropical seas. Although the adults are very wide-ranging in tropical seas, they return to the coasts where they were born to mate and lay eggs. The females come up onto the sandy beach, excavate a hole, bury their leathery-shelled eggs, and return to the ocean. When the young hatch, they scramble up out of the covering sand and make for the water. If they happen to emerge during the hours of daylight, they are usually doomed because gulls and other predacious birds carry them off before they can make it to the water. A remarkable aspect of the sea turtle's behavior is their homing ability. This ability to swim hundreds of miles through trackless seas to a particular beach on a particular small island where they happen to have been born is one of the marvels of animal navigation, rivaled only by the feats of certain birds, the Atlantic eels of Europe and America, and the salmon.

PROTECTED SANDY AREAS

As we leave the open ocean beach and go to quieter and more protected sandy flats, or offshore beyond the range of great wave turbulence, the substrate is far more stable. Many more kinds of animals can live here. Intertidal flats of sand or sandy mud are inhabited by many kinds of burrowing animals that require a more permanent sort of burrow than can be maintained in the constantly shifting sands of the open beaches.

The majority of the burrowers are either filter feeders or animals that are able to ingest the detritus that collects at the surface of the sand or mud.

Bivalve molluscs. As in other habitats, the plankton feeders tend to be dominant. On the flats, most of them are either bivalve molluscs, annelid worms, or burrowing crustaceans. All levels of the flats contain bivalves, from the plump, heavy-shelled ribbed cockles at the surface, which can be collected with a rake, to large, deeply buried clams such as the gaper clam (*Schizothaerus*) or the geoduck, which may be as much as 2 or 3 feet below the surface. In general the structure of a clam, or even merely its shell, tells much about its habits and ecology. Those clams that live well below the surface have correspondingly long siphons. The larger the siphon, the more prominent, as a rule, is the mark left on the empty shell by the line marking the mantle attachment to the shell along its edge—the indentation in the line marking the attachment of mantle muscles, known as the pallial sinus.

The soft-shell clam *Mya arenaria* and the gaper clam *Schizothaerus nuttalli* are examples of large, deeply burrowing clams with long siphons. These clams are commonly a characteristic and dominant element in their biocenoses. The soft-shell clam is widely distributed along boreal and temperate shores in both the Atlantic and the Pacific. The gaper clam is prominent along the Pacific coast of North America, and its shells also sometimes constitute the principal fossil remains in Pleistocene marine terraces.

Large clams with big siphons commonly withdraw their siphons quickly when alarmed, as by the vibrations caused by someone walking near them or, in some cases, by a shadow passing over them. As they withdraw, they may squirt water as much as 1 or 2 feet into the air.

Since the circulation of water is from the ventral siphon into the large ventral mantle chamber, through the gills into the dorsal mantle chamber, and out through the dorsal siphon and since both siphons leave the clam together at the posterior end, often even bound together in a common muscular tube, clam burrows have only one opening at the surface. The same is true of such animals as phoronids, in which the digestive system is looped, bringing the anus near the mouth, so that the mouth and food-gathering tentacles are together with the anus at the mouth of the burrow.

Annelid worms. Other animals, such as the lugworms (*Arenicola* and *Abarenicola*), the parchment worm (*Chaetopterus*), and the spoon worm (*Urechis*), and crustaceans such as the ghost shrimp (*Callianassa*), which circulate water through their burrows, obtaining either their food or oxygen, or both, from the passing stream, have U-shaped burrows with two openings at the surface. These burrows with the circulation of water through them are favorable homes for a variety of other small animals such as pea crabs, small gobies, and commensal copepods. Lugworm burrows are usually in very muddy sand near the upper reaches of intertidal flats. The incurrent end of the burrow is in a depression in the sand, whereas the excurrent opening is at the top of a small black mound. The darkness is caused by reduced mud, with hydrogen sulfide having been expelled there. Detritus coming in with the incurrent water together with some sand or mud is swallowed, and the sand is expelled as coiled castings around the excurrent opening of the burrow. As is the case with earthworms in terrestrial habitats, the work of a bed of lugworms tends to stir, cleanse, and renew the muddy sand in which they live. It has been calculated that where they are abundant they may work over almost 2000 tons of mud per acre each year.

The lugworm usually lies in the bend of the U-tube burrow. When the burrow is covered with water it stretches its head out around the entrance to feed. Around midsummer they spawn. In this case gametes are not shed into the water, but the eggs are contained in small gelatinous translucent pinkish sacs, with one end anchored in the sand by a threadlike extension. Each egg mass may contain many thousands of eggs or developing young.

Another worm abundant in some flats on both sides of the Atlantic is the trumpet worm, *Cistenides,* which constructs a conical tube of a neatly fitted mosaic of sand grains fitted and cemented onto a chitinous base. The tubes are shaped something like the shells of scaphopod molluscs and serve much the same function, being open at both ends. The small posterior end sticks out of the surface, enabling the worm to bring a respiratory current down to its gills as it forages in the soft substrate, digging with alternating vigorous shoveling motions of two groups of comblike bristles on either side of the head and sorting desired food particles with the tentacles around the mouth.

According to A. T. Watson, the young worms first secrete a membrane about themselves, and then begin to collect sand grains with their tentacles and pass them to the mouth where they are rolled about and felt. Selected grains are then fitted individually at chosen spots at the edge of the tube. As the worm grows, it continues to build the anterior end of the tube, so that, as Watson puts it, "Each tube is the life work of the tenant, and is most beautifully built with grains of sand, each grain placed in position with all the skill and accuracy of a human builder."[*]

Echiurida. Spoon worms, or Echiurida, are widely distributed in sandy mud bottoms and mudflats, often in areas similar to those in which ghost shrimp and *Upogebia* are found. The worms are thick-bodied, and indistinctly or not segmented, with a fleshy spatulate "prostomium" in front of the mouth. Like the lugworms, they live in U-shaped tubes. The circulation of water through the tubes is maintained by peristalsis-type movements of the body. By means of glands in the prostomium and anterior end of the body, they fashion a funnel-shaped mucous or slime net in the anterior part of the burrow. The water being circulated through the burrow by the peristaltic action of the body is filtered through this slime net, which the worm swallows from time to time with its entrapped detritus and small organisms. Some species are said to sweep the ciliated ventral surface of the proboscis over the mud near the entrance of the burrow and to sweep detrital particles to the mouth along ciliated tracts.

Following are some of the better known genera:

Thalassema. About a dozen species occur in both Atlantic and Pacific. No preanal bristles are present, and the proboscis is rather pointed. The sexes are quite similar in appearance.

Echiurus. A few species. Preanal bristles are present, and the body is marked with rings bearing spines. The proboscis is not bifurcated. The setae are in one or two rows. The sexes are quite similar in appearance.

Urechis. Very large, fleshy worms about 3 to 5 inches long and 1 inch or more thick, with small proboscis and a conspicuous circle of setae at the posterior end. The proboscis is not deciduous. This is a prominent member of some mudflat communities along the California coast. The sexes are quite similar in appearance.

Listriolobus. Also occurs along the Pacific coast. A smaller worm without posterior setae and with a more elongate, sometimes deciduous proboscis. The sexes are quite similar in appearance.

---

[*]From Watson, A. T. 1928. Observations on the habits and life history of *Pectinaria (Lagis) koreni* Mgr. [edited and introduction by P. Fauvel]. Liverpool Biol. Soc., Proc. Trans. 42:25-60.

Bonellia. Found along the European coasts. Displays marked sexual dimorphism, the female being a fairly large, bright green worm with a large, bifurcated proboscis that is held above the substrate and used to gather plankton. The male is minute and lives in or on the body of the female. The larvae have been said to be equipotent—becoming females if they settle by themselves, and male if they settle near a female.

Spoon worms show some affinity to annelids in that they possess a spacious schizocoel, trochophore larvae, and nephridia, but they are so different in many ways that they are usually regarded as constituting a separate phylum. Formerly they were sometimes grouped with the sipunculids and the priapulids in a phylum called Gephyrea, usually regarded as an appendage to the Annelida.

Sipunculids and priapulids. These worms are also substrate dwellers. The priapulids live in muddy substrate. Like the Echiurida they are thick, fleshy worms, with a complete digestive tract ending in a terminal anus at the caudal end. The gut is straight and wide. In most species there are one or two large caudal appendages bearing respiratory papillae, looking something like bunches of grapes. The pharyngeal region is in the form of an introvert, or eversible, proboscis, which is plump and covered with rows of small spines. Lang (1948), Hyman (1951b), and others regard the body cavity as a pseudocoel and have classed them with the assemblage termed Aschelminthes, which includes diverse forms distinguished primarily by a pseudocoel type of body cavity and a noncellular cuticular investment of the body and usually by a tendency toward constancy in the number of somatic nuclei and by a syncytial epithelium in which the nuclei are gathered into special tracts or papillae. However, according to the researches of Shapeero (1961, 1962), the priapulids actually have a true coelom lined with peritoneum and should be classed among the eucoelomate animals, perhaps as an independent phylum. Priapulids seem to prefer cool waters and are not found in tropical or subtropical regions. Some of them occur in deep water, down to over 7000 meters.

The sipunculids, or peanut worms, are a much larger group commonly found on somewhat coarser substrates, as in the sand among eelgrass roots, in fissures and cracks in rocky areas where sandy sediment has accumulated, or under rocks on sandy or

Fig. 8-48. A sipunculid worm with the introvert extended, showing tentacles in feeding position. (Oregon Institute of Marine Biology photograph.)

sandy mud substrate. In the sipunculids there is also an introvert, but the mouth, when everted, is surrounded by a circle of soft, branched tentacles with which they obtain particles from the substrate to ingest. The intestine is unique in that the distal portion of it is wound in a rather tight spiral about itself, ending in an anus in a forward position in what might be termed the shoulder region, just posterior to the base of the introvert. There are numerous genera and the group as a whole enjoys worldwide distribution.

SHALLOW-WATER BENTHIC COMMUNITIES

Since the pioneering work of Petersen (1913, 1915, 1918) and Petersen and Boysen-Jensen (1911) in the shallow Danish fjords,

Fig. 8-49. A, Another animal found on or near the surface of sand or mud in shallow water of estuaries, etc. is the cockle *Cardium corbis*. This one from South Slough, Coos Bay, Oregon, is viewed from the rear to show the short siphons. **B,** Olive snails, *Olivella biplicata*, partially buried in the sand, with the snorklelike siphons extended. **C,** *Amphitrite,* a polychaete worm that lives in a sandy mucous tube, the body buried in the substrate or in a crevice or underside of a rock. The anterior end is crowned with three pairs of much-branched red gills, and many very long extensile and contractile tentacular filaments used to find detrital particles of food and transport them to the mouth. **D,** Acorn worms, balanoglossids. Burrowing substrate-ingesting worms long thought to be related to the vertebrates. Supposed chordate characteristics include a putative notochord, pharyngeal gill slits, and a segmental enterocoelom. The tornaria larvae resemble rather closely the larvae of certain echinoderms. **E,** Ghost shrimp, *Callianassa californiensis,* inhabits rather deep burrows in the mudflats. **F,** *Upogebia pugettensis,* another, somewhat larger anomuran, shrimplike crustacean living in burrows in mudflaps or estuaries. (**A, C, D,** and **E,** Oregon Institute of Marine Biology photographs; **F,** M. W. Williams photograph.)

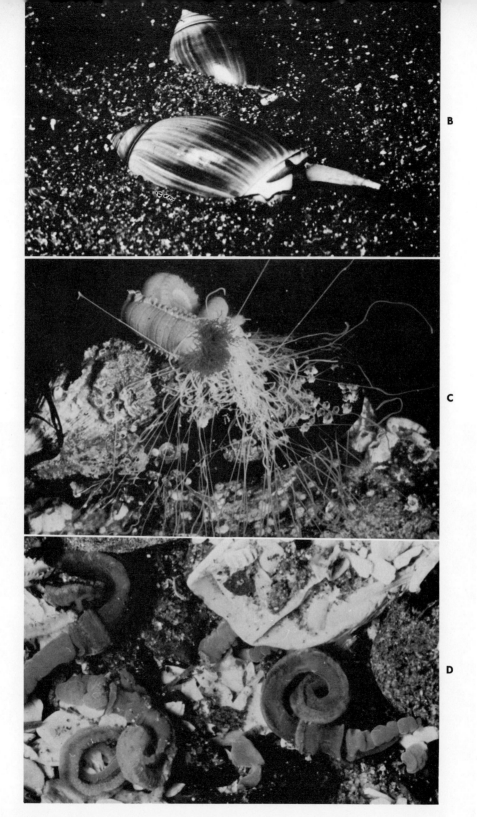

Fig. 8-49, cont'd. For legend see opposite page.

E

F

Fig. 8-49, cont'd. For legend see p. 236.

it has been recognized that the benthic animals of soft substrates are not randomly distributed with respect to each other or with respect to the environment but tend to occur in more or less well-defined associations, or **communities,** that can be identified or defined in terms of their most abundant and prominent members. Animals living in soft fine-grained substrates such as sand or mud Petersen termed **infauna,** in contrast to those living on the surface of hard substrates, which he termed **epifauna.** Through extensive quantitative sampling and statistical analysis, Petersen demonstrated in the Baltic region several distinctive infaunal communities. Such communities could usually be correlated with certain combinations of ecological factors, of which the exact nature of the substrate (particle size, stability, compaction, organic content, etc.), the average and extreme temperatures and their seasonal fluctuation, illumination, salinity, and the movements of the overlying water are important. There are also a number of other, more subtle factors that have far-reaching influence upon biotic communities. Often the composition of the communities themselves is a more reliable and readily available key to such parameters than is an examination and analysis of the environment itself.

The work of Petersen and of the biocenologists who followed him was basically a Linnaean, or classificatory, approach, in which they attempted to divide and classify the various types of environment into units and subunits in a manner analogous to the approach of a taxonomist to animal and plant groups. A good example of the culmination of this approach is the work of Gislin (1930), who attempted to exhaustively divide and classify all the communities in a small Swedish coastal region into a system of faunal units, subunits, and microunits.

This was termed the **particulate approach** by Wieser (1953, 1954, 1957, 1959). It is felt by some that, carried to its extreme, this approach results in a bewildering patchwork of units and subunits in which the content of general information is eventually lost and the labor involved in analyzing and defining subunits of the environment gets to be out of proportion to the return in an understanding of the factors underlying the observed distributions and associations. Jones (1950) and Thorson (1958) have presented good summaries of progress in the study of marine benthic communities.

The other approach to synecology is more analogous to the approach of a geneticist or physiologist to organisms and has been termed the **comparative approach.** In this approach the trends of faunal change in different types of habitat are compared, or the effects of a variable environmental parameter over a range of different situations are studied. The work of Wieser (1953, 1954, 1957, 1959) on free-living marine nematodes from different sorts of substrate exemplifies the comparative approach.

The two approaches are, of course, not mutually exclusive, any more than are taxonomy and physiology. Both, judiciously employed, are necessary to a proper understanding of biotic communities.

Communities are more than chance aggregations of species. They are relatively permanent associations. Not only are the communities of a given region fairly well defined and the order of importance of the species within a given community rather constant, but parallel communities occur in different latitudes or in widely separated regions where appropriate combinations of environmental conditions occur. Such parallel communities commonly show very similar structure and species composition but with a number of the species of one replaced by ecologically equivalent species in the other, sometimes belonging to the same genera.

Because of different growth rates, lifespans, differences in success in settlement of immature individuals from year to year, and fluctuations in the environment which do not affect all members of the community in the same way, a given community may show wide fluctuations in relative abundance of some of the species from year to year or over a longer span of time.

Communities, like individual species,

may also show the phenomenon of submergence. A community that occupies shallow inshore water in the far north tends to follow the isotherm into more southern latitudes, being found there in deeper offshore water, while warmer-water communities replace them near shore.

Mare (1942) divided the benthos into three categories—the **macrobenthos,** including all organisms too large to pass through a 1 mm. mesh sieve, the **meiobenthos,** including those which will pass through a 1 mm. mesh but be retained by a 0.1 mm. mesh, and the **microbenthos,** those forms that pass through a 0.1 mm. mesh.

The macrobenthos contains all the obvious animals and plants included in most surveys of communities. The meiobenthos includes many small copepods, nematodes, flatworms, the most minute species and immature stages of various molluscs, worms, and other animals, rotifers, gastrotrichs, gnathostomulids, the larger foraminifers, and others. Individually, such forms commonly range from 0.05 to less than 0.001 mg. in weight. The microbenthos is composed primarily of small protozoans, especially flagellates, ciliates, and amoebae, and of bacteria.

Until recently it was thought that the upper few centimeters at the immediate surface of muddy or fine, sandy substrates comprised by far the richest layer for all groups of meiobenthos and most groups of the microbenthos. This does indeed seem to be true for most groups. It is now apparent, however, that in the case of the gnathostomulids, which dominate the anaerobic layers in fine sands, the fauna becomes richer in individuals and more diverse in species below this surface layer. This situation is somewhat analogous to the recent finding that the diversity of benthic foraminiferans of the sea bottom increases with greater depth and distance from shore.

In recent years careful attention to the microfauna, to the precise nature and organic content of the sediments, to the effects of organisms such as predators belonging to the nekton, to the effects of plankton blooms and of the biological properties of the water masses passing over the communities, and to numerous other factors has added greatly to our understanding of the dynamics of infaunal communities, as well as to our understanding of the changes observed in them from one time to another and from place to place.

In most communities the bulk of the standing crop, or biomass, is present as macrobenthos. However, the number of individuals and the rates of metabolism and reproduction among the meiobenthos and microbenthos are so much greater that it is these elements of the community that may often be responsible for the greater turnover of organic matter and the greater total production over a period of time.

The dominant members of the macrobenthos are plankton feeders and detritus or substrate ingesters. Although always far less in total biomass, the carnivores are not necessarily fewer in species and may even exceed in this respect some of the more dominant categories.

Areas of extremely cold water, such as those found in the arctic or antarctic regions, tend to be characterized by communities of fewer species, but sometimes there is a very large standing crop of those species that are present. The individuals are commonly larger, longer lived, and slower growing, and a greater proportion of them show direct development. They are communities in which there is a high biomass of the dominant species but relatively slow rates of production. Changes of the composition of such communities tend to be slow. The overall result is that such communities tend to be very uniform.

Only a few examples of the many shallow benthic communities that have been recognized can be mentioned here. Some of those recognized by Petersen and other early workers along the boreal coasts of Europe and the North Sea and along the Danish coasts of the Baltic are given below.

Macoma baltica shallow-mud community

This is one of the most common communities of the boreal Atlantic waters. It is characterized by the clams *Macoma baltica* and *Mya arenaria* and the cockle *Cardium edule*. The lugworm *Arenicola marina,* the gastropod *Hydrobia ulvae,* and the amphipod *Corophium volutator* are also prominent members. This shallow-water community usually extends from the tidal zone to about 10 meters in muddy bottoms. Where there is somewhat finer sediment with some clay or silt, the bivalve *Scrobicularia plana* may occur, whereas in places where the mud is more sandy, *Cardium* tends to be more abundant. Since *Macoma, Cardium,* and *Hydrobia* have relatively long life-spans of two or more years, the rate of turnover is rather low.

Thorson (1958) demonstrated that communities similar to this but with different species of *Macoma* and comparable species of *Cardium, Mya,* and *Arenicola* occur in the Arctic, along the North American Atlantic coast, and in the Northeast Pacific.

Tellina-Donax shallow-sand community

This assemblage is found on rather exposed sandy shores from about the lower edge of the tidal zone out to about 10 meters, where it grades into the offshore sand communities. In the Atlantic boreal to temperate regions this community is dominated by the bivalves *Tellina tenuis* and *Donax vittatus.* Other characteristic animals are the polychaetes *Nephtys caeca* and often *N. hombergi,* the lugworm *Arenicola marina,* and the sand star *Astropecten irregularis.* The amphipod *Bathyporeia pelagica* is also common. In slightly deeper water with finer sediment, *Tellina fabula* may largely replace *T. tenuis,* and the heart urchin *Echinocardium cordatum* becomes abundant.

Venus community

Where the sea is slightly more open and the sandy bottoms go to a depth of about 50 meters, the boreal communities are often dominated by the bivalves *Venus gallina* or species of *Spisula,* the latter tending to become dominant where the sand is rather loose. Other bivalves characteristic of this community are *Mactra* and *Psammobia.* The polychaete *Nephtys* occurs in this community as well as in the *Tellina-Donax* complex, as does also *Ophelia* and sometimes *Pectinaria.* Other polychaetes common in *Venus* communities include *Glycera* and *Lumbriconereis.* The heart urchins *Echinocardium* and/or *Spatangus purpureus* are common, as is sometimes the sand star *Astropecten* and the gastropod *Natica.* Where the sand is somewhat compacted or firmer, *Tellina* may be present as well as the brittle starfish *Ophiothrix fragilis.*

Where substrate is coarser, composed of shell gravel and coarse sand, larger, more solid bivalves such as *Venus fasciata, Glycymeris glycymeris, Ensis arcuata,* and *Tellina crassa* tend to dominate. The heart urchin *Echinocardium flavescens* occurs here, as does also the lancet *Branchiostoma (Amphioxus).*

Syndosmya (Abra) muddy-sand community

This type of community occurs in relatively sheltered, sometimes somewhat estuarine situations, fairly rich in organic matter. The predominant animals are small, thin-shelled bivalves such as *Syndosmya alba, S. prismatica, Nucula turgida,* and *Cultellus pellucidus.* The brittle star *Ophiura texturata* and the heart urchin *Echinocardium cordatum* are common echinoderms in these situations. Polychaetes include species of *Glycera, Nephtys,* and *Scalibregma inflatum.* The abundance of small clams and polychaetes makes this community a good feeding ground for fish. Offshore these communities may become more dominated by brittle stars and grade into associations dominated by *Amphiura filiformis* and *Echinocardium cordatum.* The tall-spired *Truritella communis* is common in these communities and in places may at times dominate them. Other common gastropods are *Aporrhais pes-pelicani* and *Philine aperta.*

Farther offshore the sandy muds tend to grade into finer-grained, soft mud areas in which the brittle star *Amphiura chiajei* and the heart urchin *Brissopsis lyrifera* are conspicuous elements. The small clams *Syndosmya prismatica* and *S. nitida* and species of *Nucula* are fairly abundant, together with numerous polychaetes such as *Nephtys incisa, Glycera rouxi,* and *Lum-*

briconeris impatiens, and sedentary worms such as capitellids and maldanids. At the outer (deeper) portions of these communities the brittle star *Ophiura sarsi* may become prominent, and additional polychaetes not found prominently in the shallower zones, such as *Terebellides stroemi, Chaetozone setosa,* and *Melinna cristata,* are present.

Dendraster excentricus community

In somewhat protected shallow areas or in shallow sand beds beyond the intertidal zone along the western coast of North America occur beds of the purple sand dollar, *Dendraster excentricus,* commonly found partially buried in an oblique position in the sand with about one half to two thirds of the disc above the sand surface.

Heart urchins are also very common in the offshore sandy bottoms.

Turtle grass community

In the West Indies, around Bermuda, and in the Florida Keys in the coral reef zone on coral sands, marine grasses—the turtle grass *Thalassia testudinum,* together with lesser admixtures of manatee grass and shoalgrass—form a characteristic biocoenose. Turtle grass is a broad-leaved marine grass with stout, jointed rhizomes with heavy sheaths at the base. Usually epiphytic algae such as *Fosliella, Ceramium,* and *Aegira* are to be found growing on it. It grows on mud and sand in quiet, protected waters, where it may form continuous meadows, but it also grows where there is some wave action, often forming, in these more ex-

Fig. 8-50. Sand dollar bed, in which *Dendraster excentricus* lies typically almost half buried vertically or obliquely on edge in the sand. West Coast. (M. W. Williams photograph.)

posed situations, patches of growth conspicuously elevated above the surrounding bottom as a result of accumulations of pebbles and shell fragments among the rhizomes. Coralline algae growing among the *Thalassia* add their skeletons to the mass and give it more thickness. *Thalassia* mats of this sort represent a calcareous substrate resembling unconsolidated aeolian limestone, whereas mangroves, in contrast, usually fill in with peaty substrate tapering off into mud and soil. The seaward edge of such *Thalassia* patches is usually abrupt, not tapering, and the mat is about 15 cm. thick.

Characteristic of such patches of turtle grass are the large starfish *Oreaster* and the sea biscuit, the echinoid *Clypeaster subdepressus,* usually buried in the sand. Mas-

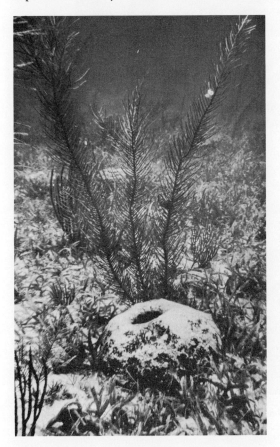

Fig. 8-51. Bed of turtle grass with loggerhead sponge, *Speciospongia vespera,* and tall gorgonian in foreground. (Chesher photograph.)

sive loggerhead sponges and growths of numerous alcyonarians such as gorgonians, soft corals, sea fans, and sea pens are common. Several species of large, predacious conchs—such as the horse conch *Pleuroploca gigantea,* which occasionally grows to 2 feet in length, the chief predator, which is the beautiful queen conch *Strombus gigas,* and the cask shell *Tonna galea*—also occur here, as do occasional octopuses, *Octopus vulgaris,* which are also found on European and Mediterranean shores. Some of the characteristic fish are the pipefish, *Syngnathus* spp., which are often difficult to see because of their body shape and posture, which mimic the turtle grass blades. The related sea horse *Hippocampus* is also characteristic as well as the cowfish *Lactophrys trigonus* and several species of corals. Long-spined active sea urchins *Eucidaris* and *Diadema,* the spiny lobster *Panulirus argus,* the mantis shrimp, the sea hare *Aplysia dactylomela,* and several species of coral are characteristic of this association. Sea turtles, the loggerhead turtle, green turtle, and hawksbill, sometimes visit the *Thalassia* beds to prey on sea biscuits, conchs, and other animals. Sea cucumbers are also abundant, including a variety of the circumtropical large black sea cucumber *Holothuria atra.*

As is commonly true of tropical communities, there tend to be more species present and greater diversity in this community than in comparable temperate or boreal communities, such as eelgrass communities, or in the various soft substrate communities already mentioned.

Somewhat similar communities are found in other tropical regions such as the Indo-Pacific, but many of the species are represented by ecologically more or less equivalent species. In numerous instances, closely related, so-called twin species are found, one in the Indo-Pacific and the other in the Caribbean region (Chapter 14).

Mississippi Delta communities

The Mississippi Delta region offers an interesting example of an area in which a number of types of bottom communities that would normally occupy larger, rather separated areas are brought together in a small area. The Mississippi River discharges huge quantities of sediments and of cold, fresh water into the warmer, saline waters of the Gulf of Mexico. The result is an unusual degree of patchiness of the sediments and nonuniformity of salinity and of temperature of the waters. There are, then, in the area many relatively small patches of sediment resulting from localized differences in temperature, salinity, and flow of the water. This provides a large

Fig. 8-52. Tropical American group of sea animals. **A,** *Diadema antillarum*, long-spined sea urchin typical of tropical waters. **B,** *Oreaster reticulata*, a large starfish, eating a dying sponge. **C,** Flamingo-tongue snail feeding on sea fan. **D,** Hermit crab feeding on plate urchin *(Clypeaster subdepressus)*, photographed at night with flash camera. **E,** Scorpion fish, camouflaged against bottom. **F,** Sea urchin, *Toxopneustes rosei*, with protective cover of debris held over it by means of the tube feet, and, **G,** with the debris removed. (Chesher photographs.)

number of ecological niches in a relatively small area, almost as varied as the conditions found in rocky substrates. Furthermore, the climate varies from freezing weather in winter to subtropical temperatures during summer. Actually, then, there

Fig. 8-52, cont'd. For legend see opposite page.
Continued.

are several communities of benthic soft-bottom animals represented in a patchy manner in a relatively small area. Areas of this sort generally show a considerably greater number of species than are found in more uniform areas of the same size, but they normally do not develop such heavy stands of any particular species.

PRODUCTIVITY OF BENTHIC COMMUNITIES

The standing crop of benthic communities can be estimated in various ways. Estimates may be based upon the numbers of animals of the dominant groups and upon the wet weight of these animals or the weight of dry organic matter present per unit of area. The latter gives the most reliable comparisons between different communities, since the great differences in size between different species, the differences in the proportion of shell material and of water content of the tissues, and other differences make comparisons based on numbers or wet weights very difficult and often misleading.

Productivity of the community is even more difficult to assess because of the great differences in the life-span among different members of the community. The microorganisms present, although not representing the bulk of the biomass, may, because of their more rapid multiplication, be responsible for the greater part of the organic production and turnover. A community with the greatest biomass, or standing crop, is not necessarily the most productive. A community in which the greatest part of the biomass is comprised of relatively large, long-lived animals may have a large standing crop but a relatively small rate of turnover. Such communities tend to be relatively stable. It must also be remembered that in benthic communities most of the actual organic production does not take place in the benthic community itself but rather in the plankton of the waters that pass over the community. What we see in production of the benthos is largely a matter of the extent and efficiency of the conversion of organic materials derived from the plankton rather than primary production.

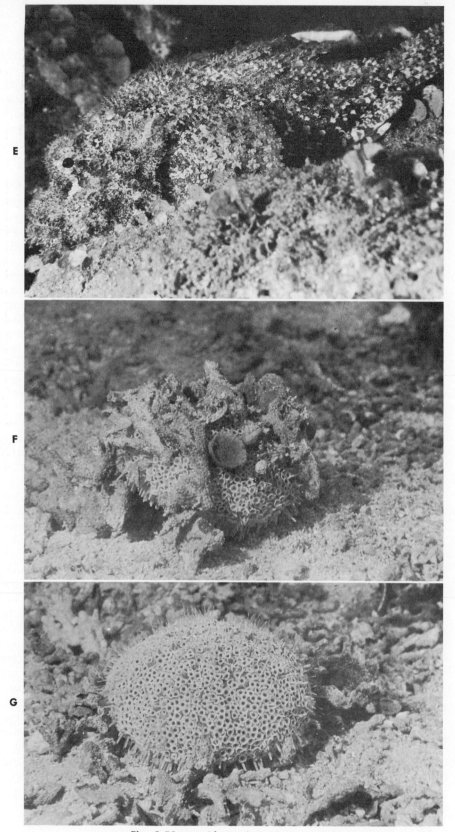

Fig. 8-52, cont'd. For legend see p. 244.

For practical purposes, reliable estimates of the production of economically significant species rather than total productivity of the community is desired.

Petersen (1915, 1918) and Petersen and Boysen-Jensen (1911) estimated that in the Kattegat area of Denmark, the benthic fauna contained about 1 million tons of "fish food" (bivalves, gastropods, polychaetes, and crustaceans), about 5000 tons of flounders, and about 75,000 tons of invertebrate predators such as starfish and brittle stars. Assuming a ratio of consumption proportional to the weights of the predators, he concluded that only about 6% or 7% of the fish food was actually consumed by fish. The rest was consumed by the invertebrate predators. It has subsequently become clear that in reality the invertebrate predators eat about four times as much in a given period of time in proportion to their weight as do the flounders, so that the fish are actually receiving only about 1% or 2% of the fish food. Furthermore, the fish cannot readily take advantage of increases in the standing crop of "fish food," since the invertebrates, with their faster reproduction, can respond to such an increase long before the fish can.

It is often found that the production, or standing crop, of given species or of groups of species in benthic communities on seemingly similar sea floors, sometimes not very distant from each other, varies to a surprising degree. In some cases the reason is fairly obvious, but other instances are found in which no ready explanation is apparent why one area should be very poor in bottom fauna or in certain elements thereof, whereas another, comparable area is quite rich.

In particularly rich beds, the bivalves alone in soft substrates sometimes reach a weight of as much as 300 gm. of dry organic matter per square meter. Usually the values are much less, ranging from only a few grams up to approximately a hundred. Dense mussel beds on rocky stretches may attain more than 3 kg. per square meter.

Part of the difficulty in interpreting the differences between areas may lie in the lack of information on seasonal and other long-term fluctuations of the biomass. The majority of quantitative investigations cover relatively short periods such as a few weeks or one summer. Until we know the magnitude of the changes from year to year and something of their causes, comparisons of different areas based only on short examinations of each are bound to contain large elements of uncertainty. The role of predators, especially of mobile ones, not themselves part of the benthos but able to range over wide areas of sea bottom, is also difficult to assess.

The settlement of larvae of different species occupying the same water mass and therefore subject to the same gross environmental conditions may nonetheless vary greatly and in directions inconsistent with each other or with any general, overall environmental influence. For example, Loosanoff and associates (1955) found that in Long Island Sound, over a period of eighteen seasons, there was no consistent agreement between the number of American oyster larvae and the number of starfish larvae settling on collectors placed in the water.

PREDATOR-PREY RELATIONSHIPS

In soft-bottom communities dominated by the brittle stars, such as *Amphiura* and *Ophiura,* the brittle stars are sometimes so numerous that every square centimeter of the bottom is searched daily for prey. How is it, then, that bivalves and other animals that also occur in these communities are able to establish themselves? Their newly settled young are in the size range utilized by these brittle stars.

The answer seems to lie in the fact that the predatory species take little or no food during their own spawning period. In *Amphiura* the gonads may become so distended that the stomach is very crowded or even histolized. There is a period of about two months when the *Amphiura* are feeding lightly or not at all. During this time many newly settled mollusc larvae will have grown beyond the size range preyed upon by *Amphiura*. Species for which the area would otherwise be satisfactory but whose larvae and young happen to settle

Fig. 8-53. Serpent stars dominating soft bottom surface. (Oregon Institute of Marine Biology photograph.)

when the *Amphiura* are actively feeding are excluded. Similar relations have been established in other predator-prey situations.

This may also be of considerable advantage to the predator. For example, a young *Mytilus* settling at the beginning of the month during which *Asterias* is spawning may grow from about 0.35 mm. to about 10 mm. during this time, increasing its organic matter 2000-fold. More slowly growing forms will increase at least 500-fold. By eating only one prey animal a month after it has settled, the starfish can obtain as much food, expressed in calories, as it might have obtained by eating 500 to 2000 of the same brood a month earlier. The crop of prey animals can then support a much larger biomass of predators without becoming depleted.

A further conservation of prey species is effected by the fact that during the planktonic larval life of the predator species a further grace period of up to three weeks is given, during which many newly settled prey individuals grow to a size at which they are safe from the attack of newly settled, extremely voracious young of the predators.

Thus the physiological rhythms of predator and prey and their relative growth rates often play an important role in determining the species composition of animal communities. Since most communities include numerous different species of both predator and prey forms and since all the species go through a series of critical size ranges during their growth, during which times they prey upon different species and are subjected to predation by different species, these relationships can become very complex and can produce fluctuations in the populations of various species not explainable by any simple change in the environmental conditions.

INFLUENCE OF THE BIOLOGICAL PROPERTIES OF THE WATER MASSES

Subtle properties of water masses may strongly influence the biology of organisms over which these water masses pass. For example, Wilson (1951a), Wilson and Armstrong (1958), and subsequently numerous other investigators have found that apparently **external metabolites,** or **exocrines,** of organisms in the water in very high dilutions exert far-reaching influence. Wilson and co-workers demonstrated that off Plymouth, when water masses containing a certain characteristic assemblage of plankters, termed *Sagitta elegans* water, were present, the water was eminently satisfactory for use in culturing the larvae of certain benthic forms. When, on the other hand, other water masses with a somewhat different assemblage, termed *Sagitta setosa* water, were present, these same larvae could not be nearly as readily cultured. The addition of some *S. elegans* water to batches of *S. setosa* water rendered them satisfactory. The two water masses could not be differentiated on the basis of the physical-chemical properties tested by conventional hydrographic techniques. In the North Sea and the English Channel, *S. elegans* water indicates at least some admixture of open Atlantic water with the coastal water masses. This mixed water, for some reason, supports a population of *S. elegans* and certain other plankters not found in unmixed waters of either the open sea or coastal water masses and is intimately bound up with the success of the larvae of various benthic forms of the region. Not all benthic species are affected the same way or to the same extent by the biological properties of the water masses passing over them, so that the effects may be highly selective.

The biological condition of the water masses is, of course, of great importance to the plankton and nekton in these water masses as well. The mixed waters indicated by the presence of such forms as *S. elegans* and *Clione limacina* also seem to be correlated with the shoals of herring, whereas *S. setosa* water is correlated with very poor fishing conditions for herring. Thus the questions of biological conditioning of the water take on great practical as well as theoretical importance. Since the herring remain associated with this "mixed water" even during the non-feeding period immediately prior to spawning, it has been suggested that perhaps it is not simply the presence of swarms of *Calanus,* their principal food, that brings the herring but that perhaps both herring and *Calanus* are drawn to this water because of its biological properties.

SPECIATION IN INFAUNAL AND EPIFAUNAL GROUPS

Most epifaunal groups, such as prosobranch and nudibranch gastropods, crabs, amphipods, etc., show large increases in the number of species as one proceeds from polar regions toward the tropics, whereas most pronounced infaunal groups, such as level-bottom bivalves, cumaceans, brittle stars, and holothurians, do not. In the far north and along the shores of Antarctica and the antarctic islands, the epifauna of the intertidal and upper subtidal depths is severely limited or eliminated by ice scouring. The more stable temperature conditions as one approaches the tropics permits greater diversification to occur, whereas the seasonal changes in temperature in the temperate and boreal regions tend to restrict the numbers of species because of the increased range of environmental extremes with which they must cope. The overall result among epifaunal groups is the development of epifauna consisting of fewer species but more individuals of those species that can live in the northern and southern regions and of more species but in less massive growths of most of the species in tropical regions.

In the case of the level-bottom infauna, these conditions are much more constant in almost all regions, and the number of types of microlandscapes that might provide habitats for new species and absorb them is comparatively restricted. Thus for infauna, the differences in conditions that lead either to restriction of the number of species or to diversification are much less between different latitudes than is the case for the epifauna.

DEEP-SEA BENTHIC COMMUNITIES

The formidable difficulties involved in sampling and photographing the benthos of the abyss, as well as the impossibility to date of making any but the most limited direct observations in this region, have greatly retarded our understanding of the communities of organisms found on the deep-sea bottom. Prior to 1872 only a few scattered collections had been made, notably those of Allman and Milne-Edwards, who described the profuse growth of hy-droids, alcyonarians, corals, worms, mus-sels, and bryozoans on a broken transat-lantic cable recovered from a depth of 2160 meters; the dredgings of O. Torell be-tween Greenland and Spitsbergen at 2500 meters; and the work of M. Sars and G. O. Sars in Norwegian waters and that of Charles Wyville Thomson and Carpenter northwest of Scotland.

The field was really opened up by the Challenger Expedition under the leader-ship of Charles Wyville Thomson, from 1872 to 1876—probably the most important and successful expedition of deep-sea ex-ploration ever undertaken. The fifty vol-umes of the *'Challenger' Reports* opened up a new era in marine biology and ocean-ography. Subsequently there have been nu-merous important expeditions in various parts of the world and important techno-logical advances, especially in the last decade, which have enormously extended and consolidated our knowledge of the deep-sea conditions and of the life of the abyss.

Two deep-sea benthic zones are com-monly recognized, the **archibenthal** and the **abyssal.** The archibenthal zone includes the continental slopes beyond the edge of the continental shelf, down to approximately 1000 meters; and the abyssal zone, the floor of the sea below this. The boundary be-tween the two is not always sharp. Probably

the best distinction is the limit of the hemi-pelagic sediments, which were derived in part from terrigenous sediments swept out over the continental slope and containing more organic matter than typical eupelagic sediments of the abyssal zone proper. Tem-peratures in these depths are low and quite constant, varying from a little below 0° C. in polar regions to highs of about 4° C. in the abyssal zone and about 9° C. in the archibenthal zone a little north of the equator. Hydrostatic pressures here reach levels at which the rates and nature of cer-tain enzymatic and chemical reactions may be affected, and the physiology of abyssal forms may differ in certain characteristic ways from that of shallow-water species.

The nutritional value, or organic con-tent, of the sediments decreases as one goes farther from shore. Murray (1895), report-ing on the dredging and trawling stations of the Challenger Expedition, compared hemipelagic and eupelagic sediments with regard to the number of species and num-ber of specimens of macrobenthos taken, omitting the South Sea stations, which were richer in both the hemipelagic and the eupelagic regions (Table 8-3).

Ekman (1953) divides the deep-sea fauna into two components: the **eurybathic** spe-cies and the **endemic deep-sea** species.

The eurybathic species are forms occupy-ing a wide range of depths and commonly found on the continental shelf as well as

Table 8-3. Numbers of species and of individual animals taken per dredge or trawl haul in representative types of marine sediments*

	Species	Specimens
Dredging		
Hemipelagic deposits	25	57
Globigerina ooze	5	6
Red clay	1.7	4.2
Trawling		
Hemipelagic deposits	32	90
Globigerina ooze	15	39
Red clay	9	30

*From Murray, J. 1895. Summary of the scientific results obtained at the sounding, dredging, and trawling stations of H.M.S. Challenger. 'Challenger' Rep. **2:**797-1608.

in the greater depths. Table 8-4 lists a number of characteristic eurybathic species, with the range of depths from which each is taken.

The endemic deep-sea species and usually the genera to which they belong are mostly limited to the deep sea and are not found on the shelf. Following are some of the most characteristic:

Porifera
  The glass sponges, class Hexactinellida: This entire group, about 375 species comprising some

Table 8-4. Some eurybathic animals and the depths from which they have been taken*

Porifera	
Thenea muricata	30 to 3440 meters
Stylocordyla borealis	2 to 3000 meters
Tentorium semisuberites	26 to 2970 meters
Polychaeta	
Lumbriconereis impatiens	At least to 3000 meters
Glycera rouxi	At least to 3000 meters
Notomastus latericeus	At least to 3000 meters
Hydroides norvegica	At least to 3000 meters
Pomatoceros triqueter	5 to 3000 meters
Amphicteis gunneri	20 to 5000 meters
Cirripedia	
Verruca stroemia	Littoral to 3000 meters
Cumacea	
Diastylis laevis	9 to 2980 meters
Eudorella truncatula	9 to 2820 meters
Isopoda	
Antarcturus furcatus	10 to 3010 meters
Lamellibranchiata	
Limopsis aurita	38 to 3175 meters
Astarte sulcata	10 to 2000 meters
Scrobicularia longicallus	36 to 4400 meters
Gastropoda	
Neptunea curta	8 to 2580 meters
Neptunea islandica	30 to 3000 meters
Puncturella noachina	8 to 2000 meters
Scissurella crispata	12 to 2300 meters
Natica groenlandica	3 to 2300 meters
Natica affinis	0 to 2600 meters
Scaphander punctostriatus	35 to 2800 meters
Asteroidea	
Pseudarchaster pareli	15 to 2500 meters
Henricia sanguinolenta	0 to 2450 meters
Ophiuroidea	
Ophiacantha bidentata	5 to 4400 meters
Ophiopholis aculeata	0 to 2040 meters
Ophiura sarsi	10 to 3000 meters
Ophiocten sericeum	5 to 4500 meters
Echinoidea	
Echinocardium australe	0 to 4900 meters
Holothuroidea	
Mesothuria intestinalis	20 to 2000 meters

*From Ekman, S. 1953. Zoogeography of the seas. Sidgwick & Jackson, Ltd., London.

Fig. 8-54. *Euplectella aspergillum,* glass sponge characteristic of deep-water soft substrates.

80 genera grouped into 15 families and 2 subclasses, is characteristically found in deep water. The few found on the upper shelf are mostly antarctic, although the bottom of Sagami Bay, Japan, famous for its *Euplectella,* is only 150 to 300 meters deep.

Some representative genera are *Aphrocallistes, Caulophacus, Euplectella* (about 15 species), *Farrea* (11 species), *Hyalonema* (about 90 species), *Monorphaphis,* and *Rossella* (mainly antarctic). Among the Demospongiae, the genera *Thenea* and *Cladorhiza* are largely deep-sea animals.

Coelenterata

Hydroids: Most are shelf forms, but a few, including the huge *Branchiocerianthus imperator* are exclusively deep-sea animals.

Alcyonaria: Gorgonians and pennatularians are well represented in deep water. Among the gorgonians, *Caligorgia, Chrysogorgia* (about 30 species), *Narella,* and *Thouarella* are almost wholly archibenthal. Twelve of the fourteen families of pennatularians, or sea pens, have archibenthal or abyssal species. The families Anthoptilidae, Funiculinidae, Kophobelenmonidae, Protoptilidae, and Umbellulidae comprise almost wholly archibenthal and abyssal forms. Chunellidae are purely abyssal.

Corals: Most deep-sea corals are solitary and are found in greatest numbers in the Sulu Sea of the Indo-Pacific. *Lophohelia* and *Amphihelia* are representative.

Zoantharians: The genus *Epizoanthus* contains both shallow-shelf and deep-sea species. Genus *Izozoanthus* is a truly deep-sea genus.

Antipatharians: All but two of the genera are represented in the deep-sea fauna, and five are exclusively so.

Ctenophora: This group is chiefly pelagic. The peculiar platyctenids, somewhat resembling polyclad flatworms superficially, are benthic, sometimes found in fairly deep water. The genus *Tjalfiellidea* is found at a depth of 475 to 575 meters off western Greenland, attached to the sea pen *Umbellula lindahli.*

Annelida: Most deep-sea polychaetes belong to genera that are also represented in the shelf faunas.

Pogonophora: This peculiar group, the tubes of which were for some time thought to be those of deep-sea annelids, seems to be limited to and characteristic of the deep-sea benthos.

Arthropods

Decapod crustacea

Brachyura: The brachyurans, or true crabs, are rather poorly represented in the deep-sea fauna, considering the size of the group and its great prominence in shelf fauna. Two families, the Homolodromiidae and the Homolidae, and the subfamily Tymolinae, with much-reduced eyes, are deep-sea forms. Single genera from other families, such as *Platymaja* and *Geryon,* and some species from genera represented chiefly in the shelf fauna, are also deep-sea forms. Some are calm-water forms with very long appendages.

Anomura: In contrast with the true crabs, the Anomura contain many deep-sea forms. The group Galatheidea contains the truly deep-sea genera *Uroptychus, Galacantha,* and *Munidiopsis,* modified for deep-sea life with reduced eyes and greatly lengthened antennae. There are over 100 species of *Munidiopsis,* most of them not occurring above a depth of 400 meters, and half not above 800 meters. The large genus *Munida* is preponderantly archibenthal. Among the Paguridea, the family Pomatochelidae, often living in *Dentalium* shells, are mostly deep-sea forms, and, except in cold regions, the Lithodidae are mostly deep-water forms. Over half of the family Axiidae are deep-water forms.

Macrura: There are numerous deep-sea Reptantia—most of the family Eryonidae, with the genera *Polycheles* and *Willemoesia* usually abyssal, and showing reduced eyes. Among the Astacura are *Phoberus, Thaumastocheles,* and *Nephropsis.* The Natantia, or shrimplike forms, have the deep-sea families Hoplophoridae and Glyphocrangonidae, as well as bathybenthal genera belonging to other groups, such as *Benthecicymus, Benthonectes,* and *Amalopenaeus.*

Amphipoda and **Isopoda:** Rather poorly represented are Amphipoda and Isopoda, most of the deep-sea genera containing only a few species. The isopod *Munnopsis*, from medium depths, shows great elongation of some of the appendages, representing an adaptation to living in the loose ooze. The genus *Nannoniscus* and the amphipod genus *Trischozoztoma* are confined to deep waters, and contain a few species each.

Cumacea: The two families of deep-sea Cumacea are both very small families—the Platysympodidae, with two monotypic genera, and the Procampylaspidae, with one genus containing only a few species. Other families generally contain some deep-sea species, and the genera *Macrocylindrus* and *Bathycuma* are purely abyssal.

Barnacles: The genera *Megalasma*, *Hexalasma*, and *Verruca* are purely deep-sea forms, and the genus *Scalpellum* includes a number of archibenthal and abyssal species.

Pycnogonids: Most genera of the Pycnogonida seem to be rather eurybathic. The genus *Pipetta* contains a few purely abyssal species from the Indo-Malayan and Antarctic regions.

Molluscs: A number of cephalopods, both squid and octopods, are characteristically demersal deep-water forms. Among the decapods the families Cirroteuthidae, Opisthoteuthidae, and Stauroteuthidae may be mentioned, comprising about 25 species. Of the octopods, the genera *Bathypolypus* and *Benthoctopus* contain about 20 species. Deep-sea species of most other mollusc groups are known, but most are members of genera or families better represented in the shelf fauna.

Echinoderms: Two of the classes of echinoderms, the Crinoidea and the Holothuroidea, are among the most characteristic and distinctively deep-sea faunal elements. Well over half the known species of crinoids are purely archibenthal or abyssal, while several other species found on the shelf occur also in deeper water. The mobile unstalked comatulids are mostly shelf forms, but even in this group the families Atelecrinidae and Thalassometridae, and the genera *Trichometra* and *Thaumatometra* of the Antedontidae, are deep-water forms. The majority of the typical stalked crinoids, including all of the families Pentacrinidae, Apiocrinidae, Hyocrinidae, Phyrnocrinidae, and Bathycrinidae, are deep-water forms. The last-mentioned family contains members found more than 5000 meters deep.

Two of the five orders of Holothuria, the Elasipoda and the Molpadonia, are restricted to deep, mostly abyssal areas. Among the other three orders, most of the family **Synallactidae**, order Aspidochirota, with about 16 genera, are abyssal. The genera *Staurocucumis* (Dendrochirota) and *Acanthotrochus* (Apoda) also are deep-sea forms. The holothurians are the most important group of deep-sea substrate ingesters. In some regions they may be responsible for a very considerable reworking and redistribution vertically of marine sediments. For example, in certain areas where there has been a deposition of volcanic ash at a given period, instead of forming a clean-cut layer of sediment marking the time of the deposition, the ash particles are found distributed through a considerable vertical accumulation of the sediments. This is thought to be the work of sea cucumbers. There is some evidence that great herds of sea cucumbers may move along the sea floor, working over an area and moving on to new areas somewhat in the manner of a herd of sheep.

The asteroids, echinoids, and ophiuroids also all have characteristically deep-sea groups, although in most cases they are not as distinctive or as well marked off from the shelf-inhabiting forms as is the case with the crinoids and sea cucumbers. Among the starfish, the long-armed **Brisingidae,** and the **Porcellanasteridae,** characterized by their high marginal plates, are mostly abyssal. The species *Albatrossaster richardi* of the latter family occurs at 6000 meters or more. Some species such as *Ctenodiscus crispatus* that occur in shallow arctic waters are archibenthal or abyssal in other latitudes. Two other families containing mostly deep-sea species are the **Benthopectinidae** and **Zoroasteridae.**

The ophiuroid family **Asteronychidae** is a deep-sea family, and most other ophiuroid families contain one or more deep-sea genera.

Among the Echinoidea the suborder Meridosternata (families **Pourtalesiidae, Urechinidae,** and **Calymnidae**) are purely deep-water forms in distribution. The Pourtalesiidae are remarkable in having a very elongate body, with the mouth sunk into a deep furrow in the anterior part. Other deep-sea families of echinoids include **Aspidodiadematidae, Echinothuriidae, Pedinidae,** and **Saleniidae.** Some of the echinothuriids are remarkable for their large size and their soft bodies, which collapse if lifted from the water.

Tunicates: Most tunicates are either part of the shelf fauna or are eurybathic. Only about one fifth of the genera—and most of these have only a few species—are typically archibenthal or abyssal. The deep-sea families are Pterygascidiidae, Hypobythiidae, and Hexacrobylidae, all rather small families.

Fish: Although most deep-sea fish are pelagic, there are a considerable number of deep-water demersal or benthic forms. Only a few groups will be mentioned here. Deep-water elasmobranchs include the genera *Centrophorus* and *Centroscyllium*. The whole group of Holocephali are preponderantly deep-water forms. *Chimaera* is archibenthal and abyssal, and *Harriotta* is purely abyssal.

Among the teleosts, the whole family Macruridae (approximately 140 species) is archibenthal or abyssal. Early writers included most of these fish in the genus *Macrurus*, which has subsequently been divided into a number of genera.

The large family Liparididae (about 125 species) is likewise comprised mostly of archibenthal and abyssal forms. The large genera *Careproctus* and *Paraliparis* occur in the shelf fauna in cold regions, but submerge to deeper water in other regions. Most genera of the Zoarcidae are archibenthal or abyssal. *Bassozetus* and *Porogadus* are purely abyssal, while *Lycodes, Lycenchelys, Neobythites,* and *Dicrolene* are archibenthal to abyssal. There are also about 15 monotypic abyssal genera in this group.

The Halosauridae and Notacanthidae (approximately 35 species) are almost wholly abyssal. Among the Pediculati, the family Ogcocephalidae contains 3 deep-sea genera, *Dibranchus, Halieutaea,* and *Malthopsis,* in addition to shelf forms. Several genera of codlike fish, Gadiformes, are preponderantly archibenthal and abyssal.

Summary

With respect to the benthic fauna of the deep sea as a whole, it can be said that it is characterized by a large number of forms that are almost completely lacking in the shelf fauna. Among these forms are several fairly high taxa, such as whole families, suborders, or orders of organisms found only in the deep sea. The fauna are poor as compared with the shelf fauna, but they contain a relatively high proportion of endemic forms, restricted to the deep sea. This is particularly true of the abyssal fauna on eupelagic bottoms. The deeper and farther from land one goes, the fewer are the eurybathic species found, and the fauna become more abyssal in character.

This is in line with Thienemann's rule that the more specialized and rigorous the environment, the poorer in species it becomes. The second part of his generalization, that although the number of species decreases there is usually a great increase in the number of individuals of those species that do inhabit such an environment, does not apply to the deep-sea fauna, since the relatively small amount of food in the abyssal regions imposes a low absolute limit on the development of the benthos.

Recent surveys using newly developed, more adequate deep-water benthos-sampling equipment, such as the anchor dredge and epibenthic sled of Sanders and Hessler (1969), have demonstrated that the benthic fauna of deep waters display a very high degree of diversity, comparable to those found in tropical shallow waters—a striking confirmation of the general rule that long-continued stability of the environment is a major factor leading to diversity of its biota.

PART IV

MARINE
ANIMALS
of
TERRESTRIAL
ANCESTRY

CHAPTER *9*

Marine birds

Marine birds are reliable indicators of the productivity of the surface waters of the oceans. Wherever they are found in great concentrations, there the waters will be teeming with life. Several different groups of birds have independently become adapted, each in its own distinctive and peculiar ways, to obtaining their sustenance from the seas. Some of them have become so completely adapted to oceanic life that they practically never come to the land except to breed. Most seabirds breed either on offshore islands or on steep cliffs facing the sea, where they are protected from most predators. These breeding grounds are often overwhelmingly crowded by the nesting populations, with thousands, hundreds of thousands, or even millions of birds crowded together in relatively small areas, using every available site. In particularly rich areas, such as around the offshore islands of Peru, the breeding islands become so heavily covered with the bird droppings, or guano, as to form the basis for a very profitable fertilizer industry.

During the nonbreeding portions of the year some oceanic birds fly many hundreds of miles, or even several thousand miles over the open ocean, far away from land. That they manage to find their way unerringly back to their nesting site, often only a very small island, across hundreds or thousands of miles of trackless ocean is one of the great marvels of animal navigation. Certain species have been transported experimentally all the way across the North Atlantic from the British Isles to North America in covered cages, so that there was no chance of their seeing anything en route. After release they were in some cases back in the same nest within a two-week period. Others have proved their capacity to fly back to Midway, two small islands in the Central Pacific, after having been similarly transported to such distant places and different directions as Washington, Alaska, and Japan.

The means by which this navigation is effected is still not clear. A recent promising suggestion is that possibly the birds are sensitive to polarized light and that the set or angles of polarization by reflection from sky and sea, which differ in each part of the world, depending on one's relation to the sun, may in some way guide them back. This would require a fine sense of time as well, since the angles of light change throughout the day and are largely reversed between morning and evening as well as undergoing some seasonal changes.

Birds that spend long periods at sea also must be adapted to meeting their water requirements from the physiologically dry, salty seawater, something which their terrestrial ancestors and relatives are unable to do. Like the teleost fish, they meet this requirement in two ways—through the evolution of special salt-excreting mechanisms and through reduction and concentration of the urine so that only the minimum required for elimination of their nitrogenous wastes is formed, most of the water being absorbed in the kidney tubules.

Major groups of marine birds

ORDER PROCELLARIIFORMES—albatross, petrels, fulmars, shearwaters

This group of birds, sometimes called the tube-nosed swimmers because of the elongation of the nasal bones covering the nasal aperture for a distance, comprises the most truly oceanic group of birds. They are long-winged, web-footed birds ranging in size from the huge wandering albatross with a wingspread of up to 11½ feet, the greatest of any living bird, to the smallest of the seabirds, certain petrels no larger than most of our familiar songbirds. Most of them nest on remote oceanic islands but spend months or even years at sea, sometimes several thousand miles from their nesting site, mostly out of sight of land. They can drink seawater with impunity, excreting the excess salt from enlarged nasal glands near the base of the beak. When captured or seriously disturbed they emit a highly unpleasant smelly, oily, musky fluid from the nose and mouth. It has been suggested that this fluid, in addition to its defensive use, serves as a means of eliminating excess vitamin A, accumu-lated because the diet is rich in vitamin A.

Some of them perform elaborately stylized courtship and recognition rituals. Incubation is shared by the parents and in the larger species, may last up to seventy or eighty days. The young of some of the largest species remain in the nest as long as eight months or more, leaving only when finally deserted by their parents. They are fed regurgitated food by both parents and become very fat.

FAMILY DIOMEDEIDAE—ALBATROSS. There are thirteen species of albatross. Nine of them are confined to the cold waters of the southern oceans and three to the North Pacific. One species is equatorial but feeds in the cool waters of the Humboldt Current in the eastern Pacific. Most species breed only every other year. The nest is either a hollow in the ground or a low mound of moss and twigs. Albatross are masters of gliding flight and can glide for hours on seemingly motionless wings. They alight on the water to feed, eating mostly fish and squid. Some of them follow ships to feed on the refuse. When taking off from the water, they must run over the surface with outstretched, flapping wings for a considerable distance to attain enough speed to become airborne. If approached by a boat while on the water, they often try to escape by swimming rather than flying because of the difficulty of takeoff.

Perhaps the most famous species is the great wandering albatross *Diomedea exulans* of the southern oceans. When mature, the male is a huge white bird with black wingtips, a black trailing edge of the wing, and mottled tail and wing coverts. The young are brown, becoming whiter at each molt. They spend as much as four or five years at sea before returning to their nesting site and are thought to sometimes circumnavigate the globe in the Southern Hemisphere during their wanderings.

The Laysan albatross, or gooney, became well known during World War II in Midway and on other Pacific islands for its persistence in occupying its nesting sites in spite of the conversion of these sites into airstrips by the military.

The only albatross of equatorial waters

is the Galápagos, or waved, albatross, which has its rookery on Hood Island of the Galápagos group and fishes in the cool waters of the Humboldt Current.

The short-tailed albatross, *D. albatrus,* the only species in the North Pacific and formerly the most abundant species to reach the offshore waters of western North America, was virtually exterminated by professional plume hunters, who raided their nesting sites.

FAMILY PROCELLARIIDAE—PETRELS, FULMARS, SHEARWATERS. These birds also are long-winged pelagic birds that spend most of their time far out at sea. They range in size from small petrels no larger than our ordinary songbirds to large fulmars and shearwaters about the size of a pigeon to that of a goose. They feed at the surface of the water on small surface animals or floating flesh.

Fulmarus glacialis, of the North Atlantic and Arctic waters, is about the size and color of a herring gull. It breeds on cliffs, eats fish and offal, and often follows whaling boats, feeding on the blubber. The eggs, oil, and feathers are valued. Related species occur in the North Pacific and in the Antarctic.

Ossifraga gigantea, the giant fulmar, sometimes called Mother Carey's goose, is almost as large as an albatross. It occurs in the southern seas and ranges northward along the western coast of North America to the United States.

Priocella antarctica, the silver-gray fulmar, is one of the most abundant birds on the pack ice throughout the circumpolar antarctic waters.

Several species of shearwaters, genus *Puffinus,* ranging in size from about that of a pigeon to that of a large gull, are widely distributed in both the Atlantic and the Pacific. They usually skim close to the waves in flight. The Manx shearwater, *P. puffinus,* is common along the eastern North Atlantic, and the muttonbird, *P. tenuirostris,* is a common Pacific species. The latter was destroyed in great numbers for oil and feathers.

Several genera of usually smaller seabirds are known as stormy petrels or Mother Carey's chickens. Most of them are dark colored with a patch of white at the rump. The largest is probably the antarctic giant petrel, *Macronectes giganteus,* about the size of a large gull.

Wilson's petrel, *Oceanites oceanicus,* to which the names stormy petrel and Mother Carey's chickens are often applied, is a small species somewhat resembling swallows in flight. It flies with rapid wingbeat and shows the distinctive white patch at the base of the tail. In rough weather the birds skim along in the troughs of the waves. They breed on barren islands in the Antarctic Ocean but fly as much as 10,000 miles to spend the summer along the Gulf Stream, going as far north as Labrador, and then return to the Southern Hemisphere in winter for the southern summer. This remarkable migration of up to 20,000 miles each year, mostly out of sight of land, enables them to utilize the warm months of the ocean in both hemispheres. They eat small squid and crustaceans found in the surface waters. Feeding flocks look as though they were dancing on the water as they face into the wind just above the water, wings held out stiffly, paddling with their webbed feet. This is the smallest of the oceanic birds and the most abundant species of bird on earth.

The genus *Prion* includes the whalebirds of the southern oceans, characterized by their peculiar broad lamellate bill. They are gregarious and often follow whaling vessels to feed on the blubber, oil, and offal.

Another rather atypical group of petrels are the diving petrels of the Southern Hemisphere, genus *Pelecanoides.* These are rather short-winged, short-tailed forms somewhat resembling auks in appearance and habits and are expert divers.

ORDER SPHENISCIFORMES—penguins

Penguins are stout-bodied, short-legged, flightless seabirds of the Southern Hemisphere, ranging from about the size of a duck to a length of up to 4 feet and a weight of 80 pounds. On land they walk clumsily with erect posture and are well known for their tuxedo-clad appearance and clownish behavior. All but one of the sixteen living species are confined to the Southern Hemisphere, most of them with rookeries in New Zealand, Australia, southern South America, and South Africa and on numerous islands in the Southern Ocean. Two species occupy the fringes of Antarctica, and one is found on the Galápagos Islands.

Penguins are gregarious, nesting in crowded rookeries and commonly swimming in flocks. They swim low in the water with little more than the head and neck exposed, diving and surfacing like porpoises. Unlike most swimming birds, they use the wings for propulsion in the water. The wings are flipperlike with only rudimentary scalelike quills, incapable of flexure but moved with a rotary motion by specially developed muscles. In swimming, the feet are used only for steering. Penguins are expert divers and swimmers and subsist on shellfish, fish, krill, small squid, and other animals they can take in the surface waters or shallow bottom.

Most species make nests of stones or bits of vegetation, in the open, and incubate their one to three eggs in the usual fashion, both parents participating in the chore. The emperor penguin, *Aptenodytes fosteri,* of Antarctica and the king penguin, *A. patagonica,* of the Falkland Islands, Kerguelen, etc. are the largest living species. They lay a single egg, incubate it by carrying it on their feet and covering it with a loose fold of skin. The emperor penguin's rookeries are usually on sea ice adjacent to the shoreline of Antarctica. The male stands on the ice through much of the antarctic winter, incubating the egg on his foot, while the female wanders off to the edge of the pack ice and fishes. She returns in the spring when the egg is ready to hatch, and both share in the care and feeding of the young. During the summer all of them go out to the edge of the pack ice to spend the summer fishing in the rich waters and fatten up for the rigors of the coming winter.

The other antarctic species is the well-known Adélie penguin, a smaller species abundant around Antarctica. It nests on offshore islands and the edge of the main-land during the summer and winters out on the pack ice. Each pair has a fixed nesting site to which it returns each year.

The smallest penguin is the blue penguin, *Eudyptula minor,* abundant around Wellington Harbor and other shoreline areas in New Zealand and said to extend along the coast of eastern Australia. It is only about 1 foot high.

A number of fossil species of penguins are known. One of them, *Palaeeudyptes antarcticus* from the Eocene of New Zealand, reached a length of nearly 7 feet.

ORDER STEGANOPODES—pelicans, cormorants, boobies, frigate birds, tropic birds

These are large fishing birds with big beaks, usually hooked at the tip, and totipalmate feet, all four toes united by a broad web. Most of them are strong fliers and constitute a characteristic element of the coastal fauna in most parts of the world. Some of them are often seen far out at sea as well.

Pelicans, family **Pelicanidae,** are noted for their particularly large beaks, from the lower half of which is suspended a large semielastic gular pouch used in catching fish. Some pelicans fish in groups, herding small schools of fish into shallow water where they can be scooped up in their large pouches. The brown pelican fishes more like a giant kingfisher, soaring about rather high above the surface and plunging into the water in a spectacular dive. Brown pelicans breed in great numbers off the west coast of South America, contributing, along with other birds, to the rich guano deposits there. The white pelican nests inland along lakes of the Great Basin from Great Slave Lake in Canada south to Texas and winters along the seacoast. Flocks of white pelicans are often seen soaring at great heights. Pelicans are awkward on land but strong, graceful flyers. They are commonly seen flying in groups, in a line, alternating a few slow strokes of their large wings with an approximately equal distance of gliding. They are voiceless or nearly so.

Cormorants, family **Phalacrocoracidae,**

are large black long-bodied birds with slender snakelike necks and moderately long bills hooked at the tip. They eat fish, which they chase and catch under the water, both the feet and the wings being used in swimming. They occur in the Pacific from Alaska to New Zealand, being most abundant in the tropics. In the Atlantic they occur along both the eastern and western coasts. Cormorants usually fly low over the water, with the neck and bill extended forward and the broad wings beating faster than the wingbeat of any gull.

In Japan cormorants have been used to catch fish, a ring being placed around their throat to prevent them from swallowing the fish.

Lack (1945) found that where two species, the shag *Phalacrocorax aristotelis* and the cormorant *P. carbo,* were living in the same area, they were not really in direct competition with each other. The shag ate mostly clupeoid fish and sand eels, whereas the cormorant subsisted chiefly on shrimps and prawns, flatfish, gobies, and other fish.

On Albemarle Island of the Galápagos, the Hancock expedition of 1933 reported that there was still a colony of several hundred flightless cormorants but that their continued existence was in danger because of man, of whom they showed no fear.

Boobies and gannets, family **Sulidae,** are large seabirds, somewhat smaller than pelicans but more powerful in flight. They occur in temperate and tropical seas, sometimes great distances from land. They feed on fish. They are great divers, often diving from heights of 60 to 100 meters, clasping their wings to their bodies shortly before entering the water, plunging in at great speed, and going 30 or 40 meters through the water before rising. In tropical waters boobies are also expert at catching flying fish.

In tropical and subtropical seas within easy flying distance from land, wherever there are boobies there will usually be some frigate birds, family **Fregatidae,** also called man-of-war birds. Frigate birds are somewhat larger than boobies but are more expert flyers. Unlike other birds of this order, the frigate birds do not actually enter the water. They plunge toward the sea with a hissing noise and seize fish or other animals in the surface water with their beaks. They are also habitual robbers of other fishing birds, especially boobies and gannets. When a booby rises from the water heavily laden with fish it has captured and swallowed during its dive under water, a frigate bird may swoop down on the booby, causing it to regurgitate the fish, which the frigate bird seizes in midair. Male frigate birds are marked by a large bright red inflatable gular pouch used in courtship and territorial display during the breeding season. The female has a plain white breast.

Frigate birds lack the oil on their feathers that makes most aquatic birds impervious to wetting. They must therefore remain dry, since if they get into the water, it will soak them so that they are unable to rise. For this reason frigate birds must stay within easy flying distance of the land, to which they return at night. On land they are awkward and must alight in trees or bushes or near an overhang or cliff from which they can launch themselves, rather than on flat ground. They have relatively very small, light bodies and short legs but very large wings with up to an 8-foot wingspread. They are probably the most expert flyers and aerial acrobats of all large predatory birds.

ORDER CHARADRIIFORMES—shorebirds, gulls, auks, terns, puffins, guillemots

This is a very large assemblage of birds. There are about 314 species, which are usually grouped into 124 genera, 16 families, and 3 suborders. Although a smaller group than the Passeriformes, with its 65 families, the Charadriiformes are probably more diverse. The various anatomical characteristics upon which the order is based are not constant throughout the group. Three large families, the **Laridae, Scolopacidae,** and **Charadriidae,** contain 72% of the total number of species.

Most members of the order are various shades of brown or gray or are black and white. In several species the bill and feet are brightly colored. There is usually little sexual dimorphism.

Differences in feeding habits are strongly reflected in the structure of the bill, which varies from the short, thick, finchlike bill in the seed-eating **Thinocoridae** to the very long slender bills of some of the **Scolopacidae,** used for probing deeply in soft sand or mud.

The eggs are laid on the ground with or without nesting material. The young are already down-covered upon hatching and in some groups can actively get about and forage. Most of the seabirds belonging to this group nest in colonies. Species living in the tropics are nonmigratory, but those nesting in the temperate, boreal, or arctic zones tend to be strongly migratory in habit.

The great majority of members of the order are associated with large bodies of water. They usually stay near the shore, but some of the seagoing members of the group are pelagic.

Suborder Lari

FAMILY LARIDAE—GULLS AND TERNS. The sea gulls, genus *Larus,* more than any other bird come to mind when one mentions marine birds. They have practically worldwide distribution, occurring along ocean coastlines except in the South Pacific between South America and Australia. Some of them are also found around inland waters. They are large, heavyset birds, usually larger than pigeons, with long wings, webbed feet, rather strong, heavy bills, and short, unforked tails. They are strong flyers. Some of them tend to congregate around fishing docks, wharves, etc. or to follow boats in near coastal waters to feed on the refuse. They are both predatory and scavengers, eating fish, shellfish, crustaceans, and refuse. Some of them often prey on the eggs or young of other birds. When excited or feeding they are noisy, large groups of them keeping up a continual and characteristic din with their short screams. Most of them are white, or white with some black or gray. The young tend to be brown or mottled. A few of the better-known species are discussed below.

The herring gull, *Larus argentatus,* is widely distributed in the Atlantic along both European and American shores. It migrates north in the spring to breed, in North America breeding from Maine and the Great Lakes northward. Tinbergen (1953) has published fascinating studies of the behavior of this species and others, which are among the classics in the early studies in ethology.

The mew, or sea mew, *Larus canus,* is the common European gull. Along the American Atlantic shores the ring-billed gull, *L. delawarensis,* is a rather small, graceful species often seen along sandy beaches, treading the sand at the waterline and searching for coquina clams. Other northern species are the glaucous gull, *L. hyperboreus,* and the great black-backed gull or cob, *L. marinus,* in which the back and upper part of the wings is slate-colored. The smallest gull is the black-headed "little gull" of Europe, only about 11 inches long. The western gull, *L. occidentalis,* is one of the several common species along the Pacific coast of North America. The adult is just over 2 feet in length, with dark mantle and black wing tips but otherwise mostly white. The feet are a pink, pale flesh color, and the stout yellow bill has a red spot near the tip of the mandible. The California gull, *L. californicus,* is often seen in large flocks, inland as well as along the coast. These may serve as strong checks on certain insect and rodent populations. In Salt Lake City, Utah, a monument has been erected to this bird, credited with saving the crops of the early Mormon settlers from ravages of the Mormon cricket.

Some of the other genera of gulls include *Bruchigavia,* of which the Australian silver gull with its coral-colored bill, feet, and legs, is one of the most beautiful species of gulls. *Pagophila* includes the beautiful ivory gull, or snowbird, a northern circumpolar species in which the adult is pure white with black feet. Another northern genus is *Rhodostethia,* which includes Ross's gull, a rather small, rare species with a wedge-shaped tail and the lower parts of rosy color. Genus *Rissa* includes a few species known as kittiwakes, in which the hind toe is short or rudimentary. The only true gulls with forked tails are the Arctic fork-tailed gull, genus *Xema,* and the Galápagos swallow-tailed gull, *Creagrus.*

The terns, subfamily Sterninae, sometimes called sea swallows, are usually smaller than the gulls and have slenderer bills and weaker feet. The tail is commonly forked. They are more graceful and dashing in their flight. They fly at some height above the water with the bill pointed down and plunge into the water to capture small fish. Most terns nest in colonies. Typical terns, genus *Sterna,* are mostly white with a black cap and a bluish-gray mantle. A few of the better-known species are discussed below.

The Caspian tern, *S. caspia,* is a widely distributed form occurring inland, as in the Great Lakes Region, as well as along seacoasts. It is the largest of the terns. The sooty tern, *S. fuliginosa,* is another widely distributed form preferring the warmer waters of the tropical coasts and West Indies. The royal tern, *S. maxima,* occurs along the southern United States and southward and is rather strikingly beautiful, with its black crest and bright orange bill. The fairy bird, or little tern, *S. minuta,* of Europe, Asia, and Africa is one of the smallest terns. The arctic tern, *S. paradisaea,* performs one of the most remarkable migrations of any bird. Its breeding grounds are in the Arctic. In the winter it migrates south to an area south of the Antarctic Circle, a distance of 22,000 miles round trip.

A few of the other genera of terns include the noddies, genus *Anous,* dark brownish, short-tailed terns found in tropical regions, said to be stupid and tame. The fairy tern, genus *Gygis,* of the South Pacific, pure white except for its black beak, is one of the most beautiful species. It is remarkable in its nesting habit. A single egg is laid in a tree crotch, or in a depression or knothole on the bare branch of a tree. It balances there through the two to three weeks' incubation despite the comings and goings of the parent and the blowing of the winds and hatches without falling. The young bird remains there until able to fly.

FAMILY STERCORARIIDAE—JAEGERS AND SKUAS. Jaegers and skuas are large strong-flying rapacious marine gulllike birds, usually dark blackish or sooty brown above and lighter below, with a hooked and cered bill. The female tends to be larger than the male. The birds are largely polar or boreal in distribution and are noted for harassing weaker birds, somewhat in the manner of frigate birds, until they drop or disgorge their prey. They also prey on the eggs and young of other birds. In the Northern Hemisphere some of the species are the jaeger, or pomarine, *Stercorarius pomarinus,* which nests in the Arctic. It is larger and darker than the parasitic jaeger, *S. parasiticus,* which also occurs in the northern seas. In both, the middle tail feathers are longer than the rest, extending well beyond the rest of the tail, those of the pomarine being rather obtuse, whereas in the parasitic jaeger they are pointed. The great skua, or sea crow, *Megalestris skua,* is a dark brown species of the North Atlantic, especially along European coasts. In the Antarctic the south polar skuas, *Catharacta skua* and *C. maccormicki,* are the boldest and most rapacious birds of prey.

FAMILY RHYNCHOPIDAE—SKIMMERS. Among the most oddly modified of the Lari in bill structure and feeding habits are the skimmers of America, southern Asia, and Africa. In these birds, related to terns, the lower mandible is strongly compressed and extends out in front beyond the upper. The pupil of the eye is vertical. The birds are partly nocturnal or crepuscular in habit and feed by flying rapidly just above the water surface with the end of the mandible in the water, skimming off small animals. The American black skimmer, *Rhynchops nigra,* is about 18 inches long, mostly black above and white below, with the base of the bill and the feet red and the end of the bill black. It occurs along the southern coasts of the United States and southward in warm regions.

Suborder Alcae—auks, puffins, auklets, murres, guillemots. These heavy-bodied, short-winged, short-tailed, short-legged, web-footed diving birds with thick, compact plumage belong in the family **Alcidae.** The legs are set far back, giving them an erect posture on land reminiscent of that of penguins. They are diving birds, eating fish, crustaceans, squid, and some larger plankton organisms. There are about 22 species, all limited to the northern Atlantic, northern Pacific, and Arctic regions. Most of them fly rapidly with rapid wing strokes and are expert swimmers and divers, using the webbed feet in swimming. Most of them nest in rookeries on cliffs and islands largely protected from terrestrial predators. The eggs are strongly tapered, reducing their tendency to roll off from the precarious nesting sites.

One of the most unique members of this group, the great auk, *Pinguinus impennus,* was a large flightless bird about 2½ feet high, sometimes called the arctic penguin. The great auk was formerly abundant on both sides of the North Atlantic as far south as Massachusetts on the American side and the northern British Isles on the European side. It was ruthlessly slaughtered by the millions for its feathers, and the eggs were also collected. The last known living specimens were killed near Iceland in 1844.

The puffins, often called sea parrots, are among the oddest-appearing seabirds. *Fratercula arctica* of the North Atlantic is black above and around the neck, sharply contrasting with the white underparts of the large white cheek patches, which include the eyes. The bill is high, laterally compressed, deeply grooved, and short; the triangular basal portion is black, bordered with yellow, and the distal portion is bright orange. The legs and feet are also bright orange, with black claws. The tufted puffin, *Lunda cirrhata,* of the Pacific is dark blackish brown all over except for the white cheek patches, the orange and yellow bill, and a large yellow plume over each eye extending back over the neck to the back. Its legs and feet are also bright orange. The horned puffin, *Fratercula corniculata,* is a Pacific species quite similar to the North Atlantic form but with a small fleshy, hornlike appendage on the eyelid.

The razor-billed auk, or scout, *Alca torda,* found on both coasts of the North Atlantic and on the northern islands, is about 16 inches long, black above and white below. The bill is laterally compressed, sharp-edged, black, and crossed by a white band. A smaller, short-billed auk, the rotche or dovekie, *Alle alle,* is common on the northern ice flows. It breeds along the coasts of the Arctic Ocean and ranges southward in the winter.

Guillemots, sometimes called sea pigeons or sea doves, are dark-plumaged birds of this family, inhabiting rocky cliffs along the seacoast. *Cepphus grylle* is the common black guillemot of the North Atlantic, and *C. columba* is the pigeon guillemot common along the North Pacific coasts.

The rhinoceros auklet, *Cerorhinca monocerata,* of the North Pacific is curious in having a deciduous horn on the bill.

Murrelets, genus *Synthliboramphus,* are rather small members of this group mostly inhabiting islands of the North Pacific but ranging south to Mexico and Japan. The species *S. wumizusume* is the crested murrelet of Japan.

Genus *Uria* includes the true murres, sometimes also called guillemots, of which there are several species.

Other seabirds

Numerous other birds are found along the beaches, in near-shore waters, and in the waters of bays and estuaries. These include numerous shorebirds, such as sandpipers, godwits, willets, curlews, and others of the Charadriiformes; many anserine birds, ducks and their relatives, some loons, grebes, and other diving birds; and a number of large wading birds such as herons, cranes, egrets, and others. Some of the Falconiformes, such as ospreys, also may fish in marine water. Most of these birds belong to groups that might better be termed aquatic rather than marine, since they also occur around inland waters and are not as distinctively modified for or adapted to specifically marine conditions as the groups discussed above.

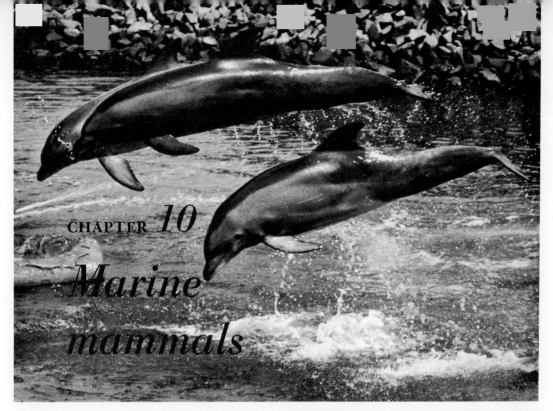

CHAPTER *10*

Marine mammals

Fig. 10-1. Bottlenose dolphins leap in unison at **Sea** World, San Diego, California. (M. W. Williams photograph.)

*T*here are three major groups of marine mammals, each of which has arisen independently from different mammalian stocks and become adapted for life in the ocean. Life in the ocean is so different from that on land that it has impressed certain general tendencies or convergencies in the evolution of all three groups. All three have diverged sufficiently from the original stocks that it is no longer readily apparent from casual examination of presently existing species to which groups of terrestrial mammals they are most closely allied.

Having become basically adapted for life in the oceans, all three groups have undergone a kind of evolutionary radiation, that is, have undergone diversification and adapted themselves to exploiting different ecological niches—different possibilities offered by the marine environments.

Common features found in all three groups are (1) modified limbs and body shape, (2) thermal regulation, (3) modified respiratory system, and (4) osmotic adaptations.

Changes in limbs and body shape enable the animal to adapt to a swimming life. These changes involved conversion of the limbs into paddlelike flippers and in some cases the loss of the hind limbs and the development of the posterior end of the body into flukes somewhat resembling a fish's tail but horizontally placed.

Thermal regulation involved the provision for heat conservation in an environment generally considerably colder than the animal's body temperature. It was achieved by a variety of means including thick skin with insulating layers of fatty, oily tissue beneath, the reduction of peripheral circulation in the skin and appendages, a metabolism producing more internal heat, and an increase in size. The larger volume-to-surface area ratio in larger animals enables them to conserve heat more efficiently than is the case with the majority of mammals.

Modifications of the respiratory system enable marine mammals to breathe efficiently without taking salt water into the lungs and to hold their breath longer than most mammals, so that they are able to make rather prolonged, deep dives.

Osmotic adaptations prevent the intake of excessive amounts of salt water, enabling marine mammals to get along without fresh water. They do not possess special salt-excreting glands as do the teleosts and the procellariiform seabirds. They seem to have accomplished the adjustment by limiting the intake of water as much as possible—by having an impervious skin and by obtaining most of their water from the fish diet, which is more dilute than seawater. The resorption of water in the kidney is very efficient, and it seems possible that the extensive fatty insulating layers beneath the skin may also play a role in storage of metabolic water as has been found to be the case in some desert animals.

These changes have, of course, resulted in these animals' becoming either very cumbersome and awkward on land or wholly unable to leave the water.

It is remarkable that in as short a time, geologically speaking, as fifty million years or so, the seagoing mammals have achieved such radical changes and become so successful in their environment. The whales have become completely adapted to marine life and never come to land, even to bear young. Some of them have evolved into the largest animals that have ever lived, considerably exceeding in bulk the largest of the great dinosaurs. The seals have produced the greatest populations of large carnivorous mammals on earth. Both of these groups seemed to hold secure and dominating positions in the world's oceans until the spectacular explosion of population and technology on the part of man in the last few hundred years.

ORDER CETACEA—whales and porpoises

Cetaceans are all large to very large marine mammals in which there has been a complete loss or extreme reduction of hair. The forelimbs are paddlelike, are used as stabilizers in swimming, and show a secondary increase in the number of bones in the digits over that found in other mammals. The hind limbs have been lost, the pelvic bones being reduced to internal vestiges. The body is. insulated by a layer of oily blubber beneath the skin, which

also aids in providing bouyancy and perhaps in the storage of metabolic water. The nostrils of all modern species are located on top of the head rather than at the fore-end, as in other mammals. The brain is large, with convoluted cerebral hemispheres. The stomach of some of them is very complex, with up to four or five chambers. Females have one pair of mammary glands, inguinal in position. The posterior end of the body is very strongly muscular and shaped into a pair of horizontal flukes resembling a fish's tail, the semi-rotary motion of which gives thrust somewhat in the same manner of a ship's propeller. It has been estimated that a 90-foot whale may generate thrust equivalent to that provided by a 520-horsepower motor. The largest whales can make bursts of speed at 20 knots and can swim all day at 10 knots or better.

Suborder Archaeoceti. The cetaceans are believed to have been derived from the earliest line of carnivorous mammals, the creodonts of the Eocene period, or possibly from certain insectivores. Whales of this early period were long-bodied species. In contrast to modern forms, they possessed differentiated, or heterodont, teeth with three incisors, one canine, four premolars, and two or three molars on either side of each jaw. Also, the position of the nostrils and the relations of the maxillae are intermediate in character between those of primitive land mammals and modern cetaceans. They are grouped together in the suborder Archaeoceti, or Zeuglodontia. They achieved a worldwide distribution, fossils having been found in Europe, America, and Australia.

From this early group the modern whales presumably differentiated during the late Oligocene or early Pliocene.

Suborder Odontoceti—toothed whales. In this group at least some teeth are present throughout life, and there is no baleen. The nostril, or blowhole, is single. The skull is very specialized. It is asymmetrical and has reduced nasals, and the flattened maxillae are above the supraorbital processes of the frontals. The sternum is composed of several elements that fuse into one piece only in the adult. Several pairs of ribs join the sternum. Although whales possess no vocal cords, they produce a variety of sounds. The Odontoceti all produce very high frequency clicks, have acute hearing, and presumably use echo location to find objects and prey in the water and perhaps to communicate as well. They also produce some squealing or whistlelike sounds within our auditory range, probably made in the larynx. This group is quite diverse and comprises seven distinct families.

The beaked whales, or bottlenose dolphins, family **Ziphiidae,** have the functional teeth reduced to one or two pairs in the lower jaw. Additional vestigial teeth are also commonly present. There is a pair of longitudinal grooves in the lateroventral throat region. A dorsal fin is present at about the posterior third of the body. The hind margin of the tail is entire, not notched. The largest existing species of this family is *Berardius bairdi,* a North Pacific species reaching a length of 40 feet. Another large species, *B. arnuxi,* about 30 feet long, is common in Antarctic waters and occurs off New Zealand and Argentina.

Another member of this group is the bottle-nosed whale, *Hyperoodon ampullatus.* The male is larger than the female, about 30 and 24 feet, respectively, and the teeth remain buried for most of the lifespan, usually only one pair erupting in old males. A peculiarity of this species is a large bony mass on each maxilla, which increases in size disproportionately with age, producing a large, gibbous swelling of the forehead in older individuals.

The dolphins, family **Delphinidae,** are a large group that includes some forms commonly called porpoises and others usually termed whales. The skull is asymmetrical, with the structural midline displaced to the left and the right bony nares larger than the left. As in all the Odontoceti, there is only one external blowhole, located on the top of the head. The snout in many species is produced into a "beak." The tail is notched in the middle. Many, but not all, have a prominent dorsal fin. There is a large mass of blubber between the blowhole and the end of the snout, known as the melon. Melon oil has been used as a lubricant.

The subfamily Delphininae contains many familiar dolphins of the short-beaked genus *Lagenorhynchus,* species of which are found in most oceans. The bottlenose dolphin, *Tursiops truncatus,* has a well-defined beak 2 to 3 inches long and twenty to twenty-two teeth in each row in both jaws and is dark brownish above and white below. It is a favorite in large oceanariums because of its intelligence and playfulness, the ease with which it can be trained to do a variety of tricks, and its built-in smile.

The cosmopolitan common dolphin, *Delphinus delphis,* is mostly black above and white below, with wavy streaks of white, brown, yellow, or gray on the sides and has forty to fifty teeth in each row. This mon in the Atlantic and Mediterranean and is another very familiar form, being common of the commonest species around the British Isles.

The subfamily Orcinae comprises the killer whales and related forms. In this group there is no well-defined beak. There is a prominent dorsal fin. The teeth are generally massive and conical, ranging from eight to nineteen in each row in the various species.

Orcinus orca, the killer whale or grampus, is perhaps the best known. The male is almost twice as long as the female, approximately 30 and 16 feet, respectively. The dorsal fin is high and falcate in the female and young, becoming triangular and up to 5 feet in height in the older males. Killer whales are very distinctively colored, with sharply defined black and white areas, black above and white below. They have a conspicuous white patch above and behind the eye and a grayish saddle behind the dorsal fin. The flippers are rounded, and in older males the flippers, dorsal fin, and tail flukes increase disproportionately in size. They are among the most voracious and dangerous predators of the seas, eating fish, marine birds, and mammals and even attacking larger whales. They are cosmopolitan in distribution.

Porpoises of the genera *Phocaena* and *Neomeris,* having one triangular dorsal fin or none, the tail notched posteriorly, and the teeth with expanded, spade-shaped crowns, are sometimes grouped into a separate family, the **Phocaenidae**. *Neomeris* includes the finless black porpoise of Asiatic coasts from India to China; these occur also in rivers, going up the Yangtze more than 600 miles to beyond Tung-T'ing Lake.

Another group of mostly tropical dolphins, genera *Steno, Sousa,* and *Sotalia,* comprise the **Stenidae**—long-snouted forms also occurring in estuaries and rivers as well as in the ocean.

The typical river dolphins, family **Platanistidae**, are very long-snouted forms with the skull less compressed in the front-to-back direction than in the Delphinidae and with all the neck vertebrae free. They occur both in South American rivers, such as the Amazon, and along Asiatic coasts and rivers from India to China.

The blackfish, caaing whale, or pilot whale, *Globicephala melas* (so named because of its high, globose forehead) is another common form, the oil of which is a fine lubricant. They are gregarious, sometimes occurring in very large schools. Schools of blackfish have on occasion become stranded on beaches after strong tides. On one occasion 147 of them were stranded on an East Lothian beach near Edinburgh, creating temporarily a major disposal problem.

Two typically arctic and circumpolar, or northern, North Atlantic forms are the narwhal, *Monodon monoceros,* peculiar for its unique dentition. It is reduced to a single pair of teeth, which only exceptionally erupts in the female. In the male the left tooth grows out as a straight, spirally marked tusk sometimes more than 9 feet long. Occasional specimens have both teeth developed into tusks. The other is the beluga, or white whale, *Delphinapterus leucas,* which is dark gray when young and becomes pure white as an adult. This is a gregarious form found in the arctic and Hudson Bay areas, sometimes as far south as the Gulf of St. Lawrence, and in the eastern North Atlantic, on rare occasions as far south as the northern part of the

British Isles. These two forms lack a dorsal fin, the neck vertebrae are free, and the flippers are broad and bluntly rounded. They constitute the family **Monodonidae**.

The largest of the Odontoceti are the great sperm whales, family **Physeteridae**. The sperm whale, or cachalot, *Physeter catodon,* may attain a length of more than 60 feet, the very massive head occupying about one third of the total length and extending forward well beyond the narrow lower jaw and mouth. The lower jaw bears twenty to twenty-six large teeth up to 8 inches long on each side. The upper teeth are more or less vestigial and variable in number. The dorsal fin is represented by a series of low humps beginning back of the middle of the body and diminishing in size posteriorly. The remarkable melon, or spermaceti organ, in front of the nares is developed as a fibrous case containing up to a ton or more of clear, colorless oil, giving this animal by far the most enormous nose on record. The blubber also yields a high-grade oil. Small amounts of spermaceti are also found in the back.

Sperm whales occur in the warm parts of all oceans. Old males sometimes wander as far north or south as the polar seas. They eat chiefly large squid, including the giant squid. Large sucker marks and scars on the head of the whales sometimes attest to the dramatic battles that must take place in the depths between the sperm whales and the giant squid.

Sperm whales can dive to greater depths than any other air-breathing animal. A cable repair ship once recovered a drowned specimen entangled in a transoceanic cable from a depth of 3,720 feet, more than half a mile. Since the spermaceti, oil, and blubber can absorb about six times as much nitrogen as an equal weight of blood, it is thought that in addition to providing insulation and bouyancy, these may help the whales in making deep dives followed by relatively rapid ascents without getting the bends. The bends is a serious difficulty from which human divers suffer if the pressures under which they are working in deep water are too suddenly released by rapid ascent. The pressure forces nitrogen from the air in their lungs into the blood. When this pressure is released faster than the nitrogen equilibrium can be restored by way of the lungs, bubbles of nitrogen form in the blood vessels, with very painful, sometimes fatal results.

The hard, parrotlike beaks of the large squid sometimes fail to pass through the digestive tract, remaining to form gradually a pathological foul-smelling dark sticky or waxy material known as ambergris. It may be spewed up by the whales and subsequently found floating on the water or cast up on a beach, or it may be taken by whalers when they cut up the whale. For a long time ambergris, added in minute amounts to fine perfumes, was the best substance known for making perfumes hold their scent. In recent years other perfume fixatives have partially replaced it, but ambergris still brings up to ten dollars an ounce.

Suborder Mysticeti—whalebone, or baleen, whales. The origin of the Mysticeti is not clear, although presumably they arose from some early group of Odontoceti, since although the teeth are lacking in the adult, numerous vestigial teeth are present during the late embryological stages, more than are characteristic for any primitive group of land mammals or zeuglodonts. In the adult they are replaced by an elastic horny substance, not true bone, termed baleen. Baleen occurs in the form of a series of triangular plates anchored in the roof of the mouth on each side, separated by brief intervals. The plane surfaces of these plates are at right angles to the long axis of the head, and the inner, lingual edges of the plates are frayed out into a fringe of bristles that combine to form a matted sieve, or strainer, for the collection of macroplankton. Baleen is more related to hair than to bone and like hair has its origin in the skin. Baleen plates are longest, up to 12 feet long, in the right whales and considerably shorter, broader, and less flexible in the rorquals. A pad of softer, hornlike substance, the intermediate substance, occurs between adjacent plates. There may also be transverse rows of small subsidiary plates on the inner edges of the larger ones.

In contrast to the Odontoceti, the skull is symmetrical, and the blowhole has double, posteriorly divergent slits. The lower jaw is large, the mandibles consisting of two very large bones, bowed outward and loosely united in front. Only the first pair of ribs join the sternum, which consists of only one piece.

Baleen whales feed by swimming through clouds of zooplankton with open mouths, like gigantic, animated plankton nets. The water is forced out at the sides of the mouth between the baleen plates, and the macroplankton organisms are caught in the fringe of the bristles along the inner edges of the plates, to be licked off by the tongue and swallowed.

Pequegnat (1958) gives some figures on the energy requirements and the amounts of food needed to meet them for the antarctic blue whale, the largest animal known. According to him, an adult blue whale weighing about 90 tons, swimming at 10 to 12 knots most of the day, and using an assumed 20% efficiency of the muscles in propelling it, requires 780,000 calories per day just for propulsion. Another 230,000 calories are required for its metabolism and maintenance of body heat —a total of more than a million calories daily. Since, according to him, the blue whale feeds only during the antarctic summer, it must take in double this amount or more to provide for the stored energy used the rest of the year—stored in the form of blubber. Thus, during the feeding season an adult blue whale requires approximately 3 million calories daily. Krill, mostly *Euphausia superba,* upon which it largely feeds, represent 460 calories per pound. Thus the blue whale requires approximately 3 tons of krill daily during its feeding season to meet its needs. Prior to the extensive antarctic whaling of this century, it is estimated that there were at least half a million blue whales, which must have meant that the blue whales alone consumed some 270 million tons of krill each summer. Since they probably do not harvest more than 20% of the crop, there must be a minimum of 1,350 million tons of krill produced annually in the antarctic waters.

The actual figure is probably well in excess of this, since several other species of whales also feed there in the antarctic summer, as well as numerous other eaters of krill, such as seabirds, fish, and squid.

Because of the short, efficient food chain for baleen whales, consisting only of phytoplankton and krill, a great deal more whale flesh can be produced than would be possible for a predator such as the killer whale, which is at the apex of a longer food chain involving at least two or three intermediary predators such as fish, squid, and seals. At each step in any food chain, there is approximately a 90% drop or greater in production because of the low efficiency of conversion of organic matter from one organism to another. Thus, even though blue whales are much larger, it is probable that there were many more blue whales, than killer whales before they were decimated by man.

According to Pequenat's estimates, the antarctic feeding grounds must produce about half a ton or more of krill per acre during the antarctic summer.

The Mysticeti are usually divided into three families. The largest family, **Balaenopteridae,** includes the rorquals and the humpback whale. They have a relatively broad, flat rostrum and short, quite rigid baleen plates, and the anteroventral half of the body is marked by parallel longitudinal ridges and grooves. A small dorsal fin is present about two thirds of the way back along the body or slightly posterior to this. The giant blue whale and several other species belong to the genus *Balaenoptera.* They range in size from the lesser rorqual, *B. acutorostrata,* found in temperate to polar latitudes of both hemispheres and not exceeding 33 feet in length, to the gigantic blue whale *B. musculus,* also known as Sibbald's rorqual, which has been known to reach 100 feet in length. The humpback, *Megaptera novaeangliae,* attains lengths up to 50 feet. It is cosmopolitan, commonly migrating along inshore waters between higher and lower latitudes. The humpback is noted for its unusually long flippers nearly one third as long as its body, as well as for

its generally dumpier form and more arched back.

The right whales, family **Balaenidae,** were so named by early whalers because they float after being killed, whereas the other whales, the "wrong whales," usually sank and were lost. In these whales the skull is strongly arched, the baleen plates are long, narrow, and flexible, and the body is not marked by the parallel grooves in the anterior region. The neck vertebrae are fused. In genus *Balaena* there is no dorsal fin. The flippers are broad, with five digits in each flipper. The head is approximately one fourth to one third of the total length. The genus *Caperea,* which includes the pygmy right whale, has a less arched head, constituting only about one fifth of the total length. A dorsal fin is present, and there are only four digits in each flipper.

Family **Eschrichtidae** includes the California or Pacific gray whale. This animal is about 45 feet long. It has no dorsal fin, the flippers are four fingered, and there are scattered hairs over the entire head and lower jaw. There are only two or three longitudinal ventral grooves on each side in the throat region. The neck vertebrae are not fused.

The gray whale is a North Pacific species, migrating south in the winter to breed in shallow lagoons off Lower California in the eastern Pacific and off Japan and Korea in the western Pacific. It spends the summer in high northern waters of the Arctic and the Bering Sea. It seems to prefer relatively shallow waters and is frequently observed during its migrations swimming quite close to shore, sometimes even going into the zone of breakers. At one time it was nearly exterminated by whaling but has responded well to protection. A greater danger at present is that the breeding grounds may be altered by man.

ORDER PINNIPEDIA—seals, sea lions, elephant seals, walrus

All three families comprising this group appeared during the Miocene, but their immediate ancestry is uncertain. All are large aquatic carnivores and are gregarious and polygamous. They haul out on shore or on the ice to have their pups. The bulls are larger than the cows and in most cases establish harems defended from encroachment by other bulls. They are awkward on land but are expert swimmers and divers. The limbs are webbed flippers, with short, upper limb segments but long, broadwebbed manus and pedes. The teeth are simpler than those of land mammals, the premolars and molars being of similar character. Most species have been extensively hunted for hides, oil, and meat. Some have been nearly exterminated. Others have been placed under international protection and are harvested now on a more or less sustained-yield basis.

The three families of the group are Otariidae, Phocidae, and Odobenidae.

The eared seals, sea lions, and fur seals, family **Otariidae** have small but well-formed ears. The hind limbs are independent, mobile, turned forward when on land, and used to facilitate movement on land. The nostrils are at the tip of the snout. The pelt bears a soft underfur. The neck is rather long.

Perhaps the most famous of this group is the northern fur seal, *Callorhinus ursinus,* of the North Pacific, which hauls out on the remote islands of the Pribilof group in great numbers each year to mate and bear young. They spend the winter roaming through the Pacific, sometimes as far as south of the equator. The Pribilof Islands are the only place at which they come to land, except for, recently, San Miguel Island off the California coast. The bulls haul out first, in June, and establish their territories, which they maintain and defend throughout the summer. Each successful bull acquires a harem as the females move in, presumably drawn to suitable areas for having their pups rather than being attracted to particular bulls. During the summer the females with pups come and go on fishing trips which are necessary in order that they may feed the young, while the bulls remain at their territorial posts, not feeding until sometime in August, by which time they have become quite emaciated. The migration patterns of the males and females during the rest of the year seem to differ, the females apparently roaming the ocean to greater distances as do also some of the younger males, whereas the older bulls seem to stay closer to home. Formerly older bulls used to come down the coast of North America at least as far as Washington and Oregon or northern California, but since the time of the last major pelagic sealing operations off the West Coast of the United States, about 1900, they remain farther north. How these far-ranging pelagic seals manage to migrate back to and find their hauling rounds in the Pribilof Islands is one of the remarkable mysteries of animal navigation.

In 1968 a small colony consisting of one bull, a harem, and about forty pups established itself on San Miguel Island off the California coast. All the pups were tagged to facilitate study of this group. The wisdom of this is questionable, since the type of tag used is one that causes considerable tissue irritation and chafing as the young animals clamber over the rocks, resulting in infections and some losses. A better, rivet type of tag, with which it is hoped this difficulty will be overcome, is under development at this writing.

The southern fur seals, genus *Arctocephalus,* comprise a few species found in southern Australia, New Zealand, southern South America, and southern Africa. They spread northward along the western shores of South America and Africa in the cool Humboldt and Benguela currents. One species was formerly abundant at Guadalupe Island in the Galápagos but was ruthlessly slaughtered and for a time believed to be extinct. The Hancock expeditions of 1932 and 1933 found some animals and reported that they seem to be making a comeback.

Two species of sea lions familiar along the West Coast of the United States are the Steller's sea lion, *Eumetopias jubata* (= *E. stelleri),* and the smaller, darker California sea lion, *Zalophus californianus.* The Steller's sea lion is more northern in its distribution, being found in the North Pacific from about the Santa Barbara Islands northward, whereas the California sea lion ranges from Vancouver Island south-

Fig. 10-2. California sea lions haul out at Point Reyes, California. (M. W. Williams photograph.)

ward to Cape San Lucas, Lower California, and the Gulf of California. Its occurrence in the northern portion of its range is sporadic and irregular. Where sea lions are not protected, considerable numbers are shot by fishermen under the impression that they compete seriously for the fish and damage gear. Analyses of stomach contents, however, indicate that they feed mostly on fish species other than those currently intensively fished and are not really in serious competition with fishermen. Their presence in large numbers in an area is a good indication of its general productivity in fish, squid, etc., rather than a threat to fishermen, and their elimination for this reason would seem wholly unjustified. The California sea lion is the species most commonly trained to do tricks in circuses.

The hair seals, family **Phocidae**, have no external ears, the hair is short and coarse, without a soft underfur, the nostrils are somewhat dorsal in position, and the hind limbs are partly fused together with the tail

and are of little use on land, being dragged along posteriorly. In swimming the body is propelled by strong side-to-side strokes of the paddlelike hind flippers, each acting alternately in the power stroke. The smallest of the seals is a freshwater species of Lake Baikal in Siberia, *Pusa sibirica*, which attains a length of only about 3 feet. The remaining 17 species are all larger marine forms, ranging in size from about 5 feet in small species to 16 to 20 feet in the huge elephant seal. There are more species, and they occur in greater numbers, in the high northern and southern latitudes. Normally only one pup is born to a cow during a given season. It is nourished by the mother for a relatively short time, during which it grows rapidly. The rich seal milk contains about 50% fat. At birth, most seals have soft, silky fur, which is later shed and replaced by coarser hairs. In a few species the molt is precocious, being partly or wholly completed before birth. The cows are impregnated soon after the birth of the pups, but implantation of the embryo may be delayed, so that in some species at least, and perhaps in all, only about seven of the eleven intervening months between copulation and birth of the young represent gestation. Seals possess a thick coat of blubber from which a high grade of oil is obtainable. Seals commonly live about thirty years or more, but the average life-span is much shorter because of the vulnerability of the very young to accident and predation.

The cosmopolitan harbor seal, *Phoca vitulina,* is perhaps the most familiar species to shoreline residents of Europe and America. It frequents coastal waters and bays, sometimes even ascending rivers for a short distance. It tends to occur in relatively small groups, never abundant enough in any one area to be commercially exploited or to warrant efforts to reduce the herds.

Two related northern species, however, the harp seal, or saddleback, *Phoca (Pagophilus) groenlandica,* and the ringed seal, *Phoca (Pusa) hispida,* are the objects of extensive commercial sealing. In the Southern Hemisphere the leopard seal, *Hydrurga leptonyx,* the antarctic Weddell seal, *Leptonychotes weddelli,* and the elephant seal, *Mirounga leonina,* have been the principal objects of exploitation. The northern elephant seal, *M. angustirostris,* which occurs along the western coast of North America, was slaughtered until, for a time, it was thought to be extinct, but in recent years it seems to be making a comeback. There is a sizable rookery at the island of Año Nuevo, north of Monterey Bay, California, and individuals have recently been seen as far north as Cape Arago, Oregon.

Coastal species such as the harbor seal are commonly rather sedentary, but some of the pelagic species, such as the harp seal, the Weddell seal, and the crabeater seal, regularly make rather extended migrations, live most of the time in the open sea, and only come out on floating ice. The crabeater seal, *Lobodon carcinophagus,* is unusual and somewhat whalelike in its manner of feeding, living on planktonic crustaceans that it strains from the seawater by passing the water through a sievelike area formed by the serrated cusps of its lateral teeth.

The third family, **Odobenidae,** includes the walrus. These animals are very large and are earless like the hair seals but have hind limbs like those of sea lions. Large males may weigh more than a ton and are very thick and heavy in the neck and shoulder region. Walrus are unique in having the upper canine teeth developed into long protruding tusks, those of the bulls considerably larger and thicker than those of the cows. These tusks are used to dig clams from the sea floor. The other teeth are simple, more or less peglike, and used to crush the shellfish. Both the larger Bering Sea walrus, *Odobenus obesus,* and the closely related Atlantic form, *O. rosmarus,* have been extensively hunted for their valuable skins, for the oil that can be obtained from the blubber, and for the ivory tusks, especially the large tusks of the male. They have become very rare now except in the far northern portions of their former ranges.

ORDER SIRENIA—manatees, dugongs, sea cows

The Sirenia are large bulky aquatic mammals superficially resembling whales in the general torpedo-like shape of the body, tapering to expanded horizontal flukes posteriorly, with the hind limbs lost and the forelimbs modified into paddles. But as far back as 1816, De Blainville recognized the basic differences between this group and the Cetacea and correctly grouped them with the ungulates related to the proboscidean stem. They are sluggish animals, frequenting shallow waters and estuaries and feeding on water plants and seaweeds. The head is proportionately smaller than in the whales, and the muzzle is broad with mobile, transversely expanded lips and coarse, heavy vibrissae, or bristles. The eyes are small, the nostrils rather dorsal in position, and the auditory openings without pinnae. The head merges indistinctly into the massive forepart of the trunk. The skin is thick and is naked or has sparse hairs. Females have one pair of mammary glands, pectoral in position.

The bones of the skull and thoracic region are exceptionally massive. There is no clavicle and no pelvic bones except a small pair of vestigial bones suspended at some distance from the vertebral column. The brain is much less convoluted and developed than that of the cetaceans.

Sirenians generally show little fear of man and are readily approached. They bear one offspring at a time, for which they show considerable affection and concern.

Although some fossil sirenian teeth seem to bear a resemblance to the teeth of early Eocene artiodactyls, and others to those of the Eocene rodent *Ischyromys,* there is much more evidence pointing to the Proboscidea as the stem group. The earliest fairly complete skeleton, *Eotheroides,* from the upper Eocene of Egypt, shows, according to Andrews, a number of resemblances to that of the early proboscidean *Moeritherium* found in the same formations. He also cited a number of anatomical details in which modern sirenians and elephants show a resemblance in spite of the great differences in external appearance and mode of life.

The middle and late Tertiary sirenians, such as *Halitherium* and *Halianassa,* were larger than modern manatees and dugongs and had a pair of large vertical tusks in the upper jaw. These tusks are present in dugongs but have been lost by the manatees.

There are three groups of recent Sirenia, the dugongs, the manatees, and the Steller's sea cows. Several species of dugongs and manatees have been named, but it is doubtful that they are really distinct species rather than races or varieties.

The dugongs, genus *Dugong (= Halicore),* occur in coastal waters of the Red Sea, on the east coast of Africa, in the Indian Ocean, the East Indies, southern Asia, and Australia. In them the tail is bilobate, the molar teeth number five or six, and the upper incisors of the male are tusklike. The upper lip is only shallowly cleft. They are more marine in habit than are the manatees and eat mostly the larger seaweeds. It has sometimes been stated that they attain a length of 20 feet, but au-

thentic specimens are much shorter, the adults usually not more than 10 feet in length. Dugong oil is unusually clear and free from bad taste. About 10 to 12 gallons can be obtained from a full-grown specimen. It contains no iodine but is said to have the same therapeutic properties as cod-liver oil. The meat is delicious, resembling beef, and is highly prized in some areas. Because of this the dugongs have been extensively hunted and practically eliminated from large areas of their former range, especially along the coasts of Asia.

Manatees, genus *Trichechus (= Manatus),* have a broad, rounded tail, rather than a bilobed one. Eleven pairs of molar teeth are formed in each jaw, but since the anterior ones are lost before the hind ones erupt, there are seldom more than six at any one time. The upper lip is deeply cleft. Manatees occur in the Caribbean region, in shallow waters of the West Indies, Florida, and northern South America and along the west coast of Africa. They ascend rivers, feeding on aquatic plants in fresh and brackish water to a greater extent than the dugongs. Adults attain a length of 8 to 15 feet and weigh about 1500 pounds. They are sluggish and are usually found singly or in small groups. Females bear one calf at a time. It is thought that the calf stays with the mother for more than a year. Occasionally a female is seen with two calves of different age. The manatee occurring along the west coast of Africa displays some skeletal differences from the Caribbean variety and has been named *T. senegalensis.*

Although the features of sirenians, observed closely, are a far cry from the type of beauty usually attributed to mermaids, the sight of dugongs in a semiupright position in the water, holding a calf with one flipper and suckling it much in the manner of a human mother, was the foundation of the mermaid myth started by early Greek sailors.

The Steller's sea cow, *Hydrodamalis gigas (= Rytina stelleri)* was discovered in 1741 by Bering and Steller near Kamchatka in the extreme North Pacific around two islands now known as Bering and Copper islands. Steller was shipwrecked there, spending several months on the islands before being rescued. Upon his return to Germany he published an account of the sea cows' habits and anatomy. They were very abundant and were the largest of the species of sirenians, attaining a length of 25 feet. He reported that they were sluggish, unsuspicious, and affectionate. They differed from other sirenians in having no teeth, the masticating surfaces being merely dense hard horny pads of the jaws. They ate the large seaweeds growing abundantly around these two islands. After the existence of this herd of magnificent large sea cows became generally known, commercial hunters quickly descended upon them, and they were exterminated by 1768, only twenty-seven years after their discovery, in spite of their remote and secluded habitat.

ORDER ARCTOIDEA—sea otter

The sea otter, *Enhydra lutris,* family **Mustelidae,** formerly occurred in considerable numbers along the Pacific coast of North America. It attains a length of about 4 feet, possesses a typical slender otterlike body with relatively short legs and large webbed hind feet. The tail is cylindrical and blunt. Sea otters are active, intelligent, alert, and playful. They swim singly or in small groups, usually near shore, along the margins of the giant kelp beds. They dive readily and eat a variety of marine animals including fish and invertebrates taken along the shallow bottom. Frequently, when they have taken a sea urchin, shellfish, or crab from the bottom, they also bring up a stone. Upon surfacing they will float on their backs, hold the prey on their chests with one paw, and hit it with the stone to break the shell.

The fur of sea otters is one of the most valuable on the market. The dark brown underfur is thick and rich, and the outer coat has somewhat coarser, gray-tipped hairs. For this, they were hunted nearly to extinction before stringent protection, together with their own scarcity, rendered hunting unprofitable. Since the late 1950's or early 1960's they seem to have been making a comeback in the Monterey region of California and to be spreading northward from there. There are also some along parts of the Canadian and Alaskan coasts.

Fig. 11-1. Green sea turtle. (Photograph by M. W. Williams, taken at Steinhart Aquarium in San Francisco.)

CHAPTER *11*

Marine reptiles

*I*n contrast to the marine mammals and birds, which appeared in the Tertiary and which, at least until the advent of man, were still undergoing evolutionary diversification and development, the marine reptiles as well as reptiles on land experienced their great flowering during the long Mesozoic era. Those found today are mere remnants of a once much greater and more diverse assemblage.

Contemporary marine reptiles

Sea turtles. The best known and most widespread contemporary marine reptiles are the sea turtles, family **Cheloniidae,** comprising five widely distributed tropical or subtropical species. Three species, the green turtle, or edible turtle, *Chelonia mydas,* the hawksbill turtle, *Eretmochelys,* and the loggerhead turtle, *Caretta caretta,* are familiar along the coasts of the southeastern United States and the Gulf states. All the sea turtles are large species with nonretractile heads. The limbs are modified into flippers for swimming. The fore limbs are especially well developed

and are the chief organs of propulsion. The carapace and plastron are completely covered with horny shields, the tortoiseshell of commerce. The green turtle feeds on marine plants. The other species are carnivorous, preying on conchs, crabs, echinoderms, and other large invertebrates. Sea turtles seem to have a strong homing instinct, sometimes performing extended migrations between their feeding grounds and the beaches on which they spawn. The females come out of the water, dig pits in the sand of the upper beach, and deposits a clutch of eggs (sometimes up to 200) about 2 inches in diameter, with a leathery shell.

The largest of the marine turtles—in fact, the largest of all living turtles—is the Atlantic leatherback, *Dermochelys coriacea,* the only member of the family **Dermochelidae.** The leatherback sometimes attains a weight of three fourths of a ton and measures up to 12 feet. They were exceeded in size only by the extinct *Archelon ischyros* from the Upper Cretaceous of North America, thought to have attained a weight of over 3 tons.

Instead of the usual carapace and plastron, the leatherback has a mosaic of small dermal bones buried in its leathery skin, completely free from the internal skeleton. The larger of these skin bones form seven rows dorsally, which appear externally as longitudinal keels running the full length of the trunk. Ventrally there are five rows of smaller bones, beneath which are vestiges of bones representing the plastron of other turtles.

In other anatomical characters *Dermochelys* resembles such Cretaceous forms as *Protostega* or *Archelon,* some of which had lost nearly all the costal plates and had only reduced neurals and marginals. It is argued that perhaps these large, soft-bodied turtles, because of changing conditions or habits over a long period of time and possibly exposure to new, more dangerous enemies such as the zueglodonts, again found a survival premium in greater protection. Having already lost their ancestral shells, they were unable to retrace the evolutionary steps leading to the shells of other Chelonia but met the need through a secondary proliferation of the marginals and other epithecal elements. If this view is correct, *Dermochelys* represents a highly specialized, rather than a primitive, line of marine turtles.

Sea snakes. Snakes are probably the most recently evolved group of reptiles, originating from lizards sometimes during the Cretaceous period. There is no known geological history of distinctively aquatic groups of snakes. Most snakes swim well, with rapid, undulatory movements of the body. Poisonous snakes, with special development of the poison glands and fangs for the injection of the poison into their victims, are known only from relatively recent geological times. It is probable that venomousness is the last important new specialization to have occurred among reptiles.

One family of contemporary snakes, the **Hydrophidae,** has become adapted to wholly aquatic marine life. They comprise a group of about 15 genera and some 50 species of highly venomous snakes related to cobras. All of them live in warm seas of the Indian Ocean and the western Pacific. The adaptations fitting sea snakes for marine life are not as striking as those found in other groups of terrestrial animals that have returned to the sea. The body and tail have become quite flattened from side to side, facilitating swimming. The broad transverse scales so characteristic of the underside of most snakes are reduced to a few vestiges or are lacking, and the snakes are viviparous, bearing their young at sea and never coming on land for any purpose whatever, except for the Laticaudinae, which lay their eggs on land above the high-water mark. The Hydrophinae bear living young at sea and are quite helpless on land. Fish comprise the major part of their diet.

Marine lizards. The only contemporary marine lizard is the Galápagos sea lizard, or marine iguana, *Amblyrhynchus cristatus,* which lives along the rocky sea beaches of the Galápagos Islands and never ventures far inland from the shore. They are large, heavyset lizards 3 to 4 feet in length and attain a weight of about 20 pounds. They are dark-colored and rather sluggish in movement on land. The tail is laterally flattened, and the feet are slightly webbed. In the water they are rapid and graceful swimmers, propelling themselves with serpentine movements of the body and tail, the legs being held motionless close to the sides. Their food consists of seaweeds.

These lizards are of interest in indicating one of the ways in which terrestrial reptiles and other animals have become aquatic. Tempted by an abundance of food in shallow waters, they ventured farther and farther to obtain it. A selective premium on adaptation to the aquatic environment has led to flattening and muscularization of the tail as a swimming organ and to the ability to remain submerged for longer periods than ordinary lizards. Darwin reported that one of the sailors of the Beagle attempted to drown one by submerging it attached to a weighted line but that it was still active when pulled up an hour later.

Although it has sometimes been stated that these lizards also seek protection from terrestrial enemies by swimming, quite the opposite situation seems to have prevailed, at least until very recently. In fact, it may be the prevalence of sharks and the total lack of terrestrial enemies on the Galápagos Islands that have been responsible for the lizards' retaining as much of a terrestrial habit as they have. Darwin reported that he was unable to scare them into the sea. It was easy, he said, to drive them down to any little rocky point overhanging the sea, but he was unable to frighten them into the water. Upon being picked up and thrown in several times, the lizard invariably returned directly to the same place. They seem to prefer to spend most of their time on the rocks along the shore, venturing into the water only to feed. Reproductively, they are also still tied to the shore, depositing their eggs in tunnels dug 12 to 18 inches deep into the sandy slopes above the high-tide line.

The mode of swimming seems to indicate that long-tailed aquatic reptiles did not use their limbs as primary organs of propulsion in the water.

Marine reptiles of the past

"I doubt not that, in a few years it may be, I shall be compelled to say that the work which I have today finished, and to which I have given so much labor, is but the first glimpse of the immense creations of ancient times."*

Order Sauropterygia

Plesiosaurs. In number and variety of species and genera and in worldwide distribution the plesiosaurs constituted, by a rather wide margin, the largest, most successful group of marine reptiles that has ever existed. They are remarkable for the great relative elongation of the neck in many species and for the extreme variation in its structure. Elongation was effected

*From Cuvier, G. L. C. F. D., Baron von. 1834-1836. Récherches sur les ossemens fossiles de quadrupèdes . . . mit Discours préliminaire 1812. Letzte Aufl. posthum. hrsg. von F. Cuvier. 12 vols. (Closing sentence of the chapter on plesiosaurs.)

both by increase in the number of cervical vertebrae over that found in other reptiles and by increase in the length of some of the elements, especially those near the base of the neck. The number of cervical vertebrae in various plesiosaurs ranged from thirteen in relatively short-necked forms to twenty-six. In the long-necked species the neck was relatively thick at the base, tapering toward the head, and the head was relatively small. In *Elasmosaurus platyurus,* the head was 2 feet long, the neck 23 feet, the body 9 feet, and the tail about 7 feet, a total of 41 feet. In the short-necked *Brachauchenius lucasi,* on the other hand, the head was 2½ feet long, the neck less than 2 feet, and the body about 5 feet. The length of the tail is uncertain, but it was not greater than 5 feet. The vertebrae were biconcave but with shallow saucerlike or nearly flat ends, quite different from the conical fishlike cavities of the strongly amphicoelous ichthyosaurian vertebrae.

The trunk was not much elongated, having only twenty-five to thirty vertebrae, and the tail was always shorter than the trunk, tapering rapidly to the end. The ribs in the cervical region were short but locked together posteriorly in a way permitting little lateral motion. Those of the trunk were attached high on the arch to the end of the stout transverse processes by a single head and ended freely below, with no attachment to a breastbone or other ventral bony structure. Their shape and position indicate that in life the body of plesiosaurs tended to be broad and somewhat flattened dorsoventrally.

The limbs were modified into broad paddles, and unlike other aquatic animals, the hind limbs were not reduced and were probably as effective as the forelimbs in swimming. The humerus and femur were broad, massive, and elongate, as in the sea turtles. Their strong, muscular rugosities suggest that the plesiosaurs possessed powerful swimming muscles. The paddles were undoubtedly more powerful than those of any other aquatic air-breathing animals. All plesiosaurs had five digits in each hand or foot but often an excessive increase in phalanges, some digits of certain species containing as many as twenty-four bones.

Some species were relatively broad-headed, with long, slender, recurved, pointed teeth in deep sockets. Others were slender-jawed and gavial-like, with smaller, more numerous teeth. The large posterior surface of the skull for attachment of the masseter muscles is evidence of powerful muscles for biting and seizing prey.

Small plesiosaurs were about 10 feet long when adult, but some of the largest species attained lengths of about 50 feet, the neck comprising about one half the total length.

The rigid nature of the jaws, united in front and incapable of lateral movement behind, indicates that they could not swallow large prey whole. The flat-headed plesiosaurs probably tore their food apart much in the manner of crocodiles, with quick, powerful jerks. The narrow-jawed species probably ate fish only, which their small, sharp, more numerous teeth fitted them to catch. Small fish were swallowed whole, and this group of plesiosaurs probably were not dangerous to larger animals.

The swimming must have been more turtlelike than fishlike, with the tail used as a rudder rather than as an organ of propulsion. They could hardly have been really swift swimmers like the ichthyosaurs and apparently seized their prey by swift downward lunges of the neck rather than by quick swimming. Many of them, like some modern crocodiles, apparently swallowed fairly large numbers of pebbles to be used as gastroliths in a muscular stomach for grinding food.

Unlike the ichthyosaurs, they appear to have been of more or less solitary habits, and some of them, at least, made journeys of considerable distance at sea. They have been found in deposits believed to be of deep-water origin and hundreds of miles from the nearest sources of the gastrolith pebbles found with their skeletons. Most species, however, probably inhabited coastal waters.

Like the other great reptiles—the mosasaurs, pterodactyls, dinosaurs, and others—they became extinct at the close of the Mesozoic era from causes not yet fully understood. As far as we can tell, it was a considerable time before their place was taken by any other creatures.

Ichthyosaurs. Like the whales among mammals, the ichthyosaurs among reptiles were the most perfectly adapted marine animals of terrestrial descent of their day. Their fossils are abundant in ancient Jurassic rocks, and we know them as a flourishing, fully developed group. But like the whales, they occupied a somewhat isolated position. So perfectly were they adapted to marine conditions that it is difficult to relate them to the earlier terrestrial forms from which they arose. Paleontologists of the nineteenth century regarded them as descended from fishes, so marked is the convergence shown.

Externally, ichthyosaurs were porpoise-like in general appearance. The body was naked, and the head was produced into a long, slender snout and closely joined to the body, with no indication of a neck. The body was somewhat torpedo-shaped—cylindrical and expanded in front, with a large thorax and abdomen, then tapering off rapidly and ending in a large, strong fishlike tail. The front paddles were close behind the head. The hind paddles were reduced in size, their function being replaced by that of the broad tail.

The vertebrae are amphicoelous and deeply biconcave, like those of fish. However, at the time when the ichthyosaurs must have arisen, probably from the very earliest Cotylosauria stem, all reptiles had biconcave vertebrae. In the ichthyosaurs, this type of vertebra persisted, as it has in the fish, because in the water such vertebrae are best suited for the quick, pliant movements of the spinal column required of most aquatic animals, although other reptile lines developed firmer backbones. In modern porpoises, which closely resemble these ancient reptiles in size, shape, and probably in habits, the small, flat-ended vertebrae are widely separated by discs of flexible cartilage.

There was no sacrum, that is, no united vertebrae in the hip region for attachment and support of the pelvis, since no such support was needed. This same adaptation is also found in cetaceans and sirenians. The long, slender ribs resemble those of fish-eating porpoises, but their articulation to the vertebrae by two attachments to the centrum and none to the arch differs from that of all other animals. The hind legs were small, and the pelvis was weak and suspended below the spinal column in the fleshy, abdominal wall. It is probable that if the ichthyosaurs had persisted to the present time, they would by now have lost the hind legs altogether, as have the cetaceans. The earlier ichthyosaurs of the Triassic (often classed as a separate family, the Mixosauridae) had a larger pelvis, longer paddles that were more leglike than in the later forms, a less well developed tail, and somewhat longer, less fishlike vertebrae. For propulsion, they apparently used their paddles more and their tail less than did their later Jurassic descendants.

The ichthyosaurs were apparently gregarious and inhabited the open ocean. Remains are often found in deep-water sediments, very complete and in a natural position, showing that they were not disturbed after the death of the animals as those of shallow-water coastal creatures usually are.

Ichthyosaurs were viviparous. Some fossils show as many as seven embryonic skeletons associated with an adult. The remains of food found in some of the fossils indicate that they fed mostly on fish, squid, belemnites, and various invertebrates. Their mouth and teeth were adapted for seizing and holding slippery prey but not for tearing or chewing their food or for swallowing very large prey. The jaws were closely united, permitting none of the expansion characteristic of snakes.

The ichthyosaurs attained a nearly worldwide distribution but died out before other contemporaneous groups of marine reptiles, such as the plesiosaurs and mosasaurs.

Mosasaurs. Mosasaurs were large aquatic lizards distantly related to the monitor lizard, *Varanus*. They appeared during the Cretaceous, enjoyed a rather rapid adaptive radiation, soon reaching its culmination in size, numbers, and distribution, and disappeared before the close of the Cretaceous.

Their skeletons have been found in Australia and New Zealand, Europe, and North America and are especially abundant and well developed in the Kansas chalk deposits.

The skull of mosasaurs was large in proportion to the total size, nearly one sixth of the total length. Both jaws were armed with numerous strong, sharp, conical, recurved teeth, and there were also additional teeth on the pterygoid bones in the back of the mouth. Unlike the teeth of other lizards, they were inserted on large tumid bony bases rather loosely attached in shallow pits. They were relatively easily dislodged and were readily replaced by new ones throughout life, so that adult mosasaurs had teeth of all sizes at the same time. The jaws had considerable freedom of movement both laterally and vertically and were rather loosely joined in front. Furthermore, they possessed a unique joint just back of the teeth, permitting movement between the front and back portions of the jaw. The large skull and heavy armament of teeth and mobile jaws indicate extremely predacious and pugnacious habits.

As in other aquatic reptiles the eyes were protected by a ring of bony plates, and the ears had a thick, cartilaginous drum in place of a simple membrane. The neck was short and strong. The body was elongate, and the tail was long and strong, laterally flattened, with the beginning of expansion toward the free end into a caudal fin. The vertebrae were procoelous like those of terrestrial lizards and snakes, giving greater strength to the spinal column but less flexibility than in other aquatic reptiles, a condition doubtless carried over from terrestrial lizard ancestors.

The limbs were adapted for swimming, the hind limbs reduced in size as in other forms that propel themselves by means of the tail. The bones of the upper arm and thigh were much shortened and broadened, and in the most specialized types, such as *Tylosaurus,* the wrist and ankle bones were mostly cartilaginous. This tendency of the ends of the long bones and the bones of the wrists and ankles to become less well ossified in purely aquatic animals is so strong that aquatic habits are suspected at once in any reptile showing them, even though the skeleton may be very imperfectly known.

The body was covered with small, overlapping scales, somewhat smaller and smoother in comparison to the size of the animal than those found on modern monitor lizards. The top of the skull was covered by horny plates, as in most lizards.

Mosasaurs ranged from about 15 to over 40 feet in length and were doubtless one of the fiercest, most voracious groups of predators that have existed. They were able to take rather large prey, which they probably swallowed whole, head first, in the manner of snakes. Their jaws and teeth were not fitted for tearing off bite-size chunks from larger animals, nor for chewing, but rather for seizing and holding prey. It is a curious thing that in a form of which the fossils are so abundant, no embryonic or very young individuals have been found. This would seem to indicate that mosasaurs were oviparous, in which case they probably laid their eggs on shores and beaches as do sea turtles and crocodiles. The waters in which mosasaurs lived contained many kinds of predacious fish, together with other mosasaurs, so it would seem highly perilous for the young, perhaps a foot or so in length, to venture into them. It is possible that the females entered estuaries and rivers to find protected places and that the young remained in such places until they were large enough to cope with conditions in the sea. Much less is known of the fauna of lakes and rivers of this period.

Thalattosaurs. The thalattosaurs were a group of small marine reptiles, only 4 to 6 feet in length, first studied by J. C. Merriam from Upper Triassic rocks in northern California. They are known only from purely marine deposits containing little or no material of terrestrial origin.

The skull displays adaptation to the aquatic environment in the elongation of the snout, with the nares well back from the end and dorsal in position. The jaws were armed with conical pointed teeth in the front, the teeth becoming rounded, rugose, and obtuse farther back. The palate was largely covered by similar teeth, which must have been used for crushing crustacean or mollusc shells—perhaps the light shells of ammonites, which were abundant in these seas at that time. The conical pointed teeth of the forepart of the jaw indicate that they probably also seized swifter-swimming prey such as fish.

The eyes show the ring of ossifications surrounding the eyeball, a characteristic of truly aquatic reptiles.

The limbs were structurally much like those of the mosasaurs. The ribs had a single articulation to the centrum, a characteristic shown by lizards, snakes, and mosasaurs and unlike those of other reptiles. The vertebrae were amphicoelous, as were the vertebrae of all primitive reptiles existing at the time of the thalattosaurs. The ball-and-socket joints of the vertebrae of present reptiles did not arise until long after the disappearance of the thalattosaurs.

The thalattosaurs apparently never achieved a worldwide distribution, nor did they enjoy a very long existence in geological terms. Apparently they were eliminated in competition with the forerunners of the plesiosaurs, ichthyosaurs, and mosasaurs, or perhaps by other marine predators such as some of the fish.

Order Thalattosuchia

The Thalattosuchia represent an early marine offshoot from the great crocodilian stem during the Upper Jurassic period. In general form and appearance they resembled what we might imagine present-day gavials would become if they should become marine in habit. They were 10 to 20 feet long, with relatively small heads drawn out into a long, slender snout as in present gavials. The eyes had a ring of sclerotic bones as in other truly marine reptiles but which is found in no other crocodiles. Their eyes were directed laterally, rather than more or less upward. The tail was long and strong, making up about half the total length and developed into a fleshy terminal fin. There was no sternum, a structure found in all other crocodiles. The forelimbs were paddlelike and attached well forward, serving as balancing organs like the pectoral fins of most fish rather than as organs of propulsion. In contrast to the legs of other marine forms propelled chiefly by the tail, the hind legs were larger than the forelegs and not very different in structure from those of gavials. The pelvis was well developed and firmly attached to the sacrum. The presence of well-formed hind legs and of a ventral armor of bony ribs seems to indicate that the marine crocodiles visited the shores periodically, probably to lay eggs. Perhaps failure to become viviparous was a limitation that may have contributed to their early extinction in competition with the truly viviparous aquatic carnivores.

Order Pterosauria

The flying reptiles, or pterodactyls, of Jurassic through Cretaceous times seem to have been the reptilian equivalent of the seabirds of today. The wings were supported by an enormous extension of the fifth, or "little," finger, which stretched out a very long but rather narrow skin plane. The thumb and first three fingers were small and clawed.

Pterodactyls ranged in size from that of moderate-sized birds such as terns to large species somewhat larger than any contemporary bird. They did not have very powerful wing muscles, and thus probably launched themselves from cliffs near the ocean and soared over the water, perhaps snatching fish from the surface as frigate birds sometimes do. Having relatively weak wings, the larger species, at least, may have been unable to take off from the water surface readily or from flat places on land.

The brain was large for a reptile, with an especially well-developed birdlike cerebellum.

All remains of pterodactyls have been found in marine sediments.

Fig. 12-1. Walrus herd on ice floes at edge of the ice pack, Cape Lisburne, Alaska. (M. W. Williams photograph.)

Adaptations of terrestrial vertebrates to the marine environment

M̲ost of the reptiles and mammals that became aquatic fish-eaters show elongation of the face, with the development of long, slender jaws armed with sharp teeth. The nostrils are commonly placed well back, sometimes dorsal, rather than at the tip of the snout, enabling them to breathe more easily while floating near the surface. Those which subsist on larger aquatic vegetation, such as the Sirenia or the Galápagos sea lizards, tend to have blunt muzzles and teeth modified for chewing and grinding. A third category, those which feed on the benthic, often hard-shelled animals such as molluscs, crustaceans, and echinoderms, are likewise often rather short-muzzled forms, with the teeth, or the hard horny surfaces substituting for teeth, as in the turtles, fitted for crushing their prey. Some of them, as the walrus, have special tusks developed for digging bivalves out of the substrate.

The plankton feeders, best exemplified by the whalebone whales, exhibit very specialized adaptations for straining the sea-water and retaining the macroplankton. To a much lesser extent, the crab-eating seal is modified in a parallel manner, straining the water between its lateral teeth.

Propulsion in water is a very different problem from propulsion on land. Land animals returning to the sea have solved the problem in two ways, each of which entails a host of secondary modifications. Some have developed the limbs as the primary organs of propulsion, whereas others depend upon sinuous movements of the body or tail to move through the water in a more efficient, fishlike manner.

Those that depend on the limbs for their movement in the water are usually relatively short-bodied, short-tailed animals with well-developed, and in some cases very much elongated, necks and relatively small heads. The hind limbs are retained, and one or both pairs become either webbed or paddle-like in form. The upper arm and thigh bones are elongate but rather massive. In all forms in which the limbs develop as paddles, the lower bones—the radius and

283

ulna of the forelimbs and the tibia and fibula of hind limbs—become greatly shortened, whereas the digits are much elongated and often contain a larger number of skeletal elements than in other animals. The ends of the long bones and the bones of the wrist and hand and of the ankle and foot tend to be cartilaginous rather than well ossified, making the appendage more flexible. The dolichosaurs, nothosaurs, and plesiosaurs all possessed relatively short, nonpropelling tails and very long, mobile necks. The turtles all have well-developed necks.

Those animals that propel themselves with their tails generally have more torpedo-like bodies, with very short necks or none at all, and often show shortening and coalescence of cervical skeletal elements. If the sinuous movements are made horizontally and the swimming is largely level, the tail tends to be laterally flattened and, in the best-adapted forms, to develop as a fishlike vertical fin. Those that have adopted vertical undulatory movements, used largely for ascending or descending rapidly, develop horizontal flukes, as in the Cetacea and Sirenia. The limbs, no longer used for propulsion, are primarily used as balancing organs. The forelimbs become paddlelike, but in most instances the hind limbs are reduced or wholly lost. The body is often elongate, and in fish-eating forms the jaws are usually long and narrow. The bones of the upper arm and thigh are greatly shortened. The spinal column becomes more flexible.

Many groups tend to lose external structures such as hair or scales and to develop a smooth skin. The bones tend to become lighter and spongier, except in some of the bottom-feeding types such as Sirenia, in which the massive bones apparently help them stay submerged in shallow waters while feeding.

In both the reptiles and mammals, some of the most completely adapted tail-propelled groups have developed dorsal fins that serve as a sort of stabilizer in rapid swimming. The ichthyosaurs among the reptiles and the dolphins among the mammals probably represent culminations of this line of evolution and show remarkable convergence in general form and presumably in their habits. Truly marine reptiles all show a ring of sclerotic bones in the eye, enabling the eye to compensate for pressure in deep water.

Marine vertebrates, of course, all must solve the problem of osmotic relationships. The reptiles and mammals lack special salt-excreting glands such as are found in teleost fish and some birds. They apparently meet the problem through restricting the intake of seawater as much as possible and obtaining much of their water requirements from the body fluids of their prey. The kidneys do a much more thorough job of resorption of water in the kidney tubules, so that the urine tends to be minimal in volume and much more concentrated than in terrestrial animals. It is possible that the fat, which serves as a thick insulating layer in homoiothermic forms, also plays a role in the storage of water, as in some desert animals.

Although some very successful groups in both the reptilian and mammalian lines, such as the sea turtles and the pinnipeds, have never cut their reproductive tie to the land, the most completely adapted marine forms in both groups, the ichthyosaurs, Hydrophinae, and cetaceans, all bear their young alive in the water and never return to the land for any purpose.

The inability to breathe under water remains as the last great barrier preventing these animals from adopting a way of life wholly similar to that of the fish.

PART V

THE
REDUCERS

Fig. 13-1. Professor Claude E. ZoBell, of the Scripps Institution of Oceanography, La Jolla, California—one of the leaders in the study of marine bacteria—shown with pressure chambers used in culturing marine bacteria under high-pressure, simulating conditions in deep water where hydrostatic pressures may be so intense as to change the rates or character of certain enzymatic processes of biological importance. (Reproduced with the permission of Dr. ZoBell.)

CHAPTER *13*

Marine bacteriology

The microbiology of the oceans long remained a practically virgin, unexplored territory. There are, in the first place, formidable difficulties and expense involved in securing adequate samples from various depths and regions in the sea. Furthermore, it is necessary to carry out cultural procedures on shipboard immediately after the samples are obtained if their interpretation is to be clear. In addition, medical bacteriology, then, somewhat later and to a lesser extent, soil and agricultural bacteriology, and finally microbial genetics and physiology have occupied the attention of the overwhelming majority of bacteriologists.

The open ocean presents special problems not only for the microbiologist but also for any minute nonphotosynthetic organisms unable to ingest particulate food, that live in it. Such organisms must assimilate dissolved organic molecules that can be transported through their cell membranes. They obtain these molecules by secreting exoenzymes into their environment, which break down larger molecules and aggregations of organic materials. These enzymes are very specific in their action, each kind of enzyme acting only on particular chemical groups and thus only on a particular substrate or a group of closely related substrates that contain the molecular configuration in question. Dissolved organic matter in the sea is too dilute to directly provide efficient feeding of this kind for the bacterial populations there. Enzymes secreted by a bacterium suspended alone in the seawater would quickly be washed away and diluted beyond useful concentration. Any molecules of suitable type formed by the action of such enzymes on particles of organic matter in the seawater would be further diluted. The return to the bacterium of usable food molecules would be practically nil. Under these conditions bacteria could not grow and reproduce.

However, all surfaces tend to attract and

hold a film of molecular and colloidal particles. Thus the living and dead bodies of all marine organisms, their shells, tests, molted exuviae, feces, and even fine clay or silica particles suspended in the sea constitute sites of concentration of available food for bacteria. In addition, many of these particles themselves are composed of organic matter that can be utilized.

For these reasons, mobility plus a sessile habit are very important for bacteria in the open ocean. They must be able to find and attach themselves to surfaces of larger particles or objects in the water. With the increased food supply available on such a surface and the increased efficiency of digestive enzymes when they are secreted against a surface that tends to hold them and their substrates together and to catalyze the reactions between them, the bacteria are able to grow and multiply. An additional advantage of adhering to surfaces is that several kinds of bacteria may be growing together, producing a greater variety of digestive enzymes, so that the variety of organic materials that can be attacked is increased. Some species that cannot directly attack the original substrate of the particle can break down some of the intermediate products resulting from the action of enzymes of other species. Decomposition then proceeds faster, in many steps, with different species of bacteria participating at different levels of the process. Many marine bacteria, upon adhering to a surface, secrete a slimy material that holds them there and in which they multiply.

As Kriss and his colleagues (1964) have emphasized, those organic substrates most suitable for bacteria will be rapidly utilized, some of them *in statu nascendi,* that is, they will be utilized as rapidly as they are produced. Thus chemical analysis of the water may fail to disclose their presence in significant amounts, even though they are being continuously produced and used on a fairly large scale. For this reason, an analysis of the populations and types of microorganisms present in a water mass may be much more revealing of the extent and nature of the biochemical processes taking place there than a chemical analysis of the water. Only the more resistant organic residues would tend to accumulate, as a sort of organic marine humus. Chemical analysis alone may, by directing attention to those compounds playing the least active roles, actually be misleading.

Of great interest in this connection is the phenomenon in which, in any sample from a large body of water held for a few days in a bottle or flask, a sudden rapid increase occurs in the population of heterotrophic bacteria. The quantity of heterotrophs in isolated water samples increases several hundred or even several thousand times. This increase is the result of a marked increase in only that part of the microbial biocenosis which uses labile organic matter as a substrate. Thus the number of bacteria capable of multiplying in peptone media increases hundreds or thousands of times, whereas the total number of bacteria present, as revealed by direct microscopic count, may be only a few times greater than at the beginning.

This invalidates the claims of bacteriologists such as Kuznetsov (1955) and Sorokin (1955, 1957), who tried to measure the multiplication of chemosynthetic bacteria by measuring the incorporation of radioactive carbon-14 added to isolated samples of water or mud. The fact that the fixation of the carbon is mainly by bacteria multiplying in the flasks through heterotrophic assimilation of CO_2 is also indicated by the fact that in samples to which readily assimilable organic substrates such as glucose are added, the amount of radioactive carbon absorbed by bacteria is several times higher.

The vigorous multiplication of heterotrophs in sample flasks is due to a double effect of the glass–water surface interface. In the first place, molecules of labile organic compounds, which may be present in the water in amounts too small for the bacteria to utilize, are concentrated at such an interface, making them more available. Even more important, however, is a denaturation of nonlabile organic matter at such a boundary, the denatured proteins being more readily hydrolyzed by bacterial enzymes. The sharp increase in labile organic matter, reflected by a marked rise in the population of heterotrophic bacteria, takes place at the expense of nonlabile organic compounds already present in the water but unavailable in that form to these bacteria.

The ocean depths are rich in such aquatic humus. Experiments with seawater stored in flasks indicates that its transformation into forms assimilable by microorganisms invariably occurs at contact boundaries between suspended particles and the water. This greatly increases the significance of the sessile habit of marine bacteria, since each particle suspended in the seawater or settling through it creates a microzone where accumulation and transformation of aquatic humus occur. The development of heterotrophs on the particle is aided, which brings about its partial or complete mineralization. This also explains the dramatic rise in the number of heterotrophs found at the surface of the sea floor, above the number found a short distance either above or below this surface.

These facts emphasize the importance of making bacteriological analyses of water samples immediately after their collection. They also have important implications for the interpretation of such conventional tests as the biochemical oxygen demand (BOD) test of waters. This method consists in agitating a water sample to saturate it with atmospheric oxygen and then pouring aliquots into two or more oxygen flasks. The amount of oxygen present at the beginning is compared with the amount present in the flasks containing water kept for five, ten, twenty, or more days, and the oxygen loss due to oxidation of organic matter in the flasks, the BOD is thereby determined in milligrams of oxygen per liter.

The BOD data merely show that the biochemical utilization of oxygen in the flasks continues at the expense of stable organic matter transformed at the solid-liquid interfaces of the flask into forms that can be broken down by the enzymes of heterotrophic bacteria.

Since no such rise in the number of heterotrophs occurs in the natural body of water from which the sample was taken, as occurs in the flasks, the results of the conventional BOD test cannot possibly be used to estimate the value or rate of oxygen demand in the sea itself, as Skopintsev had already pointed out in 1950.

Since almost all bacteria must utilize organic matter, their numbers will bear a direct relationship to the amount of organic matter present in the water. In the ocean the production of organic matter occurs in the euphotic zone through the activities of phytoplankton organisms. The distribution of bacteria in the oceans tends to parallel that of plankton, with the peaks of numbers of marine bacteria coming a little later in time and a little lower in the water than the phytoplankton peaks.

Recently it has been found that the phytoplankton of some regions, such as the Antarctic and the Arctic, exerts a powerful antibacterial effect in the bodies of animals. Organisms such as euphausiid shrimps that concentrate the phytoplankton and predators that prey on the euphausiids, or even secondarily on such predators, show marked antibiotic properties in their gut contents and tissue fluids. Thus penguins feeding on the euphausiids have been reported to be bacteriologically sterile, or nearly so. Young nestling penguins fed on regurgitated matter from their parents, and even skua gulls feeding on the young penguins, also show this condition. Their tissue fluids and gut contents have marked antibacterial properties. It is possible that this antibiotic property of the phytoplankton may play an important role in the ecology of these regions.

As mentioned in the introductory paragraphs, the overall result of bacterial activity is to break down the accumulations of organic matter represented in the bodies, exuviae, secretions, feces, skeletons, tests, and other remains of plants and animals, releasing their chemical constituents in the form of simple soluble inorganic ions and gases that can be utilized again by the plants. This process is termed the **mineralization of organic matter**. It is of the utmost importance in the cycling of available chemicals necessary for life and thus for the continuation of life on earth. In the sea the principal places for mineralization of organic matter are (1) in the upper waters, where all of the organic matter is produced, slightly below the level at which the greatest amount of phytoplankton is found because of the tendency of dead and decomposing particles to slowly settle, and (2) at the bottom of the sea at the surface of the marine mud, where organic detrital particles and larger remains that have not been completely disintegrated during settling tend to accumulate.

CHARACTERISTICS OF MARINE BACTERIA

Marine bacteria tend to differ characteristically from terrestrial and freshwater bacteria in a number of ways. Many of them require seawater or seawater media, at least for primary isolation and good growth, whereas most terrestrial forms do not do as well in such media as they do in media made with fresh water. Not only the salt content of the water but also, apparently, trace quantities of other elements or compounds of biological origin are important, since, as a rule, synthetic seawater prepared from purified chemicals and distilled water, although nutritionally adequate, osmotically similar, and chemically similar insofar as all major ions are concerned, is not usually as satisfactory for cultivation of marine bacteria as is natural seawater.

Since phytoplankton cells store very little reserve carbohydrate and produce much less cellulose than do terrestrial plants and practically no lignin and since the carbohydrates formed are used almost as fast as they are formed in the metabolism of the plant cells, the proportion of saccharolytic (carbohydrate-utilizing) to proteolytic (protein-utilizing) bacteria in the oceans is much less than on land. In the sea, one of the organic residues formed abundantly is **chitin,** the hardened cuticle of the exoskeleton of arthropods and some other animals. This differs from cellulose in being a nitrogenous material rather than a carbohydrate, but its formation and tendency to accumulate in the seas may be compared to that of cellulose on land. In both cases there is an abundantly formed organic residue, resistant to attack by the majority of species of bacteria. The chitinolytic, or chitin-attacking, bacteria of the sea play a role somewhat comparable to that of the cellulolytic, or cellulose-attacking, bacteria on land.

In general, populations of marine bacteria do not survive unusual environmental stresses such as heat or changes in their chemical environment as well as do populations of terrestrial bacteria. Their environment normally varies within narrower limits, and resistant stages are usually absent.

Because most marine bacteria normally grow under cool or cold conditions, there is a greater proportion of types that prefer a cool environment. This makes preservation of such marine products as fish or shrimp by refrigeration less effective than is the case with meat products of terrestrial origin, since many of the bacteria associated with these products actually prefer to grow at about ordinary refrigeration temperatures. In the sea the proportions of spore-forming bacteria and of cocci in the total bacterial population are much less than on land. There are more gram-negative, polar flagellated, nonsporing rods and spirals.

It should not be thought that because the marine humus is not readily attacked by the typical heterotrophic forms which grow so well in peptone media, there are no organisms at all which attack it directly. Direct microscopic techniques, such as studying the growth that occurs on slides or other surfaces suspended at various depths in the water, have revealed a number of types of microbial cells of unusual morphology, most of which have not been successfully grown on laboratory media. Some appear to be obligate periphyton, since they appear only on overgrowth slides. The lists of species isolated from marine sources do not yet fully reflect, even as to morphological types, the full diversity of forms present in the sea. Only with further development of physiological-ecological methods and the use of carefully designed enrichment media will we begin to approach a detailed picture of the microbiological biocenoses of the oceans.

The problem of the taxonomy of marine bacteria also presents grave difficulties. Microbial taxonomy is beset with unusual difficulty in any case because of the impossibility of knowing the totality of characteristics revealing the biological potentialities of the forms under study and of including even the most common range of mutational variation in their descriptions. In the case of bacteria found in the sea, there is the added problem of knowing whether or not they are closely related to terrestrial species that have already been described (often very inadequately). Even though the majority of terrestrial bacteria washed into the sea in vast numbers daily do not long survive, a number of species may be capable of living in either environment. Whether or not a number of forms now known only from the sea also occur in terrestrial environments is uncertain because no one has made much of an effort to look for them there. To describe all the bacteria found living in the sea as new species simply because they occur in the oceans is no real solution to the problem. In the present state of our knowledge, it is perhaps just as well to use species names in a rather inclusive sense. If a marine bacterium is not clearly different from a previously described terrestrial form in morphological or physiological characteristics, it may be regarded as a variant, race, or ecotype of the named species.

The large group of filamentous, branching Actinomycetales, so abundant and prominent terrestrially, seems to be lacking in the oceans except where washed in from adjacent land. On the other hand, the peculiar group of filamentous, nonbranching, botryose organisms known variously as the Krasil'nikoviae or the Calamobotryocidales are known only from the oceans, in which they are apparently cosmopolitan.

BACTERIA ON THE FLOOR OF THE OCEAN

At the substrate-water interface of the floor of the ocean the accumulation of organic detrital particles settling down from above and probably also the surface effect of denaturation of dissolved organic humus support a large microbial population. Here the final stages of unfinished mineralization of organic residue take place. Below the surface the bacteria quickly use up all the free oxygen, and their metabolism of organic matter results in an accumulation of hydrogen sulfide, so that the subsurface and deeper layers are anaerobic, black, and smell of hydrogen sulfide. The number of bacteria declines sharply downward from the surface, and the proportion of anaerobic bacteria increases. It is a curious fact, however, that even from the deeper layers of anaerobic mud, a considerable number of aerobic bacteria can be cultured. It may be that they have remained in a state of suspended animation ever since they became buried below the surface of the marine mud.

Since the rate of accumulation of marine sediment is on the order of 1 mm. per century, these deeply buried aerobic bacteria must have existed there in this state for hundreds of thousands of years in some cases. At the surface of marine sediments, where they occur in the greatest numbers, they exercise their strongest effects on pH and thus on the processes of calcification and lithification. Their growth on the particles of sediment, often accompanied by secretion of cementing materials, may cause

Table 13-1. Aerobic and anaerobic bacteria per gram of sediment*

Depth in ocean floor (cm.)	Aerobic bacteria per gram	Anaerobic bacteria per gram
0–10	62,000,000	8,900,000
40–50	91,000	23,000
240–250	2,000	900
500–510	580	26

*From ZoBell, C. E., and D. Q. Anderson. 1936. Vertical distribution of bacteria in marine sediments. AAPG Bulletin 20(3):258-269. Reprinted by permission of the American Association of Petroleum Geologists.

Table 13-2. Percentage of elements of organic origin in marine sediment*

	Carbon	Hydrogen	Oxygen	Nitrogen	Phosphorus
Marine sapropel	52	6	30	11	0.8
Recent sediment	58	7	24	9	0.6
Ancient sediment	73	9	14	3	0.3
Petroleum	85	13	0.5	0.4	0.1

*From ZoBell, C. E. 1946. Marine microbiology. By permission of The Ronald Press Co., New York.

a cementing together of these particles and thus bring about a change in the character of the ooze and hasten its consolidation and lithification.

Table 13-1, from ZoBell and Anderson (1936), shows in round figures the numbers of aerobic and anaerobic bacteria cultured from a sample of marine sediment core taken at 2230 meters (over a mile) below the surface of the water.

The activity of the microorganisms in the marine mud also gradually alters the proportions of chemical elements of organic origin in the sediment, making them more petroleum-like, as shown in Table 13-2. The relative proportions of oxygen, nitrogen, and phosphorous are reduced, and with the loss of these, the proportion of carbon and hydrogen increases.

It is thought that in shallow seas where vast amounts of organic matter may collect, the activities of marine bacteria in the sediments bring about the initial changes leading to the formation of petroleum. After these altered marine deposits become deeply buried in sediments and subjected to great pressures and perhaps high temperatures, they finally become converted into petroleum deposits.

OTHER EFFECTS OF MARINE BACTERIA

Marine bacteria are important in numerous other ways in addition to their primary role in the mineralization of organic matter.

Many are ingested along with minute detrital particles by filter feeders such as bivalve molluscs and copepods, and are also eaten by small animals such as protozoans, rotifers, and nematodes, thus serving directly as food for many animals.

Where very numerous, they may affect the pH of the water. This in turn will affect such processes as the deposition or solution of calcium carbonate and the rate of sedimentation of fine particles in the water, such as silt and clay carried out to sea from river mouths. Their tendency to adhere to and multiply on the surfaces of particles in the water may directly increase the tendency of such particles to agglomerate and settle. Marine bacteria may thus be of some geological importance.

In localized areas where the organic content of the water is above average and the weather very calm, they sometimes deplete the oxygen in the surface waters, causing the death of many animals there.

The settlement and growth of marine bacteria on all submerged surfaces constitute a first step in the colonization of these surfaces by marine organisms, making them more favorable for the settlement and growth of larvae of barnacles, mussels, hydroids, sponges, bryozoans, and algae. Thus they play a role in the settlement of the so-called "fouling organisms" of ship bottoms, underwater pipes, etc. They do this by rendering such surfaces less smooth and shiny, by changing the pH at the surface-water interface, by themselves providing food for some of the larvae, and perhaps by affording some protection from any toxic elements that might be put in paints for the protection of such surfaces.

Marine bacteria may also contribute some of the biological compounds, traces of which seem to be important in "conditioning" the seawater for the growth of other forms. The dissolved metabolites may also be liberated from larger aggregates of organic matter during their decomposition by bacteria.

Marine bacteria are directly important to man in causing spoilage of marine food products and deterioration of nets, ropes, and cables and in initiating the fouling of all submerged surfaces such as boat bottoms.

Marine bacteria thus play a primary role in influencing the physical, chemical, biological, and geological processes in the oceans and ocean floors.

DISTRIBUTION OF HETEROTROPHIC BACTERIA IN THE OCEANS

In recent years extended expeditions, particularly those of Kriss and his associates under the sponsorship of the Soviet Union, have done much to extend and clarify our understanding of the distribution, number, kinds, and activities of bacteria in the oceans, on a worldwide basis. Kriss and associates (1964, 1967) described some 186 kinds of heterotrophic bacteria, 15 yeasts, 12 mycobacteria, and 2 actinomycetes taken at all depths and in all major ocean regions. They were able to distinguish characteristic tropical species and associations of bacteria and other species and groups characteristic of cold polar waters, to trace the movements of major water masses and currents by analysis of their contained bacteria, and to arrive at some idea of the turnover of organic matter in various types of ocean waters through quantitative study of the populations of heterotrophic bacteria they contained. Unlike most larger organisms, many bacteria are not seriously limited by differences in hydrostatic pressure and can be found in any depths where nutrient conditions and temperature permit their growth.

Tropical waters were, in general, found to contain a much larger quantity of bacteria than did temperate, boreal, and polar waters, although not necessarily more kinds. It was also found that as a rule the bacteria typical of cold waters possessed a wider range of biochemical abilities (that is, they could utilize a greater variety of organic substrates) than did those characteristic of tropical waters. No ready explanation for this difference was apparent, although it was speculated that perhaps the comparatively high concentration of readily assimilable allochthonous organic matter in tropical waters, which causes the high density of heterotrophs in these regions, obviates the necessity for enzymatic reactions that would ensure total utilization of organic material. In the high latitudes where the water is poorer in food reserves, adaptations may have occurred in the heterotrophs, allowing them to use the organic matter more economically by breaking it down and converting it more thoroughly.

In all regions the proportion of heterotrophic bacteria able to metabolize inorganic nitrogen sources (that is, growing well on media containing mineral sources of nitrogen) was strikingly high—86% to 99% of the cultures.

The percentage of actively proteolytic species and of species fermenting carbohydrates varied in different regions, being highest in the North Atlantic and arctic regions and lowest in the equatorial and South Pacific regions. In the Indian Ocean the percentage of active cultures was intermediate, being higher in the antarctic regions and lower in the equatorial portions.

In order to bring out more clearly certain features of the distributions of bacteria, the data of Kriss and co-workers were analyzed according to the method of Mc-Connaughey (1964) as modified by McConnaughey, Moestafa, and Lestari (unpublished). Treated in this manner, the 215 kinds fall into 115 groups, which in turn associate as 72 communities. These 72 communities, in turn, form two clearly distinct major associations, one containing 24 species and varieties, centered in the tropics, and the other being a looser association composed of two vaguely separated portions and containing 182 kinds, found primarily in cooler waters and higher latitudes. Nine species, most of them taken in only one haul each, were not associated with any other groups.

The circulation of waters in the oceans, of course, carries tropical bacteria into high latitudes, and vice versa, in certain regions and at certain times and depths. When a

Table 13-3. Most frequently taken species of heterotrophic marine bacteria and other microorganisms*

Num-ber	Species or variety	Stations	Group	Com-munity	Association
1	*Bacterium agile* Jensen, var. B	120	1	1	Warm
2	*Bacterium agile* Jensen, var. A	89	1	1	Warm
3	*Pseudomonas sinuosa* Wright, var. A	61	2	2	Cool
4	*Bacterium agile* Jensen, var. D	52	1	1	Warm
5	*Bacterium candicans* (Frankland), var. B	52	3	1	Warm
6	*Bacterium candicans* (Frankland), var. A	46	1	1	Warm
7	*Bacterium agile* Jensen, var. C	32	4	3	Warm
8	*Pseudomonas sinuosa* Wright, var. B	30	5	1	Warm
9	*Bacterium parvulum* Conn, var. B	27	6	4	Cool
10	*Bacterium agile* Jensen, var. F	25	7	5	Intermediate
11	*Pseudomonas sinuosa* Wright, var. C	24	8	6	Cool
12	*Micrococcus albicans* Trevisan, var. E	24	9	7	Cool
13	*Bacterium candicans* (Frankland), var. D	21	10	3	Warm
14	*Bacterium agile* Jensen, var. E	15	11	8	Warm
15	*Pseudobacterium biforme* Eggerth	14	12	9	Cool
16	*Pseudomonas sinuosa* Wright, var. E	14	13	10	Cool
17	*Pseudomonas sinuosa* Wright, var. F	12	14	11	Intermediate
18	*Bacterium agile* Jensen, var. H	11	15	12	Warm
19	*Bacterium agile* Jensen, var. K	11	16	13	Intermediate
20	*Bacillus catenula* Migula var. D	11	17	14	Cool
21	*Micrococcus radiatus* Flügge var. B	10	18	15	Cool
22	*Bacterium candicans* (Frankland) var. C	9	19	16	Cool
23	*Bacillus idosus* Burchard, var. A	9	20	14	Cool
24	*Rhodotorula mucilaginosa* (Jörg)	9	21	16	Cool
25	*Debaryomyces rosei* (Kudryavtsev)	9	22	17	Cool
26	*Pseudomonas sinuosa* Wright, var. G	8	23	18	Cool
27	*Micrococcus radiatus* Flügge, var. H	8	24	19	Intermediate
28	*Pseudomonas liquida* Frankl, var. B	8	25	3	Warm
29	*Rhodotorula glutinis* (Fres)	8	26	17	Cool
30	*Bacterium parvulum* Conn, var. E	7	27	1	Warm
31	*Pseudomonas sinuosa* Wright, var. D	7	28	15	Cool
32	*Pseudobacterium variabilis* (Distaso), var. B	7	29	20	Cool
33	*Sarcina subflava* Ravenel, var. A	7	30	21	Cool
34	*Micrococcus albicans* Trevisan, var. D	7	30	21	Cool
35	*Bacillus catenula* Migula, var. I	7	31	22	Cool

*Adapted from Kriss, A. E., I. E. Mishustina, N. Mitskevich, and E. V. Zemtsova. 1964. The microbial population of oceans and seas. [Transl., 1967, by K. Syers; edited by G. E. Fogg.] Edward Arnold (Publishers) Ltd., London.

water mass carries its microbial population into regions far removed from their normal distribution, it remains identifiable by its microbial content, as well as by other, more often studied characteristics, as long as the bacteria continue to live, which is probably about as long as the water mass itself maintains its distinctive identity.

The grouping of the first 35 species and varieties, all of them taken at seven or more different stations, is illustrated in Table 13-3. This can be rearranged and simplified by grouping the species as follows:

Species centered in warm waters	No. of stations
Bacterium agile var. B	120
Bacterium agile var. A	89
Bacterium agile var. D	52
Bacterium candicans var. B	52
Bacterium candicans var. A	46
Bacterium agile var. C	32
Pseudomonas sinuosa var. B	30
Bacterium candicans var. D	21
Bacterium agile var. E	15
Bacterium agile var. H	11
Pseudomonas liquida var. B	8
Bacterium parvulum var. E	7

Intermediate species*

Bacterium agile var. F	25
Pseudomonas sinuosa var. F	12
Bacterium agile var. K	11
Micrococcus radiatus var. H	8

Species centered in cool waters

Pseudomonas sinuosa var. A	61
Bacterium parvulum var. B	27
Pseudomonas sinuosa var. C	24
Micrococcus albicans var. E	24
Pseudobacterium biforme	14
Pseudomonas sinuosa var. E	14
Bacillus catenula var. D	11
Micrococcus radiatus var. B	9
Bacterium candicans var. C	9
Bacillus idosus var. A	9
Rhodotorula mucilaginosa	9
Debaryomyces rosei	9
Pseudomonas sinuosa var. G	8
Rhodotorula glutinis	8
Pseudomonas sinuosa var. D	7
Pseudobacterium variabilis var. B	7
Sarcina subflava var. A	7
Micrococcus albicans var. D	7
Bacillus catenula var. I	7

---

*The intermediate species include forms that were taken about equally with warm-water and cool-water types, and in all latitudes, or that clearly centered in intermediate latitudes.

PART *VI*

ZOOGEOGRAPHY
of the
SEAS

*B*iological provinces or zoogeographical regions are characterized most importantly by the endemic organisms they contain. Both the number and proportion of endemic species and the taxonomic rank of endemic groups are important. In general, any higher taxonomic group represents a longer evolution than most of its present species or subgroups. If an entire genus is endemic to a given region, this fact has more weight than the endemism of an equal number of species of a more widely distributed genus. Similarly, an endemic family or order carries more significance than a species or genus.

Endemism may result either from failure to spread beyond the region of origin or from restriction of a formerly greater range. A high degree of endemism in a given region generally denotes the relative isolation of that region for a prolonged period of time, the degree of endemism attained being roughly proportional to the degree and duration of the isolation. Obviously, given isolating factors do not affect all groups of organisms to the same extent. Something that may completely block the spread of certain types of animals or plants may have no significance at all with respect to the spread of other groups.

Zoogeographical research results in the delimitation of a hierarchy of regional units and subunits analogous to the hierarchy of taxonomic units resulting from systematic research. However, in both cases the final aim is not the creation of a hierarchy of units in itself but the understanding of the evolutionary and geographical history that these systems reflect.

In assessing the fauna within a zoogeographical system, it is advantageous, when possible, to use mathematical values for the various elements, combining these into more comprehensive figures by summation to arrive at values for the degrees of endemism or nonendemism of various groups and for the degrees of affinity or independence of the region in comparison with other regions, so as to arrive at conclusions as objectively as possible. Ekman (1940) laid the basis for such a procedure. In order to use methods of this sort satisfactorily, it is, of course, first necessary to have adequate faunistic and floristic information, which can be provided only by detailed surveys.

The distribution of organisms is the result of interaction between the physiological properties of the organisms, the quality or characteristics of the environment, and the history of both these components of the

297

ecosystem. Throughout time the faunas and floras have been continuously changing as the result of infinitely complex interactions between animate and inanimate nature. The biogeographical conditions that have emerged at any particular time are the result of all that has gone before and cannot be well understood out of their historical context.

In the seas, the distribution of organisms is complicated by the dimension of depth. The epipelagic organisms and the faunas and floras of the continental shelves show very different patterns of distribution than do the animals of bathypelagic or abyssal depths. The most significant transition between what is termed the shelf fauna and the deep-water archibenthal or abyssal fauna lies at the "mud line," the upper limit of loose mud sediment, as was recognized by John Murray in his reports on collections made by the Challenger expedition. This line may or may not coincide with the edge of the continental shelf in different regions, but in a general way it seems to do so.

Fig. 14-1. Mossy chiton, *Mopalia muscosa.* (R. Buchsbaum photograph.)

CHAPTER *14*

Organisms of the continental shelves and neritic waters

For the epipelagic and shelf organisms, the temperature of the surface waters is of crucial importance in determining their distribution. Most organisms are living much closer to the upper limits of their thermal tolerance than to the lower, since most biochemical reactions are faster and more efficient in this part of the range. An organism is usually more quickly incapacitated by moving from its normal environmental temperature into warmer surroundings than by moving into correspondingly cooler ones. Furthermore, where warm and cool waters meet, the warmer, less dense water tends to remain at the surface and the cooler water to sink below it. The cool-water organisms, instead of invading the warm surface waters, are more likely to be carried down or to swim down with water more to their liking, showing the phenomenon of submergence, or else to be killed by the warmer water.

The result is that warm tropical waters are relatively free from incursions of organisms from the colder waters to the north and south. Life itself and perhaps the majority of major groups of organisms appear to have originated in warm tropical waters, whereas the colder waters to the north and south and in the depths have been populated as a result of movement of organisms from warmer waters into them, together with the organisms' adaptation to life in cooler waters. Once adapted to life in cold waters, organisms have less tendency to recolonize the warm waters, both because of the greater physiological risk in moving from cooler to warmer waters and because of the competition from the forms already filling the ecological niches there.

Furthermore, the colder-water regions are subject to progressively greater ranges of environmental stress and seasonal changes in temperature and illumination. The colder and darker periods bar species adapted only to warmer waters, and

299

warmer periods bar species adapted to still colder waters. The growing season becomes progressively more abruptly seasonal, and the annual fluctuations in conditions tend to favor those species best able to take advantage of the growing seasons through rapid growth. The surface waters of cold regions tend to be more fertile in the spring because of the annual winter turnover and mixing of waters from different depths.

In the tropics, on the other hand, conditions are far more stable. There is far greater opportunity for ecological succession to set in. Instead of the great seasonal blooms involving a relatively small number of favored species, there is a more steady state of equilibrium characterized by far greater diversity but much less tendency to produce an immense biomass of particular species. Because of its relative isolation and greater diversity, the tropical zone of the world as a whole displays a high degree of endemism. Many genera, families, and even higher systematic groups are wholly, or almost wholly, confined to the tropics. Some are circumtropical, appearing in all major divisions of the zone. Others are limited to one or two of the main divisions. The circumtropical species form only a very small fraction of the total enormous number of species found in the tropics.

TROPICAL AND SUBTROPICAL REGIONS

Two great biocenoses, or associations of organisms, characterize and are limited to shallow tropical waters around the world. They are the coral reef and the mangrove swamp. They are of the utmost zoogeographical importance because they occupy dominant positions in many parts of the tropics and because they provide the habitat for special faunas of great variety. No other faunal groupings can compare in variety and beauty with those of the coral reefs.

The subtropical waters to the north and south of the tropical zone proper contain a more or less thinned-out tropical fauna, lacking the more stenothermal warm-water forms but having most of their constituents in common with the tropics, and a far smaller number of species in common with the temperate fauna on the other side. Although the subtropical waters do contain some endemic elements, these elements are fewer in number and lower in taxonomic rank than the endemic constituents of tropical waters.

The main divisions of the tropical waters are the result of long-standing barriers to the distribution of shallow marine fauna in the form of continents lying across the equator in a north-south direction and in the form of the vast East Pacific barrier, that great stretch of very deep water lying uninterruptedly between outer Polynesia and Hawaii on the west, and the American continents on the east. These barriers divide the tropical belt into three major components: the **Indo-West Pacific**, the **tropical Atlantic**, and the **tropical East Pacific.**

INDO-WEST PACIFIC

The Indo-West Pacific region is centered in the great Malay Archipelago off Southeast Asia, extends west across the northern part of the Indian Ocean to the eastern coast of tropical Africa and the Red Sea, and east through Polynesia in the South Pacific to Hawaii, the Marquesas Islands, and the Tuamotu Archipelago. North and south it extends from the northern coasts of Australia to the Korea Strait and Japan.

The homogeneity of this vast region is best attested to by those groups of animals that are most readily transported great distances by currents and that have therefore attained distribution throughout, but limited to, this region. The fish, more than any other group, fit this category. Even strongly coastal species can be transported by currents over wide stretches of pelagic regions in either larval or adult condition without harm. Thus chance plays a larger part in the distribution of fish than is the case for most other elements of the shelf fauna. Gunther, as long ago as 1886, noted that the number of species of tropical fish ranging from the east coast of Africa through the East Indies to Polynesia was very large but that these same fish did not appear in any great number of species in the tropical East Pacific along the coasts of the Americas. The fish fauna of the Indo-West Pacific is very much richer than that of the other main tropical regions, and contains all the families and the majority of genera of the tropical marine fish fauna of the world (except for 4 small families).

Another group of animals characteristic of the region are the sea snakes, family Hydrophidae, with 15 genera and more than 60 species ranging over the whole of this region, but only one species occurring outside the region, along western Africa. The dugongs, genus *Dugong (= Halicore)* are also limited to this region.

Among the invertebrates, the swimming crab *Neptunus pelagicus* has achieved a similar distribution, as has also the coconut crab, *Birgus latro,* distributed during the pelagic zoeal stage. Several stomatopods with long-lived pelagic larval stages have also achieved wide distribution in the region, as has the spiny lobster *Palinurus japanicus.* Of the crabs of the Red Sea, no less than 30% are also found in Hawaii.

All the families and subfamilies of crinoids, except one monotypic West Indian family, are represented in the Indo-West Pacific. Of the 14 families and approximately 70 genera of comatulids, 6 families and 45 species are exclusively Indo-West Pacific. Thirty-eight species of other endemic echinoderms inhabit the entire extent of the Indo-West Pacific region, and several others are limited to parts of the region.

The faunistic center of the region is the Malay Archipelago, the world's greatest archipelago, which contains extensive areas with less than 200 meters' depth. It is to be expected that such a region would be a great center for the development of shallow-water fauna and flora. The Malay Archipelago and surrounding seas are extraordinarily rich in species. The faunas of the islands of the Central Pacific are, for the most part, derived from this rich fauna but show a progressive diminution in the number of the species characteristic of the faunal center and some increase in endemic species of their own with increasing distance. The Hawaiian Islands on the extreme northern and eastern boundaries of the region, although still showing strongest Indo-Pacific affinities, lack a rather large number of the groups characteristic of the faunal center and have a fairly large proportion of endemic species of their own, with a small contingent of forms derived from the western American fauna.

A similar diminution of characteristic Malayan species and groups and an increase in the number of local endemic species can be noted in a western direction to the Red Sea and eastern Africa.

The northern Australian coastal fauna is poor in endemic echinoderms and closely related to the Indo-Malayan fauna. The Great Barrier Reef region off the northeastern coast of Queensland is more independent than northwestern Australia, presumably because until the late Quaternary the area of the shallow Torres Strait was above the surface, constituting a barrier isolating this area from the heart of the Indo-Malayan area, so that it remained more closely related to the eastern coast of New Guinea.

The Red Sea is of special interest because it is known to have been formed only about a million years ago at the end of the Pliocene or the beginning of the Quaternary period. It is a sea of exceptionally high salinity because of the intense evaporation between the African and Arabian deserts. Its temperature is relatively high, as much as 21° to 25° C. to as deep as 200 meters. Of about 430 species of decapod crustaceans known from the Red Sea, some 31% of the Macrura and 33% of the Brachyura are endemic. For crinoids the high figure of 70% endemic species has been given, but this may be partly because of imperfect knowledge of the fauna of the tropical Indian Ocean floor. In any event, the case of the Red Sea demonstrates that the rather short geological period of 1 million years is sufficient for a fairly large number of species to undergo evolutionary changes sufficient to differentiate them as new species.

TROPICAL ATLANTIC

The West Indies, the world's second largest archipelago, occupy a faunistic relation to the Atlantic somewhat akin to that of the Malay Archipelago of the western Pacific. In each case, warm tropical currents crossing the ocean over vast distances from the east bathe the region in warm waters. Although not possessing as rich a fauna as the Malay Archipelago, the West Indies do possess a rich and varied shelf fauna, especially as compared with other Atlantic regions. There are extensive coral reefs, but they are not as rich in coral species and genera as those of the Indo-Pacific. The endemic fauna is richer in species and genera than is that of any other part of the Atlantic. Of the stenothermal warm-water crabs, approximately 70% of the species and 17% of the genera of the area are endemic. All 26 of the Atlantic genera of reef corals are present in the West Indies. There is a rather large number of endemic fish genera, most of them, however, with few species. Some of the world's finest mangrove swamps have developed in the region of the Florida Keys and along the southwest coast of Florida.

In contrast, the tropical eastern Atlantic fauna is rather poor, comprising a markedly smaller number of species and genera. The communication between the tropical faunas of the eastern and western Atlantic has been quite restricted in recent times, as shown by the contrast in the percentage of amphi-Atlantic species in comparison with the number of genera. Of the crabs and echinoderms, which are the best-known groups, considering only the stenothermal warm-water forms, there is a total of about 300 crabs representing 115 genera, and 160 species of echinoderms representing 62 genera. Of these, only about 8% of the species of crabs, and 15% of the species of echinoderms, are amphi-Atlantic, while the corresponding figures for the genera are 22% and 42%. Of the reef corals, only 5 of the Atlantic genera occur along the western African coast.

Curiously enough, the West Indian fauna shows a closer relationship with the fauna of the Indo-West Pacific than it does with that of the eastern Atlantic, which is in the same ocean with it. If only present-day geography is considered, this would be almost inexplicable, considering the complete barriers to communication between these tropical faunas, on both the east and the west. The relationship with the Indo-Pacific is particularly evident at the generic level. For example, in the case of the crabs, 36.5% of the genera are also represented in the Indo-Pacific. For the echinoderms, approximately 65% are common. Of the 26 genera of reef corals, 21 are also found in the Indo-West Pacific. The number of examples of other groups showing similar relationships is considerable. With respect to species, however, the relationship is slightly closer with the eastern Atlantic fauna. This discrepancy between the relationships of genera and of species means that in very recent times, as long as the present species have been in existence, communication across the Atlantic has been easier than between the Atlantic and the Indo-Pacific but that at an earlier date, within the lifetime of present-day genera but not within the time of the majority of present Atlantic species, there was open communication between the Atlantic and the Indo-Pacific. This is further indicated by the presence along the eastern Atlantic coasts of a number of additional fossil species and genera of Indo-West Pacific origin, and evidence that there was formerly a much closer relationship between the tropical faunas of the eastern and western Atlantic. In more recent times the tropical fauna of the eastern Atlantic was largely destroyed during the ice ages and replaced by a more temperate fauna, the tropical fauna being largely restricted to the West Indies region.

Recent geological discoveries show that prior to and during the Mesozoic era, not only was there open communication in an east-west direction completely around the world in the tropical regions, dividing the land masses into northern and southern groups, but the major continental platforms themselves were in a very different relationship to each other. They formed two great land regions—a northern mass termed Laurasia comprising parts of what are now North America, Greenland, and northern Eurasia and a southern mass termed Gondwanaland, comprising what are now South America, Africa, India, Australia, and Antarctica, with the Tethys Sea communicating from east to west between them. This open communication presumably existed for some time after continental drift had begun to tear apart these great land masses and to raft them into their present positions during the Cretaceous and Tertiary. During the Cretaceous and early Tertiary periods the Atlantic coastal waters of northern African and Europe contained a rich tropical fauna, continuous at first with that of the Indo-Pacific. This fauna has now largely disappeared from these regions and to a considerable degree from the entire Atlantic. The present tropical Atlantic shallow-water fauna, chiefly centered in the West Indies, can be regarded as a relic from this ancient, more extensive tropical Atlantic fauna, which at a still earlier date was an extension of, and was continuous with, the fauna of the Indo-Pacific.

TROPICAL EASTERN PACIFIC

The Isthmus of Panama forms an insurmountable barrier to the spread of marine animals between the tropical Atlantic coast and the tropical American Pacific coast. The faunas of these areas are somewhat less closely related than are those of the West Indies region and the eastern Atlantic. Nevertheless, the tropical eastern Pacific fauna shows a closer relationship to that of the West Indies than it does to that of the Indo-Pacific. Again, like the relationship between the West Indies fauna and the Indo-Pacific fauna, this is most evident at the generic level, indicating that the present barrier has been in place long enough for considerable speciation to have taken place, but not so long as to have excluded present-day genera.

Among the crabs endemic to tropical American waters, for example, about 2% of the species but 18% of the genera are common to both sides. For the echinoderms the figures are about 0.3% and 10%, respectively. However, the Pacific side has about 38% of the total number of endemic Amer-

ican crabs and 41% of the echinoderms, in both cases more than are found on the Atlantic side. Those species that are common to both sides appear to be very ancient species that have not changed in morphology since the time the barrier was formed. Among other groups as well, amphi-American genera are more numerous than amphi-American species. Another indication of the close relationship between the two faunas is the occurrence of a number of so-called twin species, species that are very much alike, belonging to the same genus, one on each side of the barrier. Gunther, McMurrich, and others, on the basis of zoogeographical resemblance between the two fauna, postulated a direct connection between the two in the past, which was later confirmed by geologists, who demonstrated an ancient channel across Central America. The relatively small number of amphi-American species shows that this communication ceased to exist before most of the species reached their present state of development. The age of the present barrier, then, is greater than the average age of the species, but the existence of several pairs of twin species shows that the connection must have been of comparatively recent date in geological terms.

These conclusions agree well with geological findings that the last direct connection existed during the Lower Pliocene. During the Eocene, Oligocene, and Miocene, the two oceans had direct connections for considerable periods. The western American tropical fauna is, then, in the main, of Atlantic origin. The deep, broad East Pacific Barrier has evidently been an effective barrier for a longer time than has the Isthmus of Panama.

The relationship of the tropical western American fauna to the tropical West Indies fauna is considerably stronger than its relationship with the temperate western American faunas on either side of it.

SUMMARY

During the Mesozoic and Cretaceous periods there was a broad sea continuous around the tropical belt of the world, dividing the land masses into northern and southern components. Only the eastern Pacific deep-water barrier interrupted the general continuity of the tropical fauna. During the Miocene and Pliocene periods, the geographical situation was being drastically altered by the gradual continental drift, by the formation of the Central American isthmus, and by the land bridge between Asia and Africa. The ancient Tethys Sea was divided into the present Indo-Pacific, Mediterranean, Atlantic, and American Pacific components, largely isolated from each other. Climatic changes occurring at about the same time brought about a decided cooling, especially in the Atlantic, which decimated the former tropical fauna, particularly in the eastern Atlantic.

The relationships between the present tropical faunas of the Indo-Pacific, the Atlantic, and the eastern Pacific reflect the ancient Tethys period, while their differences represent the effects of isolation and of different climatic regimes in more recent times. The comparative poverty of the Atlantic tropical fauna today as compared with that of the Indo-Pacific does not necessarily mean that the Indo-Pacific was always the greatest and the only important center of origin and distribution of tropical faunas nor does it mean that the Atlantic is poorer in tropical fauna for the same reason Hawaii is—that it is on the periphery of the area of influence. The former Atlantic tropical fauna seems to have been as rich as that of the Indo-Pacific, but it has been decimated by climatic changes.

TEMPERATE TO COLD NORTHERN REGIONS
MEDITERRANEAN-ATLANTIC AREA

The western Mediterranean and neighboring portions of the Atlantic Ocean from about Cape Blanco on the African coast to the English Channel form a faunistic region with rather ill-defined boundaries or transitional zones. The Strait of Gibraltar is not a very important zoogeographical boundary. The southern portion of this Atlantic area is known as the Mauretanian region, and the northern, as the Lusitanian. The Lusitanian region extends northward, merging into the boreal region.

There are a number of endemic elements, some restricted to one or another portion of the region, and others found throughout. The well-known precious coral *Corallium rubrum* occurs in the Mediterranean and in the Mauretanian region southward as far as the Cape Verde Islands, as does the sea pen *Pennatula rubra*. Ten of the 18 squid species of the Sepiolidae are endemic here, most of them exclusively in the Mediterranean. About 40% of the echinoderms collected on the shelf in this region are endemic, perhaps 15% to 16% of them restricted to the Mediterranean. Among the endemic pelagic fauna are the sardine *Sardinia pilchardus* and the anchovy *Engraulis encrasicholus*. Only about 6% of the genera of shelf fauna are endemic. A considerable similarity has been noted by various zoogeographers between the Mediterranean and Japanese faunas, with a number of species being either the same or closely related twin species between the two regions. This resemblance was explained through the discovery of the extent of the ancient Tethys Sea, in much the same manner as were the similarities between the West Indian and Indo-Pacific faunas.

The water of the Mediterranean is somewhat more saline than most ocean waters because of the high rate of evaporation in the region and the fact that it does not receive a major freshwater inflow from large river systems except the Nile. Surface waters tend to flow into the Mediterranean from the Atlantic, and somewhat deeper strata flow out from the Strait of Gibraltar, bearing warm, highly saline water that strongly influences the temperature and salinity of neighboring Atlantic areas and that bears members of the Mediterranean-Lusitanian fauna, coming to the surface at certain times of the year near the entrance of the English Channel or along the seaward sides of the British Isles. This warm saline deep water from the Mediterranean makes the temperature-depth relationships of this part of the Atlantic somewhat abnormal, there often being no considerable fall in temperature until a considerable depth is reached.

During the Tertiary the main connection between the Mediterranean and the Indo-Pacific region was north of the present Red Sea area through southwestern Asia. It was many times broader, and lasted longer than the brief Miocene union of the eastern Mediterranean and northern Red Sea region, which was not at that time open to the Indian Ocean. In the Middle Pliocene there was, for a brief time, a connection with the Mediterranean, but it was not important in faunal exchange, since the Nile disgorged great masses of fresh water into the gulf, forming a freshwater barrier and the connection was rather brief. Since then there had been no direct connection between the Mediterranean and the Red Sea until the completion of the Suez Canal in 1869. The faunas of the two areas, the eastern Mediterranean and the northern Red Sea, were very distinct with respect to species. Only a few circumtropical species and perhaps 4 or 5 others were common to the two areas. In 1924 a study was made of the canal by a group from Cambridge, and these studies were later continued by others. It was shown that most of the species of animals in the canal came from the Red Sea because strong tidal currents from that end carried them up the canal as far as the Bitter Lakes, whereas the northern and central parts of the canal lacked such currents. Fish, because of their greater ability to swim, entered the canal from both ends. For many years the high salinity of the Bitter Lakes, which the canal traverses, proved to be a barrier to effective migrations through the canal. The salinity of the lakes varied from $46^0/_{00}$ to $50^0/_{00}$ at the surface and was even greater in deeper water. In recent years, however, continued dilution has rendered this barrier less effective. A number of Red Sea species have entered the eastern Mediterranean, and there has been some exchange of fish in both directions.

SARMATIC INLAND SEA

During the Eocene and Oligocene the Tethys Sea included large areas of central and southeastern Europe as well as the Mediterranean and parts of Northern Africa and Western Asia. During the Upper Miocene this sea diminished, the western part becoming shallow and the eastern part more and more brackish. Stenohaline marine forms such as corals and echinoderms, brachiopods and cephalopods, and dogfish died out and were replaced by an extensive brackish-water fauna, probably one of the most extensive brackish-water faunas ever developed. This great Eastern European–Western Asian portion of the former Tethys Sea became separated from the Mediterranean and is known as the Sarmatic Inland Sea. At its greatest extent it covered much of the Balkan Peninsula, parts of southern Russia, most of Hungary, and the areas of the present Black Sea and Caspian Sea. It flooded the valley of the Volga as far north as Kazan and large parts of the lower Kama and Ural river valleys. The water continued to become more brackish. The western end became connected with the Mediterranean during the Lower Quaternary; in the eastern part the Caspian became separated from the Black Sea during the Middle Pliocene but for a long time made connection with Lake Aral. During the second interglacial period the water of this Caspian-Aral Sea was about 80 meters higher than the level of the present Caspian Sea. The fauna of this Sarmatic Caspian-Aral basin was largely brackish and mixed to a considerable extent with freshwater forms. At present only remnants of this formerly richest-known brackish-water fauna are left in the Sea of Azov, in various Black Sea estuaries and lagoons, and in the Caspian Sea. The Caspian Sea fauna is almost purely of Sarmatic origin. That of the Black Sea, however, now is composed largely of a limited number of species of Mediterranean origin that can stand the low salinities of the Black Sea and a smaller Sarmatic component mostly limited to more brackish lagoons or estuaries. In both the Black Sea and the Caspian, inadequate circulation has resulted in stagnation of the deep water and the accumulation of hydrogen sulfide, so that the benthos is limited to shallow marginal zones. In the Black Sea the benthos extends down to about 100 to 150 meters, occupying approximately 23% of the Black Sea floor. In the Caspian Sea the hydrogen sulfide does not come as near the surface, and the benthos extends to a depth of 400 meters. The Caspian Sea has been isolated from the world's oceans for about 30 million years and has thus preserved its highly individual character.

THE NORTH ATLANTIC

Although the North Atlantic is one of the most thoroughly studied of all ocean regions, the lack of clear faunistic boundaries resulting from the gradual change in water temperatures in a north-south direction, the lack of effective physical barriers, and the fact that the wide seasonal fluctuations in temperature allow vegetatively eurythermal but reproductively stenothermal organisms both from colder and warmer areas to live in the region, reproducing at different times when the water temperatures suit them best, make it hard to define the borders of this region in a precise manner.

It is best defined in terms of endemic boreal species and their centers of abundance and spread rather than by negative characters such as the northern or southern limits of occurrence of a selected number of species.

The northern European boreal region is centered in the North Sea and extends northward off the coasts of Norway, dominated by the relatively warm North Atlantic Current, a mixture of waters mainly derived from the Gulf Stream but with an admixture of colder water from eddies off the cold East Greenland Current and a cold outflow from the Baltic through the Skagerrak. This current continues northward and branches, one branch continuing into the southern part of the Barents Sea where it bars southward extension of truly Arctic species and the other branch forming an extensive eddy south of Spitsbergen in the Greenland Sea. Arctic waters approaching the European coasts are, then, always of a mixed character.

On the American side, the coast south of Cape Cod has a peculiar mixed character because of the warm offshore Gulf Stream but colder inshore waters derived in part from the Labrador Current, especially in winter. The waters of the New England coast, Newfoundland, and the Gulf of Saint Lawrence are essentially boreal in character but with a strong arctic admixture.

Many endemic species occur on both sides of the Atlantic, as do many additional nonendemic species also found in these waters, so that a manual of the organisms of the boreal region of either side of the Atlantic serves fairly well on both sides. The current patterns are such, however, that shallow-water American forms are more readily transported towards Europe than the reverse. The result is that more of the American boreal species are amphi-Atlantic, whereas the European boreal contains a somewhat larger number of species peculiar to the European side. Both hydrographically and faunistically, the European boreal is more clearly definable.

The boreal waters support some of the world's most productive fisheries. A few of the outstanding examples are the herring *Clupea harengus,* the cod *Gadus morrhua,* the haddock *Melanogrammus aeglefinus,* and the plaice *Pleuronectes platessa.*

Some of the more familiar invertebrates are the softshell clam *Mya arenaria,* the mussel *Mytilus edulis,* the barnacle *Balanus balanoides,* the lugworm *Arenicola marina,* the periwinkle *Littorina litorea,* the starfish *Asterias forbesi* and *Solaster endeca,* the sea cucumber *Cucumaria frondosa,* and many others.

One of the outstanding features of pelagic boreal waters are the great seasonal swarms of the red copepod *Calanus finmarchicus,* or herring feed. Although this species is not confined to the boreal Atlantic, undergoing submergence and appearing in other latitudes and in the North Pacific, the immense aggregations appearing in or near surface waters are characteristic of the Atlantic boreal and subarctic waters.

BALTIC SEA

Extending inland from the North Sea between the Scandinavian countries and the rest of Europe, the Baltic, with a surface area of 422,000 km.[2], constitutes the world's largest brackish-water basin at the present time. The salinity varies both geo-

Table 14-1. Decrease in numbers of marine species in the Baltic Sea, from the Kattegat to the gulfs of Bothnia and Finland*

	Kattegat	Belt Sea	Upper Baltic	Gulfs of Bothnia and Finland
Sponges	27	16	0	0
Sea anemones	18	4	2	0
Polychaetes	160*	70	15	4
Amphipods	132	36	13	5
Bivalves	87	34	24	5
Echinoderms	35	8	2	0

*Adapted from Ekman, S. 1953. Zoogeography of the seas. Sidgwick & Jackson, Ltd., London.

graphically and seasonally according to the circulation of marine waters and the influx of fresh water. In general it decreases with the distance inland, reaching minima in the gulfs of Bothnia and Finland. Near the entrance to the Baltic, the so-called Belt Sea, salinities vary around $15^0/_{00}$, plus or minus $5^0/_{00}$. At the upper end, in the gulfs of Bothnia and Finland, they drop to about $2^0/_{00}$.

Stenohaline marine forms are, of course, excluded; the number of species of marine animals and plants decreases toward the interior, and the number of brackish-water and freshwater forms increases. Parts of the gulfs of Bothnia and Finland are frozen over for five or six months or more each winter. The decrease in the number of marine species from the mouth of the Kattegat to the upper part of the Baltic proper between Darsser Ort–Gedser and Bornholm is shown in Table 14-1. Very few of them extend into the gulfs of Bothnia and Finland and then only occasionally into the lower parts of these bodies.

A number of North Sea fish are found at times in the Baltic as far as the mouths of the gulfs. The most important is a brackish race of herring, *Clupea harengus membras,* which is of importance economically for the whole region. The widespread euryhaline jellyfish *Aurelia,* the clam *Mya arenaria,* a recent immigrant, and a few other marine forms are distributed fairly well throughout the extent of the Baltic proper.

Euryhaline freshwater and brackish-water species show reversed gradients of numbers of species, being most numerous in the gulfs of Bothnia and Finland, with some of them extending their ranges some distance into the upper parts of the Baltic proper. Such species, for example, are the rotifers *Keratella cochlearis, Notholoca longispina,* and *Asplanchna priodonta,* some of the planktonic freshwater and brackish-water crustaceans such as *Bosmina,* sludge worms such as *Tubifex,* some pond snails such as *Lymnaea ovata* and *Theodoxus fluviatilis,* and a number of other animals including such fish as the stickleback *Gasterosteus,* the whitefish *Coregonus,* the bream *Abramis,* and the perch *Perca fluviatilis.* The most euryhaline and widely distributed fish in the Baltic are the Atlantic salmon *Salmo salar* and the eel *Anguilla anguilla.*

The Baltic has not always been in communication with the North Sea. During the ice ages it was sometimes glaciated, and at times it was an inland lake or was in communication with the high arctic. A number of relict high arctic species such as the circumpolar *Cottus quadricornis,* the mysid *Mysis oculata,* the amphipod *Pontoporeia affinis,* and a few others occur in the upper reaches of the Baltic; fossil beds of *Portlandia arctica* and other high arctic bivalves, together with remains of the Greenland seal *Phoca groenlandica,* attest to former high arctic fauna of the region.

ARCTIC OCEAN

The Arctic Ocean is a great, nearly land-locked basin. It communicates with the North Atlantic over relatively shallow sills, the Nansen's Rise, between 1000 and 2000 meters deep, east of Greenland, and a similar ridge across the Davis Strait, west of Greenland. Its communication with the North Pacific through the Bering Strait is much more restricted.

The zoogeographical bounds of the marine arctic do not parallel the lines of latitude however, since in both the North Atlantic and the North Pacific cold currents flowing southward out of the Arctic Ocean along the western shores and warmer water flowing northward along the eastern shores cause a marked north-south skewing of temperature relationships. Thus, arctic conditions extend southward along the Atlantic coasts of Greenland and northern Canada because of the eastern Greenland and Canadian currents and along the Asian coast of the Pacific because of the Oyashio Current. Boreal conditions extend northward along the Norwegian coast to the Barents Sea and along the Alaskan coast some distance beyond the Bering Strait.

The Arctic Ocean itself is divided into two major zones, each more than 4000 meters deep, by the Lomonosov Ridge, which runs from just north of northwestern Greenland, through the area of the North Pole to the New Siberian Islands off central Siberia. Both the Eurasian Basin and the Canadian Basin are, in turn, divided into two parts by secondary ridges paralleling the Lomonosov Ridge. In the Eurasian Basin this is a continuation of the Mid-Atlantic Ridge and rift system; in the Canadian Basin it is simply a rise variously known as the Alpha Rise, Marvin Ridge, or Fletcher's Rise. The continental shelves along most of the Eurasian side of the Arctic Ocean are very broad, extending out as much as 700 km. or more, forming a series of relatively shallow seas. Beginning just north of Norway and going east, these are the Barents Sea, Kara Sea, Laptev Sea, East Siberian Sea, Chukchi Sea, and the Beaufort Sea north of Alaska and northwest Canada.

The coastal waters and the surface layer of the Arctic Ocean are strongly influenced by drainage from the surrounding land, the melting of ice, and the influx of North Pacific water of somewhat lower salinity than that of the Atlantic. The result is a layer of water about 100 to 300 meters deep, with lowered salinity (below $34^0/_{00}$) and, for the most part, with temperatures below the freezing point. Beneath this is a somewhat thicker layer of Atlantic water coming from between Greenland and Spitsbergen. This is slightly warmer, commonly 0.5° to 1.5° C., and more saline, about $34^0/_{00}$. It forms a stratum extending down to approximately 700 meters. In the depths of the basins, the temperatures are again negative and the salinity nearly $35^0/_{00}$. These layers are not everywhere as distinct as was formerly thought. The patterns of inflow and outflow from the Arctic Ocean are fairly stable at present.

The Arctic Ocean is at least partially covered by ice at all times, the extent of ice coverage varying from about one tenth of the surface to complete freezing, depending on the season and the climatic cycle. One-year-old ice is seldom more than 2 or 3 meters thick, but the ice of large ice islands rafted from the shelf ice along northeastern Canada or Ellesmere Island may be more than 12 meters thick. The salt leaches out of sea ice in one season, so that when it melts, the ice dilutes the surface water. Where winds and storms sweeping across the ocean push the ice against the shore, extensive areas of pack ice may be compressed, broken up, and refrozen as a very rough, irregular surface that is difficult to negotiate. The scouring by ice also keeps the intertidal rocks of many areas bare.

Curiously enough, it now appears that the Arctic Ocean was relatively open and ice free during early stages of the great glacial ice ages, but was largely frozen over during the late glacial and most of the interglacial periods, such as the present. This, according to the theory of Ewing and Donn (1956), is the result of a sort of cyclic exchange of great volumes of water between the Atlantic Ocean, the Arctic Ocean, and continental masses of ice. According to this theory, during interglacial periods the glaciers melt and retreat, and their water is added to the oceans, gradually raising the sea level. The increased exchange of water between the Arctic Ocean and the North Atlantic would have the effect of cooling the Atlantic water and of warming the Arctic Ocean, finally causing it to become ice free. A relatively warm, ice-free polar sea whipped by arctic winds would contribute great amounts of water vapor to the air, resulting in blizzards and storms, which would deposit, over thousands of years, a blanket of ice over the northern portions of nearby continents. Eventually the withdrawal of water from the oceans, forming these ice sheets, would reach a critical point where the exchange of water over the shallow sills between the North Atlantic and the Arctic Ocean would no longer be sufficient to warm the polar sea. The Arctic Ocean would freeze over, cutting off the supply of water vapor and putting an end to the massive deposition of continental ice. The reduced exchange between the Atlantic and the polar sea would result in a warming of the Atlantic waters, and a gradual recession of the ice sheets would begin, again contributing to a rising of sea levels. This theory accounts nicely for known changes in sea levels as shown by marine terraces, etc., and has subsequently been further bolstered by archeological findings along the shores of the Arctic Ocean, which seem to indicate that it was, in fact, ice free during early parts of the last ice age.

Since their elaboration of this theory of ice ages in 1956, Ewing and Donn have modified it in certain important aspects, especially with respect to the later stages of the growth of the ice sheets and the termination of ice ages. In its new form the theory may be restated as follows:

The migration of the geographical poles to regions of thermal isolation (or perhaps the thermal isolation of the geographical poles by migration of the continents) in the Tertiary brought about the development of a strongly zoned world climate with high latitude temperatures at the threshold for glaciation. Continental glaciation probably began in Antarctica, where it has persisted to the present, contributing to a general cooling of both hemispheres.

Northern Hemisphere glaciation was initiated in the arctic latitudes by increased precipitation caused by an ice-free Arctic Ocean. However, after the glaciers reached a certain critical size, their refrigerating effect was sufficient to cause them to become self-perpetuating in North America and Europe, where they could draw upon abundant moisture provided by the North Atlantic and, in western North America, the North Pacific. In Siberia, the central desert limited glacial growth because of lack of moisture. The freezing of the Arctic Ocean marked the cessation of glacial growth in that region. The size of the Siberian ice sheet probably represents the approximate size the glaciers would have attained elsewhere if their growth had been fed chiefly, or only, by moisture provided by the Arctic Ocean.

As the glaciers grew in extent, they finally reached the point where warming from the south was in dynamic equilibrium with their own refrigerating effect. By this time, or sometime thereafter, the continued cooling of the oceans lowered their vapor pressure, by as much as 50% over much of the North Atlantic and to a smaller but significant extent elsewhere. This reduced, beyond the equilibrium value, the atmospheric moisture available for building glaciers, and a retreat began. Retreating glaciers leave very wet soil in the regions of their margins, which greatly increases the absorption of radiation, the albedo of wet soil being half that of dry soil. This warming effect would be enhanced through heat absorption by dirty, stagnant, marginal ice. Once the retreat began, the warming effects continued, and the general level of atmospheric moisture remained low because of a significant lag between the beginning of the retreat and the warming of the surface waters of the oceans. The oceans were still being fed great quantities of nearly ice-cold fresh water, which tended to remain at the surface because of its lower density.

By the time the temperatures of the ocean surfaces had increased sufficiently to provide a significant increase in moisture, the ice sheets had been reduced beyond the critical size at which, by their refrigerating effect, they could benefit from this and had retreated to regions dependent chiefly upon the Arctic Ocean for moisture. Thus they were unable to build themselves back. The retreat continued at an accelerated pace. Only on Greenland, which was surrounded by water and had sufficient moisture to keep its ice pack built up, was this not the case.

The critical point is that both the initiation and the termination of ice ages were caused by changes in precipitation rather than by initial changes of temperature. The growth and decline of the glaciers were responsible for, rather than caused by, changes in world temperature. The lag between the onset of glacial retreat and the warming of the oceans allowed the retreat to proceed beyond the point of no return, until such time as the Arctic Ocean itself is again ice free, providing moisture to initiate a new cycle of glaciation.

Probably the reason the ice ages began when they did and were not a chronic feature of the world's history throughout earlier geological ages lies in the different relationships of the continental masses and oceans during earlier periods. It was not until continental drift rafted the continents into their present spatial relationships that the poles were in positions of thermal isolation and the thermostatic effect of the exchange between the Arctic Ocean and the North Atlantic could cause these tremendous climatic oscillations.

Arctic Ocean fauna

The fauna of the Arctic Ocean includes a rather high proportion of endemic species, most of which are circumpolar in distribution. About 80 species of fish belonging to 45 genera occur in the Arctic Ocean. Of these, 48 species, or 60%, are known only from arctic waters or in a few cases also as relict species present also in such places as the upper Baltic. Of the 45 genera, 18, or 40%, or endemic.

Numerous endemic invertebrates are also present, including the amphipod *Onismus,* with 8 purely arctic species; *Pseudalibrotus,* with 4 arctic species, only one of which occurs also in northern boreal waters; and *Acanthostepheia,* with 3 arctic species. The isopod *Mesidothea* is purely arctic, although *M. entomen* still occurs as an arctic relict

in the Caspian and the Baltic. The bivalves *Portlandia arctica* and *Yoldia hyperborea* are common in the polar sea.

A group of species living only in the coldest waters can be distinguished as "high-arctic species." Others preferring slightly warmer water near the subarctic waters are known as "low-arctic species." Some species occur throughout the Arctic in a "pan-arctic" distribution.

Several low-arctic and some pan-arctic species that occur on the shelf areas in the Arctic show the phenomenon of submergence and occur also in deeper water in boreal latitudes.

The arctic fauna is, in general, poorer in species than the faunas of lower latitudes, but those which do live there sometimes occur in great numbers in accordance with the general rule that extreme environments, where adequate nutrition is available, will support small numbers of species but immense numbers of individuals of those species that can exploit them.

The short arctic summer is a season of intense growth and reproduction. Planktonic animals, such as *Calanus,* that may have two or even three broods a year in more southern latitudes generally have only one in the arctic. There is also a tendency for them to grow to slightly larger size as individuals.

THE NORTH PACIFIC

The continental shelves form a great uninterrupted arc around the North Pacific from Lower California to Japan, since the Bering Strait is in shallow water. That portion of the shelf on the Asiatic side of the Bering Sea north of the Aleutian Islands to the Kamchatka Peninsula, Kurile Islands, and Sea of Okhotsk is arctic or subarctic in character because of the cold Oyashio Current. South of the Aleutians, however, the north-flowing Japan Current, which is the Pacific equivalent of the Gulf Stream, turns eastward and crosses the North Pacific as the North Pacific Current, bring its warm waters to the American shores at about the latitude of Oregon and Washington. Here it diverges, sending one part, the Alaska Current, as a great eddy northward to the Gulf of Alaska, where it is deflected to the west. The southern branch flows parallel to the coast as the California Current to about Point Conception, where the eastward trend of the shore leaves the current farther out to sea, still trending in a southern direction and turning west off Lower California as the North Equatorial Current.

On the American side, conditions similar to the boreal of Europe extend along the entire shelf region from Alaska to Lower California. Upwellings of deep water near shore along the coasts of California and Lower California keep the shelf waters cooler than they would otherwise be, while the waters from the North Pacific Current moderate the water temperatures of the Pacific Northwest, the Canadian coast, and the Gulf of Alaska. Some of this warm eddy in the Gulf of Alaska also passes north between the easternmost Aleutian Islands along the coast of Alaska to the Bering Strait, so that the American side is noticeably warmer than the Siberian. The overall result is that surface temperatures over the shelf all the way from the Gulf of Alaska to the southern part of Lower California are remarkably uniform, usually varying between 10° and 20° C. For most of this stretch, the temperatures usually range between 13° and 18° C. Many of the more eurythermal species of the shelf fauna are distributed the whole distance, or nearly so. In the northern part of the North Pacific, these conditions do not follow the main shelf but rather are found only along the narrow southern edge of the Aleutian chain. At the western end of the Aleutians lies a deep-water region, forming a physical and faunistic boundary between the American and Asiatic shelf fauna.

Just as the cold Labrador Current (derived largely from the arctic Canadian Current) brings cold water south along the western coast of the Atlantic while the North Atlantic Current (derived largely from the Gulf Stream) moderates the temperatures along most of the eastern North Atlantic coast, so do the cold Oyashio and Liman currents flowing south along the western Pacific to northern Japan, while the North Pacific Current moderates the coastal temperatures of most of North America. On the western sides of both oceans, then, there is a much more abrupt transition between warm temperate conditions and cold high boreal or subarctic conditions, but on the eastern sides there are long stretches of relatively mild, cool temperate or boreal conditions.

There is a high degree of endemism along the North American Pacific coast, roughly half the species of marine organisms being endemic to this region, in contrast to the European coasts of the Atlantic, to which not more than a fourth of the resident species are endemic. There are also more endemic genera and families.

The American North Pacific is the world's leading area with respect to the number and variety of starfish, exceeding even the richest areas of the East Indies. Of the approximately 92 species, about 60% are endemic. Approximately 31% are also found in arctic or subarctic regions, and only about 8% are common to more southern regions. Most of the thirteen endemic genera of starfish are monotypic, but *Pisaster* contains 3 species and *Mediaster,* 2. *Pisaster ochraceous,* the most abundant and conspicuous intertidal species, comes in several strikingly different color phases, all often found together, and ranges along the whole Pacific coast of North America from the Gulf of Alaska to Lower California. One of the most striking of the monotypic genera is *Pycnopodia,* the very large, soft-bodied, "twenty-rayed starfish," *P. helianthoides,* being a characteristic element of the lowest intertidal and subtidal rocky areas. In addition to the endemic genera, several more widely distributed genera such as *Henricia, Solaster, Pteraster,* and *Dipsacaster* have more species in this region than anywhere else.

Among the molluscs, octopus, chitons, and buccinid gastropods are best developed in the American North Pacific.

The crab family Lithodidae has 13 genera and 26 species in this region, none in the subtropical waters to the south, and 2 on the antiboreal South American coast. Other crustacean genera especially well represented in the North Pacific are *Pandalus* with about 10 species, *Spirontocaris* with more than 40, and *Crangon* with about 17.

Among the fish 18 of the 20 genera of surf fish, Embiotocidae, are endemic to the American North Pacific, the other 2 being found in Japan.

The giant kelps, Laminariales, consist of some 30 genera, mostly endemic to this region, although a few genera such as *Laminaria* and *Alaria* are much more widely distributed.

The entire Pacific coast from Alaska to Lower California is not, however, a wholly uniform region from a faunistic standpoint. Subjective analyses by early astute biologists such as Dana (1853), Dall (1921), Bartsch (1921), and others, and later statistical analyses by Schenk and Keen (1936) and, still later, by Newell (1948) have demonstrated that there are faunal provinces or subregions. The most clear-cut faunal break, or transitional zone, occurs at Point Conception, California, where the California Current leaves the coast. Another is at about Forrester Island north of Vancouver Island.

Newell, on the basis of statistical analysis of the molluscan distributions, divides the coast into three broad provinces north of the warm subtropical **Panamic Province:** the **Californian Province,** extending from Cape San Lucas along the outer coast of Lower California north to Point Conception in southern California; the **Oregonian Province,** from Point Conception north to the Forrester Island; and the **Bering Province,** north from Forrester Island to the Bering Strait.

Newell also made interesting contributions to the methodology of analysis and emphasized a point frequently overlooked by others concerned with the problem of analysis of faunal provinces, that is, the discrepancy between the distributions of short-range species (rare species) and the distributions of long-range, abundant species. This is the fact that the short-range, rare species tend to cluster in ecotonal situations at the borders of biotic communities or provinces. With respect to the numbers of species present in various parts of it,

a biotic community or province, then, is typically structured more like a doughnut than a discus. The greater part of its extent is characterized by a relatively small number of species, some of them having an extended range and a tendency to occur in dense stands wherever local ecological factors permit, whereas the border areas, or transitional zones, are characterized by larger numbers of species, some of them having a very limited range.

Thus, on the local level at least, a high proportion of endemic species may indicate a transitional area rather than an ancient center of the regional fauna. The endemism of species that are abundant and dominant in a region is probably of more critical importance from a zoogeographical standpoint than the endemism of an equal number of rare species, although the latter may well indicate the borders of such regions.

SUMMARY

When the Northern Hemisphere is considered as a whole, it becomes apparent that the major faunistic break is between the warm tropical-subtropical regions and the cool-to-cold temperate-boreal–arctic. All the families of organisms and between 50% and 75% of the genera found in arctic regions are also present in the boreal and temperate areas, whereas only 38% of the families and about 8% of the genera characteristic of temperate regions are also found in the tropics and subtropics.

TEMPERATE TO COLD SOUTHERN REGIONS

Whereas the shelf regions of the Northern Hemisphere are almost continuous around their northern portions, those of the Southern Hemisphere are more limited in extent and are separated from each other by extensive tracts of deep sea. The temperate regions of southern and eastern Africa and Australia are rather closely linked with tropical regions already considered, but those of the western coasts of Africa, South America, and Antarctica are more isolated. The relations of the antiboreal and antarctic regions are totally different from those of the boreal and arctic regions. Instead of a nearly landlocked polar sea, there is an ocean-surrounded polar continent, still in the grip of a prolonged ice age. The Southern Ocean is broadly and directly contiguous with all of the major oceans—Pacific, Indian, and Atlantic—and is continuous without interruption around the world.

Surface circulation of waters is characterized by the **West Wind Drift,** a great circumpolar current that washes the shores of all temperate oceanic islands of the Southern Ocean and the southern parts of South America, Africa, and Australia. Where partially intercepted by the southern tips of South America and Africa, it gives off north-flowing branches of cold current along the western coasts of these continents—the **Humboldt** or **Peru Current** along South America and the **Benguela Current** along Africa. The shelf areas on the western sides of these continents are for the most part rather narrow, and the extensive upwelling along the coasts contributes to the great productivity of the waters of these currents and strongly influences the coastal climate.

Immediately adjacent to the antarctic continent there is an eddy, the **East Wind Drift.** The antarctic waters of the southern part of the West Wind Drift are moving in a northeastern rather than a strictly eastern direction because they are continually being replaced near Antarctica by upwelling from intermediate depths. These Antarctic waters meet and submerge below warmer, less dense antiboreal water (so-called subantarctic water) at the **Antarctic Convergence.** Surface temperatures in the Antarctic Convergence are about 3.5° to 4.5° C. in summer, and between 1° and 2° C. in winter. To the south they become colder, and to the north, warmer. The Antarctic Convergence lies, in general, between the latitudes of 50 and 60 degrees south. Its position is not absolutely fixed. It forms the most important faunal boundary of the surface waters of the Southern Ocean.

Approximately 10 degrees to the north of the Antarctic Convergence is another convergence, the **Antiboreal Convergence** (often termed Subtropical Convergence),

lying roughly along the 14° to 15° C. summer isotherm or the 10° to 12° C. winter isotherm. The Antiboreal Convergence does not touch the continental coasts, however, and so does not significantly influence their shelf faunas. It is of importance for the epipelagic oceanic faunas. Because of the general eastern direction of flow of southern temperate and antiboreal surface waters, groups of epipelagic, or shelf-dwelling, organisms having their centers of origin or dispersal in this zone in general distribute themselves to the east of such centers, but because of the extreme distances over open ocean between shallow areas, only shelf organisms with exceptionally long-lived pelagic larval stages or in which the adults are able to endure such long journeys would be able to make it to the next available habitats. Thus in these regions faunas originating in southern Australia might be expected to be quite well represented in New Zealand but have only a few outlying members near the southern end of South America and practically none in southern Africa. A rich southern African fauna, however, might well have some outlying members in Australia. Similarly, a southern South American shelf fauna might be expected to be sparsely represented in the islands of the South Atlantic and perhaps in southern Africa but not in Australia.

AUSTRALASIA

The biogeography of the Australasian region is complicated by its position extending from warm regions in the north to cold in the south, by the rather complicated pattern of currents, some of which display seasonal reversals, and by the relative isolation of the region from other biogeographical areas for a long period.

Australia has derived most of its intertidal fauna from the tropical Indo-Pacific. To the south, as one encounters cooler waters, there is a considerable thinning out of the number of species able to survive. Because of its isolation from other land masses, there is not much recruitment of a compensating cool-water intertidal biota, especially since the current systems in the area tend to favor the spread of warm-

Australia–New Zealand region
Biotic provinces recognized by Knox, 1963
for intertidal zone
(Composite, adapted from four maps in Knox, 1963)

```
······· = Warm, tropical, subtropical coastal water
▪▪▪▪ = Warm temperate coastal water
▬▬▬ = Transitional waters
║║║║ = Cold temperate water
⟶ Warm currents
- - ⟶ Cold currents
⟸⟶ Warm most of year but
     cooler and reversed at times
```

New Guinea

Dampierian Province

Great Barrier Reef Province

Solanderian Province

Coral Sea

Trade Wind Drift

Northern Territory

Queensland

Western Australia

South Australia

New South Wales

West Australian Province

Flindersian Province

Victoria

Maugean Province

Tasmania

Philippian Province

Norfolkian Province

Aupourian Province

Kermadecian Province

Peronian Province

Tasman Current

Cookian New Zealand Province

Moriorian Province

Antiboreal Convergence
(often termed Subtropical Convergence)

Forsterian Province

West Wind Drift

Antarctic Convergence

Kerguelian Province

Antipodean Province

Cold, subantarctic water

Kerguelian Province comprises Kerguelen
Island in southern Indian Ocean and
Macquarie Island shown here.
Antipodean Province comprises Auckland, Camble,
Bounty, and Antipodes Islands.
Moriorian Province comprises Chatham Island.
Philippian Province comprises Lord Howe Island.
Norfolkian and Kermadecian Provinces comprise Norfolk
and Kermadec Islands, respectively.

Fig. 14-2. The Australasian region.

water species into the cooler areas rather than the reverse.

A summary of some of the important features of the region is shown in Fig. 14-2. Knox (1963) recognizes no less than sixteen biological provinces for the shallow-water marine areas—seven along the shores of Australia proper, three in New Zealand, and six small provinces for various islands or isolated groups in the region whose shore biota differs recognizably from those of any of the other provinces and from each other. These provinces are very briefly outlined below.

Australian provinces

The **Dampierian Province** extends along the northern shore of Australia west of Torres Strait to the Houtman Abrolhos Islands off western Australia. This is a tropical-subtropical region, with variable east-west trending currents along the northern coast and north-south currents in the western part. The waters come in part from the north through the Malay Archipelago, in part from the east from the Coral Sea, and in part from the northeast part of the Indian Ocean. The coastal waters are warm, ranging from about 24° to 29° C. along much of the northern coast and from 22° to 27° C. in the western part. Much of the biota is pantropical in character; a smaller portion derived from species endemic to the Malay Archipelago and northern Australia. It is still questionable whether Torres Strait marks a sufficiently distinct faunal boundary between the Dampierian Province and the Solanderian Province along the northern coast of Queensland to merit their designation as separate biotic provinces. It seems probable that for some time in the Pleistocene, the Torres Strait was closed, so that these areas were temporarily separated from each other. Because the northern coasts are less well known than any other regions of the Australian shelf, conclusions as to its distinctness are still doubtful. There do seem to be a number of endemic Dampierian species, especially toward the northwestern portion of the province, at least among the echinoderms, which have been better studied than most groups of the area.

The **Solanderian Province** of the rocky mainland coast of northern Queensland is a second tropical-subtropical area closely related to the northern Dampierian and overlapping the northern end of the temperate Peronian Province between Double Island Point and Point Vernon, between latitudes 25 and 27 degrees south. The coastal waters are warm, about 21° to 28° C. The zonation, or sequence of dominant intertidal organisms, differs in characteristic ways from that of the more southern Peronian Province. The upper littorinid zone is dominated by *Nodilittorina pyramidalis,* and the lower littorinid (*Melarhaphe*) zone, by *Melarhaphe melanacme.* Below this the small barnacle *Chthamalus malayensis* is dominant down to about the high-watermark of the neap tide or slightly below it. A conspicuous oyster zone is formed below the barnacles by *Crassostrea amasa,* extending downward to approximately the mean low level of spring tides. From the lower edge of this oyster zone a larger barnacle, *Tetraclita squamosa,* extends to the algal zone beginning slightly below the lower level of low neap tides. The common chiton of this province is

Liolophura giamardi. The common limpets are *Patelloida saccharina, Cellana conciliata,* and *Chizacmea heteromorpha.*

The **Great Barrier Reef Province** parallels the Solanderian, offshore, and is also bathed with warm waters. It is dominated by reef-building corals and shows an affinity with eastern New Guinea and the western Pacific more strongly than with the Indonesian region. There are, of course, many faunal elements in common with the nearby mainland coast, but there are also characteristic differences. Endean (1957) concluded that the echinoderm faunas of the Great Barrier Reef and the Queensland coast were distinct entities; the mainland species had more elements in common with the East Indies, and the Barrier Reef complex showed more relation to New Guinea and the western Pacific. The intertidal zonation of Heron Island, a typical Barrier Reef island, differed in several respects from that on the mainland. The littorinids were much less abundant and lower in the intertidal. There was no upper barnacle (*Chthalamus*) zone, and the oyster zone was less well developed. The large barnacle was represented by a different species, *Tetraclita vitiata,* and the algal zone of the mainland was largely replaced by a lithothamnion-zoanthid-coral zone. The commonest chiton is *Acanthozostera gemmata.* The only limpet is *Penepatella inquisitor.*

The **Peronian Province** of New South Wales overlaps the Solanderian Province to the north in southern Queensland and the cold-water Maugean Province to the south along the coast of Victoria. The water temperatures run from about 17° to 20° C. There are several noticeable differences in the intertidal zonation. The lower littorinid (*Melarhaphe*) zone is dominated by *Melarhaphe unifaciata.* The upper barnacle zone is dominated by *Chthalamus antennatus.* Two of the most conspicuous differences are that the zone occupied by oysters (*Crassostrea*) in the north is here dominated by the barnacle *Tetraclita rosea* and that the zone occupied by the barnacle *T. squamosa* in the north is here occupied by calcareous tubes of the serpulid worm *Galeolaria caespitosa.* The algal zone contains many more tunicates, becoming, in effect, an algal-ascidian zone. Typical echinoderms include *Asterina inopinata, Halopneustes pycnotilus, Centrostephanus rodgersii,* and *Heliocidaris tuberculata. Heliocidaris erythrogramma* and *Patirella gunii* occur both here and in the Flindersian Province of South Australia and southern West Australia.

The **South Australian,** or **Flindersian, Province** extends from western Victoria, along the shores of South Australia and most of the southern shoreline of Western Australia, where in the western part it intergrades with the West Australian Province. The exposed capes are of Paleozoic rock. Coastal waters range from about 13° to 19° C. The intertidal zonation is marked by a supralittoral band of the littorinid *Melarhaphe unifaciata,* the lower part of which overlaps a prominent black band of *Calothrix fasciculata.* Most of the littoral zone is dominated by barnacles—*Chamaesipho columna* in the upper part, followed by the surf barnacle *Catophragmus polymerus* on exposed slopes, and *Balanus nigrescens,* together with coralline algae, *Galeolaria caespitosa,* and limpets. The algal zone is dominated by species of the brown alga *Cystophora,* with some *Sargassum bracteolosum* and *Ecklonia radiata.* The Peronian barnacle *Tetraclita rosea* is absent.

Around the shores of Tasmania and adjacent shores of Victoria is a province having still cooler water, the **Maugean Province,** which is rather closely related to both the Flindersian and the Peronian provinces. The water temperatures are mostly between 11° to 15° C. One conspicuous difference is the dominance of the giant fucoid seaweed *Durvillea potatorum* in the seaweed zone. Among the mat of corallines in the lower littoral, large chitons, genus *Peronoplax,* are dominant, and a large ascidian, *Pyura stolonifera,* is abundant. In slightly deeper water the kelp *Macrocystis angustifolia* becomes dominant.

The **West Australian Province** rounds the southwestern corner of the continent, intergrading with the western ends of the Dampierian Province on the north and the Flindersian Province on the east. The water temperatures here are mostly about 17° to 20° C. The supralittoral zone commonly has an upper band of blue-green algae, below which are the littorinid *Melarhaphe unifaciata* or *Tectarius rugosus.* In the midlittoral zone the large barnacle *Balanus nigrescens* tends to dominate, along with the limpets *Notacmea onychitis* and *Patellanax peroni* and the chiton *Clavarizona hirtosa.* In the algal zone, *Ecklonia radiata* and species of *Cystophora* are abundant, and many of the rocks are covered with a lithothamnion coralline algal encrustation. The abalone *Haliotis roei* is dominant here. In general the West Australian Province differs from the Flindersian Province in lacking a number of the characteristically cooler-water forms, and from the Peronian in lacking a number of the characteristic species of that similarly situated province of the eastern coast, whose places are taken by endemic western species. The long stretches of cooler waters along the South Australian coast and of warmer water to the north have apparently been rather effective barriers to active interchanges between the biotas of these two Australian temperate areas.

New Zealand provinces

As shown in Fig. 14-2, New Zealand is situated in such a way relative to Australia, and to the oceanic currents in the region, that it would be expected to recruit biota rather strongly from Australian centers of distribution, but that exchange in the reverse direction is likely to be minimal. Its position, lying obliquely across the latitude of the Antiboreal Convergence, results in rather different water temperature conditions in the northern and southern areas.

Knox (1963) recognized three marine biotic provinces—a cold-water Forsterian Province, of the southern part of South Island and Stewart Island; an intermediate Cookian Province, embracing the northern two thirds of South Island and approximately the southern half of North Island; and finally the warmer-water Aupourian Province, comprising the northern portion of North Island.

The **Forsterian Province** is a mixed subantarctic and cold temperate region differing in several respects from the Cookian Province with which it intergrades. It possesses a strong endemic element, especially among the algae, and numerous more widespread subantarctic species are also well represented. The endemic species include *Hymenena semicostata, Brogniartella australis,* and *Apophloea lyallii*. Among the prominent subarctic species, *Pachymenia lusoria, Delesseria crassinerva, Myriogramme crispata,* and *Codium dimorphum,* together with such circumpolar antarctic species as *Adenocystis utricularis, Ballia scoparia, Chaetomorpha darwinii, Schizoseris davisii,* and others, are well represented. The echinoderm fauna strongly intergrades with that of the Cookian. Of the more than 60 species of shelf echinoderms known from the region, only about 10% are regarded as typical Forsterian species. Subantarctic species include *Trachythyore amokurae* and *Oncus brevidentis*. Characteristic Forsterian molluscs include the limpet *Cellana strigilis* var. *redimiculum* and the siphonariid *Kerguelenella stewartiana*. The supralittoral zone is characterized by the blue-green alga *Verrucaria* and an encrusting red alga, *Hildenbranchia,* in the most southern areas. The littorinids *Melarhaphe cincta* and *M. oliveri* are abundant. Throughout the midlittoral zone, bands of algae are superimposed on the barnacles and mussels. *Chamaesipho columna* is the dominant barnacle. *Bostrychia arbuscula* extends through the barnacle zone, and below this is a band of *Apophloea lyallii*. The leafy red alga *Pachmenia lusoria* forms a band in the lower midlittoral, and patches of the mussels *Modiolus neozelandicus* and *Mytilus edulis aoteanus* also occur. In some areas algae grow so densely as to largely exclude barnacles in the midlittoral zone.

The **Cookian Province** comprises mixed cold temperate waters of the central regions of New Zealand, with some gradation in temperatures and differences in biota along the north-south extent of the province, as well as some characteristic differences between the biota of the east coast and that of the west coast. On the east coast the intertidal comprises a supralittoral zone with an upper band of lichens, below which the littorinid snails *Melarhaphe cincta* and *M. oliveri* are dominant. The upper littoral is marked by a barnacle zone comprised of *Chamaesipho columna* and, in the northern portion, *C. brunnea*. In areas of strong wave action, *Elminius plicatus* is found with *Chamaesipho*. A zone of mussels, species *Modiolus neozelandicus,* may mix with the lower part of the barnacle zone, and below these occur *Mytilus edulis aoteanus*. Tubes of the serpulid worm *Pomatoceros cariniferus* may form dense encrustations in localized protected areas. Limpets are abundant throughout the midlittoral zone, sometimes extending into the upper littoral, especially species of *Cellana, Patelloida, Sypharochiton,* and *Siphonaria*.

Along exposed coasts the lower littoral is dominated by large fucoids, *Durvillea antarctica* and *D. willana,* with a variable undergrowth of corallines and *Lithothamnion*. In more protected places species of *Cystophora, Carpophyllum,* and *Xiphophora* tend to replace the *Durvillea*.

One of the major differences between the east coast and the west coast in the Cookian Province is the absence of the large brown algae from the lower littoral and upper sublittoral zones of the west coast. The horse mussel, *Modiolus,* is found in more prolific growths, sometimes covering entire upper and midlittoral zones and extending into the lower littoral.

The **Aupourian Province** of northern New Zealand is an area of transitional warm temperate waters. The supralittoral zone is characterized by an upper band of yellow and grey lichens, below which the littorinids *Melarhaphe cincta* and *M. oliveri* dominate the rocks, with black patches of blue-green algae and the lichen *Lichina pygmaea*. *Chamaesipho brunnea* is the main barnacle of the upper littoral, giving way to *Chamaesipho columna* and *Elminius plicatus* in the midlittoral zone, along with the limpets *Cellana ornata* and *C. radians,* and the whelks *Lepsiella scobina* and *Lepsia haustrum*. Bands or patches of the algae *Apophloea sinclairii* and *Bostrychia arbuscula* occur among the barnacles.

In sheltered positions the oyster *Saxostrea glomerata* dominates the lower midlittoral zone. In other places the serpulid *Pomatoceros cariniferus* occurs at this level. The brown algae *Xiphophora chondrophylla* and *Carpophyllum maschalocarpum* dominate the lower littoral, with undergrowth of corallines and various smaller red algae. Below these the large brown algae *Lessonia variegata, Ecklonia radiata,* and species of *Carpophyllum* and *Sargassum,* and the red algae *Pterocladia lucida, Melanthalia abscissa,* and *Vidalia colensoi* occupy the upper sublittoral zone.

Island provinces

The smaller provinces designated by Knox in this region are three island groups north of New Zealand in relatively warm waters and three to the south in cold water.

The northern islands are Lord Howe Island, off New South Wales (**Philippian Province**), Norfolk Island, directly north of New Zealand (**Norfolkian Province**), and the Kermadec Islands, northeast of New Zealand (**Kermadecian Province**). Lord Howe Island marks the southernmost occurrence of coral reefs.

The southern islands are Macquarie Island, regarded as an extreme eastern outpost of the **Kerguelian Province;** three island groups and an island south of New Zealand, the Auckland Islands, Campbell Island, the Bounty Islands, and the Antipodes Islands, grouped together as the **Antipodean Province;** and finally the Chatham Islands, southeast of New Zealand, constituting the **Moriorian Province.** One of the most striking endemics of these southern islands is the monotypic penguin, genus *Megadyptes.*

SOUTHERN SOUTH AMERICA

Along the west coast of South America the north-flowing Humboldt Current, plus extensive upwelling, combine to keep the surface temperatures of coastal waters low along almost the whole extent of the continent and to make the waters off Peru and northern Chile extraordinarily productive. The fact that there is an uninterrupted continental shelf extending from the North Pacific to southern South America has also strongly influenced the coastal fauna. A number of groups from the north show submergence in the tropical regions and reappear in the southern fauna, either as the same or as closely related species. Others, which made the crossing at some time in the past, are not now found in the tropics but have representatives both north and south. The phenomenon of north-south **bipolarity** is probably better shown along the west coasts of the Americas than anywhere else.

Southern South America and a number of oceanic islands in the region have a colder-water antiboreal biota distinct from the more temperate regions to the north. The antiboreal region extends north to about the level of Chiloe Island on the west, but its extent along the southeastern coast of South America is less clear because this region is perhaps the least well known of any major coastal region.

In the antiboreal region winter surface temperatures (August) range from 4.5° to 9.5° C., and summer temperatures (February) from about 8° to 16° C. The temperate regions show winter temperatures of 10° to 16° C., rising as high as 19° to 20° in summer. There is a rather abrupt transition in the north to subtropical and tropical waters where the Humboldt Current turns west to become the South Equatorial Current.

Approximately every seven years a disastrous change in current patterns brings warm water of the Equatorial Countercurrent farther south than usual in the winter along the coast of Ecuador, converging with and overlying part of the Humboldt Current. As these waters mix, the exceptionally rich plankton and other animals in the northern portion of the Humboldt Current are destroyed wholesale. Large amounts of organic matter decomposing in the water and sinking to the bottom produce sufficient hydrogen sulfide to poison the water and blacken the paint of ships passing through the area. Many of the guano-producing birds die, and others leave the area. The coastal weather is adversely affected, producing heavy rains causing serious erosion of the relatively unprotected soils. This phenomenon is known as "El Niño."

The antiboreal fauna of South America has a high degree of endemism, about 52% of the echinoderms and 73% of the fish being endemic. The proportion of endemic genera is much less—about 8% for echinoderms. The largest proportion of nonendemic elements are antarctic or subantarctic elements, the largest contingent being 20 species, or 16% of the total shared with southern Australia, southern New Zealand, and the oceanic islands to the south of these. This is a relatively high figure, considering the vast distances separating South America from these regions.

A few of the most abundant or characteristic elements of the South American antiboreal fauna include the starfish *Anasterias antarctica* and *A. pedicellaris,* endemic to the area, and the antarctic *Cycethra verrucosa.* The antarctic ophiuroids *Astrotoma agassizii, Gorgonocephalus chilensis, Ophiomyxa vivipara, Ophiomitrella falklandica, Ophiactis asperula,* and *Ophiocten amitinum,* the latter two being especially abundant, are also characteristic. Sea urchins include 6 endemic monotypic genera, one of them, *Tetrapygus,* extending northward along the west coast as far as Peru. The endemic genus *Austrocidaris* comprises 3 species.

Among the fish the connection with the Antarctic seems to be weaker, only about 3% being common to both regions, compared with 38% among echinoderms. The fish fauna of antiboreal South America seems to occupy a more independent position. Some of the characteristic endemics of the region are 2 genera—*Cottoperca,* with 3 species, and *Notothenia,* with 8 species—and seventeen other species of the Nototheniiformes. Six genera of Zoarcidae, including *Austrolycus,* with 4 species, and *Crossostomus,* with 3, are also endemic.

Some examples of forms showing bipolarity include, among the fish, the families Zoarcidae, Liparididae, and Scorpaenidae. Among the invertebrates, several genera of the crustacean family Lithodidae, such as *Lithodes* and *Paralomis,* and numerous echinoderm genera, including *Flor-*

ometra, Antelaster, Ctenodiscus, Diplopteraster, Echinus, Solaster, Porania, Bathybiaster, and *Ophiocten,* show this phenomenon. The genera *Henrica* and *Pteraster,* common in both Europe and North America, show submergence in the tropical regions.

SOUTHERN AFRICA

The biotic zonation around the southern extremity of Africa is complicated by the meeting here of warm southwest-flowing currents from the Indian Ocean with the cold waters of the West Wind Drift. The Mozambique Current, called the Natal Current farther south, flows southward along the eastern coast of Africa, keeping the shore temperatures essentially subtropical, about 20° C., to about Algoa Bay. South of Cape Agulhas, the southernmost tip of Africa, where this warm current is also known as the Agulhas Current, it meets and is split up by the cold West Wind Drift. Some of the cold water from the West Wind Drift wedges itself between the Agulhas Current and the mainland, flowing close to the southern coast as far as Algoa Bay. At Cape Agulhas the immediate coastal waters may be several degrees warmer than the offshore waters, which farther out give way to waters of subtropical temperatures. The patterns vary somewhat with the seasons and with vagaries of the local weather. Here, more than in most areas, it becomes clear that in the zoogeography of the oceans, the concern is with certain kinds of water rather than with fixed areas, boundaries, and locations.

The western coast is bathed by cold, north-flowing waters of the Benguela Current and further cooled by upwelling, a situation analogous in many respects to that of western South America. Water temperatures here are in the neighborhood of 14° to 15° C. in the south, increasing northward and becoming subtropical beyond the region of Cape Frio. The southwest African temperate region from Cape Frio to St. Helena Bay is usually termed the **Namaqua Biotic Province,** following W. Michaelsen (1915), who contributed much to our early understanding of the biology of this region. Being a region of high illumination, with cool coastal currents fertilized by upwelling, the waters here are very rich in phytoplankton and are very productive. Dinoflagellates in particular become seasonally enormously abundant. "Red tides" are a regular recurrent feature between Walfish Bay and Lüderitz Bay in the spring, often causing high mortality among the fish and invertebrates of the area. They occur sporadically also in other areas along the southwest African coast. The organic matter resulting from these mass mortalities accumulates on the sea floor, so that a nearly azoic benthic region smelling strongly of hydrogen sulfide and containing the bodies and skeletons of animals killed during the red tides has formed. It is approximately 30 km. wide and extends down to depths of 140 to 150 meters. During the rest of the year pelagic organisms and fish repopulate the upper strata of water, but the benthos is almost extinguished.

The region is generally unfavorable to a high development of the benthos because of a sandy sea floor and lack of protection from the waves. In the calmer, deeper offshore water decomposition of algae and other organisms contributes to the stagnation of the water and the smothering of benthic organisms. Along the immediate shoreline and in the intertidal zone, on the other hand, the fauna is very rich. The water movements prevent the deposition of sapropelic mud, and the organisms have the advantage of nutrient-rich waters with abundant plankton. The fauna in general shows affinity with that of the rest of the Atlantic Ocean rather than with that of the Indian Ocean, and numerous species appearing here also occur along the European coasts. Examples of such species are the brittle star *Ophiothrix fragilis,* the tunicate *Leptoclinides capensis,* the molluscs *Venus verrucosa* and *Cerithium vulgatum,* and the polychaetes *Nephthys hombergi* and *Sabella pavonina.* Other characteristic species of the intertidal fauna are the molluscs *Mytilus meridionalis, Cominella delalandei,* and *Patella granatina.*

A number of echinoderms characteristic of the North Atlantic are also represented here by "twin species" or varieties. Some of these are *Astropecten irregularis* var. *pontoporaeus, Hippasteria phrygiana* var. *capensis, Marthasterias glacialis* var. *rarispina* and var. *africana, Brissopsis lyrifera* var. *capensis,* and *Echinocardium flavescens* var. *capense.* Similar examples could be cited for other groups.

It seems possible, as Darwin postulated, that temperature decreases during the ice ages may have made possible a continuity of distributions of the northern and southern regions that has been interrupted by the rewarming of tropical warm waters since the last glacial epoch. This cooling effect seems to have been especially strong in the eastern Atlantic.

Another group of species of southern and southwestern Africa seems to have been derived from distant regions of South America or the southern Atlantic islands by way of the West Wind Drift. Still others have a circumpolar distribution, being found in the colder parts of the Australasian region, in southern South America, and in southern and southwestern Africa. A few of these widely distributed forms, carried around the world by the West Wind Drift, include the lugworm *Arenicola assimilis,* the crustaceans *Squilla armata* and *Jasus lalandii,* and the ascidian *Corella eumvola.* Fish of the genus *Callorhynchus,* described from southern Africa, South America, and southern Australia and New Zealand, may be merely widely separated varieties of a circumpolar species.

THE SOUTHERN, OR ANTARCTIC, OCEAN

The Southern Ocean is more than the confluence of three great oceans, the Pacific, Atlantic, and Indian. It has an identity of its own, even though its northern boundaries cannot be precisely defined. Perhaps the most logical way to determine the boundary, if there must be a boundary, would be to include all the waters involved directly in the West Wind Drift, which would place the northern limit at the Antiboreal Convergence. Biologically, however, the Antarctic Convergence is the sharpest line of demarcation between typically antarctic surface waters and those of other oceans. The strongest flow in the West Wind Drift coincides with the Antarctic Convergence.

Unlike the other oceans, the Southern Ocean is continuous around the world, the narrowest part being the 1000 km. separating the southern tip of South America from the Antarctic Peninsula (formerly the Palmer Peninsula). Elsewhere it is more than 2000 km. from the coast of Antarctica to any continental landmass.

While the Southern Ocean occupies more than 75 million km., or 22% of all the ocean area of the world, its heat content is only 10% of the oceanic total. The heat exchange between Antarctica and the rest of the world profoundly influences the nature and patterns of the world's climate, and the temperatures and current patterns of the world's oceans.

Near the shores of Antarctica, especially in the embayments forming the Ross, Amundsen, Belingshausen, and Weddell seas, large areas are covered by ice much or all of the year. Cold water from the coastal areas sinks to the bottom, forming deep-water abyssal slow-moving currents trending northward along the low contours of the ocean floors, detectable in the Atlantic even north of the equator. The sinking water is replaced by upwelling from deep waters just above the north-trending bottom water, moving south as a countercurrent and upwelling near Antarctica as the **Antarctic Divergence.** Upon reaching the surface layers, most of this water is caught up in the West Wind Drift, and begins to move in a northeast direction until it reaches the Antarctic Convergence, where it sinks beneath the warmer antiboreal water, forming a layer of antarctic intermediate-depth water moving down and to the north between the Antarctic Convergence and the Antiboreal Convergence. The surface antiboreal waters between the two major convergences are, in general, moving in a southeast direction, comprising the northern portion of the West Wind Drift.

Along much of the shoreline of Antarctica, especially in the seas, there are large eddies, so that the coastal water is moving west rather than east.

Fossils discovered in Antarctica ever since the first explorers ventured ashore near the turn of the century demonstrate clearly that this continent has not always been covered with ice. In fact, the succession of life there has been shown to be much the same as in the other major Southern Hemisphere continents until well beyond the time of the appearance of seed plants relatively recently in the geological history of the world. Fossil trees with well-defined rings indicating rapid growth in a temperate climate prove this beyond doubt. This is another facet in the growing body of evidence that Antarctica, rather than having always been an isolated landmass, may, until about the end of the Mesozoic era, have been at the heart of a great southern continent comprising South America, Africa, Antarctica, Australia, and perhaps part of southern Asia. The splitting up of this former supercontinent, Gondwanaland, and the rafting of Antarctica to the geographical pole of the planet, making this pole a thermally isolated region rather than part of the world ocean, seems to have been the precipitating cause of the glaciation that still holds the continent in the grips of a prolonged ice age. The former closer association of the lands now comprising the continents of the Southern Hemisphere would also go far to explain the apparent worldwide distributions of some of the ancient groups of animals and plants without the necessity of hypothesizing numerous and improbably long land bridges as was frequently done by earlier paleogeographers.

Antarctic Ocean fauna

The high fertility of the surface waters of the Antarctic resulting from continuous large-scale upwelling, the extreme nature of the environment, and the short growing season through the antarctic summer combine to produce the world's most dramatic outbursts of phytoplankton growth, dominated by rapidly reproducing species of diatoms. The zooplankton, too, tends to be dominated by a relatively small number of species of rapidly reproducing forms.

One species in particular, the red euphausiid shrimp *Euphausia superba,* commonly called krill, tends to dominate large areas of surface waters during the antarctic summer. Krill is the major link between the phytoplankton and the larger animal life in the Antarctic. One of the larger members of the macroplankton, it is a little over 2 inches long, large enough to be preyed upon directly by many kinds of fish, squids, birds, and even mammals. It thus provides for these animals an unusually efficient one-step food chain enabling a much larger biomass of large animals to be supported than would be the case if most of them preyed upon animals of intermediate size. The world's largest animals, the great whalebone whales—the blue whale, finback whale, humpback whale, and others—congregate here during the antarctic summer. Whalebone whales are basically immense animated plankton nets having the proper mesh to capture macroplankton organisms. It has been estimated (Pequegnat, 1958) that a mature blue whale must eat about 3 tons of krill daily during its feeding season. It is estimated that, prior to recent intensive whaling in the region, at least half a million whales congregated in the Antarctic each summer, which would mean that at a conservative estimate at least 1350 million tons of krill, about 1000 pounds per acre, were produced to support them, since the whales certainly did not consume more than 20% of the total annual production. Actually, the figure must be very much higher because of the hordes of other predators that also feed upon the krill.

Krill is an ideal food for predators such as the whales. The euphausiids congregate in immense shoals in the surface waters, so that only a minimum of energy is required to fish them in great quantities. Furthermore, they are a nutritious high-protein food.

Hardy and Gunther (1935), studying the distributions of phytoplankton, krill, whales, and dissolved phosphate (a plant nutrient) in the region of South Georgia Island, made the interesting discovery that these distributions are closely correlated and that from a knowledge of any one of them, the distributions of the other three could be accurately derived. Clouds of phytoplankton are inversely correlated with high phosphate content of the water (because most of the phosphate has been used up in producing the phytoplankton). High phosphate content of the water, shoals of krill, and the presence of blue whales were positively and very significantly correlated. Hardy was even able to predict, solely from records of the distribution of whales, during a year in which their distributions in the region were unusual, what the content of as-yet-unopened bottles of plankton from a plankton survey made the same year would be.

The life histories of the krill and some of the other members of the macroplankton are nicely timed to the circulation of waters in the Antarctic Ocean. During the summer the actively reproducing adults are drifting in a generally northeast direction with the surface waters, until they and the eggs and young come to the Antarctic Convergence. Here they are carried below the surface. The larval stages sink into the intermediate depths in which the water is drifting south. The waters at the depths in which they are drifting (being a thicker layer of water moving the same distance and delivering the same volume of water in a given time) are moving more slowly than the surface waters, so that the young euphausiids begin coming to the surface as newly transformed adults the second summer following their submergence. Thus, by changing their level in the water, populations of drifting planktonic organisms are able to maintain themselves in a favorable region in spite of the continual movement of the surface waters out of the region. The West Wind Drift has caused a general circumpolar distribution of most pelagic antarctic organisms, but even so, there are areas where localized conditions result in greater-than-average concentrations.

The krill are fed on directly by whalebone whales, the crabeater seal, Adélie penguins, hordes of flying birds such as petrels, fulmars, and terns, and hordes of small fish and squid.

The small fish and squid form the principal food of larger fishes, emperor penguins, fishing birds such as the shag, and most seals, such as the Weddell seal, Ross's seal, and the leopard seal. The skua feeds on young penguins, steals fish from other fishing birds, and takes some small fish and squid. Sperm whales feed principally on squid. Leopard seals prey on penguins, fish, and other seals; the voracious killer whales attack any large animal, including other whales larger than themselves.

The benthos is supported, as everywhere, by mixed plankton in areas shallow enough so that they are in depths occupied by phytoplankton and by thinned-out zooplankton and organic detritus from above in deeper water. The unusual richness of the Antarctic Ocean enables it to support a luxuriant benthos, some groups such as sponges being better developed here than anywhere else. At the shoreline the ice keeps the rocky intertidal and subtidal free of attached animals and large algae, so that the margin of the sea is barren. In many areas this barrenness extends downward approximately 10 meters below the level of the floating ice, to the lowest level at which ice crystals will form on a submerged object; the formation of ice crystals and their detachment as they grow perhaps prevent the establishment of small, early stages of attached benthos. Below this level there is an abrupt colonization of the rock with a rather rich benthos.

Although feeding is richly provided for in the Antarctic, nesting sites for birds are at a premium. The rookeries tend to be densely populated with thousands of birds, which use every available site during their nesting seasons.

There is a high level of endemism in the fauna of the Antarctic Ocean. According to Ekman (1953), about 73% of the 256 species of echinoderms known from the region are endemic to the Antarctic Ocean, and approximately 27% of the genera, several of them monotypic are endemic. Among the fishes, 90% of some 78 species and 65% of the 37 genera are listed as endemic. The family Notothenidae is an especially noteworthy group of endemic antarctic fish.

One of the most extreme and curious adaptations to the cold, oxygen-rich antarctic waters occurs in certain fish of the family Chaenichthyidae, related to the notothenids. These so-called **icefish** attain a length of 2 feet or more in some species and are large-headed, thin-bodied, rather sluggish fish with no red blood cells and no hemoglobin. About 16 species are known, all of them living in cold, stable, food-rich antarctic waters, where they do not have to expend great amounts of energy obtaining food. The oxygen dissolving into their colorless blood plasma from the surrounding water is sufficient to maintain their sluggish metabolism. Examination of kidney and spleen tissues, where blood is formed in other fish, shows no evidence of the formation of blood cells. Thus their anemia is probably total, at least in some species. The blood volume is, however, unusually large for fish of their size and weight, enabling them to transport more dissolved oxygen than would otherwise be the case. Only in stable, cold, oxygen-rich, food-rich waters, such as those of various antarctic areas, does it seem that such an anemic race of fish could even begin to evolve.

In comparison with the Arctic Ocean, the Antarctic Ocean is far richer in its fauna, particularly the benthic elements, and displays a much greater degree of endemism, probably because its formation took place long before that of the north polar sea. Its relationships to other oceans create the conditions of massive interchange of waters, with extensive upwelling of water from intermediate depths, that make the surface waters exceptionally productive. The greater productivity of the surface waters is, in turn, reflected in higher productivity at all depths than would otherwise be the case.

Although the Arctic Ocean has a vastly greater extent of shallow-water shelf areas, which are directly continuous with richly populated continental shelves of both sides of the boreal Atlantic and also with those of the northern Pacific and although for the most part the shores of Antarctica slope off into deep water much more abruptly and the Antarctic is isolated by great distances, by the direction of the surface current systems, and by great differences in ecological conditions from other regions with shallow-water shelf faunas, nonetheless the shelf fauna of the Antarctic is much richer in species as well as in biomass than is that of the Arctic. Clearly Antarctica has been a center for the development of a marine fauna for a very long time and has had an uninterrupted cold climate ever since at least the early Tertiary.

CHAPTER *15*

Distributions of the deep-sea benthos

GENERAL FEATURES

Some of the general features of the deep-sea benthos have already been discussed in Chapter 8. Here we will be concerned only with the question of the extent to which the deep-sea benthos displays regional differences of zoogeographical significance. This is difficult to assess accurately because of the formidable problems of securing adequate samples. The fact that very large areas are often of relatively uniform character somewhat offsets the difficulty. Nevertheless, the amount of data and of direct and indirect observation at the floor of the deep seas is so meager compared to that related to shelf biota that generalizations must, of necessity, be broader and less precise.

Since, as a rule, temperature decreases and salinity slightly increases with depth, the very deep oceanic basins tend to be somewhat isolated from each other by tem-

Fig. 15-1. *Pheronema giganteum* F. E. Sch. (actual size about 26 cm. long), one of the glass sponges. (From Ijima, I. 1926. *In* Reports of the Siboga Expedition, 1899-1900. Vol. 6. The Hexactinellida of the Siboga Expedition. E. J. Brill, Leiden, Netherlands.)

perature barriers along the rises between them and sometimes by temperature differences of their contained water. When the waters of such basins also differ in other respects such as oxygen content or biological "conditioning," the differences become reinforced, making exchanges of the more sensitive and stenothermal elements of their benthos unlikely. For benthic organisms the ridges and rises between adjacent basins may present barriers much as do mountains on land.

In the deep sea, **eurybathic** species ranging over wide limits of depth, like cosmopolitan species in the superficial layers, are common but contribute less to zoogeographical characterization of a given region than do **stenobathic** forms. Archibenthal species, such as most Pogonofera, found chiefly on

325

the continental slopes, have wider geographical distributions as a rule than do strictly shelf-dwelling species, but not as wide as the abyssal forms. The East Pacific Barrier, for example, is effective for archibenthal as well as shelf-dwelling species, but it is of little significance for truly abyssal ones.

Early impressions that the deep-sea fauna of the world is relatively uniform were based on insufficient material and on failure to distinguish between bathypelagic species, which show a much greater tendency to be cosmopolitan, and bathybenthal species. It is now clear that the Atlantic abyssal fauna is quite distinct from that of the Pacific and Indian oceans, especially at the species level. The polar seas also have somewhat isolated and distinctive abyssobenthal faunas. The upper temperature limit for the strictly polar abyssal faunas seems to lie at about 0° C. The deep polar basins containing water at or below this temperature are relatively isolated from other abyssal areas, especially in the north, although Baffin Bay, which is separated by shallow areas from the major arctic basins, does have an arctic abyssal fauna.

The finding of some archaic types of animals known previously only as fossils or most closely related to animals known as fossils, during early deep-sea explorations, led for a time to the concept of the abyss as a sort of refuge for primitive, outmoded types of animals that had long since disappeared from other areas because of competition with more recent species or because of the greater environmental changes to which the shallow waters are sometimes subjected. It was hoped that the study of abyssal faunas would go far to fill in gaps of our understanding of the evolution and relationships of various groups. To some extent it has done so. Many stalked crinoids have fossil records going back as far as the Eocene, and some genera, even to the Jurassic. The recent discovery of *Neopilina,* a representative of the primitive molluscan group Monoplacophora, thought to have been long extinct, is another case in point.

However, by no means all the "living fossils" of the sea are abyssal, as shown by the spectacular discovery of the coelacanth fish *Latimeria* in the Indian Ocean off eastern Africa and the Comoro Islands in water not more than about 600 meters deep. Furthermore, archaic types do not dominate the abyssal fauna to a greater extent than they do the shelf fauna. Such familiar archaic types as *Lingula, Limulus, Heliopora, Nautilus, Branchiostoma,* and *Cestracion* are all shelf-dwelling forms, or they live in the upper strata of water. The abyss also contains many types that are more specialized and have more recently evolved than their nearest relatives, now living on the shelf.

The shelf faunas of the world seem to have originated in the tropics, from which they spread into northern and southern cooler regions, becoming progressively adapted to living in colder waters. Most of the archibenthal and abyssal faunas, even of warm regions, appear to have been derived from the northern or southern shelf faunas after their adaptation to cool or cold temperatures. Elements of these cold-adapted shelf faunas subsequently descended into deeper waters, both directly in the regions in which they found themselves and by submergence along cool isotherms beneath the warmer subtropical and tropical waters. This general pattern of dispersal is especially true for the benthos and to a lesser extent for bathypelagic and abyssal nekton.

ISOLATED SEAS

The **Mediterranean** is a special case in many respects. The Gibraltar Ridge prevents the exchange of all but the superficial layers of water with the Atlantic, so that the cool, deep, well-oxygenated Atlantic water is prevented from entering. There is a flow of surface water into the Mediterranean and of subsurface, more saline water out over this sill. Strong evaporation increases the salinity sufficiently to cause the warm surface water to become denser and to sink, so that the deep water in the Mediterranean is warmer than correspondingly deep water in the adjacent Atlantic. The temperature difference, the difference in salinity, the outflowing current of subsurface waters, and the height of the sill of the Gibraltar Ridge all combine to prevent the entry of Atlantic deep-water species into the Mediterranean. Thus there is no stenothermal cold-water abyssal fauna in the Mediterranean; there is only an archibenthal abyssal fauna, which diminishes in richness toward the eastern part of the sea. Because of the unusual temperature distribution, littoral and sublittoral species extend their ranges deeper than they do in most areas, and the archibenthos extends all the way to the bottom.

The **Red Sea,** of more recent origin than the Mediterranean, displays similar conditions even more intensified. It is only 2800 meters deep in its greatest depth and is separated from the Indian Ocean by a sill no more than about 100 meters in depth at its lowest point. Thus the deep water of the Indian Ocean is excluded. Evaporation here is even more intense than over the Mediterranean, and the sea is shallower, so that the temperatures and salinities along the bottom are even higher than in the Mediterranean. Bottom temperatures are about 21.5° C., and the salinity is about 40.5⁰/₀₀. Many groups of stenohaline animals, such as sponges, echinoderms, and ascidians, found in the shallower waters disappear before reaching the bottom. The deeper parts of the bottom are colonized chiefly by gastropods and decapod crustaceans of types that in other regions live on the shelf or are, at most, archibenthal. For the Red Sea as a whole, at the species level there is a rather high degree of endemism, indicating that approximately one million years of relative isolation is sufficient to produce numerous changes at the species level. The level of endemism of higher taxa is low, however, the fauna being clearly related to that of the Indo-Pacific.

The **Sea of Japan,** at present, has a depth of some 4000 meters, but it is separated from the Pacific by sills no deeper than 165 meters. During comparatively recent periods of the ice ages, it is thought to have undergone a history of elevations during the glacial periods and subsidence during the interglacial periods, which at times converted it into a great freshwater lake. This relatively recent history and its present isolation from the deep waters of the Pacific account for the remarkable fact that in spite of its very considerable depth and its nearness to regions with well-developed abyssal faunas, the Sea of Japan lacks a specifically abyssal fauna, the deeper parts of the benthos being colonized only by a number of eurybathic species, some of which have descended from the shelf to about 3500 meters.

PART *VII*

TIME
and
MARINE LIFE

CHAPTER 16

Thoughts on the role of the seas in the origin and early evolution of life

Here our approach must of necessity be more indirect and speculative rather than observational and descriptive. The origins of life on earth have receded so far into the remote past as to wholly preclude any direct attack on the problem. Furthermore, it was not until living things had already become very widespread, highly organized, and diversified that they began to leave recognizable traces. Most of the traces left by the life of past ages are still locked up in ancient rocks or have been altogether destroyed.

The best we can do is to study the hierarchies of living things as we know them today, from the simplest viruses to the most complex animals, study the properties of living matter as such, and study the progression through time of living things as we find it recorded in fossil-bearing rocks. By extrapolating backward from the various trends observed in the evolution of living matter and extrapolating forward from what we can deduce to date with respect to abiotic chemical and or-

ganochemical evolution, we can try to arrive at some sort of conceptual meeting ground with respect to the conditions under which such an amazing system could have arisen, and the period of time involved.

More and more in recent decades the conviction has grown that life is not something set apart from, and, as it were, foreign to, the material universe, merely using some of the materials in a mysterious way for its purposes. Rather, life is just as much an inherent property of the universe as is any other manifestation of the matter-energy-time complex. According to this view, given suitable conditions for a sufficiently prolonged period of time, matter will inevitably manifest the properties of life. Suitable conditions are, so far as we know, exceedingly rare in the universe. The universe is, however, so inconceivably vast that it seems probable that there are millions of planets supporting some forms of life. The complexity of living things is so extraordinarily great and their evolution involves such long periods of time and is so chancy

Fig. 16-1. "Animal" gardens 60 feet below the surface of the Caribbean Sea near St. Johns, The Virgin Islands. National Park Service. (M. W. Williams photograph.)

and irreversible that it is most unlikely that any of the same species of animals or plants would evolve independently on two different planets. In this sense it is probable that man stands alone. Any planet on which life arose and flourished for long periods would, of necessity, evolve groups of organisms ecologically and possibly structurally comparable to those found on earth.

KINDS OF ORGANISMS

Living things as we know them today can be arranged into five major groups, or kingdoms, based on different orders of complexity or basically different modes of nutrition: the **noncellular organisms**, the **akaryotic cellular organisms**, the **Protista**, the **Metaphyta**, and the **Metazoa**.

NONCELLULAR ORGANISMS

The noncellular organisms are the **viruses** and **episomes** as we know them today. In its active, so-called vegetative stage, a virus consists of a single strand or circle of nucleic acid immersed in the chemically rich environment of a cell of some higher organism. There it takes advantage of the host's metabolic machinery, replicates itself at the expense of chemicals present in the host cell, and finally directs the synthesis of specific protein, which eventually coats the strands of virus nucleic acids individually, forming virions that are able to survive outside the host cell and to infect new cells of the same type. Some viruses, notably the bacteriophages, may vary this cycle by joining the nucleic acid ring of the host cell, replicating synchronously with the replication of the host nucleic acid, and becoming to all intents and purposes a gene, or functional unit, of the host nucleic acid.

Another category of entities, even simpler than the viruses, are small rings of nucleic acid known as episomes, which exist independently of the host nucleic acid in the cytoplasm and replicate there. Episomes must be passed from infected cells directly to noninfected cells, usually by conjugation or direct contact of the cells, since they do not direct the formation of specific proteins and become complete virions. Like the virus nucleic acid, episomes in some cases show a definite relationship to the genome of the host cell and may behave as if they were part of the cell's own genetic mechanism. Both viruses and episomes may, on rare occasions, exhibit the phenomenon of crossover with host-cell genetic material, exchanging one or two genes of their own for one or two genes of the host cell's nucleic acid.

Some biologists prefer not to regard precellular stages in the evolution of life, or the viruses and episomes, as organisms—reserving the term for those forms that display the cellular level of organization. It is to be noted that both episomes and viruses exhibit properties of life only when immersed in the extremely rich biochemical environment of a host cell. They show a definite relationship to the nucleic acid of the host cell, are quite specific as to what kinds of cells they can infect, and consist only of nucleic acid or of nucleic acid and a specific protein coat, or at most a small number of specific proteins, in the case of the virion of bacteriophage. Another major difference between the viruses and cellular organisms is that in viruses only the nucleic acid replicates directly, whereas the protein is synthesized through redirected activities of the host cell's metabolic machinery, after which the complete virion is assembled. The cells of cellular organisms, on the other hand, duplicate themselves as complete units.

It seems probable that viruses and episomes as we know them are not holdovers from the primitive precellular ancestral forms of life but are, rather, detached bits of genetic material of cells of cellular organisms continuing to function and evolve independently of the genome of the cells in which they live. Nonetheless, they may, by analogy, give us some insight into the probable order of structural complexity and environmental needs of ancient precellular entities.

There is a very big gap between the postulated precellular entities, presumably little more than gigantic macromolecules of nucleic acid replicating in a medium rich in the organic chemicals necessary for building their structure, to even the simplest of cellular organisms, with its hundreds of proteins and enzymes, its carbohydrates, lipids, and high-energy phosphate compounds, its intricate membrane system, and its digestive, respiratory, and excretory processes, all of which are organized into a functioning whole that reproduces as a complete organized unit, controlled in some way by the intimate internal ordering of coded units in the nucleic acid that constitutes its genome. Once the cellular level of organization was achieved, it was enormously successful and has continued to grow, proliferate, differentiate, and cover the earth with a living mantle that is largely responsible for the physical and chemical character of the earth's surface, its atmosphere, and its waters.

AKARYOTIC CELLULAR ORGANISMS

The simplest of the cellular organisms do not have a nucleus with a definite nuclear membrane, nor do they have a number of chromosomes. The nucleic acid consists apparently of a single very long strand of nucleic acid that is structurally a closed circle, although it may be complexly folded and not give the appearance of a circle in electron micrographs. The akaryotic cellular organisms comprise the bacteria, actinomycetes, and blue-green algae. The latter differ from the first two groups in having developed a mechanism of photosynthesis. Certain specialized groups of sulfur bacteria are also photosynthetic but by a different mechanism that does not involve the production of free oxygen as a by-product.

Another significant evolutionary jump, although less difficult to visualize than the transition from noncellular to cellular organisms, occurs between the akaryotic cellular organism and the karyotic, or nucleated, types, in which the cells have well-organized nuclei surrounded by nuclear membrane and cell division is by a process of mitosis in which discrete units of DNA are condensed as a definite number of chromosomes. The organisms in which the cells are nucleated comprise three groups: the Protista, the Metaphyta, and the Metazoa.

THE PROTISTA

The Protista are organisms with nucleated cells that have remained at the cellular grade of construction. There is one phylum, Protozoa, an immense assortment of species commonly divided into two subphyla—the Plasmodroma, with monomorphic nuclei similar in character to those of the cells of other animals and plants, and the Ciliophora, with peculiar dimorphic nuclei, a micronucleus and a macronucleus, which sets them rather sharply apart from other groups. Among the Plasmodroma, the class Flagellata contains both plantlike photosynthetic forms, often classed with the algae, and holozoic animallike forms. There is no sharp break between the photosynthetic flagellates and the higher plants, such as occurs between the holozoic protozoans and the higher animals, probably because the mode of obtaining food is much the same in all plants but the shift from ingestion at the cellular level to ingestion into a communal stomach by way of a mouth posed a more difficult evolutionary obstacle, with fewer possible transition stages.

THE METAPHYTA

The plant kingdom, or the Metaphyta, comprises all higher plants, the multicellular photosynthetic forms, and some related forms such as fungi, which are saprophytic.

THE METAZOA

The animal kingdom, or the Metazoa, comprises all the multicellular animals.

ORIGIN AND EARLY EVOLUTION OF LIFE

The basic common thread running through the entire series of living things is the presence of gigantic, self-replicating macromolecules of nucleic acid. We must then look to the formation of nucleic acid as the transitional step between chemical and biological evolution. So far as can be determined, such molecules need a fluid or semifluid environment extremely rich in very complex organic compounds in order to replicate themselves. A major stumbling block for early theorists in connection with the origin of life was to account for such a pool of complicated organic compounds in a world without life, when the only sources known for such compounds were living organisms. For some time only an extremely long period of slow, fortuitous chemical evolution could be postulated, with any organic molecules formed accumulating over the ages because there were no bacteria or other organisms to destroy them. These were visualized as accumulating in shallow seas warmed by the sun and gradually polymerizing into more complex units until the conditions for the formation and functioning of self-replicating molecules had been met. The theory was not very satisfying and could be formulated only in the broadest and vaguest terms.

Several developments in the last two decades have done much to clarify possible mechanisms of chemical evolution and to demonstrate that it probably proceeded much faster than had formerly been thought possible. Oparin first gathered together the evidence for abiotic biochemical evolution in a plausible form.

The deduction by Urey and others, from geochemical and astronomical evidence, that the primitive atmosphere of the world was a reducing atmosphere without free molecular oxygen provided a much more plausible setting for this biochemical evolution without oxidation and destruction of most of the more delicate types of organic compounds as fast as they were formed. Finally, the demonstration by Urey's student, Miller (1953), that an anaerobic mixture of gases similar to that postulated for the primitive atmosphere would, in the presence of ionizing radiation, quickly generate organic compounds by condensation of carbon radicals and their polymerization greatly shortened the time period that had been thought necessary for biochemical evolution. Miller, starting with such a mixture of gases circulating over a vessel of warm water and exposed to electrical discharges, came up with several types of organic compounds including amino acids and polymerized groups of amino acids, in a week's time.

From this it was postulated that the primitive anoxic earth's atmosphere, bombarded by intense ultraviolet light not damped by an ozone layer as at present, was the site for the formation of a variety of organic molecules. Some of these would find protection from destruction by this same ultraviolet light by entrapment in spaces between clay particles on shores of estuaries, so that a reserve could be built up upon which could be based further steps in the biochemical evolution.

Perhaps an even more likely site for the accumulation of such molecules, protected from the ultraviolet light but still within the sphere of influence of much of the visible light, would be along the thermoclines in the primitive oceans. Tropical thermoclines would represent the most extensive, relatively stable regions where such molecules could collect, prevented from sinking by the increased density of the water below, gently stirred and mixed by internal waves at the interface between the two layers of water, and also many of them perhaps carried up in convection currents such as those of the Langmuir circulation, concentrated in the zones of convergence, and polymerized into larger aggregates.

The shoreline would seem a less likely place for successful biochemical evolution to proceed very far because of the time factor. Shorelines change much faster. Storms, depositions of new sediment, drying, burning by ultraviolet light, changes in sea level or local changes in the level of the land, etc. would continually be undoing the work of perhaps many years of biochemical evolution in such areas, but along the thermoclines in the open ocean, the process could go on uninterrupted for thousands of years, during which the concentration of organic materials would be increasing rapidly because of their formation in the atmosphere and entrapment in the sea. Thus the upper part of the oceans, particularly at the level of the thermocline, would become an increasingly concentrated broth of organic compounds, and the polymerization of these compounds into ever more complex units would be favored.

For details as to how such polymerizing organic compounds would build themselves into increasingly complex and individualized units (blobs of polymer jelly, termed coacervates, with the evolution of enzymes in the interstices between them or proteinoid microspheres), the evolution of increasingly efficient mechanisms of synthesis of their component molecules (the walling off of functional units by single or double monomolecular lipid films forming eobionts, eventually with internal as well as external membranes, leading to compartmentalization and diversity of function within the single proto-organism), and the formation of ribonucleic acids, of deoxyribonucleic acid, and eventually of cells, the references indicated under "Origin and Early Evolution of Life" in the Annotations to the Bibliography should be consulted.

Even the so-called prebiological or chemical stages in the evolution of life probably did not represent simply chance reactions but were, rather, to a large degree self-guiding and directional processes partaking in some sense in a hereditary process. Simons' principle that hierarchic systems will evolve more rapidly than nonhierarchic systems of comparable size and complexity, and the principle that a hereditary organization occurs with a higher probability than a nonhereditary organization resulting in the same degree of complexity, probably apply here as well as to systems recognized as living.

Once the stage had been reached where organisms capable of reproducing themselves had been formed, their increase in the relatively rich, predator-free primitive seas must have been very rapid until the time came that so much of the organic soup of the primitive sea was incorporated into their bodies that it became too thin to support further rapid increase of the biomass. As this critical point was approached, the rapid increase in the biomass would have slowed down, perhaps rather abruptly, and diversity would have begun to increase. Mutants that, during the phase of rapid growth of the population, never attained an appreciable proportion of the population might continue to increase, while the formerly rapidly growing parent populations became stabilized in numbers or even began to decrease. Ecological succession would set in. Any forms that had evolved the process of photosynthesis, building up new organic matter from inorganic substrates by using absorption of energy from the visible spectra, would have been especially and selectively favored and would have increased rapidly. These may well have been in the form of bacteria-like blue-green algal cells. Likewise, any cells that had developed the capacity to ingest other cells as a source of organic nourishment would now be especially favored. The dichotomy between holophytic and holozoic modes of nutrition would become accentuated, whereas the relative importance of saprozoic nutrition would have diminished.

Whether life originated in shallow near-shore waters such as estuaries and lagoons or along the tropical thermoclines in the ocean, the latter would appear to be the earliest places where the free oxygen evolved as a by-product of photosynthesis could have accumulated sufficiently to make possible the change from anaerobic fermentative metabolism to the much faster, more vigorous oxidative metabolism characteristic of most organisms today. Near the shore the oxidation of the great amounts of mineral substances still in a reduced state would use up the oxygen as fast as it was produced for a much longer period of time than in the upper strata of the ocean along the thermocline.

The early oceans at this stage, then, probably had a layer of photosynthetic blue-green algalike cells concentrated along the thermocline, with perhaps also coatings of these or similar cells growing on the substrate in shallow areas where they could obtain light. The origin and early evolution of holozoic forms using oxidative metabolism probably took place along the thermoclines, as Weyl (1968) has pointed out. Such forms might have been able to utilize the mat of growth on the substrate in shallow areas at times and places when more oxygenated seawater swept over them or to penetrate marginal areas for short periods to feed on the presumably lush protected growth in shelf areas if they could swim back into more oxygenated water when necessary. Evolution of both plant and animal forms could then have proceeded for a long time and attained a high degree of diversification before the animals were able to colonize shallow-water shelf areas, marginal lagoons and estuaries, etc. to any great extent. Only when enough oxygen had accumulated so that most of the exposed reduced mineral surfaces and suspended particles in the water had been oxidized, could colonization occur.

In the evolution of early animals along the thermocline, density would have been a critical factor. The development of forms with dense, compact muscular systems or heavy skeletons or shells would have been precluded. Within the bounds of this limitation, however, a very considerable diversity could have been achieved well before extensive colonization of the shallow regions became possible. According to Weyl, the period of this colonization was probably during late Pre-Cambrian or early Cambrian times. The fact that considerable evolution and diversification of invertebrates had already occurred prior to this time helps account for the relatively sudden appearance of a great variety of shallow-water types. Once in the shallow water where increased density is no longer a liability but often an advantage, the evolution of all sorts of skeletal and muscular systems, without regard to density, would have occurred. Indeed, the study of Cambrian fossils does show a progression from early Cambrian forms with lighter, more fragile shells or exoskeletons to types in the late Cambrian with heavier, better developed ones. It was only with the evolution of forms with heavier bodies and hard parts that fossils began to be preserved on a sufficiently large scale that we are likely to find any of them today.

Fig. 17-1. Floor of Indian Ocean, with mid-ocean ridges, fracture zones, major basins, continental shelves, and extensive mud aprons deposited by great rivers.

CHAPTER 17

Sea floor spreading and continental drift

The theory of continental drift was first espoused by Wegener (1912) and others on rather circumstantial evidence such as the coincidences in shape of the major continents, which are such that one can visualize them fitted rather closely together like the pieces of a jigsaw puzzle. To this evidence were added curious coincidences between the faunas and floras of such continents as would have been joined together in this manner at the time of origin of the major groups of animals and plants of today. Geological evidence in the form of similarities and continuities of geological formations that would fit together in a logical manner in such an arrangement of the continents lent further plausibility to the idea, as did similarities in the fossil faunas and floras of Mesozoic age or older.

Many scientists, felt, however, that the major ocean basins and continental platforms had probably held much the same general relationships to each other throughout the known history of the world, and isostatic changes in the level of land or of areas of the sea floor, together with changes in sea level and the formation of land bridges during certain periods between some of the continents, could account for the coincidences noted by the proponents of the continental drift theory more easily than the postulated migrations of continents. The difficulties of imagining the physical causes and mechanisms that would account for the supposed movements of whole continents seems to pose a greater problem than the finding of alternative explanations for the coincidences noted. In other words, the theory seemed to raise more serious and difficult problems than it solved.

During the past two decades, however, an impressive array of new evidence from several different disciplines has again brought the theory of continental drift into prominence and seems to have demonstrated its essential correctness.

In the first place, a great worldwide system of ridges and rifts showing tensional features, as if the earth were gradually being pulled apart along this system, has been discovered. The Mid-Atlantic Ridge

337

extends the full length of the Atlantic Ocean, passes through Iceland, and continues north into the Arctic Ocean where it bisects the Eurasian Basin. In the Pacific the East Pacific Rise continues the system from the antarctic region to the Gulf of California, from the head of which it continues northward as the San Andreas Fault. Another branch extends north and west from the Antarctic portion of the Pacific system into the Indian Ocean where it runs north into the Red Sea, finally turning back as the Great Rift Valley in Africa. In many parts of the world this rift system has a longitudinal central valley or trench, with the mountains or rises arranged more or less symmetrically along either side of it. It is also characterized by the presence of major fracture zones running at right angles to the main rift system, along which horizontal displacements, sometimes of many miles, have taken place, such as might be expected if lateral movements of great blocks of the earth's crust were taking place, not all at exactly the same rate, along the course of the rifts (Fig. 17-1).

In other parts of the world there are mountain systems running somewhat parallel courses that show compressional features as if they had been pushed up by horizontal pressures in the earth's crust. Along the western rim of the Pacific, for example, is a great system of offshore deep trenches, followed on their western sides by island chains or continental margins showing compressional characteristics.

It has also been determined that these rift and ridge systems are the principal sites of tectonic activity of the world. The majority of all recorded earthquakes are centered along them, as is most volcanism.

Evidence that the volcanic rock found near the axis of the mid-ocean ridges is younger than that found at some distance to either side and that the rocks on either side are progressively older the farther they are from the central rift has come from a study of magnetic anomalies. When rock or sediments are formed, magnetic properties of incorporated magnetic elements are oriented in a definite manner with respect to the magnetic poles of the earth, forming as

it were, permanent compasses oriented according to the position of the magnetic poles at the time the rock was formed. From the study of terrestrial formations, it has become clear that at certain times in the past the magnetic field of the earth collapsed and then was reconstituted. When this happens it appears that the magnetic poles may be reconstituted either in their original positions or in a reversed position. Several such reversals have been discovered and dated by the known ages of the strata in which they occur. Similar correlated magnetic reversals have also been found in some deep-sea sediment cores.

Along either side of the mid-ocean ridges, symmetrically placed patterns of magnetic reversal have been found corresponding to the magnetic reversals already known from their vertical locations in strata and cores from other regions. Along the sides of the mid-ocean ridges, however, these patterns of magnetic reversal are laid out as horizontally disposed bands paralleling the axis of the ridge systems. This would seem to indicate that new rock has been forming along the axis of the ridges and has been transported laterally in both directions. The distances between successive magnetic reversals, together with the known dating of the reversals with which they have been correlated, give us a means of estimating the rate of this lateral spreading of the ocean floor. The rate, for the most part, seems to be about 1 to 4 cm. per year. Where the spreading is relatively fast, as along the East Pacific Rise, the resulting submarine topography is one of gently rising and undulating hills. Along much of the Mid-Atlantic Ridge, where the rates of spreading are somewhat slower, much steeper, more rugged mountain ranges have been formed, paralleling the axis of the rift on both sides, apparently because much more new lava is poured out and piled up along the axis of the rift before it is carried away by the lateral movements of the earth's crust.

Other lines of evidence pointing to the same conclusions lie in the lack of, or thinness of, eupelagic sediments near the axis of the ridge systems and their increasing thickness laterally. The thinner sediments contain only the recent top layers of the marine sediment, but the thicker sediments at greater distances from the axis contain progressively older sediments as well.

Even the history of continental faunas seems to bear out the idea. The fact that, in a much shorter geological period of time, mammals and birds have each diversified into about twice as many orders as did the reptiles over a much longer time span has been attributed to the fact that during the period of the origin and early diversification of the reptiles, there were only one or two great land masses. Hence there was less isolation of different populations of the more primitive reptiles. In the case of the mammals and birds, however, the continents drifted apart shortly after the origin of the groups, isolating different populations of their more primitive members, each of which proceeded to undergo adaptive radiation to about the same extent that the whole group would have done had the populations not been isolated.

The crustal movements seem to be caused by gigantic convection currents in the mantle beneath the earth's crust. The material being brought to the surface at the mid-ocean ridge systems should not be thought of as pushing the rest of the sea floor away laterally. Rather, widespread movements of great segments of the earth's crust, which floats on the more plastic material of the mantle, are pulling the crust apart along these lines, and the new crustal material originating there can be thought of as a sort of compensatory upwelling. In some regions land masses are included with the parts of the crust being rafted. Along the western Pacific the corresponding downwelling occurs just to the east of the main continental mass, producing the deep trenches where the mantle turns downward and pulls the crustal material with it back into the mantle. The pressures of this downturning current, which perhaps in part is deflected downward by the continental platform and which perhaps meets another, similar current of the mantle flowing in the opposite direction beneath the continental platform, produce the island chains and mountain ridges to the west of the trenches, which show such marked compressional characteristics.

There is some evidence that after great trenches such as those along the eastern sides of the Pacific island chains are formed, they eventually become filled with sediments and, for reasons not yet clearly understood, are later raised, forming new island chains. New deep trenches are formed to the seaward. The shallow seas between the earlier island chains and the mainland eventually also fill with sediments, thus extending the continental border.

Be that as it may, it is now abundantly clear that the earth's crust is not as stable as had formerly been supposed. Present continents have not always occupied their present relative positions. There has been, at least since the end of the Mesozoic period, an extensive exchange of materials between the earth's crust and the mantle, involving for the most part crustal material of the floors of the oceans where the crust is thinnest. The major continental platforms, although they have undergone changes in position and, of course, erosional and tectonic changes of their surfaces and edges and even perhaps been split apart, have nonetheless existed as recognizable entities for a far longer time than have the present floors of the oceans. The Atlantic Ocean is a relatively new ocean, formed in the gap as the American continents and the great Eurasian-African land masses were rafted apart.

This explains why deep-sea sediments have not formed layers approaching the thickness calculated for them on the basis of known rates of sedimentation and assuming relatively stable oceanic basins since at least Cambrian times. In a way, this is rather disappointing to paleontologists. It had formerly been felt that, since the sea floors are relatively quiet regions protected from most of the rapid erosional and other changes so noticeable on and near land, they must have, locked in their deeper sediments, a far more complete and far better record of the past world history than could be found on land. Now, just as our technology is developing to the point where we can look forward to obtaining good samples from these ancient oceanic sediments, we find that they do not exist—that the present floors of the oceans, for the most part, date only from about the end of the Mesozoic era and that all their priceless, more ancient sediments have slipped away forever beneath the earth's crust, returning to the mantle whence they came and from which, even now, new ocean floor materials are rising and spreading out along the mid-ocean ridges.

For very ancient fossils and geological history of the earth and its oceans, we must then continue to rely upon sediments formed where ancient seas flooded areas of continental platforms, laying down sediments that have been outside this great circulation of crustal and mantle materials but that have been subjected to the vicissitudes of terrestrial environments.

PART VIII

The *IMPACT* of *MAN*

Never before in the history of this planet has its thin life-supporting surface been subjected to such diverse, novel, and potent agents. I believe that the cumulative effects of these pollutants, their interactions and amplification, can be fatal to the complex fabric of the biosphere.... I believe that continued pollution of the earth, if unchecked, will eventually destroy the fitness of the planet as a place for human life.... Many of the pollutants—carbon dioxide, radio-isotopes, pesticides, and excess nitrate—are invisible and go largely unnoticed until a lake dies, a river becomes foul, or children sicken."

Barry Commoner. 1966. Science and survival. The Viking Press, Inc., New York.

Fig. 18-1. Manner of fishing with a flying lure at Banda. (From Weber, M. 1902. *In* Reports of the Siboga Expedition, 1899-1900. Vol. 1. Introduction and description of the expedition. E. J. Brill, Leiden, Netherlands.)

CHAPTER *18*

Utilization of marine resources

*E*ver since the dawn of man's history the seas have served as a source of foods and other products for coastal populations, supplementing the use of terrestrial resources that have always been the principal sustenance for our species. Ancient Indian kitchen middens along the Pleistocene marine terraces of the Pacific coast attest to the use of *Cryptochiton, Kathrina,* clams, fish, and in some places marine mammals.

As man's numbers and technical proficiency increased, so too did the utilization of marine resources. However, our utilization of ocean resources has remained relatively primitive, corresponding to the preagricultural gathering and hunting phases of man's development on land. Although the total amount of photosynthesis in the oceans is comparable to, or perhaps a little greater than, that on land, phytoplankton is difficult to harvest in significant amounts, the effort expended exceeding the return. In the seas we have confined ourselves, with a few minor exceptions, to skimming off the animals at the higher trophic levels rather than managing them on an efficient basis,

which might ensure continued returns. Meanwhile, our methods of gathering and hunting have greatly improved in both scale of operations and efficiency. The typical response to decreased returns due to overfishing has been to increase the fishing effort and the efficiency of fishing methods.

During the past century mankind has entered into the terminal phases of the logarithmic population expansion that characterizes any population of organisms confronted with an open environment—that is, an environment in which the carrying capacity for that species vastly exceeds the population of the species. This phase of logarithmic population growth is invariably followed either by a period of equilibrium in which the numbers become more or less stabilized or by a period of decline that may be more or less sharp depending upon the extent to which the increase in the species destroys vital features of its own environment. In the case of man the situation has been greatly exacerbated by rapid technological developments just as he was coming into this phase of his biological history, which, in effect, opened up the environment repeatedly and also suddenly lowered death rates, resulting in a currently much greater than normal terminal population increase. The discovery of and expansion into the New World continents, together

343

with the rapid technological developments that have enabled man to exploit levels of his environment hitherto unavailable to any animal, have enabled him to continue to expand his population and exploit his environment to a degree never before even remotely approximated by any large animal throughout the world's history.

The rapidity with which basic changes in his way of living have taken place has made man also the most bewildered and confused species of animal ever to exist on the earth. It has endowed many individuals who have been in a position to benefit from this rapid expansion with the naïve Alice-in-Wonderland attitude that such rapid expansion is a normal and desirable state of affairs and that our technology can be depended upon to continue to open up more and more new facets of our environment for exploitation as we deplete the old ones.

Suddenly now, we begin to realize that this is not the case—that we have now reached a major ecological crossroads, that a high standard of living, or perhaps any living at all, will depend upon our ability to bring the world population of mankind into a balanced equilibrium at a much lower numerical level than the present 3.5 billion. Such a change will, of course, entail complete reorganization and change in our economics, politics, religion, and philosophy, which, in the Western world at least, are geared to a steady state of expansion.

In such a grave crisis, it is natural that most people first turn to the old approach that served us so well in the past and begin to look about for new facets of the environment to open up and exploit, for new technological developments that will, like a magician, pull more rabbits out of the hat for us to eat.

Thus, all the great powers and major economic interests are now turning eagerly to the oceans as the last great reservoirs of untapped resources and are also busy with plans to decimate the few remaining terrestrial wild areas, such as the Amazon Valley and the jungles of western New Guinea, equatorial Africa, and Borneo, to squeeze what can be squeezed out of them.

At this stage of our social and biological evolution when populations are on a runaway collision course with disaster and there are no effective plans in sight to limit them, other than by widespread starvation or nuclear war, and in the present state of international economic and political anarchy when the major effort of all great powers is devoted to sterile, wasteful military posturings and confrontation, the discovery now of any further really large new, untapped resources could be a tragic development for mankind rather than a blessing.

Exploitation of new ocean resources can provide constructive outlets for the energies of a rather large number of people, but the overall effect could be to continue to divert men's attention from the directions in which possible solutions to the human dilemma actually lie. The resources discovered will be seized upon by the presently powerful economic and military interests to increase their power, increasing the gap between the haves and the have-nots. The bulk of any such resources will be wasted by the military instead of being used in rational ways. It could simply mean that if the time ever comes when man does bring his populations into planned and viable equilibrium with his environment, that environment will have been already much more thoroughly impoverished, and man will have to exist at a far lower level than would otherwise have been the case. If he doesn't bring his populations into a viable equilibrium, such discoveries will simply act as temporary palliatives, enabling the present disastrous increases to continue a little further and making the scale of starvation or of killing by war or pestilence just that much greater but not much farther off in the future.

The only certain element in this clouded picture of man's immediate future is that, barring nuclear war, populations will continue to increase rapidly for a few more years, after which they will be greatly reduced or exterminated.

The first question that arises in connection with increased utilization of marine resources during the terminal agonies of population increase is: To what extent can

Table 18-1. Ocean cropping*

Commercial fish catch for 1964 (including invertebrate fisheries, but not including sports fishing and unrecorded fishing in many areas of the world)	$\sim 5.2 \times 10^7$ tons fresh weight
Commercial fish catch in terms of carbon	$\sim 3.9 \times 10^6$ tons
Annual carbon productivity of the oceans	$\sim 5.3 \times 10^{10}$ tons
Maximum possible carbon productivity	$\sim 1.6 \times 10^{11}$ tons

*Assuming a three-link food chain with 5% efficiency of transfer, an annual cropping of 3.9×10^6 tons of carbon as fish requires 3.2×10^{10} tons of carbon as living organisms as an ecological basis. Conclusions about possible expansions of ocean fisheries must be based on estimated annual carbon fixation. The figure for maximum possible productivity is merely a hypothetical figure based on an assumed 100% utilization of incident light, with no limitations imposed by amounts and rates of turnover of phosphorus and nitrogen.

present ocean fisheries be expanded to provide additional food? Table 18-1, modified from one supplied by Dr. Vishniac and updated to 1964, should be studied in conjunction with Tables 1-2, 1-3, and 1-4.

From Table 18-1 it can be seen that the hope of feeding our expanding populations by a severalfold expansion of ocean fisheries is not within the realm of possibility. At most the fisheries can be only not quite doubled before reaching the limit imposed by the total annual productivity of the oceans.

Actually, the potential level of expansion of ocean fisheries is less than that indicated by the tables because a considerable fraction of ocean production goes into species not readily exploited on a commercial basis and because of the fairly extensive unrecorded fishing that makes the actual present cropping somewhat higher than that shown.

In other words, we are already cropping the ocean fisheries dangerously close to the upper limits of biological possibility. Only wise and careful management practices will enable us to maintain even the present levels of cropping without destroying more species most useful to us, as has already happened to some.

Present international conventions making the oceans into regions of free-for-all competitive exploitation by peoples of all countries, in whatever manner their whims and those of their governments dictate, are not conducive to good management practices. Limited international cooperation in con-

servation has been obtained on a voluntary basis in connection with a few fisheries clearly threatened with almost immediate destruction. The work of the International Whaling Commission is an example. The Food and Agricultural Organization of the United Nations has also done much good work, especially in bringing together and making available comprehensive studies of the biology of important species. These efforts need to be greatly expanded, the conservation regulations made more enforceable, and the regulatory agencies put under the charge of competent biologists whose primary concern is the preservation of the ecosystem rather than the extraction of the maximum annual profits from it and who are not directly dependent in their employment upon the industries they are supposedly regulating.

More efficient use of so-called trash fish and parts of fish commonly wasted, for the manufacture of high-protein fish flour, offers the possibility of a significant augmentation of our benefit from fisheries without a correspondingly large increase in the present level of fishing, as does also increased use of canned, frozen, and other forms of preserved fish.

The avenues open for obtaining greatly augmented supplies of food from the oceans lie in large-scale "farming" of certain species and in the harvesting of organisms low in the trophic hierarchy. One such organism, for example, is the antarctic krill. The size and shoaling habits of this immensely abundant, nutritionally rich crustacean should make it a favorable object for commercial exploitation. Now that we have very largely eliminated the great baleen whales, perhaps man will come to occupy their ecological niche, so far as the krill are concerned. Commercial farming of bivalve molluscs, which feed directly upon the plankton, can probably also be very considerably expanded, as well as the farming of certain crustaceans and edible seaweeds.

Farming and fishing the seas for food will increasingly have to compete with other aspects of exploitation of the seas that may not be compatible, such as underwater mining, oil drilling, and perhaps military uses, as well as to cope with the generally increased destruction through pollution and alteration. Most of the best fisheries are in shelf areas, relatively near land, where they are subjected to a maximum of such influences. In still other cases, even though the adult fish may occur well offshore, the eggs or young are dependent upon nearshore environments highly susceptible to destruction.

In localized areas with sandy or muddy substrate, the production of fish can sometimes be much increased by artificial reefs, providing habitats for rock-dwelling species of invertebrates and seaweeds, and attracting desirable species of fish. The creation of suspended artificial reefs from systems of floating platforms or buoys farther offshore may also offer interesting possibilities. This method might be combined with a system of fish traps that would make harvesting the fish easy and perhaps serve as sites for large-scale farming of useful bivalves such as mussels or races of oyster adapted to the open-ocean salinities.

Another way in which the oceans may make a greater contribution to man may be through furnishing fresh water on a large scale in certain areas. A variety of schemes for desalinization have been proposed, most of which suffer from the fact that the energy required to accomplish the desalinization plus the disposal problem created by excessive quantities of concentrated brine or salt, produced as a by-product, make them impracticable except in very localized situations where fresh water is at a high premium. It has even been suggested that huge icebergs might be towed to the vicinity of metropolitan areas such as Los Angeles to be used as sources of fresh water and to provide a cooling effect.

One of the more interesting proposals that may be feasible in warm areas with moist trade winds most of the year was made by Gerard and Worzel (1967). They propose to use the power of the wind, by means of big windmills, to pump seawater up through large-diameter pipes from below the thermocline, at about 9° C. This cold water would be run through very large condensers located onshore, where they would intercept the humid wind. According to their calculations, a condenser 200 meters long by 10 meters high should deliver about a million gallons of fresh water per day. There would be no harmful brine or salt incrustations, as in desalinization schemes. The deep marine water used could be discharged into a nearby lagoon, where it would increase the fertility of the water, raising its productivity and supporting various schemes of aquiculture. The big condenser would act as an air conditioner for any settlement to the lee of it, supplying more pleasant cooler, dryer breezes than those which were customary in the region. It might even be possible to recover some power by using the water descending from the condenser to turn a turbine or perhaps by taking advantage of the temperature difference between the deep water at 9° C. and the surface waters at 20° C. to drive a turbine by boiling propane at surface-water temperatures and then cooling and recovering the propane by using the cooler deep water. The cycling of the propane would obviate any need for continual supply of more fuel. There has even been a suggestion that if such an installation were built on one of the larger Pacific coral atolls and the used deep water discharged into the lagoon, the lagoon might become fertile enough to support the farming of whalebone whales or in any case to support a greatly augmented fishery.

CHAPTER *19*

Man's impact on marine environments

Fig. 19-1. Crews near Santa Barbara, California, trying in vain to save the beaches from oil pollution by casting great quantities of hay into the water to soak up the oil. Later burning of some of this oil-soaked hay contributed to air pollution. (Wide World Photos, Inc.)

Since man is a terrestrial animal, the effects of his extraordinary proliferation and recent developments in technology have been most marked on land. Here we are involved in the greatest, most rapid redistribution and change in the faunas and floras of the world that has ever occurred. Very large numbers of the dominant elements of the flora and fauna of vast regions have been suddenly eliminated and replaced by domesticated types of crop plants and ornamentals and by domestic animals molded and shaped according to the whims and desires of man and bearing little resemblance to any "naturally occurring species." Plants and animals have been introduced on a large scale into regions where they did not formerly occur, with concomitant upsets and repercussions for the entire ecosystems of those regions. Dry regions have been irrigated, wet regions drained, water-table levels changed, and great areas sprayed and dusted with long-lasting poisons that have entered the geochemical cycle of the entire world, producing sometimes devastating, sometimes subtle effects that we can hardly begin to assess as yet. Erosion and pollution have so changed the waters that the rivers and streams of entire continents have taken on the character of open sewers, carrying off so much in the form of plant nutrients, soil, sewage, and industrial wastes that they can only be regarded as hemorrhages of the land rather than part of its normal circulation. Fossil fuels are being burned at such a rate that the character of

347

the atmosphere is being radically changed in a manner that threatens to get entirely out of hand and perhaps eliminate the present world biota altogether. Radioactive materials are being created in an altogether new order of magnitude. The level of background radioactivity has already been appreciably raised throughout the world. The proliferation of nuclear weapons and of nuclear industrial plants and laboratories is such that it is no longer just a high probability, but a virtual certainty, that either by accident or by design, large regions will be rendered uninhabitable and the levels of radioactivity throughout the world raised to such heights that they will exert profound effects on whatever life, if any, remains.

The changes that have taken place in the oceans to date are not nearly so dramatic or obvious as those on land, but they are important, and since our technology is now making possible large-scale industrial utilization of marine resources, they will increase rapidly in scope and effects in the immediate future. Many of these changes are secondary effects of changes taking place on land. Others are the result of the direct intervention of man in the oceans. We shall consider only a few of the changes very briefly, as examples.

OVEREXPLOITATION

Depending upon the particular circumstance and the nature of the resource, the results vary all the way from a slow decline in the abundance and availability of the resource to its abrupt and permanent elimination or extinction. Examples are legion. The Steller's sea cow was completely wiped out only twenty-seven years after its discovery. The great auk, or arctic penguin, was likewise exterminated. The elephant seal, the sea otter, the white albatross, and many other once-abundant sea animals have been brought to the verge of extinction. Several species of seals have been eliminated over the greater part of their former ranges. The great fish, such as swordfish, sailfish, and tarpon, are threatened.

In fisheries the usual signs of over fishing are a decline in the average size of the fish of a given species that are taken and an increase in the intensity or efficiency of the fishing effort necessary to take the same catch. The normal response to a declining fishery is just such increases in fishing effort, together with improvements in the efficiency of the gear or fishing methods and expansion of the area fished, if possible. Extensive trawling for demersal fish in limited areas such as the Grand Banks or the banks of the North Sea, which have long been celebrated fishing grounds, can be especially destructive because the big trawls are destructive to the entire benthic community upon which the fish depend. When World War I forced a four-year halt in fishing operations in the North Sea, the fish were both larger and more numerous at the resumption of fishing than they had been just before the war, demonstrating that the fishery was already on the decline at that time.

The history of the whaling industry is typical and may serve as an example. Serious whaling began in the North Atlantic in the sixteenth century. The Atlantic right whale was the first species to come under heavy attack. This was an abundant, large, black, oil-rich, 60-foot whale, termed the "right whale" because when harpooned and killed, the carcass would float, whereas certain other whales, "wrong whales," would sink and be lost. By the end of the seventeenth century the Atlantic right whale had practically disappeared. Whalers turned their attention to the Greenland right whale, a species with abundant and valuable baleen as well as oil and meat. Soon these, too, were nearly exterminated. About 1862 a Norwegian whaler invented the harpoon gun with grenade, with which whalers could kill even the gigantic blue whale. Power-driven catcher boats with pumps for pumping air into the carcasses immediately after the kill came into use, making all species of whales easy prey. Whaling continued for a time west of Greenland, around Baffin Bay, but by the early part of the twentieth century the industry was dead in the North Atlantic. The large whales had all but vanished from the Northern Hemisphere. Whalers, with much improved equipment, now turned to the great antarctic summer feeding grounds. Modern equipment includes huge factory ships that serve as mother ships for fleets of smaller whaling boats. Helicopters and radar devices are used to spot the whales. Each whaling boat quickly slaughters whale after whale, pumps air into the carcasses so that they float, marks them with flags or even mounts radio transmitters on them, and then, at the end of the day's work, comes back, affixes lines to them, and tows the whole group to the factory ship. During the first half of the twentieth century whalers slaughtered more than five times as many whales as had been killed by all the famous nineteenth-century whaling fleets combined. By 1932 it was obvious to all that the whaling industry was in serious trouble and would soon be out of business if this continued. The International Whaling Commission was set up with the function of obtaining as reliable data as possible on the numbers of different kinds of whales and setting quotas for each species every year, with a view to conserving threatened species. The Commission is handicapped by inability to enforce its quotas or even to assign quotas for particular nations. It simply sets an overall quota for each species, after which representatives of the countries concerned negotiate the division of the quota among their whaling fleets. It seems to be increasingly difficult for them to reach agreement. Furthermore, not all the nations subscribe to the authority of the Commission. Some whalers who abide by the quotas during the summer in the Antarctic also take whales at other times during their migrations or at their breeding grounds.

The gigantic blue whale, the largest animal that ever lived, is now on the verge of extinction. It is now supposedly under complete protection, but this action was probably taken too late. The blue whale ranges over the whole open sea, from the Antarctic to the equator. It appears that the population has been reduced to such a low level that too few of them are able to find mates and reproduce often enough to maintain the species.

The gray whale, which feeds in the Bering Sea and migrates to Lower California to bear its young in the warm, shallow lagoons along the desolate coasts there, has had a somewhat different history. It too was at one time nearly exterminated but, after being protected for several years, has made a rather good comeback. The principal threat to it now is not from whalers but from alterations in Scammon's Lagoon, the principal breeding ground, due to dredging and other activities of a Mexican salt company there.

One of the most dramatic cases of the sudden failure of a great fishery occurred in the Pacific sardine fishery centered at Monterey Bay, California. During its heyday, that fishery supported a large fleet of fishing boats and a number of large canneries located along the edge of the bay near Pacific Grove, known as Cannery Row. During the mid-1930's over half a million tons were taken annually. In the late 1930's and early 1940's there was a sharp decline, and the catches became more irregular. By 1947 the fishery was clearly threatened, and abundant money suddenly became available from both private and governmental sources for research programs on the biol-

Fig. 19-2. Pacific sardine. Supported a once-great fishery at Monterey, California. Actually, the Pacific "sardine" is a species of pilchard, *Sardinops caerulea.*

ogy of the sardine. The California Co-operative Sardine Research Program was organized in 1949, eventually involving most of the interested institutions and agencies along the Pacific coast. By 1952 the canneries had begun to close down, and soon the industry was dead. Most of the equipment was sold to firms operating in South America, and part of Cannery Row was converted into a West Coast branch of Ward's Biological Supply House.

Data developed through the efforts of the Cooperative Sardine Research Program seem to indicate that in this instance the causes of the catastrophic decline were complex, involving extensive hydrographical changes in the North Pacific as well as the activities of man. During the years in which the sharpest decline occurred, average water temperatures of the region were from 1° to 3° cooler than during the good years of the fishery. The sardines shifted their spawning southward, off Baja California, and also decreased sharply in abundance.

The hatching time of sardine eggs is retarded from fifty-four hours at 17° C., which is about optimum for them, to seventy-three hours at 15° C. This increases significantly the period during which the eggs are subject to predation. The cooling also slows the growth of the young fry, increasing their period of greatest vulnerability. G. Murphy (1961) estimated that the prolonged period of vulnerability caused by a 3° decrease in temperature could decrease survival by a factor of ten.

Furthermore, zooplankton was much more abundant in the water during the cool years. Many zooplankters are predators of the fish eggs and young fry, so that predation pressure on these vulnerable stages actually increased at the very time their period of vulnerability was being prolonged due to the slowing of the rates of development.

Intensified fishing effort during the period when the decline was beginning also diminished the adult reproductive stock at the same time that their replenishment was falling off because of the greatly decreased survival through the early stages.

In 1957 oceanic temperatures in the North Pacific again began to rise, rather abruptly. Southern species of fish and other organisms began to appear farther north than before. Episodes of attack by sharks farther north were reported. The coastal sea level rose approximately 5 inches, and there was a slight upsurge in sardine catch. The abruptness of this general rise in ocean temperatures and level shows that mechanisms of interaction between ocean and atmosphere can produce large-scale effects more rapidly than had been supposed and in ways not yet well understood.

Changes in "ocean climate," even of only a few degrees, may have marked effects on the distributions and abundance of pelagic organisms. Fishing pressure by man may, for heavily fished species, greatly increase the probability that slight ecological changes will have catastrophic effects for the species. Clearly, problems of conservation and management in oceanic fisheries are not simple, nor are the results always predictable. A good, easily readable discussion of this subject is given by Hedgpeth (1962, 1968) in his revised editions of Ricketts and Calvin's *Between Pacific Tides*.

CONSERVATION—TOO LITTLE AND TOO LATE

The following is the usual progression of events when natural resources are threatened by overexploitation: While a resource is abundant, those engaged in exploiting it resent and resist the "meddling" of anyone who wants to study it carefully, to recommend less wasteful ways of exploiting it, or to suggest regulations for conserving it. Only when the industry is clearly and imminently threatened do they call upon and support the activities of such persons, often too late. At such times the larger operators, who can afford to do so and who have the greatest investment to lose if the industry fails, become very public spirited and conservation conscious, eager to set the industry up on a sustained-yield basis. Smaller operators who cannot stay in business under suddenly more stringent conservation regulations are forced to sell out at a loss. In the event that enough is salvaged to enable the industry to continue its operations, the large operators, now on a sustained-yield basis or more nearly so, in all sincerity claim and are credited with far-sighted, public-spirited vision for having adopted or submitted to the regulations that alone enable them to continue operation—regulations which should have been put into effect when the industry was first opened up and which these same operators may have fought tooth and nail as long as it was profitable for them to do so. Because of their great wisdom, farsightedness, public spirit, and experience, they go on to maintain that it is they, not some outside regulatory body, who should have the decisive voice with respect to planning and regulating further resource exploitation in their areas of operation.

Since it is now becoming rapidly clear that all major resources, without exception, are becoming threatened, we can hope for a great upsurge of interest in conservation and in social regulation of major industries. The lack of any international authority capable of enforcement, the distrust and rivalry between countries, the newness of many of the enterprises, and the fact that much of the damage they do will be under water, where few will be aware of it, will make the proper planning and regulation of ocean industries exceedingly difficult to achieve.

CHANGES IN THE ABUNDANCE AND DISTRIBUTION OF ORGANISMS

The removal of any important species from the scene or its relegation from a position of dominance to one of scarcity is never an isolated phenomenon. Every such change also has far-reaching effects throughout the entire ecosystem. Changes in the relative abundance of organisms closely linked to it in the food cycle follow immediately, and from these, still further changes. Some species may be eliminated, and others may be caused to proliferate as never before. Parasites whose life cycles involve some of these species may also be greatly affected. So complex and intertwined are the relationships within the ecosystem that any changes within it or changes in the pressures upon it have repercussions throughout the entire system that are difficult to predict.

An example of the far-reaching effects of changes in the abundance of species is the current explosive growth in the populations of the formerly rare coral-eating *Acanthaster planci,* the well-named "crown of thorns" (Fig. 8-1, p. 164). This starfish has already destroyed 100 square miles of Australia's Great Barrier Reef and most of the coral along a 24-mile stretch of the coast of Guam and is decimating corals at places as widely scattered as Borneo, Fiji, Palau, Saipan, Wake, and Midway.

If this devastation continues unchecked, it could destroy the entire system of marine life upon which millions of islanders depend for their protein food, subject the islands to erosion from which the reefs now protect them, and remove one of the world's most far-flung, gigantic biocenoses, unsurpassed for variety and beauty and for zoogeographical and economic importance. The effects would be worldwide—a staggering loss from both the economic and aesthetic standpoints.*

Although the reasons for the great proliferation of *Acanthaster* have not been precisely pinpointed as yet, most biologists concerned about the problem feel that the most probable explanation lies in the elimination of some planktonic microfeeding species that formerly fed on the larvae. This could have been caused by DDT or by radioactive residues from atomic tests. The reduction or elimination of triton, a large predatory gastropod that preys on adult *Acanthaster,* by shell collectors has also been mentioned as a contributing factor in some areas. Widespread dredging, channel blasting, and other projects that have been disrupting the coral habitats on a constantly increasing scale ever since World War II provide abundant new surface areas of killed coral on which the larvae of *Acanthaster* can safely settle. The increased devastation of corals in the area by enlarged populations of *Acanthaster*

*A recent news item since this went to press indicates that narcotic agents in Australia have arrested drug addicts who are getting kicks by smoking preparations of dried *Acanthaster* organs. It would be ironic if such a great demand for this should develop that a new widespread wave of drug addiction and smuggling might play a significant role in saving the threatened Pacific coral reefs.

then permits this process to become self-augmenting, leading to rapid destruction of the reef.

Reefs that have been destroyed have never been found to come to life again.

Through both deliberate and inadvertent introduction of organisms into areas where they did not previously exist, an accelerating process because the mobility of man has increased so sharply, we are now involved in one of the fastest, most extensive distributional revolutions in the history of the earth. Like the elimination of key species, the successful introduction of new species into an area also has far-reaching, often unpredictable effects.

So-called fouling organisms—the communities of barnacles, mussels, algae, bryozoans, hydroids, tube worms, tunicates, and other, smaller associated organisms that grow on ship bottoms, along with boring organisms such as shipworms and gribbles—have been carried all over the world. As a result, many of them have established themselves in new areas, sometimes displacing elements of the indigenous biota. Among those most noticed by man have been the wood-boring species. It is probable that man is chiefly responsible for both the abundance of these forms and their worldwide distribution. Prior to the advent of man the marine environment contained very little wood—only the occasional trees washed into the sea by rivers or toppling into the margin of the sea from forested sea cliffs. Its occurrence in most ocean areas was so rare and so fortuitous that it could hardly have supported widespread thriving populations of specifically wood-boring marine animals. Their occurrence must have been very limited. Man, with his extensive lumbering, his wooden-hulled ships, his pilings and other wooden structures built along the margins of the seas, has for the past two or three thousand years provided a vastly augmented, steadily increasing supply of wood in the marine environment, together with free transportation for wood borers. What were previously very minor species of limited distribution suddenly were presented with an open environment and have become important pests on a worldwide scale. Shipworms have been important pests for all early commercial and naval vessels with wooden hulls from before the time of the Roman galleys. In the early 1700's they damaged the dikes of Holland so seriously as to gravely threaten the Netherlands with inundation. The first careful study of the group was the treatise of Sellius (1733), written at that time, in which he proved that shipworms were molluscs. He did basic anatomical work on them and reviewed all that was known about them. Another, more recent spectacular case of shipworm damage occurred after introduction of the worms into San Francisco Bay sometime before the First World War. The worms spread unnoticed and rapidly among the thousands of pilings of piers, wharves, railroad trestles, marine warehouses, and other structures along the bay front. Many of them began to collapse by the 1920's, causing millions of dollars worth of damage. After a symposium on marine borers held at Friday Harbor in 1957, a new worldwide testing program was instituted in which studies are being made from samples collected on a worldwide basis and methods of protecting wooden structures are being systematically evaluated.

The construction of canals has provided another means by which various species have become redistributed. The Suez Canal from the Mediterranean to the Red Sea, completed in 1869, was for a long time not an effective avenue for interchange of faunas because of the high salinity of the Bitter Lakes, through which it passes. Over the years this salinity has been decreased through interchange with waters by way of the canal and has been reduced from about $68^0/_{00}$ to $45^0/_{00}$, so that in recent years passage of several species has become possible. Most of the exchange has been from the Red Sea into the Mediterranean because tidal movements from the Red Sea carry organisms far up the canal from that end but not in the reverse direction at the Mediterranean end. Over 150 species of Red Sea animals now occur in the eastern Mediterranean, nine of them commercially exploited.

The Welland Canal permitted the Atlantic sea lamprey to invade the Great Lakes. About 100 years after its construction the lampreys underwent an explosive increase in the Great Lakes that destroyed the valuable whitefish and trout fisheries. The governments of the United States and Canada have spent about $16 million on control during the last twelve years, and of course the financial loss to the fishing industry was far greater.

At present there is under consideration by the United States a plan to construct a new sea-level canal through the Isthmus of Panama, possibly using nuclear explosives to assist with the earth moving. This is viewed with alarm by many biologists who feel that it may set off a chain reaction of major ecological adjustments extending across the oceans. It may also change the physical environment especially along the Pacific side by allowing warm Caribbean surface water to spread out over a large area now cooled by upwelling produced by offshore winds, producing effects comparable to El Niño along the coasts of Ecuador and Peru (p. 319). It is felt that in any case a major intervention of this sort in the world ecosystem is too grave a responsibility for unilateral decision by one country and that an adequately financed international commission of marine biologists and ecologists should be set up first to survey the region carefully, to carry out experiments on the mixing of species from both sides of the isthmus, and to monitor the changes that occur if and when the canal is constructed. Such a commission should be multidisciplinary, independent of any single government or governmental agency, university, or private concern, and should be well funded. It should be in operation for several years before the final decision to build such a canal is made.

The commission itself should make the decision—not simply be advisory—and if they decide to build the canal, it should also have general supervision of the actual construction, including such decisions as whether or not to use nuclear explosives. Military considerations should be allowed no role whatsoever in making the decision or in determining the characteristics of the canal. **The only acceptable reasons for deliberate large-scale interventions in the world ecosystem are great potential benefits to mankind as a whole.** Adequately financed, completely independent international commissions of competent scientists, empowered to make the decisions in question and to supervise their implementation, offer a more rational approach to such problems than do unilateral decisions made by competitive heads of state and military leaders.

It would be far better to put up with some delay and inconvenience in shipping for a few years than to blunder into a large-scale ecological disaster. Even if the ecological results of such a canal should be beneficial rather than disastrous, it would be better to have the biological and oceanographical studies made, so that a maximum return of scientific knowledge and understanding could be gained from the venture. Furthermore, this procedure would set a first example in deliberate cooperative international planning and decision-making based on the best scientific knowledge obtainable. This kind of attitude and cooperation is sadly needed and could in the long run be of far greater benefit to mankind than the canal itself.

The possible use of nuclear explosives during the construction of the canal is another facet of the problem with grave implications that should not be decided hastily or unilaterally. Whether or not to use such methods should not be decided by physicists and engineers or politicians interested in making the tests for their own sake but by biologists whose primary concern is environmental quality, after careful study in each instance where this is proposed.

POLLUTION

There are five principal types of pollution of major importance in connection with marine environments: (1) changes in estuaries, bays, lagoons, and coastal habitats through pollution from the land and through dredging, construction, shipping, and other activities, (2) worldwide dissemination of long-lived pesticides, (3) oil pollution, (4) worldwide contamination by radioactive materials, and (5) changes in the earth's atmosphere that may affect the carbon dioxide–oxygen–carbonate balance in a manner destructive to the entire world biota. We shall consider each of these classes of change very briefly simply to point up something of the nature and scale of the changes being made.

Even ordinary pollution with only local municipal sewage and wastes may have marked effects on the coastal biota. For example, Table 19-1, compiled from data presented by Leighton and Boolootian (1963) in connection with an entirely different sort of study, compares the growths of marine algae and eelgrass at two comparable sites on opposite sides of Santa Monica Bay, California, both of which formerly

supported flourishing growths of these plants. Point Dume has remained relatively free from local pollution, while the Flat Rock area has been subjected to municipal pollution.

From Table 19-1 it can be seen that every species of marine plant noted underwent noticeable diminution in amount or was wholly eliminated, with the sole exception of the formerly rare coralline alga, which came to dominance in the polluted area. Any such marked change in the local flora would, of course, entail great changes in the abundance and species composition of the fauna, quite apart from the direct effects of the pollution on the animals.

Bays, estuaries, and coastal lagoons not only support important fisheries in their own right, close to the land and readily available, but also serve as nurseries for the eggs and very young stages of numerous other fish that are usually found offshore as adults. The young stages of the fish and also of invertebrates such as crabs, bivalve molluscs, and other valuable animals are often particularly sensitive to alterations in their environments.

Extensive deforestation and agricultural activity, even far inland, results in erosion that greatly increases the sediment load and adds materials dissolved from fertilizers and insecticides to the water of most rivers and streams. Sewage and industrial wastes from towns and factories or mills add a variety of metallic ions, detergents, organic matter, acids, and other poisons to the water, as well as reducing its oxygen content. When the waters from such overloaded, partially sterilized rivers come to their estuaries, the excessive sediment in the water tends to smother gill-breathing animals and to choke nonselective sediment feeders, and the metallic ions, acids, or insecticides, sometimes even in only trace quantities, may wipe out sensitive embryonic and larval stages of many animals. The deposition of excessive sediment may smother much of the infauna.

Impoundment of water by huge inland dams for flood control, by irrigation projects, and by hydroelectric developments also radically alters the stream flow and the character of the estuaries into which they

Table 19-1. Effects of municipal pollution in Santa Monica Bay

Point Dume	Flat Rock
Very abundant species	
Gigartina canaliculata	Scattered occurrence
Rhodoglossum affine	Rare
Common but with discontinuous distribution	
Pelvetia fastigiata	Rare
Egregia laevigata	Rare
Gigartina leptorhynchos	None
Gigartina spinosa	None
Polysiphonia pacifica	Rare
Phyllospadix scouleri	Scattered occurrence
Of scattered occurrence	
Eisenia arborea	None
Macrocystis pyrifera	None
Dictyota flabellata	Rare
Cystoseira osmundacea	None
Gelidium purpurascens	Rare
Pterocladia pyramidale	Rare
Zanardinula cornea	None
Bossea sp.	Rare
Enteromorpha sp.	Rare
Ulva sp.	None
Rare at Point Dume	
Coralline alga	Very abundant
Codium fragile	None

discharge, as well as interfering with the spawning migrations of anadromous fish such as salmon. Dams and irrigation projects also cause rivers to drop most of their heavier sediments far inland, so that beaches that had formerly received most of their sand from these rivers may be stripped of sand for miles by alongshore transport (p. 219) and may receive too little replacement sand. Added to this situation is the fact that most bays and estuaries are the sites for development of towns and ports. The patterns of circulation are changed by dredging and by the building of roads, causeways, bridges, boat basins, etc. Not infrequently, large parts of an estuary are filled in to make more land for real estate developments. Of the twenty-three estuaries along the coast of California, all but three have already been biologically ruined, and these remaining three are on the way out. Even estuaries remote from actual direct operations of man suffer.

DISSEMINATION OF PESTICIDES

Woodwell, Wurster, and Isaacson (1967), in a study of East Coast estuaries, state that long-term, detailed observations show substantial reduction over the past decade in the populations of shrimp, amphipods, summer flounder, blue crab, spring peeper, Fowler's toad, woodcock, and other DDT-sensitive species, even in estuaries and marshes remote from direct human disturbance. They also state that the analysis of the water means little in assessing the threat, since water has the least DDT of the major components of the ecosystem. Often, although the water may show only traces of DDT, or even when it is undetectable, the animals may contain nearly lethal concentrations. It is concentrated progressively in animals higher in the trophic scale. Table 19-2 is extracted from a larger table showing progressive increase in DDT levels in animals at the Carmans River estuary in the vicinity of Long Island, New York.

It has also been found in the past few years that DDT influences the calcium metabolism in some subtle way. Birds of prey in particular, which accumulate the highest levels, have eggs with thinner, more fragile

Table 19-2. Accumulation of DDT in trophic levels*

Source	DDT in ppm
Water of estuary	0.00005
Zooplankton	0.040
Shrimp	0.16
Flying insect (Diptera)	0.30
Fundulus	1.24
Tern	2.8 to 5.17
Cormorant (immature)	26.4
Ring-necked gull (immature)†	75.5

*Based upon data from Woodwell, Wurster, and Isaacson (1967).
†In the ring-necked gull, the levels approach those found in animals dying of DDT poisoning.

shells, have more losses from breakage, and are declining rapidly in numbers.

The plankton of estuaries is also extremely sensitive to some commercial herbicides, especially substituted urea compounds such as Monuron, Diuron, Neburon, etc., and in some cases cannot tolerate them in concentrations as low as 0.5 part per billion.

Estuaries and coastal waters bear the brunt of the load of insecticides and poisons washed out of fields and woods into the streams. For example, it is estimated that the Mississippi now carries 10,000 kg. of pesticides into the Gulf of Mexico annually. The San Joaquin River in California discharges 1900 kg. per year into San Francisco Bay.

Tests on Barbados Island indicate that the easterly trade winds carry about 600 kg. of DDT per year into that distant part of the tropical Atlantic (Risebrough and associates, 1968). PCB, a toxic compound widely used in the manufacture of plastics, paints, and other products, is also found now in marine fish and birds along the coast of Southern California and also appears to be progressively concentrated at higher trophic levels. Radioactive substances commonly behave in a similar way.

People who became upset about the addition of a little fluorine to the drinking water, which at least has been shown to have some beneficial effects, should become very much disturbed over the wholesale addition of all these known poisons, insecticides, herbicides, industrial sewage products, and radioactive compounds to the world's atmosphere and waters, so that now not only every man, woman, and child alive but every organism on earth is forced to involuntarily partake of them. What the long-term effects will be no one can say, but they will undoubtedly be very profound. Most of these substances are an entirely new addition to the world's ecosystem, developed only within the past three or four decades. They have been found in animals all over the world, even from such remote areas as northern Alaska, Antarctica, the tip of Lower California, and Barbados Island.

OIL POLLUTION

A recurrent problem of ever-growing magnitude—pollution by oil—is a grave threat to coastal ecology everywhere. Because of its striking and devastating effects along resort beaches and in boat basins, as well as along less conspicuous stretches of the coast, this situation has received much publicity. It deserves much more. Oil and crude oil, spread out as a surface layer over the water, wash up along the shores, coating everything with a sticky tarry or oily film, fouling the gills and respiratory passages of all intertidal animals so that they choke to death, and killing the surface plankton. Seabirds whose feathers are fouled by the oil are unable to fly and either drown or die slow, miserable deaths along the beaches. The fur and respiratory passage of marine mammals are fouled, and, of course, an oil-fouled coastline is ruined for a long time for fishing or recreation.

Oil pollution comes from many sources—some of it from oily wastes from oil fields or refineries near the coast, producing local problems in their areas; some from tankers and ships; and some from offshore oil wells.

The tonnage of crude oil carried in tankers is steadily growing, and ships of unprecedented capacity are being put into use. When one of these has an accident or is sunk off a coastal region, major ecological catastrophy ensues.

Some companies have caused considerable local pollution by cleaning their tankers near the coasts and rinsing them with seawater. Such practices, together with occasional accidents through negligence, became such a nuisance, especially along the Atlantic coast of the United States, that in 1924 Congress passed the Oil Pollution Act, authorizing the government to make efforts to remove spilled oil from the water or the shorelines and to sue the responsible companies for costs and damages. Under this act approximately a hundred prosecutions a year were filed, resulting in a considerable improvement in the practices of oil firms, although by no means wholly solving the problem. In 1966 some amendments to this act were included in the Clean Water Restoration Act of that year, originally intended to strengthen the government's hand in such cases. However, because of pressuring from oil interests, during the House-Senate conference on the bill the definition of "discharge" was changed to "any grossly negligent, or willful spilling." Proving "gross negligence" or "willful spilling" is almost an impossibility, and no one has been prosecuted under the new law.

During World War II many tankers were sunk, especially in the Atlantic. The Coast Guard estimates that over 200 million gallons went down in the United States flagships alone. Since then, whenever an oil slick appears offshore in the ocean, it has been customary to blame it on leakage from the sunken tankers. This is doubtless true in some cases, although it will be a decreasing source as time goes on. Recent developments in chemical analysis of oils will probably enable oil slicks to be precisely identified in the future and compared with samples obtained from tankers operating in the region.

A few examples will serve to indicate the variety of types of oil accidents occurring today.

In the early summer of 1965 one of our warships stationed in the Mediterranean not far from the French Riviera accidentally discharged several thousand gallons of oil into the water. A sailor assigned to flush out and clean the water tanks opened the wrong valves. This embarrassing accident caused great damage to the expensive and crowded resort beaches along the French Riviera and hundreds of thousands of dollars' loss to local resort businesses. This was especially distressing to the Navy because the French were at the time trying to get rid of American forces stationed in France and this accident added much to the "Yankee go home" sentiments already being expressed both officially and unofficially and in signs along the highways. The fact that the Navy assigned all available personnel to help try to clean up the beaches and even shipped up boatloads of fresh, clean sand from as far away as Naples, at U.S. taxpayers' expense, did little to mitigate the irritation aroused by this *faux pas*. Fortunately, this accident caused little real ecological damage, since the French Riviera is a particularly barren strip of gravelly beaches already rendered almost biologically sterile by the vast overdevelopment that has occurred there. It would have been hard to choose a better place to have such an accident, since it engendered maximum publicity and awareness of the problem without doing much harm.

A far more serious accident was the Torrey Canyon disaster in March, 1967. The Torrey Canyon was a huge tanker filled with crude oil from Kuwait, which went aground on rocks about 15 miles off the Cornwall coast of England. This accident resulted in the spillage of some 117,000 tons of crude oil. The problem was complicated by the massive attempts to save the beaches by dispersing the crude oil slick with detergents. More than 20,000 tons of at least twelve kinds of detergent were used. This proved to be a great mistake because the detergents were water soluble and toxic, ensuring the death of almost the entire biota beneath the oil slick, as well as that on the surface. They also failed to keep the oil from the beaches. One of those most used, BP 1002, contained a highly toxic surface-acting organic-solvent stabilizer responsible for massive mortalities. The combination of crude oil and detergents wiped

out the intertidal plants and animals for miles and killed all larval forms in the offshore waters. This, together with the persistent gummy coating of rocks and beaches, will impede recruitment of new biota for a long time. The detergents further delayed recovery by killing off the bacteria capable of degrading the oil. Persons interested in this problem should read the report on this disaster by scientists from the Plymouth Laboratory, edited by J. E. Smith (1968).

As this is being written another gigantic thick tarry crude oil slick, already covering approximately 800 square miles of ocean in the Santa Barbara Channel of Southern California, is being fed at the rate of some 21,000 gallons per day from a Union Oil Company offshore drilling rig. Thousands of seabirds, at least three whales and probably more, and many fish have been killed; about 35 miles of coastline have been blackened, with most of its intertidal biota wiped out. Crews from the oil company are frantically pumping bargeloads of thick cementlike mud into the well and adjoining strata in the hope of plugging it, their work impeded by heavy swells. Conservationists, resort owners, owners of marinas, and businessmen dependent upon the tourist trade in the Santa Barbara area are up in arms. The seal, sea lion, sea elephant, and marine bird rookeries in the channel islands are threatened. Direct and indirect damage, together with the loss of oil, will probably run far in excess of whatever wealth the channel wells might have produced.

This last case is not an isolated instance but only one of a rapidly growing series. Similar blowouts have occurred off Australia, in the Gulf of Mexico, and in other areas. Offshore oil exploration and drilling is very rapidly increasing all over the world and poses a major world problem in control. The Santa Barbara case is very instructive in a number of ways.

When it was first proposed to drill in the Santa Barbara Channel, some geologists and conservationists and some persons concerned about the effect of ugly oil drilling rigs and spillages on the aesthetic appeal of the Santa Barbara region raised objections, but these objections were dismissed as unrealistic, academic, and out of touch with the economic facts of life. The public was assured by both governmental and oil company "experts" that there was no danger of pollution, that careful studies had demonstrated both the feasibility and safety of the operation, and that the operations were to be so carefully done and rigidly supervised that there was a wide margin of safety. The prospects of great wealth gushing from the sea, of oil lease fees and increased tax revenues, carried the day with both political and industrial leaders. The Secretary of the Interior now has proclaimed a moratorium on all further drilling in the channel on the grounds that not enough is known of the geology of the area as yet and this must be thoroughly studied first.*

RADIOACTIVE CONTAMINATION

Radioactive substances arise from military testing of nuclear bombs, from industrial and research nuclear reactors, and to a limited extent from natural causes such as bombardment of the earth by cosmic rays. They behave in many ways like the insecticides and other poisons in distributing themselves throughout the biosphere and in that they are very unevenly concentrated from the environment by different organisms, with a tendency to concentrate in animals of the higher trophic levels.

All the elements can be made radioactive, some of them in several isotopic forms. In general the atoms of an element in the radioactive state behave chemically in the same manner as do normal atoms of that element. They are therefore incorporated

---

*Perhaps the geology was better understood at the time when those who wished to delay drilling, pending further studies, were overruled and the leases were issued in the first place. Since this writing, the oil continued to leak, although at a much reduced rate, for several months. Pumping was finally resumed on the theory that the best way to overcome the danger of further massive oil releases in the channel was to relieve the pressure in the oil-bearing strata by pumping out the oil. During this operation one of the pipes, at the same installation, broke—resulting in another massive release of oil.

into the intimate molecular structure of all organisms in amounts that are proportional to the ratios of radioactive to normal atoms of the element in question available in the environment and dependent upon the degree to which the organism in question utilizes or concentrates that element. Each is further characterized by a half-life decay period, representing the time during which half the atoms present will decay. These half-life periods range from as short a time as 10^{-7} seconds to 10^6 years. Elements that have moderate to long half-life periods and that are either required in the bodies of organisms or are so closely related chemically to other required elements that they are incorporated along with them, are especially dangerous.

The most serious effects produced are alterations in the genetic material of the cells, causing heritable mutations, most of which are deleterious. For this reason the effects are most serious in actively proliferating tissues such as epithelium, bone marrow, or germ cells and least serious in tissues that are not proliferating such as adult nerve cells. For the same reasons growing embryos are more likely to suffer serious damage than adult organisms.

In proliferating somatic tissues the incorporation of radioactive elements may lead to somatic mutations, producing in localized areas of the body the cells that differ in some respect from normal, or it may lead to the escape of the cells from regulatory control by the body and their proliferation as cancer or leukemia. On the other hand, it may injure the cells in such a way as to prevent their normal proliferation, as in the case of white cells arising in the bone marrow, leading to leukopenia—a serious diminution of white cells that makes a person or animal much more susceptible to infections.

In germ cells, mutations are, of course, expressed as heritable changes in the offspring or in extreme cases as failure of the zygote to develop into a viable offspring at all. In human populations deleterious mutations accumulate to a much greater extent than in most animals because the individuals bearing them are sedulously nursed along with all our medical skill instead of being eliminated, while our healthiest young men are casualties of war.

There is no such thing as a safe threshold of radioactivity below which there will be no damage. Any dosage of ionizing radiation, no matter how small, can induce a mutation if it happens to be on target. The chance for damage and the degree of damage, of course, increase proportionally to the dosage. However, low levels of exposure spread over very large populations will produce just as many mutations as high levels spread over a smaller population.

Since a large proportion of nuclear weapons testing has been done in remote areas of the seas, and since the oceans have also been used to some extent as disposal sites for excess dangerous radioactive materials produced as by-products of nuclear reactors both on land and on ships and submarines, and because most of the radioactive materials in particulate form in the atmosphere eventually fall into the oceans or are washed into the oceans from the land, the oceans of the world have received a disproportionate amount of radioactive materials. Many marine organisms concentrate some of these isotopes to an astonishing extent and pass them along to other animals in the food chain, including man.

For a summary and analysis of this aspect of environmental pollution, the reader is referred to Polikarpov (1966), *Radioecology of Aquatic Organisms,* and to publication 651 of the National Academy of Sciences–National Research Council (1957), entitled *Effects of Atomic Radiation on Oceanography and Fisheries.* The general conclusion from these studies and others is that further additions of radioactive substances of any kind to the world's oceans are inadmissible.

ATMOSPHERIC POLLUTION

According to a report by Eshleman (1969) on recent studies of the atmospheres of other planets, the earth may be in grave danger from extensive atmospheric pollution and burning of fossil fuels. During the past 100 years the burning of fossil fuels has increased the percentage of atmospheric carbon dioxide by 15% and is now using up 15% of the biological production of oxygen annually. Smog, jet contrails, deforestation, and pollution of the earth's waters, together with the increased carbon dioxide in the atmosphere, tend to raise the temperature of the atmosphere. A critical level may be reached (we do not yet know what this critical level is) where the increased temperature would drive more water vapor and carbon dioxide into the atmosphere from the oceans, producing in turn an enhanced greenhouse effect that would end only when the earth's atmosphere had become a heavy, hot, suffocating, carbon dioxide–rich mixture unfit for most of the present biota of the world and leading to the greatest mass extinction since the beginning of life on earth. As Eshleman says, "It would seem that our relative ignorance of the possible global effects of small changes in the composition and temperature of the earth's atmosphere should give us serious and immediate concern about man's use of the atmosphere as a garbage dump."*

CONCLUDING THOUGHTS

It does no good to predict the end of the world or of mankind because if it happens there will be no one around to appreciate the accuracy of the prediction or to do anything about it. On the other hand, until it does happen, the presence of man will be regarded as proof of the falsity of the prediction. So let us assume for the moment that we will not in the immediate future eradicate mankind in a nuclear holocaust or suffocate and cook ourselves by setting off an uncontrollable atmospheric greenhouse effect. Let us look at the overall trends in the present development of the world ecosystem, given the continued presence of man for at least several more decades.

The overall result of the rise of man has been to contribute greatly, and at an ever-accelerating pace, to the instability of the ecosystem. It is safe to say that for the next few years, at any rate, the uncontrolled

---

*From Eshleman, V. R. 1969. The atmospheres of Mars and Venus. Sci. Amer. **220** (3):78-88.

population explosion will continue to contribute to this instability. We will continue to destroy the few remaining wild regions, to introduce deadly poisons and radioactive substances into the biosphere in increasing amounts and variety, and to burn fossil fuels. Every such change in the ecosystem entails the extermination of numerous species of animals and plants. We are in the midst of one of the great periods of mass extinction of species, a marked contraction in the numbers of kinds of living things on earth. Every increase in the instability of the ecosystem has the general effect of reducing the diversity of the living things inhabiting it. A comparison with conditions as seen in long, stable and in unstable environments in different regions of the world is instructive. Instability in the environment leads to less diversity or conversely to more uniformity in the fauna and flora. There is not necessarily any reduction in biomass of living matter produced—just less diversity.

The same thing is happening to man himself, and to his cultures. The unprecedented increase in the mobility of man and in his technology has led to a general weakening and destruction of nontechnical cultures all over the world and to the spread of Western gadgets and the ways of life that they entail. Cities the world over are becoming full of cars, gasoline stations, department stores and supermarkets, smog, and decaying ghettos in the center and expanding suburbs for the well-to-do. Radio and television, telephones, and airplanes link all peoples so closely that what is done anywhere in the world has immediate repercussions throughout all of society. Whether we like it or not, human society is rapidly becoming one entity, and exhibiting greater and greater uniformities throughout. The evolution of diverse cultures independently of each other has become a thing of the past. Because of the uncontrolled population increases, the general level of human life will be low and poverty-stricken. It will continue this way longer, and recover to a lesser degree, inso-

far as we discover and use up the remaining untapped resources of the world at this stage of our development.

We have already passed a point of no return where the ratio between the needs of our expanding population and the resources necessary to support it are such that in order simply to exist we are forced to destroy our environmental capital faster than it can be renewed. There is no hope in our time or in that of our children of avoiding an accelerated general environmental deterioration and a further lowering of the quality of life for our species. This is difficult for many of us in the United States or Western Europe to believe, since through our rapid exploitation of a virgin continent, our new technology, and our worldwide economic imperialism, we are, though only about 12% of the world's population, utilizing well over half the total production of the world and have thereby produced momentarily the most affluent society that has ever existed. While doing this, we have brought about an ever-accelerating destruction of the ecological base upon which all humanity depends for its existence.

Man will continue to crowd out and exterminate the larger animals, high in the trophic scale, that compete with him for food without themselves contributing an equivalent amount of food for man. The large carnivores, and eventually most of the large herbivores as well, will continue to become rarer or will be exterminated while man turns more and more to intensive "farming" of the lands, the continental shelves, and perhaps even of the open sea.

A world lacking basic diversity in its cultures, with reduced diversity in its faunas and floras, and crowded with poverty-stricken people, will be a less interesting and a less viable world. Unless and until mankind finds and utilizes acceptable ways of bringing his population into equilibrium with the ecosystem at a much lower numerical level than the present world population, he is condemned to inhabit a social and biological slum.

Glossary

(For clarification of other terms, please consult the Index.)

abioseston nonliving particles suspended in the water.

aboral away from the mouth; at the opposite pole of the body from the mouth.

abyssal extremely deep; the deeper regions of the sea; depths greater than 4000 meters.

abyssobenthal on or near the bottom in the abyssal zone.

abyssopelagic in the water well above the bottom in the abyssal zone.

aciculum in polychaete annelids, a large seta, larger than and distinct from the ordinary setae, and almost completely buried in the fleshy parapodium.

acraspedote without a velum; the condition of "true" jellyfish, or Scyphozoa, as contrasted with hydrozoan, medusae.

actinotroch larva a larval form in Phoronidea, somewhat related to trochophore larvae but bearing ciliated arms.

actinula a larval form in certain medusae in which the planula larva develops arms but does not settle down as a fixed polypoid form before becoming a medusa; an intermediate growth stage in coelenterates without a polypoid generation.

adductor muscle a muscle that pulls together or closes the valves or shells of an organism—as the adductor muscles of barnacles or bivalve molluscs.

adradial in the tetramerous radial symmetry of most medusae, the sector midway between the main radial (perradial) canals; adradial canals are additional canals occupying this position in some medusae.

aesthetask sense organs on the antennae of certain copepods.

agar-agar gelatinous solidifying agent obtained from certain red algae such as *Gelidium,* widely used for solidifying bacteriological media and for other purposes.

agglomerate objects or structures closely crowded into a cluster.

Agulhas Current a warm current, continuation of the south-flowing Mozambique Current from the eastern side of Africa, around the southern end of the continent to Cape Agulhas.

air bladder in many teleost fishes, a gas-filled sac originating as an outgrowth from the pharynx and serving as a hydrostatic organ.

akaryotic without a definite nucleus with a nuclear membrane.

akineton plankton without any power of self-movement, such as many spores or planktonic eggs.

Alaska Current a north-flowing, then west-flowing eddy from the North Pacific Current in the Gulf of Alaska.

algal film the growth of small to microscopic algae, mixed with other small organisms, which forms on all surfaces exposed to marine waters in the euphotic zone. A source of food for many ani-

361

mals, especially chitons and many gastropods that graze upon it, using a filelike lingual ribbon, the radula, to scrape it off rock surfaces into the mouth.

alima larva a second larval stage of certain Stomatopoda, following the pseudozoeal stage.

allochthonous acquired from elsewhere—nonnative.

allogenetic plankton carried into the region by currents, wind, etc. but normally living and breeding elsewhere.

allogenous detritus detritus carried into a region from some other area.

allopatric with separate, mutually exclusive areas of geographical distribution.

alongshore current (commonly termed longshore current) the movement of sand and other loose particles in a given direction along the coast because of the prevailing angle of the wave trains to the coastline.

Alpha Rise a rise in the floor of the Arctic Ocean, in the Canadian Basin, roughly paralleling the Lomonosov Ridge.

alveolus a pit or cavity in a surface, as in the lorica of various tintinnids. Differs from a pore in not passing all the way through.

ambergris a foul-smelling pathological waxy material formed largely from squid beaks in the digestive tract of certain toothed whales such as the sperm whale. Has been valued as a stabilizer added in minute quantities to fine perfumes.

ammocoetes larva the larval stage of lampreys, remarkable for its resemblance to the Cephalochordata.

amphiblastula larva a larval stage developing from the fertilized egg cell in certain sponges, comprised of small flagellated cells in one hemisphere, which eventually give rise to the choanocytes, and bulkier nonflagellated cells in the other hemisphere, which give rise to the other cells in the sponge body.

anabolism the constructive processes in metabolism; the building up of the compounds and structural elements that comprise the tissue or substance of an organism. *See* catabolism.

anadromous living part of the time in fresh water, part in salt water, as certain fish such as the salmon.

anal fin in fish, a midventral fin behind the anus.

ancestrula in Bryozoa, the first zooecium of a colony; the zooecium formed by the larva upon settling.

angstrom unit a minute unit of distance useful in describing such things as light waves, etc., one one-hundred millionth of a centimeter.

animal exclusion theory the theory that the patchiness and the inverse relationship in surface distributions of phytoplankton and of zooplankton is caused in part by failure of the zooplankton to make it to the surface waters when impeded in their diurnal vertical migration by dense clouds of phytoplankton.

anisogamous with some visible differentiation between gametes. Usually used in cases in which the gametes are not greatly different but are not so much alike as to be very difficult or impossible to distinguish, as in the isogamous condition.

annulus the transverse groove bearing the horizontal flagellum in dinoflagellates.

antapex the end opposite the apex; the base.

Antarctic Convergence an important zoogeographical and hydrographical feature in the Southern Hemisphere where northeast-drifting surface waters from the Antarctic meet and sink beneath southeast-drifting warmer waters. It lies, for the most part, in or near the 50- to 60-degree latitude band south.

Antarctic Divergence the region near the shores of Antarctica where water of intermediate depths rises to the surface in extensive upwelling, replacing the surface waters that are drifting in a northern and eastern direction in the West Wind Drift.

antenna in arthropods, a preoral sensory appendage of the head. Antennae are paired. In Crustacea there are two pairs, the first pair commonly called antennules, whereas the term antennae is reserved for the second pair. Feelers.

antennal scale the short, flattened exopodite, or outer branch, of the second antennae in mysids and various decapod crustaceans.

antennules in Crustacea, the first pair of feelers.

anteriad toward the anterior end.

antherozoid a swimming male gamete in certain algae and other primitive plants.

Antiboreal Convergence the second greatest zone of convergence in the Southern Hemisphere, located approximately 10 degrees north of the Antarctic Convergence. Often called the Subtropical Convergence.

antizoeal stage the first larval stage of various Stomatopoda.

aperture an opening. In shelled organisms such as Foraminifera, the main or largest opening in the test.

apophysis in Radiolaria, a horizontal outgrowth at right angles to the longitudinal axis of the radial spines.

appendix interna in Euphausiacea, a median process uniting the two pleopods of a given somite; also termed the stylambis.

archibenthal on the continental slope; between the relatively shallow continental shelf and the main ocean bottom, or abyssal plain.

arcuate arched or curved.

areolation the crowding of a surface with small alveoli, or pits, so that it forms a sort of network, or meshwork.

article in Arthropoda, one of the segments of a jointed appendage.

aspinose without spines, as in some Radiolaria. In aspinose pores, the opening of the pore is not armed with a spine.

atoll an isolated ring of coral islands and reefs surrounding a large central lagoon.

atrium an enclosed space in an animal body, lined with ectodermal epithelium and morphologically outside the animal and surrounding the pharyngeal region, into which water from the pharynx flows and into which the excretory and reproductive organs usually empty; opens to exterior by the atrial opening, or atriopore. Found in tunicates, cephalochords, etc. The term is also used for the thin-walled chamber (s) of the heart, which receives blood from the respiratory organs or the body, and from which blood goes into the ventricle.

auricularia larva the free-swimming, ciliated larva of certain holothurians. It possesses earlike, blunt extensions of the body rather than long, slender larval arms.

autochthonous native, indigenous.

autogenetic plankton plankton organisms that live and breed in the region in which they are found.

autopelagic plankton surface-living plankton.

autotrophic literally "self-feeding." Said of organisms, such as plants, that synthesize needed organic foodstuff from inorganic substrates. *See* heterotrophic.

auxiliary cell in cetain red algae, a cell to which the zygote extends a tubular extension, the ooblast, through which it migrates to the auxiliary cell. This cell participates in the formation of the fruit of the sexual plant.

auxospore in diatoms, an enlarged cell (in some cases, at least, resulting from union of gametes), which restores size to that line of cells in which the size was diminished at each cell division during the last few preceding divisions.

avicularia beaklike zooids in some Bryozoa colonies, such as *Bugula,* thought to function in keeping the colony free from debris or from larvae of other organisms that might settle on it.

axenic said of a culture of some organism that contains no other organisms.

axopodium in Protozoa, a pseudopodium containing an axial filament.

baleen the horny material comprising the plates in the mouth of the whalebone whale, through which water is strained in feeding; used to capture macroplankton. Also termed whalebone.

barrier reef a coral reef parallel to, but at some distance from, the shore.

basipodite in biramous crustacean appendages, the basal portion of the appendage before the branching.

bathypelagic living in deep water below the euphotic zone but above the abyssal zone, from about 100 meters to about 4000 meters; free in the water, not living on the bottom.

bends a painful and dangerous condition created in divers or in any air-breathing animal in deep water by too rapid a release of the hydrostatic pressure upon rising, causing formation of bubbles of nitrogen gas in the blood.

Benguela Current the major north-flowing cold current along the western coast of Africa.

benthic living in or on the bottom.

benthos the organisms living in or on the bottom.

bifurcate forked; divided into two branches.

bilateral symmetry organization of the body so that the longitudinal axis of the body divides two equivalent sides, right and left, and two non-equivalent sides, dorsal and ventral.

binary fission a form of asexual reproduction in which an organism divides into two more or less equivalent parts, each of which continues to live as a whole organism.

biocenosis a community of organisms found in a given area or region or habitat.

biological clock inherent physiological rhythm in organisms that usually manifests itself in changes in behavior or physiology timed to certain recurrent changes in the normal environment of the organism and that may continue for a time even when the environmental stimulus to which the organism is timed has been removed.

bioluminescence the production of light by living organisms.

biomass the weight of the standing crop of living organisms, or of selected kinds of living organisms, in a given unit of the environment at a particular time or over a given time period.

bioseston living particulate matter in the water, and organic particles derived directly from living organisms. *See* abioseston.

biosphere that portion of the surface of the world, its waters, and its atmosphere permeated with living organisms or their products.

biotic community the group of species of organisms living in, and characteristic of, a given habitat.

biotope a specific habitat.

bipinnaria larva a larval stage of various starfish in which the ciliated bands are drawn out into arms.

bipolarity a term describing the geographical distribution of species, or closely related forms, in both the Northern and Southern hemispheres but not in the intervening equatorial belt.

biramous two-branched, as various crustacean appendages.

black zone a band of growth of microscopic blue-green algal cells and other minute microplants at the extreme upper limit of the intertidal zone, in the splash zone. The black zone represents the transition from essentially terrestrial to essentially marine habitats.

bonification estimation of fertility and standing crop.

boreal a term descriptive of cool or cold temperate regions of the Northern Hemisphere. The corresponding region of the Southern Hemisphere is sometimes termed antiboreal.

brachiolarian larva the last stage in starfish larvae just prior to metamorphosis into the starfish form. Brachiolarian larvae develop additional larval arms, not involving the ciliated band, in connection with fixation to the substrate.

bract in Siphonophora, thick gelatinous structures of various shapes that aid in swimming and in the protection of the colony. Bracts have been regarded as modified medusoids or, more recently, as merely modified tentacles.

buccal mass in molluscs and various other organisms, the mouth and pharynx when heavily muscularized and used as an organ of prehension and mastication.

budding asexual reproduction in which a small portion of an organism grows out from and eventually develops into another individual or another equivalent part of a colony.

buffer salts or other substances that are to some degree amphoteric, able to remove either hydrogen or hydroxyl ions from solution, thus retarding changes in pH of the solution when either acids or bases are added and making for a greater constancy in the pH.

cadophore In Thaliacea of the family Doliolidae, a posterior dorsal extension of the body that carries the buds in the aggregate phase of the life cycle.

Cambrian the earliest of the geological ages from which a good fossil record has been preserved, beginning approximately 500 million years ago and lasting approximately 80 million years.

Canadian Current a south-flowing cold current from the Arctic Ocean along the northern coast of Canada.

carapace in Crustacea, a fold of skin and exoskeleton of the dorsal body wall, covering the head, part or all of the thorax, and in decapods forming lateral gill chambers.

carinate having a keel or ridge.

carotenoid yellow or orange pigments chemically related to carotene. Carotenoid pigments are lipid-soluble and are an important coloring matter in many marine organisms.

carpogonium unicellular female sex organ in many red algae.

carposporangium in red algae, a spore-bearing organ on the gonimoblast filament, derived from the zygote.

carpospore a spore resulting from division of the zygote in red algae.

catabolism the breakdown of organic molecules during metabolism.

caudal posterior, or toward the tail.

central capsule in Radiolaria, the inner portion of the protoplast, bounded by a membrane and containing the nucleus or nuclei.

centric in diatoms, having the valves arranged radially symmetrical around a central point. A member of the order Centrales.

cephalic pertaining to the head or cephalon. Toward the head.

cephalosome in copepods, the apparent head when viewed dorsally, when this consists of the head plus one or more fused thoracic somites.

cephalothorax in Crustacea, the head and thorax when united under a single carapace.

cercaria a larval stage of digenetic trematodes. The cercaria leaves the mollusc host and either directly, or indirectly by way of a metacercarial stage, invades the vertebrate host, where it matures into the sexual adult worm.

chela in Crustacea and some other arthropods, a pincer formed when the terminal articles of a limb is opposable to a process from the penultimate article, as in crabs. An appendage ending in this manner is termed a chelate appendage.

chemoautotrophic deriving metabolic energy from the oxidation of inorganic chemicals occurring outside the organism, a condition found in a few minor groups of bacteria, the nonphotosynthetic autotrophic forms. Chemotrophic; chemolithotrophic.

chitin a polymer of glucosamine that forms an impervious stable substance synthesized by many animals, such as arthropods, as an external covering. Impregnated with calcium salts, it may form a firm, rigid exoskeleton.

chlorinity grams of chlorine and bromine ions per thousand grams of seawater. Determined by simple titration with silver nitrate. Used in calculating salinity.

chloroplast in plants, a cytoplasmic structure, plastid, containing chlorophyll. The site of photosynthesis.

chromatophore in plant cells, plastids containing yellow or brown pigments in addition to chlorophyll, giving them a color other than clear green.

cicatrix in fossil Nautiloidea, a scar on the initial shell chamber, thought to represent the point of attachment of an embryonal shell.

cinereous ash gray, the color of wood ashes.

cingulum in dinoflagellates, the horizontal groove around the cell.

circular canal in coelenterate medusae, a tube of the gastrovascular cavity, lying at the periphery of the bell, encircling it, and connected to the stomach by the radial canals.

cirrus a slender, tentacle-like process. In ciliate protozoans a fused group of cilia.

claspers in various animals, appendages or organs of a male, used to hold the female and facilitate copulation.

clavate club-shaped.

cleaning symbiosis a relationship between different species of animals in which one, usually a smaller species, picks parasites, bits of infected tissue, etc. from the body surface, mouth, or gills of a larger animal.

clone genetically similar individuals derived by asexual reproduction from one individual or, in Ciliata, from a pair of exconjugants.

coccolith the tiny calcareous concretion, or plate, of the Coccolithophoridae, often forming a significant part of calcareous marine sediments.

coelom a body cavity lined with mesodermal epithelium, the peritoneum.

coenenchyme the fleshy cellularized mesogloea of alcyonarians, usually traversed by gastral tubes, solenia, connecting the gastral cavities of the polyps.

coenocyte a tissue in which the cytoplasm is continuous rather than divided into individualized cells around each nucleus.

coenosarc the living tissue in hydroid colonies, joining all the polyps.

collarette in Chaetognatha, a thickened ectodermal area in the neck region of the body.

colloblast adhesive cell on the tentacles of Ctenophora of the class Tentaculata. Used in the capture of prey.

columella in gastropod shells, the central axis of shell around which the whorls are spiraled.

comb plates the flagellated platelets in Ctenophora responsible for swimming.

commensalism an association between two species in which one of the pair is benefited, whereas the other seems to be neither benefited nor harmed.

community see biotic community.

compensatorium in certain Bryozoa, a water-filled invagination of ectodermal epithelium from which the water can be expelled as the animal retracts into its zooecium, making room for the retracted body.

conceptacle in brown algae of the subclass Cyclosporeae (the Fucales), a round cavity borne in the swollen tip of the branched thallus and containing the reproductive organs.

concrescent growing together, coalesced.

conjugation in the Ciliata, a sexual process in which there is a temporary cytoplasmic connection between two individuals. While the connection lasts there is an exchange of haploid nuclear elements, each of which fuses with a corresponding nucleus of the other individual, reconstituting a diploid condition. Then the individuals separate. The descendants of a pair of exconjugants are all genetically alike and constitute a clone.

constant plankton perennial holoplankton in a given region.

consumer an organism that feeds on already formed organic matter. All "animals," and a few other forms.

convergence a zone where surface waters meet and are carried down beneath the surface. *See* divergence.

copepodid an immature copepod, the stage occurring between the last naupliar stage and the mature adult. Copepodids look much like adult copepods but are not sexually mature.

coracidium a larval stage in certain pseudophyllidean tapeworms in which the egg is surrounded by a ciliated or flagellated embryophore and swims freely in the water.

corallite a cup-shaped structure in coral skeleton in which the polyp sits.

cordate heart-shaped.

cormidium in Siphonophora, a group or colony of individuals borne together on the coenosarc of the stalk. In some species the cormidia are easily detached and are usually found separately in plankton hauls.

coronal constriction in jellyfish of the order Coronatae, a circular constriction of the bell.

coronal pores in some Radiolaria, small pores surrounding the parmal pores. *See* parmal pores.

cortical shell in some Radiolaria, the portion of the skeleton forming an outer framework.

costae ridges of a shell.

coxa in Arthropoda, the basal segment of one of the joined appendages; in biramous appendages, the first segment or article of the basipodite.

craspedote having a velum, a small shelf of tissue forming an inward-directed ring from the edge of the bell in hydromedusae.

crenulate an edge having a series of small indentations, giving it a scalloped appearance.

critical tide levels the average levels attained by each of the tides of the daily tide cycles during the year.

cruciate; cruciform in the form of a cross.

crystalline style in bivalve molluscs and a very few other filter-feeding molluscs, a rodlike gelatinous structure in the first part of the intestine, or in a diverticulum thereof, which is rotated by ciliary action and gradually worn down anteriorly by rubbing on the gastric shield in the stomach. It functions to stir the stomach contents and slowly release digestive enzymes incorporated into it.

ctenidia the typical gills of most aquatic molluscs.

cyst a resistant wall or shell secreted around an organism, or around certain stages in the life cycle of an organism, to protect it during periods unfavorable to ordinary life activities.

dactylozooid in coelenterates, a modified, mouthless, often rodlike polyp, heavily armed with nematocysts, which serves to touch and catch prey or to defend the colony.

decussate crossed or intersected.

deep-scattering layer an intermediate-depth zone, above the bottom, which reflects sound from echo-sounding equipment, giving a record of a "false bottom." It is formed by the aggregation of certain small animals at particular depths during the day. At night, as the animals usually swim upward into water nearer the surface, the deep-scattering layer tends to rise and become more diffuse, and individual layers within it lose their identity.

delta an extension of low land formed by deposition of sediments in the mouth of a river. Deltas are usually fan-shaped, with the river forming numerous channels through them.

demersal said of a swimming organism that prefers to spend most of its time on or very near the bottom—as the flatfish, octopuses, and many shrimps.

density the weight per unit volume.

detritus particulate matter, especially that of organic origin, floating in the water or settling to the bottom.

dextral to the right. In gastropods, having the aperture of the shell on the right side when the shell apex is upward and the aperture is facing the observer (on the observer's right). *See* sinistral.

diameter in diatoms, the greatest width.

dichotomous branching into two parts at each division.

digenetic trematodes flukes, trematode worms in which the sexual adult stage is found in vertebrates and the asexual larval stages in molluscs, usually gastropods.

digitiform finger-shaped.

diploblastic formed of two layers of cells, an outer epithelium and an inner gastrothelium; the grade of construction of simpler coelenterates such as *Hydra*.

distal the part of an appendage or other structure farthest from the point of attachment. Away from the point of origin.

divergence a zone where water from below is upwelling and spreading out over the surface. *See* convergence.

diversity in zoogeography, an expression that is a function of both the number of species present in a given region or habitat and also the relative proportions of their numbers. For example, given two faunas, each with 500 species—if, in one of them, two or three species were overwhelmingly dominant and the others relatively rare, whereas in the second many of the species appeared in considerable numbers, the second fauna would exhibit greater diversity. Diversity tends to be correlated with stability of ecological conditions, other things being equal.

doliolarian larva a short-lived free-swimming nonfeeding larva of certain shallow-water crinoids, such as *Antedon*.

drill carnivorous gastropod, such as oyster drill and whelk, which uses its radula to drill a hole through the shell of its prey.

ecdysis the molting of a cuticle or exoskeleton, as in the Arthropoda and the Nematoda.

echo-sounding equipment acoustical equipment designed for determining depth of water by bouncing sound waves off the bottom and recording the time required for the echo to reach a recording instrument. Fathometer.

ecosystem the entire system of organisms and the physical and chemical aspects of their environment with which they interact.

ectocrine external secretions or hormones that affect other organisms.

ectoderm the embryonic cell layer of the Eumetazoa forming the outer epithelium, lining of the mouth, and various other structures.

endemic found only in a given region.

endite in Crustacea, an inner lobe or cuticular extension from certain basal segments of the appendages—as the endites of amphipod and isopod thoracic limbs that form the brood pouch in the females, or the gnathobases of appendages of trilobites and of *Limulus*.

endoderm the embryonic cell layer from which the stomach and its diverticula and glands are formed in Eumetazoa.

endogenous detritus detritus originating in the locality where it is found.

endogenous plankton plankton originating in the region where it is found.

endopodite in Crustacea, the inner ramus of a biramus appendage. Also endopod.

endopsammon the microscopic fauna of sand and mud.

endoskeleton a skeleton produced within the flesh of an animal and typically remaining embedded there.

endostyle a glandular ciliated groove along the ventral side of the pharynx in Urochorda, Cephalochorda, and lamprey larvae.

Entomostraca a collective term for all the lower subclasses of Crustacea, most of which are of small size and hatch from the egg as nauplius larvae.

ephyra the earliest free-swimming stage of scyphozoan medusae that have just been released from the polyp, or scyphistoma. Also ephyrula.

epicone in dinoflagellates, the portion of the cell above or anterior to the horizontal groove, or cingulum.

epifauna animals living on the surface of firm substrates. *See* infauna.

epiphyte a plant growing on another plant; sometimes extended to all organisms growing as sessile forms attached to plants.

epiplankton plankton occurring near the surface, in the upper 200 meters.

epipodite a laterally directed extension from one of the basal segments of an arthropod limb.

episome a small bit of nucleic acid in the cytoplasm of a cell, replicating independently of the cell's own genome and transferable to another cell of the same or closely related type by direct contact.

epitheca the upper or larger half of the frustule of a diatom, or the plates above the cingulum in peridinian dinoflagellates.

epizoon an organism growing attached to the external surface of an animal.

eretmocaris stage postlarval stage in certain decapod Crustacea, characterized by possession of long eyestalks.

erichthus larva the second larval stage in Stomatopoda.

estuary the brackish-water areas influenced by tides, where the mouths of rivers come to the sea.

euhalabous plankton, especially phytoplankton living in water of 30 $^0/_{00}$ to 40 $^0/_{00}$ salinity.

euphotic zone the upper strata of water that receive sufficient light to support photosynthesis.

euplankton organisms that spend most or all of their lives as plankton.

eurybathic able to tolerate a considerable range of depths.

euryhaline able to tolerate a wide range of salinities.

eurythermal able to tolerate a wide range of temperatures.

eutrophic plankton plankton in areas of especially high productivity.

exogenous plankton plankton originating in some other locality than in which it was found.

exopodite in Crustacea, the outer branch, or ramus, of a biramous appendage. Also exopod.

exoskeleton in Arthropoda, the hardened cuticular covering of the body, serving for protection of soft parts and as sites for muscle attachments.

extracapsular cytoplasm in Radiolaria, the cytoplasm peripheral to the central capsule.

exumbrella the top, or convex outer surface, of medusae. The aboral surface.

false bottom the deep-scattering layer.

fascicle a bundle or cluster of threadlike, rodlike, or bristlelike structures bound together or arising from the same place, as a fascicle of setae.

fenestra a small opening, as in a shell.

ferruginous the color of iron rust.

filament a threadlike structure.

filiform in the shape of a thread.

flagellum a whiplike structure. In Protozoa of the class Flagellata, the characteristic locomotor organ.

fluke a parasitic flatworm belonging to the digenetic trematodes.

foot in ordinary usage, the terminal portion of a walking appendage that contacts the substrate. In molluscs, the ventral surface of the body, variously modified in different groups for creeping or digging or to serve as prehensile arms.

fossette a small hollow or pit.

fracture zone in geology, a zone along which there has been displacement. In the oceans, major fracture zones are found at right angles to the mid-ocean ridge system.

frustule the siliceous shell of a diatom.

fucoxanthin a brownish xanthin pigment found in seaweeds.

furcilia larva the last larval stages in euphausiids.

fuscous brown-colored and tinged with gray; swarthy.

fusiform tapering toward both ends, as in many pennate diatoms.

gametangium in algae, a special structure in which gametes are borne.

gamete one of the germ cells, a sperm or an egg.

gametophyte in algae, the generation giving rise to gametes.

gasterozooid in Spihonophora and in Doliolidae, individuals of the colony specialized for feeding.

geniculate elbowed, with a kneelike bend, as in the antennae of various isopods and other Crustacea.

genital somite; genital segment in copepods, the somite or fused pair of somites bearing the genital openings, usually the first segment of the urosome.

gill a general term for special structures evolved for underwater respiration.

gill arch the tissue of the pharynx, which supports the gills; it is usually supported by cartilage or bone, in fish, between the gill slits.

gill rakers comblike extensions from the inner side of one or more gill arches in plankton-feeding fish that retain the plankton as water passes through the gill slits.

girdle the connecting bands of the two halves of diatom frustules.

glabrous having a smooth surface.

globigerina ooze a calcareous sediment on the floor of extensive areas of the ocean, containing, among other things, great numbers of foraminiferan shells of the genus *Globigerina*. Because of the solubility of the shells under great pressure, this ooze is not found in the deepest parts of the ocean.

gnotobiotic literally "known biology," said of cultures grown in a chemically defined medium without other organisms or with only known organisms present in the culture.

gonangium in hydroids, the modified polyp from which medusae buds are developed.

gonodendron in Siphonophora, processes in the gonophores bearing the sex cells.

gonophore in Siphonophora, modified medusoid individuals or reproductive zooids bearing the sex organs.

gonozooid in colonial Coelenterata and tunicates, individuals bearing gonads.

greenhouse effect possible self-sustaining increase in temperature water vapor, and carbon dioxide in the atmosphere; heat absorption resulting from increased temperature and carbon dioxide from burning of fossil fuel and pollution.

gross production the total of the organic matter being produced in a given unit of time in the system studied, without any correction for respiration or catabolism. *See* net production.

haliplankton marine or inland saltwater plankton.

haptic clinging tightly, as the Gnathostomulida, which are strongly haptic, clinging to the sand grains.

hekistoplankton flagellated nannoplankton.

helix a coiled spiral, as in gastropod shells.

hemiplankton organisms spending only part of their lives in the plankton; meroplankton.

heterogamous having differentiated gametes, sperm, and egg cells.

heterotrophic the mode of nutrition of organisms that cannot synthesize new organic matter from inorganic substrates. Requiring already formed organic matter. This includes all animals and most bacteria.

hinge joint in Copepoda, the major articulation between the prosoma and the urosome.

holdfast the root or disclike anchoring portion of the thallus of a seaweed.

holopelagic pelagic throughout the entire life cycle.

holophytic having plantlike nutrition, synthesizing new organic matter from inorganic substrates. Photosynthetic.

holoplankton organisms that are planktonic throughout their entire life cycle.

holozoic having animal-like nutrition. Fed by ingestion of organisms of other organic matter.

hyaline transparent or semitransparent, with a glassy or gelatinous consistency.

hydrocoele an embryonic coelomic vesicle in echinoderms and in tornaria larvae of Hemichordata, communicating to the outside by a water pore.

hydroecium in Siphonophora, the extended part of the swimming bell, covering the proximal part of the stem of the colony.

hydrotheca in Hydrozoa, the cuplike extension of the perisarc around the feeding polyp, into which it can withdraw. Present in members of the subclass Calyptoblastea.

hyphalomyraplankton plankton of brackish water.

hypocone in Dinoflagellata, the lower or posterior portion of the cell back of the cingulus.

hypodermis in Arthropoda, the epithelial cells beneath the cuticular exoskeleton.

hypoplankton demersal plankton; plankton taken close to the bottom.

hypotheca in diatoms, the smaller or lower valve of the shell; in armored dinoflagellates the lower plates below or back of the cingulus.

infauna animals living in and on soft bottom.

injector organ in certain burrowing polychaetes, such as Opheliidae, a muscular sac back of the prostomium that withdraws or injects coelomic fluid of the head coelom.

interradial in medusae, the portion midway between the primary radial (perradial) canals of the gastrovascular cavity.

isogamous having gametes that show no obvious differentiation into distinctive male and female gametes. Gametes that look alike.

isolume a theoretical warped plane or level in the water connecting points with equal light intensity.

isotherm a theoretical warped plane or line in the water connecting points of equal temperature.

Japan Current the north- and northeast-flowing current offshore from Japan, also called the Kuroshio Current.

karyotic having nucleated cells.

knephoplankton plankton found between 30 and 500 meters' depth.

krill shoals of euphausiids, which serve as whale food; in Antarctic waters especially, the red *Euphausia superba.*

Kuroshio Current the major warm northeasterly current off the shores of Japan, sometimes termed Japan Current.

Labrador Current a cold south-flowing current along the coast of eastern Canada.

lambert a unit of illumination; the luminance produced by a uniform diffuser with a total flux of one lumen per square centimeter.

lamelliform flat and leaf-shaped.

lamina a leaflike structure or object; a thin layer.

laminate formed of thin layers.

Langmuir circulation the system of cells of convergence and divergence that forms in open water at right angles to the direction of prevailing wind. If the water is flowing at a different angle, the small rips, or zones of convergence, formed may not be directly crosswind in direction.

Laurasia the northern member of the two supercontinents supposed to have existed prior to Tertiary times, composed of what is now North America, Greenland, and northern Eurasia.

laver the green alga, *Ulva,* dried and sold for food.

lecithotrophic said of larvae that are nourished from nutrients in the egg rather than by feeding on plankton.

leptopel large organic molecules and aggregates of molecules of approximately colloidal size dissolved or suspended in the water.

light saturation the amount of light at which a given kind of plant cell will carry out photosynthesis most actively.

Liman Current a cold south-flowing current in the western North Pacific.

linear forming a line, as in an end-to-end arrangement of objects. In graphic representations, usually taken to mean forming a straight line on the graph rather than a curve.

lingual ribbon the radula in molluscs.

lithocyst a cystlike hydrostatic organ in the margin of the disc in coelenterate medusae.

lithostyle a sensory club or tentaculocyst on the free margin of the disc of a jellyfish, usually containing a calcareous concretion with a cushion of sensory cilia. Lithocysts are a modified form of lithostyle.

littoral zone the intertidal zone. Some writers include the entire shelf area.

lophophore the horseshoe- or double-spiral–shaped ridge of tissue, and its supporting structures, surrounding the mouth and bearing ciliated tentacles in the Bryozoa, Brachiopoda, and Phoronidea.

lorica a secreted noncellular protective case, as in Tintinnidae.

lumen (1) an opening; (2) a unit of illumination equivalent to one meter candle or lux.

lunate crescent-shaped.

lunule an impressed area just below the beak in the shell of a bivalve mollusc.

macrobenthos the larger organisms of the benthos, more than 1 mm. in length.

macronucleus the larger vegetative nucleus in ciliate Protozoa, which does not undergo mitosis during division nor participate in nuclear exchange during conjugation.

macroplankton plankton organisms ranging in size from about 1 mm. to 1 cm. in length.

magnetic anomaly a reversal in the direction of magnetic orientation in rocks or sediments, indicating a reversal of polarity of the earth's magnetic field.

major articulation in copepods, the articulation between the prosoma and the urosome.

mantle in molluscs, the dorsal body wall, covering the visceral hump and secreting the shell in shelled forms.

mantle cavity the space between the mantle and the rest of the body in molluscs, in which the mantle has grown down around the body at any point. The mantle cavity commonly contains the ctenidia and the openings of the digestive, reproductive, and excretory systems.

manubrium in jellyfish, an ectodermally lined protrusion from the surface of the subumbrella, bearing the mouth.

marine humus the less digestible organic residues that accumulate in solution or suspension in the seas.

marine terrace a flat, platformlike area showing the position of a former beach, or area cut back by the sea.

maxillae in Crustacea, the two pairs of appendages immediately back of the jaws, or mandibles. The first pair are often termed maxillulae, the term maxillae being used for the second pair.

maxillipeds in Crustacea, the first two or three pairs of thoracic appendages back of the maxillae, when turned forward and modified to aid in the handling of food.

medullary shell in Radiolaria, the part of the shell forming an inner framework.

medusa the sexual stage of Hydrozoa and Scyphozoa, usually in the form of a free-swimming "jellyfish."

megaplankton plankton organisms larger than 1 cm. in length or diameter.

meiobenthos small benthic organisms that will pass through a 1 mm. mesh sieve but be retained by a 0.1 mm. mesh.

melon in toothed whales, a mass of oily blubber between the blowhole and the end of the snout.

membranelle in ciliates, a group of cilia fused together to form a vibratile plate or membrane.

meroplankton organisms occurring in the plankton only periodically, or only for a portion of their life cycle.

merus in Crustacea, the third segment of the leg in the thoracic legs.

meshalobous phytoplankton, or plankton in general, living in brackish water with salinity ranges of 5 $^0/_{00}$ to 20 $^0/_{00}$.

mesogloea the gelatinous noncellular or secondarily cellular material between the outer epithelium and the gastrothelium in many coelenterates. It is particularly well developed in the medusae of Hydrozoa and Scyphozoa.

mesopelagic intermediate depths below the euphotic zone but above the completely aphotic zone. The dysphotic depths. Organisms found in these depths.

mesoplankton planktonic organisms between 0.5 and 1 mm. in size.

metabolite a chemical produced as a result of the metabolism of organisms.

metagenesis the alternation of sexual and asexual generations, as in many hydroids.

metanauplius in Euphausiacea, a postnaupliar larval stage in which appendages in addition to the original three pairs of naupliar limbs are present but before metamorphosis to the calyptopis stage has occurred.

metapleural folds in Cephalochordata, ventrolateral fin folds in the pharyngeal region of the body. Probably metapleural folds of early Chordata gave rise to the paired fins of fish.

metasoma in Copepoda, the thoracic somites behind the cephalosome and in front of the major articulation.

microbenthos benthic organisms small enough to pass through a sieve with a 0.1 mm. mesh.

micron one one-thousandth of a millimeter.

micronucleus in Ciliata, the small nucleus (or nuclei), which is comparable to the nucleus of other cells. It undergoes mitosis during cell division and undergoes reduction divisions to the haploid state prior to conjugation.

microplankton plankton organisms in the size range from 0.5 mm. down to 60 μ.

mid-ocean ridge a system of rifts and parallel mountain ranges or hills found in all major oceans, thought to be sites of the upwelling of new ocean floor material from the earth's mantle, from which the ocean floors are gradually spreading out laterally.

millilambert one one-thousandth of a lambert. A very small unit of illumination used in expressing bioluminescence of some plankters.

miracidium the larval stage that emerges from the egg in digenetic trematodes and carries infection to the molluscan host.

mitraria larva a modified trochophore larva of some annelid tube worms, having two bundles of very long setae extending below the larva.

mixotrophic able to exist by either autotrophic or heterotrophic nutrition, as conditions necessitate.

monomyarian in bivalve molluscs, having only one adductor muscle.

monoton plankton very heavily dominated by one species.

mutualism a relationship between two species in which both benefit, in many cases becoming completely dependent on the relationship, as in the case of some cellulose-eating forms such as termites and the microbial fauna of their intestines.

naked dinoflagellates members of the suborder Gymnodinioidea that do not have protective plates on the cell surface.

nannoplankton very small plankton organisms in the size range from 5 to 60 μ. Plankton organisms too small to be retained by a plankton net.

natatory having the function of swimming.

nauplius a larval stage in many of the lower groups of Crustacea, having only three pairs of appendages and a single median eye.

nectocalyx in Siphonophora, one of the members of the colony modified as a swimming organ.

nectophore in Siphonophora, a nectocalyx, or swimming bell.

nekton aquatic, free-swimming organisms whose swimming activity largely determines the direction and speed of their movement, including fish, cephalopods, some larger crustaceans, etc.

nematocyst the typical stinging cell of coelenterates.

neritic relating to the waters over the continental shelves.

net production gross production minus the amount lost in the same time by catabolic processes.

neuston organisms living in the surface film at the top of the water.

nodule in pennate diatoms, part of the complex system, including the raphe and nodules, that enables them to move.

North Atlantic Current the eastward continuation of the Gulf Stream across the North Atlantic.

North Equatorial Current the west-flowing major current of the North Pacific.

notochord in Chordata, a stiff rod of turgid vacuolated cells extending longitudinally dorsal to the digestive tract, between it and the dorsal nerve cord, and serving as an axial skeletal support. In adults of most vertebrates, it is replaced by the centrum of the vertebral column.

nucleic acid macromolecules characteristic of all living things and containing the coded information by which their genetic characteristics are determined.

oceanic plankton plankton characteristic of the open ocean beyond the continental shelves.

ocellus in Arthropoda, a simple eye. Also used for light-sensitive pigmented structures in other groups, as in medusae.

oecia in colonial Bryozoa, special reproductive zooids containing egg cells.

offlap in geology, the situation where terrestrial deposits are laid down over marine deposits, as when the sea level becomes lower relative to the land in a given region.

oikoplast in Larvacea such as *Oikopleura*, the cell or cells that secrete the gelatinous house.

oligohalobous plankton plankton found in water with salinity of less than 5 $^0/_{00}$.

oligotrophic plankton plankton of seas with a low level of nutrients, where light penetrates deeply into clear, relatively barren water.

onlap in geology, the condition where marine sediments are deposited over terrestrial sediments, as when sea level becomes higher than before, relative to the land in the region.

oostegite the thin cuticular endites from the thoracic legs in female isopods and amphipods, forming the brood pouch.

ooze soft deposits on the floor of the ocean.

otocyst a balancing organ comprised of a small vessel filled with fluid and containing a solid particle or particles that impinge on sensory hairs or cilia.

ovisac in copepods, the egg sac (s) secreted by the ovaries in which the eggs are retained prior to hatching.

Oyashio Current a cold current from the Arctic, in the western North Pacific.

pallial line in bivalve molluscs, a line on the shell showing the attachment of mantle muscles to the shell.

pallial sinus in bivalve molluscs that have a well-developed siphon, an emargination in the pallial line, in the posterior end of the shell where the siphon musculature is attached.

palp in arthropods, the exopodite of mouthpart appendages, sensory in function.

palpon in Siphonophora, a simple mouthless zooid serving as an accessory digestive zooid.

pantomictic plankton plankton not dominated by any particular species.

parameter one of the constants by which a population or environment is defined.

parapyle in Radiolaria of the order Phaeodaria, the two lateral pore fields in the central capsule.

parmal pores in some Radiolaria, the primary pores through the shell, surrounding each radial spine.

parthenogenesis development of an egg without fertilization by a sperm.

pectin a complex carbohydrate produced in plants.

pedalium in certain jellyfish, an expanded gelatinous peduncle bearing one or more marginal tentacles or sense organs.

peduncle in some jellyfish, an extension of the central part of the subumbrella, bearing the manubrium at its end.

pennate a diatom with the valves bilaterally or asymmetrically arranged on a longitudinal axis. Members of the order Pennales.

pereion in Crustacea, the thorax.

pereiopods thoracic appendages, in Crustacea.

perforate provided with holes or pores.

periodic plankton organisms that regularly appear in the plankton only at certain times, as at dusk.

perradius in jellyfish, the axes of the disc in which the primary radial canals lie.

person in colonial coelenterates, urochordates, etc., one of the individuals of the colony.

pervalar axis in diatoms, the axis through the center of the valves or through the length of the diatom cell.

Peru Current the cold north-flowing current along the western side of South America.

phaeoplankton plankton of the upper 30 meters.

phenology study of the quality and quantity of seasonal changes in plankton.

Phi scale a scale used in expressing the relative sizes of particles.

photogenic capable of generating light, as photogenic tissue in luminescent organs.

photokinesis aggregation in a particular region of the spectrum; or movement with speed that varies with variation in the light intensity.

photoreceptor a light-sensitive sense organ.

photosynthesis the synthesis of organic matter from inorganic substrates, using light as a source of energy.

phototaxis response to stimulus by light; movement toward or away from a light source.

phototropism positive phototaxis; attraction toward light.

phragmocone in extinct cephalopods of the group Belemnoidea, a chambered portion of the shell.

phyllosoma larva the flattened, transparent planktonic larva of spiny lobsters.

phytoplankton photosynthetic planktonic organisms such as diatoms and dinoflagellates.

pilidium larva a helmet-shaped planktonic larva of some nemerteans.

plankter a planktonic organism.

planktobiont an organism living solely in the plankton. A holoplanktonic organism.

plankton the floating and drifting organisms in the water whose movements from place to place are due to the movements of the water rather than to their swimming.

plankton spectrum the qualitative and quantitative composition of the plankton population at a given time and place.

planktophile living mainly as plankton, although able to live otherwise.

planktoxene only occasionally in the plankton.

pleon in Crustacea, the abdomen.

pleopod in Crustacea, an abdominal appendage, especially those anterior to the uropods.

pluteus larva the ciliated planktonic larva of sea urchins and other Echinoidea.

pneumatophore in Siphonophora, the float.

podobranch in Crustacea, a gill attached to the basal segment of one of the thoracic legs.

polymictic plankton with several species present in great numbers.

polyp in Coelenterata, the stage commonly developing from a planula larva. In Hydrozoa and Scyphozoa, the polyp is usually a sessile form multiplying asexually, whereas the sexual form is commonly a free-swimming medusa or an attached reduced medusoid. In Anthozoa, there is no medusa stage, and the gonads are borne on the mesenteries in the polyp.

polythalamous many-chambered, as in the shells of some Foraminifera.

polytroch having several ciliated bands, as in some trochophore larvae.

pore canal in echinoderm and hemichordate larvae, a canal from the hydrocoele to the exterior.

pore field in Radiolaria, an area on the central capsule bearing pores.

porosity an expression relating the size and number of pores in a given porous structure to the area. It is a function relating the total area represented by pores to that of the pore-bearing structure. (In ordinary usage in laboratories, the porosity of something, such as a filter, refers only to the size of the pores.)

postlarva in Crustacea, a young individual of essentially adult morphology but not yet sexually mature.

prevalent plankton a species that constitutes more than half the plankton in a given place.

primary consumers organisms that eat plants. Herbivores.

primary production the creation of new organic matter from inorganic substrates.

producers organisms that carry on primary production. Plants.

proloculum in Foraminifera, the initial, or first, chamber of the test.

proostracum in extinct cephalopods of the group Belemnoida, a dorsal platelike portion of the shell.

prosoma in Copepoda, the forepart of the body in front of the major articulation.

proteolytic capable of degrading proteins. Said of enzymes that digest proteins.

protoplast the protoplasm within a cell wall; the protoplasm of a cell.

prototroch in trochophore larvae, the ciliated band in a preoral position.

proximal toward the origin of an appendage.

pseudochitinous superficially like chitin but chemically different.

pseudocoelom a body space, as in the Aschelminthes, between the gut and the body wall, not lined by a peritoneum and derived from the blastocoel rather than from an internal split in the mesoderm.

pseudoplankton organisms not normally planktonic, carried into the plankton by turbulence.

pustule a vacuole-like structure in dinoflagellates.

pyrenoid a structure in the plastids of plants concerned with the synthesis of starch.

radial canals in coelenterate medusae, tubes of the gastrovascular system passing from the centrally located stomach to the periphery of the disc.

radioles in many annelid tube worms, the feathery processes extending from the anterior end, used for food-gathering and also as sensory structures.

radula the tonguelike lingual ribbon of most molluscs except bivalves.

ramus a branch; in Crustacea, one of the branches of a biramous appendage.

raphe the longitudinal slit in the side of pennate diatoms.

raptorial pertaining to structures such as claws or pincers used to seize prey or enemies. The term is also used to describe organisms that actively seize their prey, as raptorial birds.

reducers organisms such as bacteria whose life activities result in the mineralization of organic matter.

reticulopod pseudopodia forming an anastomosing network, as in Foraminifera.

reticulum a network.

rhopalia in Scyphozoa, marginal sense organs that are reduced and modified tentacles containing an endodermal statolith.

rhopaloneme a type of nematocyst found only in Siphonophora.

rostrum in Crustacea, a forward prolongation of the carapace beyond the head.

salinity the equivalent amount of sodium chloride, usually expressed as parts per thousand, which would give to the water the same osmotic properties as those which result from the total amount of dissolved salts in the water; loosely, the saltiness of the water.

salt pump a physiological mechanism that concentrates an ion against the osmotic gradient.

saprophytic a mode of nutrition found in nongreen plants and many bacteria that require organic food absorbed through the body surface. If the substrate furnishing this food cannot be absorbed directly, it is partially broken down or digested outside the organism by exoenzymes before absorption.

saprozoic a somewhat similar mode of nutrition in some animals such as tapeworms, differing in that the subsequent metabolism of the organic food is of a different type than that found in plants.

scapulet in jellyfish of the group Rhizostomeae, outgrowths of the oral arms near the bell, bearing supplementary mouth openings. Scapulets are not found in all rhizostomes, only in certain families and genera.

sculpture elevated or depressed marks on a surface, as on a shell.

scutum one of the paired movable calcareous plates of a barnacle.

scyphistoma the sessile polypoid stage in the life of some Scyphozoa.

secondary production the sum of the organic matter produced as a result of predation on primary producers and their products. Net secondary production would be this figure less the loss due to catabolism.

sessile with respect to organisms, those which are fixed to the substrate. With respect to structures that are sometimes stalked or raised, the condition when they are not so stalked or raised.

seston particulate matter in the water.

setae bristlelike stiff cuticular structures, as in the parapodia of polychaetes, on the head of chaetognaths, or on the appendages of various crustaceans, etc.

sibling species species that are very much alike but that occur in separated areas or separate ecological niches. Twin species.

single-species blooms plankton blooms very much dominated by one species.

sinistral left-turning, with shell aperture on the left. *See* dextral.

siphonoglyph in most Anthozoa except corals, a ciliated canal at one or both ends of the gullet serving as a means of passing a current of water in and out of the gastrovascular cavity.

somatocyst in many siphonophores, the part of the gastrovascular canal lying in front of the proximal part of the stem.

spermaceti in sperm whales, the clear high-grade oil found in the melon, or spermaceti organ, in the rostrum.

spermatophore a packet of sperm within a complicated case secreted by the male, as in copepods and cephalopods.

spicules needles or variously shaped structures of calcium carbonate or siliceous material found in the flesh or tissues of certain organisms, as in sponges.

spongin the fibrous horny flexible material making up the skeleton of various Demospongiae. In the order Keratosa, of which commercial sponges are representative, it comprises the entire skeletal framework of the sponge. Other groups also have spicules or may have only spicules.

statocyst a lithocyst; a small vesicle containing a statolith, serving as a balancing organ.

statolith in jellyfish, a small calcareous concretion within a lithocyst.

stenobathic having a limited vertical range, or a range of depth tolerance.

stenohaline having a limited range of salinity tolerance.

stenothermal having a limited range of temperature tolerance.

stratosphere in oceanography, nearly uniform masses of cold water in the high latitudes and of cold bottom water in middle and low latitudes.

styliform having the form of a dagger, or stiletto.

subchelate a chela in which the dactyl simply presses back upon or opposes a somewhat enlarged penultimate segment of the appendage, rather than opposing a corresponding process of that segment.

subumbrella in jellyfish, the lower or inner side of the bell, usually the concave side, bearing the mouth or manubrium.

swimmeret in higher Crustacea, the biramous appendages of the abdominal segments; pleopods.

symbiosis literally a "living together," a term used by some biologists to include strictly parasitic, commensal, and mutualistic relationships between organisms but by others restricted to mutualistic relationships.

sympatric with overlapping geographical distribution.

syncytium multinucleate; protoplasm containing several nuclei without corresponding division into cells.

telotroch in trochophore larvae, a circlet of cilia in front of the anus.

telson in higher Crustacea, the caudal end of the body behind the last true somite. The telson contains the anal opening, and together with the uropods comprises the tail fan.

tentaculocyst a lithostyle.

tentilla lateral branches of the tentacle of gastrozooids in Siphonophora.

tergum in Arthropoda, the skeletal plate covering the dorsal part of each free segment.

theca a case, lorica, or shell surrounding an animal.

torsion in Gastropoda, the twisting of the visceral hump during larval life, bringing the mantle cavity and gills to a forward position.

transapical the width in a diatom shell.

trophosphere in oceanography, the upper layer of oceanic waters in middle and low latitudes where there are strong currents and warm temperatures.

tubulus in some Radiolaria, a wide radial cylindrical extension of the skeleton.

tychoplankton not regularly part of the plankton; brought up from the bottom by turbulence.

ultraplankton planktonic organisms less than 5 μ in length or diameter.

uropod in higher Crustacea, the last pair of abdominal appendages, flattened and, together with the telson, making up the tail fan.

zoospore a flagellated plant spore capable of swimming.

zygote a fertilized egg cell.

Bibliography

This list comprises only a representative sampling of the literature of marine biology—it is by no means a complete listing.

Abbott, R. T. 1954. American seashells. D. Van Nostrand Co., Inc., Princeton, N. J.

Adolph, E. F. 1925. Some physiological distinctions between freshwater and marine organisms. Biol. Bull. 48:327-335.

Adolph, E. F., and P. E. Adolph. 1925. Regulation of body volume in fresh-water organisms. J. Exp. Zool. 43:105-149.

Agassiz, A. 1906. Reports on scientific results of the expedition to the Eastern Tropical Pacific, in charge of Alexander Agassiz, by the U. S. Fish Commission Steamer "Albatross" from Oct., 1904, to March, 1905. Part V. General report of the expedition. Mem. Mus. Comp. Zool. Harv. 33:xiii, 1-75.

Agassiz, E. C., and A. Agassiz. 1865. Seaside studies in natural history: marine animals of Massachusetts Bay. Ticknor & Fields, Boston.

Ahlstrom, E. H. 1952. Oceanographic instrumentation. Nat. Acad. Sci. Pub. 309:36-50.

Aix, W. W. von. 1962. Introduction to physical oceanography. Addison-Wesley Publishing Co., Inc., Reading, Mass.

Alexander, W. B., B. A. Southgate, and R. Bassindale. 1932. Salinity of water retained in the muddy foreshore of an estuary. J. Mar. Biol. Ass. U. K. 18:297-358.

Allee, W. C. 1931. Animal aggregations. The University of Chicago Press, Chicago.

Allee, W. C., A. E. Emerson, O. Park, T. Park, and K. P. Schmidt. 1949. Principles of animal ecology. W. B. Saunders Co., Philadelphia. 837 p.

Allen, E. J. 1919. A contribution to the quantitative study of plankton. J. Mar. Biol. Ass. U. K. 12:1-8.

Allen, E. J., and E. W. Nelson. 1910. On the artificial culture of marine plankton organisms. J. Mar. Biol. Ass. U. K. 8:421-474.

Allen, W. E. 1920a. Quantitative and statistical study of plankton of the San Joaquin River and its tributaries in and near Stockton, Calif., in 1913. Univ. California Pub. Zool. 22:1-292.

Allen, W. E. 1920b. Some work on marine phytoplankton in 1919. Trans. Amer. Microbiol. Soc. 40:177-181.

Allen, W. E. 1921. Problems of floral dominance in the open sea. Ecology 2:26-31.

Allen, W. E. 1922a. Quantitative studies on marine phytoplankton at La Jolla in 1919. Univ. California Pub. Zool. 22:329-347.

Allen, W. E. 1922b. Quantitative studies on inshore marine diatoms and dinoflagellates of Southern California in 1920. Univ. California Pub. Zool. 22:369-378.

Allen, W. E. 1923a. Tidewater collections of marine diatoms taken at half-hour intervals near San Diego, Calif. Univ. California Pub. Zool. 22:413-416.

Allen, W. E. 1923b. Marine diatoms and dinoflagellates caught by aid of Kofoid bucket in 1922. Univ. California Pub. Zool. 22:435-445.

Allen, W. E. 1924. Surface catches of marine diatoms and dinoflagellates made in 1923 by U.S.S. "Pioneer" between San Diego and Seattle. Univ. California Pub. Zool. 26:243-248.

Allen, W. E. 1930. Methods in quantitative research in marine microplankton. Bull. Scripps Inst. Oceanogr. (Tech. Ser.) 2:319-329.

Allen, W. E. 1939. Cutthroat competition in the sea. Sci. Monthly 49:111-119.

Allen, W. E. 1946. "Redwater" in La Jolla Bay in 1945. Trans. Amer. Microbiol. Soc. 65:149-153.

Alvarino, A. 1957a. Estudio del zooplancton del Mediterráneo occidental. Campaña del Xauen en 1954. Bol. Inst. Esp. Oceanogr. 81:1-26.

Alvarino, A. 1957b. Zooplancton del Atlántico ibérico. Bol. Inst. Esp. Oceanogr. 82:1-51.

Amos, W. H. 1952. Marine and estuarine fauna of Delaware. Univ. Delaware Mar. Lab. Annu. Rep. 1:14-20.

Amos, W. H. 1954. Biological survey of the Delaware River estuary. Univ. Delaware Mar. Lab. Annu. Rep. 2:21-31.

Amos, W. H. 1955. Guide to aquatic animals of Delaware and to the waters in which they live. Thesis. University of Delaware. 140 p.

Amos, W. H. 1956. Selective bibliography concerning the hydrography and biology of the Delaware River estuary. Univ. Delaware Mar. Lab. Ref. 56-57. 12 p.

Amos, W. H. 1959. The living wreck. Estuarine Bull. 4:8-10.

Amos, W. H. 1960. Life in bays. National Audubon Society, Nature Program. Doubleday & Co., Inc., Garden City, N. Y. 64 p.

Amos, W. H. 1966. The life of the seashore. McGraw-Hill Book Co., New York. 231 p.

Andel, T. H. van. 1968. Deep-sea drilling for scientific purposes: a decade of dreams. Science 160:1419-1424.

Andersen, H. T. [ed.]. 1969. Biology of marine mammals. Academic Press, Inc., New York.

Anderson, A. E., E. C. Jonas, and H. Odum. 1958. Alteration of clay minerals by digestive processes of marine organisms. Science 127:190-191.

Andree, K. 1920. Geologie des Meeresbodens. Borntraeger, Leipzig.

Andrews, C. W. 1906. A descriptive catalogue of the tertiary vertebrata of the Fayûm, Egypt. Sirenia Rep. Brit. Mus. (N. H.) pp. 197-218.

Annandale, N. 1922. The marine element in the fauna of the Ganges. Bijdr. Dierk. Amsterdam (Feest-Num. W. Weber):143-154.

Apstein, C. 1907. Das Plankton im Colombo See auf Ceylon. Zool. Jahrb. Abt. Syst. 25:201-244.

Apstein, C. 1910. Das Plankton des Gregory-Sees auf Ceylon. Zool. Jahrb. Abt. Syst. 29:661-680.

Arber, A. 1920. Water plants: a study of aquatic angiosperms. Cambridge University Press, New York. Reprinted, 1963, by J. Cramer. Hafner Publishing Co., Inc., New York.

Armstrong, F. A. J. 1954. Phosphorus and silicon in seawater off Plymouth during the years 1950-1953. J. Mar. Biol. Ass. U. K. 33:381-382.

Armstrong, F. A. J. 1957. Iron content of seawater. J. Mar. Biol. Ass. U. K. 36:509-517.

Armstrong, F. A. J. 1968. Inorganic suspended matter in seawater. J. Mar. Res. (Thompson Anniversary vol.) 17:23-34.

Arnold, A. F. 1901. The sea beach at ebb tide. D. Appleton-Century Co., New York.

Atkins, W. R. G. 1922. Influence upon algal cells of an alteration in the H^+ ion concentration of seawater. J. Mar. Biol. Ass. U. K. 12:789-791.

Atkins, W. R. G. 1926. The phosphate content of sea water in relation to the growth of algal plankton. J. Mar. Biol. Ass. U. K. 14:447-467.

Atkins, W. R. G. 1945. Autotrophic flagellates as a major constituent of the oceanic phytoplankton. Nature (London) 156:446-447.

Atkins, W. R. G. 1953. Seasonal variation in copper content of seawater. J. Mar. Biol. Ass. U. K. 31:493-494.

Atkins, W. R. G., and H. H. Poole. 1940. A cubical photometer for studying angular distribution of submarine daylight. J. Mar. Biol. Ass. U. K. 24:271-281.

Atkins, W. R. G., and H. H. Poole. 1958. Cube photometer measurements of angular distribution of submarine daylight and total submarine illumination. J. Cons. Perm. Int. Explor. Mer 23:327-336.

Aurich, H. J. 1949. Die Verbreitung des Nannoplanktons in Oberflächenwasser von der Nordfriesischen Küste. Ber. Deut. Komm. Meeresforsch. 11:403-405.

Austin, T. S. 1957. Summary of oceanographic and fishery data, Marquesas Islands area, Aug.-Sept., 1956 (Equapac). Special report: Fisheries no. 217. Fish and Wildlife Service, U. S. Department of the Interior, Washington, D. C. 186 p.

Austin, T. S., and V. E. Brock. 1959. Meridional variations in some oceanographic and marine biological factors in the Central Pacific. International Oceanographic Congress Preprints. American Association for the Advancement of Science, Washington, D. C.

Baer, J. G. 1951. Ecology of animal parasites. University of Illinois Press, Urbana.

Bainbridge, R. 1949. Movements of zooplankton in

diatom gradients. Nature (London) 163:910-912.

Bainbridge, R. 1953. Studies on the interrelationships of zooplankton and phytoplankton. J. Mar. Biol. Ass. U. K. 32:375-445.

Bainbridge, R. 1957. Size, shape, and density of marine phytoplankton concentrations. Biol. Rev. 32:91-115.

Bakus, G. J. 1964. Effects of fish-grazing on invertebrate evolution in shallow tropical waters. Occasional Papers, Allan Hancock Found. Pub. 27:1-29.

Bakus, G. J. 1966. Some relationships of fishes to benthic organisms (algae, invertebrates) on coral reefs. Nature 210:280-284.

Bakus, G. J. 1969. Energetics and feeding in shallow marine waters. Int. Rev. Gen. Exp. Zool. 4:275-369.

Bal, D. V., and L. B. Pradhan. 1952. Records of zooplankton in Bombay waters during 1944-1947. J. Univ. Bombay N. S. [B] 20:75-80.

Banning, G. H. 1933. Hancock Expedition to the Galápagos Islands, 1933 (general report). Bull. Zool. Soc. San Diego, no. 10.

Banse, K. 1955. Über das Verhalten von meroplanktischen Larven im geschichteten Wasser. Kieler Meeresforsch. 11:188-200.

Banse, K. 1956. Über den Transport von meroplanktischen Larven aus dem Kattegat in die Kieler Bucht. Ber. Deut. Komm. Meeresforsch. 14:147-164.

Barbour, T. 1912. A contribution to the zoogeography of the East Indian Islands. Mem. Mus. Comp. Zool. Harv. 44:1-203.

Barlow, J. P. 1955. Physical and biological processes determining the distribution of zooplankton in a tidal estuary. Biol. Bull. Mar. Biol. Lab. Woods Hole 109:211-225.

Barnes, C. A., and T. G. Thompson. 1938. Physical and chemical investigations in the Bering Sea and portions of the North Pacific Ocean. Univ. Washington Pub. Oceanogr. 3 (2):35-79.

Barnes, H. 1959a. Apparatus and methods of oceanography. Vol. I. Chemical apparatus and methods. John Wiley & Sons, Inc., New York. 341 p.

Barnes, H. 1959b. Oceanography and marine biology: a book of techniques. The Macmillan Co., New York.

Barnes, R. D. 1968. Invertebrate zoology. 2nd ed. W. B. Saunders Co., Philadelphia.

Barnet, L., and the editors of Life. 1960. The wonders of life on earth. Time, Inc., New York.

Bartsch, P. 1921. Ocean currents, the prime factor in the distribution of marine molluscs on the west coast of America. Special Publications, Berenice P. Bishop Mus., no. 7, pp. 505-526.

Bary, B. McK. 1959a. Biogeographic boundaries: the use of temperature-salinity-plankton diagrams. International Oceanographic Congress Preprints. American Association for the Advancement of Science, Washington, D. C.

Bary, B. McK. 1959b. Species of zooplankton as a means of identifying different surface waters and demonstrating their movements and mixing. Pacific Sci. 13:14-54.

Bascom, W. 1961. A hole in the bottom of the sea. Doubleday & Co., Inc., Garden City, N. Y.

Bates, M. 1960. The forest and the sea. Random House, Inc., New York.

Bayer, F. M., and B. O. Harding. 1969. The free-living lower invertebrates. The Macmillan Co., New York.

Bé, A. W. H. 1968. Shell porosity of recent planktonic Foraminifera as a climatic index. Science 161:881-884.

Beanland, F. L. 1940. Sand and mud communities in the Dover Estuary. J. Mar. Biol. Ass. U. K. 24:589-611.

Beebe, W. 1928. Beneath tropic seas. G. P. Putnam's Sons, New York.

Beebe, W. 1935. Half a mile down. John Lane, London. 344 p.

Beklemishev, K. V. 1957. Spatial interrelationships of marine zooplankton and phytoplankton. Trudy Inst. Okeanol. 20:253-278.

Belser, W. L. 1959. Bioassay of organic materials in seawater. International Oceanographic Congress Preprints. American Association for the Advancement of Science, Washington, D. C.

Bentley, J. A. 1960. Plant hormones in marine phytoplankton: zooplankton and seawater. J. Mar. Biol. Ass. U. K. 39:433-444.

Berger, W. H. 1967. Foraminiferal ooze: solution at depths. Science 156:383-385.

Bernal, J. D. 1967. The origin of life. The World Publishing Co., Cleveland.

Bernard, F. 1939. Étude sur les variations de la fertilité des eaux méditerranéennes. J. Cons. Perm. Int. Explor. Mer 14:228-241.

Bernard, F. 1948. Recherches sur le cycle de Coccolithus fragilis Lohm. flagellé dominant des mers chaudes. J. Cons. Perm. Int. Explor. Mer 15:177-188.

Bernard, F. 1955a. Densité du plancton vu au large de Toulon depuis la bathyscaphe F.N.R.S. III. Bull. Inst. Oceanogr. Monaco 1063:1-16.

Bernard, F. 1955b. Étude préliminaire quantitative de la repartition saisonnière du zooplancton de la baie d'Alger. Bull. Inst. Oceangr. Monaco 1065:1-28.

Bernard, F. 1958. Données récentes sur la fertilité élémentaire en la Mediterranée. Rapp. P.-v. Réun. Cons. Perm. Int. Explor. Mer 144:103-108.

Bernstein, T. 1934. Zooplankton des nordlichen Teiles des Karischen Meeres [Russian and German summary]. Trans. Arch. Inst. Leningrad 9:3-58.

Berrill, N. J. 1966. The life of the ocean. McGraw-Hill Book Co., New York.

Berrill, N. J., and J. Berrill. 1951. The living tide. Dodd, Mead & Co., New York.

Berrill, N. J., and J. Berrill. 1959. 1001 questions answered about the seashore. Dodd, Mead & Co., New York.

Bert, P. 1871. Sur les phénomènes et les causes de la mort des animaux d'eau douce que l'on plonge dans l'eau de mer. C.R. Acad. Sci. (Paris) 73:372-385, 464-467.

Bert, P. 1873. La mort des animaux d'eau douce que l'on immerge dans l'eau de mer. C. R. Soc. Biol. (Paris) 23:59-61.

Bert, P. 1883. Sur la cause de la mort des animaux d'eau douce que l'on plonge dans l'eau de mer, et réciproquement. C. R. Acad. Sci. (Paris) 97:133-136.

Bigelow, H. B. 1914a. Explorations in the Gulf of Maine, July-Aug., 1912, by the U. S. Fisheries schooner "Grampus." (Oceanography and notes on plankton.) Mem. Mus. Comp. Zool. Harv. 58:393-419.

Bigelow, H. B. 1914b. Oceanography and plankton of Massachusetts Bay and adjacent water, Nov., 1912, to May, 1913. Mem. Mus. Comp. Zool. Harv. 58:393-419.

Bigelow, H. B. 1915. Exploration of the coastal water between Nova Scotia and Chesapeake Bay, July-Aug., 1913. Mem. Mus. Comp. Zool. Harv. 59:149-360.

Bigelow, H. B. 1917. Exploration of the coastal water between Cape Cod and Halifax in 1914-1915 by the U. S. Fisheries schooner "Grampus." (Oceanography and notes on plankton.) Mem. Mus. Comp. Zool. Harv. 61:365-442.

Bigelow, H. B. 1922. Exploration of the coastal water off the Northeast United States in 1916 by U. S. Fisheries schooner "Grampus." Mem. Mus. Comp. Zool. Harv. 65:85-188.

Bigelow, H. B. 1926a. Plankton of the offshore waters of the Gulf of Maine. Bull. Bur. Fish. (Washington) 40 (2):1-509.

Bigelow, H. B. 1926b. Physical oceanography of the Gulf of Maine. Bull. Bur. Fish. (Washington) 40 (2):511-1027.

Bigelow, H. B. 1938. Plankton of the Bermuda oceanographic expeditions. VIII. Medusae taken during the years 1929-1930. Zoologica 23:99-189.

Bigelow, H. B., L. C. Lillick, and M. Sears. 1940. Phytoplankton and planktonic protozoa of off-shore waters of the Gulf of Maine. I. Numerical distribution. Trans. Amer. Phil. Soc. 31:149-191.

Bigelow, H. B., and M. Sears. 1939. Studies of the waters of the continental shelf, Cape Cod to Chesapeake Bay. III. Study of zooplankton. Mem. Mus. Comp. Zool. Harv. 54:183-378.

Blaxter, J. H. S., and R. I. Currie. 1967. Effect of artificial lights on acoustic scattering layers in the ocean. In N. B. Marshall [ed.] Aspects of marine zoology. Academic Press, Inc., New York.

Blegvad, H. 1914. Food and conditions of nourishment among communities of invertebrate animals on or in the sea bottom in Danish waters. Rep. Dan. Biol. Sta. 22:41-78.

Blegvad, H. 1916. On the food of fish in Danish waters. Rep. Dan. Biol. Sta. 24:17-72.

Blegvad, H. 1925. Quantity of fish food in the sea bottom. Rep. Dan. Biol. Sta. 31:27-56.

Blegvad, H. 1929. Mortality among animals of the littoral region in icy winters. Rep. Dan. Biol. Sta. 35:49-62.

Boden, B. P., and E. N. Kampa. 1953. Winter cascading from an oceanic island and its biological implications. Nature (London) 171:426-427.

Boden, B. P., and E. N. Kampa. 1967. Influence of natural light on vertical migrations of an animal community in the sea. In N. B. Marshall [ed.] Aspects of marine zoology. Academic Press, Inc., New York.

Bogorov, B. G. 1934. Investigation of nutrition of plankton-consuming fishes. Bull. Inst. Mar. Fish. Oceanogr. U.S.S.R. 1:1-18.

Bogorov, B. G. 1939. The particularities of seasonal phenomena in the plankton of Arctic seas and their significance for ice forecastings. Zool. J. U.S.S.R. 18:735-747.

Bogorov, B. G. 1941. Biological seasons in the plankton of different seas. C. R. Acad. Sci. (U.S.S.R.) 31:404-407.

Bogorov, B. G. 1946. Peculiarities of diurnal vertical migration of zooplankton in polar seas. J. Mar. Res. 6:25-32.

Bogorov, B. G. 1956. Unification of plankton research. Coll. Int. Biol. Mar. Sta. Biol. Roscoff. Ann. Biol. 33:299-315.

Bogorov, B. G. 1957. Standardization of marine plankton investigations. Trudȳ Inst. Okeanol. 24 (R).

Bogorov, B. G. 1958a. Estimates of primary production in biogeographical regionalization of the ocean. Rapp. P.-v. Réun. Cons. Perm. Int. Explor. Mer 144:117-121.

Bogorov, B. G. 1958b. Perspectives in the study of seasonal changes of plankton and of the number of generations at different latitudes. In A. A. Buzzati-Traverso [ed.] Perspectives in marine biology. University of California Press, Richmond.

Bogorov, B. G., and C. W. Beklemishev. 1955. On the phytoplankton production in the western

North Pacific [in Russian]. C. R. Acad. Sci. (U.S.S.R.) **104**:141-143.

Bogorov, B. G., and M. E. Vinogradov. 1955. Some essential features of zooplankton distribution in the northwestern Pacific [in Russian]. Trudȳ Inst. Okeanogr. **23**:113-123.

Bossanyi, J. 1958. Preliminary survey of small natant fauna in the vicinity of the sea floor off Blyth, Northumberland. J. Anim. Ecol. **26**:353-368.

Boughey, A. S. 1968. Ecology of populations. The Macmillan Co., New York.

Bowen, E. J. [ed.]. 1965. Recent progress in photobiology. Academic Press, Inc., New York.

Bowman, R. I. [ed.]. 1964. Proceedings of the Galápagos International Scientific Project of 1964. University of California Press, Richmond.

Boyko, H. [ed.]. 1965. Salinity and aridity. W. Junk, Publishers, The Hague, The Netherlands. 300 p.

Braarud, T. 1935. The Ost Expedition to the Denmark Strait, 1929. II. Phytoplankton and its conditions of growth. Hvalråd Skr. **10**:1-173.

Braarud, T. 1961. Cultivation of marine organisms as a means of understanding environmental influences on populations. In M. Sears [ed.] Oceanography. Pub. no. 67. American Association for the Advancement of Science, Washington, D. C.

Braarud, T., K. R. Gaarden, and J. Grontoed. 1953. The phytoplankton of the North Sea and adjacent waters in May, 1948. Rapp. P.-v. Réun. Cons. Perm. Int. Explor. Mer **133**:1-89.

Braarud, T., and A. Klem. 1931. Hydrographical and chemical investigations in the coastal waters off More and in the Romsdalsfjord. Hvalråd Skr. **1**:1-88.

Brahmachary, R. L. 1967. Physiological clocks. Int. Rev. Cytol. **21**:65-89.

Brand, T. von. 1934. Methods for determination of nitrogen and carbon in small amounts of plankton. Biol. Bull. Mar. Biol. Lab. Woods Hole **69**:221-232.

Brand, T. von, and N. W. Rakestraw. 1941. The determination of dissolved organic nitrogen in seawater. J. Mar. Res. **4**:76-89.

Brand, T. von, N. W. Rakestraw, and C. E. Renn. 1937-1947. Decomposition and regeneration of nitrogenous organic matter in seawater. I-VI. Biol. Bull. Mar. Biol. Lab. Woods Hole **72**:165-175, **77**:285-296, **79**:231-236, **81**:63-69, **83**:273-282, **92**:110-114.

Brandhorst, W. 1958. Nitrate accumulation in the northeast tropical Pacific. Nature (London) **182**:679.

Brandhorst, W. 1959. Nitrification and denitrification in the eastern tropical North Pacific. J. Cons. Perm. Int. Explor. Mer **25**:3-20.

Brandt, K. 1898. Beitrage zur Kenntnis der chemischen Zusammensetzung des Planktons. Wiss. Meeresuntersuch. Abt. Kiel **3**:43-90.

Brandt, K., and C. Apstein [ed.]. 1901-1942. Nordisches Plankton. (Zoology, 7 vols.; botany, 1 vol.) Lipsius & Tischer, Publishers, Kiel and Leipzig. Reprint, A. Asher & Co., Amsterdam.

Brandt, K., and E. Raben. 1919. Zur Kenntnis der chemischen Zusammensetzung des Plankters und einiger Bodenorganismen. Wiss. Meeresuntersuch. Abt. Kiel **19**:175-210.

Breeder, C. M. 1948. Field book of marine fishes of the Atlantic coast. G. P. Putnam's Sons, New York.

Brongersma-Sanders, M. 1957. Mass mortality in the sea. In J. W. Hedgpeth [ed.] Treatise on marine ecology and paleoecology. Vol. I. Ecology. Geological Society of America, New York.

Brotskaja, V. A., and L. A. Zenkevich. 1939. Quantitative evaluation of the bottom fauna of the Barents Sea. Trudȳ Vses. Nauchno-issled. Inst. Morsk. Rȳb. Khoz. Okeanogr. **4**:99-126.

Brouardel, J., and L. Fage. 1953. Variation, en mer, de la teneur en oxygène dissous au proche voisinage des sediments. C. R. Acad. Sci. (Paris) **237**:1605-1607.

Brown, F. A. 1958. Studies of the timing mechanisms of daily, tidal, and lunar periodicities of organisms. In A. A. Buzzati-Traverso [ed.] Perspectives in marine biology. University of California Press, Richmond. (Brown and his coworkers have also published several papers on endogenous rhythms of fiddler crabs and other organisms, reference to which can be found in this paper or that of Piffendrigh.)

Bruce, H. E., and D. W. Hood. 1969. Diurnal inorganic phosphate variations in Texas bays. Pub. Inst. Mar. Sci. **6**:133-165.

Bruce, J. R. 1928. Physical factors on the sandy beach: tidal, climatic, and edaphic. J. Mar. Biol. Ass. U. K. **15**:535-552.

Brunelli, G. 1941. Note sul plancton della Laguna Veneta. Archo. Oceanogr. Limnol. **1**:55-56,

Bruns, E. 1958. Ozeanologie. Band I. V. E. B. Verlag der Wissenschaft, Berlin. 420 p.

Bruun, A. F. 1957. Deep-sea and abyssal depths. In J. W. Hedgpeth [ed.] Treatise on marine ecology and paleoecology. Vol. I. Ecology. Geological Society of America, New York.

Bruun, A. F., S. Greve, H. Mielche, and R. Sparch [ed.]. 1956. The "Galathea" deep-sea expeditions, 1950-1952. The Macmillan Co., New York.

Bsharah, L. 1957. Plankton of the Florida Current. V. Bull. Mar. Sci. Gulf Carib. **7**:201-251.

Buchanan, J. B. 1968. The bottom fauna communities across the Continental Shelf off Accra, Ghana (Gold Coast). Proc. Zool. Soc. (London) **130**:1-56.

Budker, P. 1959. Whales and whaling. The Macmillan Co., New York.

Buljan, M. 1959. Utjecaj šum na produkcija priobalnim vodama [Influence of the forest upon the production in coastal waters]. Morskov. Rẏb. Barst. **11** (9):193-194.

Burckhardt, G. 1899. Vorläufige Mittheilung über Planktonstudien an Schweizer Seen. Zool. Anz. **22**:185-189.

Burkenroad, M. D. 1931. Notes on the sound-producing marine fishes of Louisiana. Copeia 1:20-28.

Burkholder, P. R. 1959. Vitamin-producing bacteria in the sea. International Oceanographic Congress Preprints. American Association for the Advancement of Science, Washington, D. C.

Burkholder, P. R., and L. M. Burkholder. 1956. Vitamin B_{12} in suspended solids and marsh muds collected along the coast of Georgia. Limnol. Oceanogr. 1:202-208.

Buzas, M. A., and T. G. Gibson. 1969. Species diversity: benthic Foraminifera in the western North Atlantic. Science **163**:72-74.

Buzzati-Traverso, A. A. [ed.]. 1958. Perspectives in marine biology. University of California Press, Richmond.

Callaway, R. J. 1957. Oceanographic and meteorological observations in the northeast and central North Pacific, July-Dec., 1956. Special science report no. 230. Fish and Wildlife Service, U. S. Department of the Interior, Washington, D. C.

Calvin, M. 1961. Chemical evolution. Condon Lecture, Oregon State System of Higher Education. University of Oregon Press, Eugene.

Calvin, M., C. Heidelberger, J. C. Reid, B. M. Tolbert, and P. F. Yankwich. 1949. Isotopic carbon. John Wiley & Sons, Inc., New York.

Candeias, A. 1930. Estudas de plancton ova Baia Sesimbra. Bolm Soc. port. Ciênc. nat. **1** (3):11-22.

Carey, C. L. 1938. Occurrence and distribution of nitrifying bacteria in the sea. J. Mar. Res. 1:291-304.

Carrington, R. 1957. Mermaids and mastodons. Holt, Rinehart & Winston, Inc., New York.

Carrington, R. 1960. Biography of the sea. Basic Books, Inc., New York.

Carson, R. 1951. The sea around us. Houghton Mifflin Co., Boston. Rev. ed., 1961. Oxford University Press, New York.

Carson, R. 1955. The edge of the sea. Houghton Mifflin Co., Boston.

Carson, R. 1962. Silent spring. Houghton Mifflin Co., Boston.

Carter, L. J. 1968. Continental shelf: scramble for federal oil-lease revenues. Science **160**:1431-1432.

Carter, L. J. 1970. Galveston Bay: test case of an estuary in crisis. Science **167**:1102-1108.

Caspers, H. 1957. The Black Sea and the Sea of Azov. *In* J. W. Hedgpeth [ed.] Treatise on marine ecology and paleoecology. Vol. I. Ecology. Geological Society of America, New York.

Castro, J. de. 1952. The geography of hunger. Little, Brown & Co., Boston. 336 p.

Chapin, H., and W. Smith. 1952. The ocean river. Charles Scribner's Sons, New York.

Cheng, C. 1941. Ecological relations between the herring and the plankton off the northeast coast of England. Hull Bull. Mar. Ecol. 1:239-254.

Chesher, R. H. 1969. Destruction of Pacific corals by the sea star *Acanthaster planci*. Science **165**:280-288.

Chichester, C. O. [ed.]. 1965. Research in pesticides. Academic Press, Inc., New York.

Childress, J. J. 1968. Oxygen minimum layer: vertical distribution and respiration of the mysid *Gnathophausia ingens*. Science **160**:1242-1243.

Chow, T. J., and T. G. Thompson. 1952. The determination and distribution of copper in seawater. I. J. Mar. Res. **11**:124-138.

Chow, T. J., and T. G. Thompson. 1954. Seasonal variation in the concentration of copper in the surface waters of San Juan Channel, Washington. J. Mar. Res. **13**:233-244.

Chu, G. W. T. C., and C. E. Cutress. 1954. Human dermatitis caused by marine organisms in Hawaii. Proc. Hawaii Acad. Sci. **29**:9.

Clarke, A. C. 1960. The challenge of the sea. Holt, Rinehart & Winston, Inc., New York.

Clarke, G. L. 1936. Light penetration in the western North Atlantic and its application to biological problems. Rapp. P.-v. Réun. Cons. Perm. Int. Explor. Mer **101**:3-12.

Clarke, G. L. 1939. The relation between diatoms and copepods as a factor in productivity of the sea. Quart. Rev. Biol. **14**:60-64.

Clarke, G. L. 1940. Comparative richness of zooplankton in coastal and offshore areas of the Atlantic. Biol. Bull. **78**:226-255.

Clarke, G. L. 1941. Observations on transparency in the southwestern section of the North Atlantic Ocean. J. Mar. Res. 4:221-230.

Clarke, G. L., and H. R. James. 1939. Laboratory analysis of the selective absorption of light by seawater. J. Opt. Soc. Amer. **29**:43-55.

Clarke, M. R. 1966. A review of the systematics and ecology of oceanic squids. Advances Mar. Biol. 4:91-300.

Clements, F. E., and V. E. Shelford. 1927. Concepts and objectives in bio-ecology. Yearbook of the Carnegie Institute, 1926.

Cleve, P. T. 1901. Plankton from the Indian Ocean and the Malay Archipelago. K. svenska Vetensk-Akad. Handl. 35 (5):4-58.

Cloud, P. E. 1952. Facies relationships of organic reef. Bull. Amer. Ass. Petrol. Geol. **36**:2125-2149.

Clowes, A. J. 1938. Phosphate and silicate in the Southern Ocean. 'Discovery' Rep. **19**:1-20.

Clowes, A. J. 1950. An introduction to the hydrology of South African water. Investigational report no. 12. Division of Fisheries. Union of South Africa, Pretoria. 42 p.

Coe, W. R., and W. E. Allen. 1937. Growth of sedentary marine organisms in experimental blocks and plates for four successive years at the S.I.O. pier. Bull. Scripps Inst. Oceanogr. (Tech. Ser.) **4**:101-136.

Coker, R. E. 1947. This great and wide sea. University of North Carolina Press, Chapel Hill.

Cold Spring Harbor Symposia on Quantitative Biology. 1960. Vol. 25. Biological clocks. Cold Spring Harbor, New York, Biological Laboratory, Cold Spring Harbor.

Cole, H. A. 1956. Benthos and the shellfish of commerce. *In* M. Graham [ed.] Sea fisheries, their investigation in the United Kingdom. Edward Arnold, Publishers, Ltd., London.

Coleman, J. S. 1933. Nature of intertidal zonation of plants and animals. J. Mar. Biol. Ass. U. K. **18**:435-476.

Coleman, J. S., and F. Segrove. 1955. Tidal plankton over Stoupe Beck Sands, Robin Hood's Bay. J. Anim. Ecol. **24**:445-462.

Collier, A. 1953. The significance of organic compounds in seawater. Trans. N. Amer. Wildl. Conf. **18**:463-472.

Collier, A. 1958. Some biochemical aspects of red tides and related oceanographic problems. Limnol. Oceanogr. **3**:33-39.

Collier, A., S. Ray, and W. B. Wilson. 1956. Some effects of specific organic compounds on marine organisms. Science **124**:220.

Colman, J. 1940. On the faunas inhabiting intertidal seaweeds. J. Mar. Biol. Ass. U. K. **24**:129-183.

Commoner, B. 1966. Science and survival. The Viking Press, Inc., New York. 150 p.

Compton, I. 1957. The living sea. Doubleday & Co., Inc., Garden City, N. Y.

Conover, S. A. M. 1956. Oceanography of Long Island Sound, 1952-1954. IV. Phytoplankton. Bull. Bingham Oceanogr. Coll. **15**:62-112.

Cooper, L. H. N. 1933. Chemical constituents of biological importance in the English Channel, Nov., 1930, to Jan., 1932. I. Phosphate, silicate, nitrite, and ammonia. J. Mar. Biol. Ass. U. K. **18**:677-728.

Cooper, L. H. N. 1934. Variation of excess base with depth in the English Channel, with reference to the seasonal consumption of calcium by plankton. J. Mar. Biol. Ass. U. K. **19**:747-754.

Cooper, L. H. N. 1935a. Iron in the sea and in marine plankton. Proc. Roy. Soc. [Biol.] **118**:419-438.

Cooper, L. H. N. 1935b. The rate of liberation of phosphate in seawater by the breakdown of plankton organisms. J. Mar. Biol. Ass. U. K. **20**:197-200.

Cooper, L. H. N. 1937. The nitrogen cycle in the sea. J. Mar. Biol. Ass. U. K. **22**:183-204.

Cooper, L. H. N. 1938. Phosphate in the English Channel, 1933-1938, with a comparison with earlier years, 1916 and 1923-1932. J. Mar. Biol. Ass. U. K. **23**:181-195.

Cooper, L. H. N. 1948a. The distribution of iron in the waters of the western English Channel. J. Mar. Biol. Ass. U. K. **27**:279-213.

Cooper, L. H. N. 1948b. Phosphate and fisheries. J. Mar. Biol. Ass. U. K. **27**:326-336.

Cooper, L. H. N. 1951. Chemical properties of the seawater in the neighborhood of the La Badie Bank. J. Mar. Biol. Ass. U. K. **30**:21-26.

Cooper, L. H. N. 1952a. Processes of enrichment of surface water with nutrients due to strong winds blowing onto a continental slope. J. Mar. Biol. Ass. U. K. **30**:453-464.

Cooper, L. H. N. 1952b. Factors affecting the distribution of silicate in the North Atlantic Ocean and the formation of North Atlantic deep water. J. Mar. Biol. Ass. U. K. **30**:511-526.

Cooper, L. H. N. 1952c. The physical and chemical oceanography of the waters bathing the continental slope of the Celtic Sea. J. Mar. Biol. Ass. U. K. **30**:465-510.

Cooper, L. H. N. 1956. On assessing the age of deep oceanic water by carbon-14. J. Mar. Biol. Ass. U. K. **35**:341-354.

Cooper, L. H. N. 1961. Vertical and horizontal movements in the ocean. *In* M. Sears [ed.] Oceanography. Pub. no. 67. American Association for the Advancement of Science, Washington, D. C.

Cooper, L. H. N., and A. Milne. 1938-1939. Ecology of the Tamar estuary. Parts I-V. J. Mar. Biol. Ass. U. K. Vols. 22 and 23.

Cooper, L. H. N., and D. Vaux. 1949. Cascading over the continental slope of water from the Celtic Sea. J. Mar. Biol. Ass. U. K. **28**:719-750.

Coppleson, V. M. 1959. Shark attack. Angus & Robertson, Ltd., London.

Corlett, J. 1953. Net phytoplankton at ocean weather stations "I" and "J." J. Cons. Perm. Int. Explor. Mer **19**:178-190.

Corner, E. D. S. 1961. On the nutrition and metabolism of zooplankton. I. J. Mar. Biol. Ass. U. K. **41**:5-16.

Costello, D. P., M. E. Davidson, A. Eggers, M. H. Fox, and C. Henley. 1957. Methods for obtaining and handling marine eggs and embryos. Woods Hole, Mass., Marine Biology Laboratory. 247 p.

Cousteau, J. Y., and F. Dumas. 1953. The silent world. Harper & Brothers, New York.

Cowen, R. C. 1960. Frontiers of the sea. Doubleday & Co., Inc., Garden City, N. Y.

Cowey, C. B. 1956. A preliminary investigation of the variation of vitamin B$_{12}$ in oceanic and coastal waters. J. Mar. Biol. Ass. U. K. 35:609-620.

Cowlers, A. J. 1938. Phosphate and silicate in the Southern Ocean. 'Discovery' Rep. 19:1-120.

Cowles, R. P. 1930. A biological study of the offshore waters of Chesapeake Bay. Bull. Bur. Fish. (Washington) 46:277-381.

Creitz, G. I., and F. A. Richards. 1955. The estimation and characterization of plankton populations by pigment analysis. III. A note on the use of "Millipore" membrane filters in the estimation of plankton pigments. J. Mar. Res. 14:211-216.

Cromwell, T. 1953. Circulation in a meridional plane in the central equatorial Pacific. J. Mar. Res. 12:196-213.

Cross, J. C., and H. B. Parks. 1937. Marine fauna and seaside flora of the Nueces River basin and the adjacent islands. Bull. Texas Coll. Arts Ind. 8:1-14.

Curl, H. C. 1960. Primary production measurements in the north coastal waters of South America. Deep-Sea Res. 7:183-189.

Currie, R. I. 1953. Upwelling in the Benguela Current. Nature 171:497-500.

Currie, R. I. 1958. Some observations on organic production in the Northeast Atlantic. Rapp. P.-v. Réun. Cons. Perm. Int. Explor. Mer 144:96-102.

Cushing, D. H. 1951. Vertical migration of planktonic Crustacea. Biol. Rev. 19:3-22.

Cushing, D. H. 1953. Studies on plankton populations. J. Cons. Perm. Int. Explor. Mer 19:3-22.

Cushing, D. H. 1954. Some problems in the production of oceanic plankton. Document VIII. Presented to Commonwealth Oceanographic Conference, 1954. Plymouth, Mass.

Cushing, D. H. 1955. Some experiments on the vertical migration of zooplankton. J. Anim. Ecol. 24:137-166.

Cushing, D. H. 1958. The estimation of carbon in phytoplankton. Rapp. P.-v. Réun. Cons. Perm. Int. Explor. Mer 144:32-33.

Cushing, D. H. 1959a. On the nature of production in the sea. Fish. Invest. London (ser. 2) 22 (6):1-40.

Cushing, D. H. 1959b. The seasonal variation in oceanic production as a problem in population dynamics. J. Cons. Perm. Int. Explor. Mer 24:455-464.

Cuvier, G. L. C. F. D., Baron von. 1834-1836. Recherches sur les ossemens fossiles de quadrupèdes. ....Mit Discours préliminaire 1812. [Letzte Aufl. posthum. hrsg. von F. Cuvier.] 12 Bde.

Daisley, K. W., and L. R. Fisher. 1958. Vertical distribution of vitamin B$_{12}$ in the sea. J. Mar. Biol. Ass. U. K. 37:683-686.

Dakin, W. J. 1934. The plankton calendar of the continental shelf of the Pacific coast of Australia at Sydney, compared with that of the Irish Sea. (James Johnston Memorial vol.) Liverpool University Press, Liverpool, pp. 164-174.

Dakin, W. J., and A. Colefax. 1933. Marine plankton of coastal waters of New South Wales. I. The chief planktonic forms and their seasonal distribution. Proc. Linn. Soc. N. S. W. 58:186-222.

Dakin, W. J., and A. Colefax. 1935. Observations on the seasonal changes in temperatures, salinity, phosphates and nitrates, nitrogen, and oxygen of the ocean waters on the continental shelf off New South Wales, and the relationship to plankton production. Proc. Linn. Soc. N. S. W. 60:303-314.

Dakin, W. J., and A. N. Colefax. 1940. The plankton of the Australian coastal waters off New South Wales. I. Univ. Sydney Publ. Zool., Monograph no. 1, pp. 1-211.

Dales, K. P. 1966. Symbiosis in marine organisms. In S. M. Henry [ed.] Symbiosis. Vol. I. Academic Press, Inc., New York.

Dall, W. H. 1921. Summary of the marine shell-bearing molluscs of the northwest coast of America. Bull. U. S. Nat. Mus., no. 112, pp. 1-217.

Daly, R. A. 1910. Pleistocene glaciation and the coral reef problem. Amer. J. Sci. Ser. 4:30.

Dana, J. D. 1853. Crustacea. U. S. Explorer Expedition. II., vol. 14. Philadelphia. 1618 p.

Darwin, C. 1897. The structure and distribution of coral reefs. D. Appleton & Co., New York. 344 p.

Darwin, C. 1916. The voyage of the "Beagle." E. P. Dutton & Co., Inc., New York.

David, P. M. 1961. The influence of vertical migration on speciation in the oceanic plankton. Syst. Zool. 10:10-16.

David, P. M. 1967. Illustrations of oceanic neuston. In N. B. Marshall [ed.] Aspects of marine zoology. Academic Press, Inc., New York.

Davis, C. C. 1948. Effect of industrial copper as pollution upon the plankton in the Baltimore Harbor area. Pub. Chesapeake Biol. Lab. 72:1-12.

Davis, C. C. 1955. The marine and freshwater plankton. Michigan State University Press, East Lansing.

Davis, F. M. 1923a. Quantitative studies on the fauna of the sea bottom: Dogger Bank. Fish. Invest. London (ser. 2) 6 (2):1-54.

Davis, F. M. 1923b. Quantitative studies on the fauna of the sea bottom: southern North Sea. Fish. Invest. London (ser. 2) 8 (4):1-50.

Davis, J. H. 1940. Ecology and geological role of mangroves in Florida. Carnegie Inst. Washington Pub. 517:303-412.

Dawson, E. Y. 1956. How to know the seaweeds. William C. Brown Co., Publishers, Dubuque, Iowa.

Dawson, E. Y. 1966. Marine botany. Holt, Reinhart & Winston, Inc., New York.

Dawydoff, C. 1936. Observations sur la faune pélagique des eaux indochines de la mer de Chine meridionale. Bull. Soc. Zool. Fr. **61**:461-484.

Day, J. H. 1951. Ecology of South African estuaries. I. Review of estuarine conditions in general. Trans. Roy. Soc. South Africa **33**:53-91.

Day, J. H., N. A. H. Millard, and G. J. Broekhuysen. 1953. Ecology of South African estuaries. IV. The St. Lucia system. Trans. Roy. Soc. South Africa **34**:129-156.

Day, J. H., N. A. H. Millard, and A. D. Harrison. 1952. Ecology of South African estuaries. III. Knysna: a clear, open estuary. Trans. Roy. Soc. South Africa **33** (3):367-413.

Deacon, G. E. R. 1933. General account of the hydrology of the South Atlantic Ocean. 'Discovery' Rep. **7**:173-238.

Deacon, G. E. R. 1937. Hydrology of the Southern Ocean. 'Discovery' Rep. **15**:1-124.

Debenham, F. 1960. Discovery and exploration. Doubleday & Co., Inc., Garden City, N. Y.

Deevey, G. B. 1948. Zooplankton of Tisbury Great Pond. Bull. Bingham Oceanogr. Coll. **12**:1-44.

Deevey, G. B. 1952a. Survey of zooplankton of Block Island Sound, 1943-1946. Bull. Bingham Oceanogr. Coll. **13** (3):65-119.

Deevey, G. B. 1952b. Quantity and composition of the zooplankton of Block Island Sound, 1949. Bull. Bingham Oceanogr. Coll. **13** (3):120-164.

Deevey, G. B. 1956. Oceanography of Long Island Sound, 1952-1954. V. Zooplankton. Bull. Bingham Oceanogr. Coll. **15**:113-115.

Defant, A. 1958. Ebb and flow: the tides of earth, air, and water. The University of Michigan Press, Ann Arbor.

Degener, O. 1949. Naturalists' South Pacific Expedition: Fiji. Waialua, Hawaii. Published by the author. 303 p.

Delsman, H. C. 1939. Preliminary plankton investigations in the Java Sea. Treubia, Dec., 1939.

Denton, E. J., and T. I. Shaw. 1962. Buoyancy of gelatinous marine animals. J. Physiol. (London) **161**:14-15.

Devereux, I. 1967. Temperature measurements from oxygen isotope ratios of fish otoliths. Science **155**:1684-1685.

Dietz, R. S. 1948. Deep-scattering layer in the Pacific and Antarctic oceans. J. Mar. Res. **8**:430-442.

Dietz, R. S. 1962. The sea's deep-scattering layer. Sci. Amer. **207**:44.

Digby, P. S. 1967. Pressure sensitivity and its mechanism in the shallow marine environment. *In* N. B. Marshall [ed.] Aspects of marine zoology. Academic Press, Inc., New York.

Dittmar, W. 1884. Report on researches into the composition of ocean water collected by H.M.S. Challenger. 'Challenger' Rep. (phys. chem.) **1**:1-251.

Dobrin, M. B. 1947. Measurement of underwater noise produced by marine life. Science **105**:19-23.

Donn, W. L., and M. Ewing. 1966. The theory of ice ages. III. The theory involving polar wandering and an open polar sea is modified and given a quantitative basis. Science **152**:1706-1712.

Doochin, H., and F. G. W. Smith. 1951. Marine boring and fouling in relation to the velocity of water currents. Bull. Mar. Sci. Gulf Carib. **1**:196-208.

Doty, M. S. 1946. Critical tide factors correlated with vertical distribution of algae and other organisms along the Pacific coast. Ecology **27**:315-328.

Doty, M. S. 1956, 1959. Current status of carbon-14 method of assaying productivity of the ocean. University of Hawaii Annual Report for 1956, 1959.

Doty, M. S., and M. Ogun. 1957. Evidence for a photosynthetic daily periodicity. Limnol. Oceanogr. **2**:37-40.

Doudoroff, P., and M. Katz. 1950. Critical review of literature on the toxicity of industrial wastes and their components to fish. I. Alkalies, acids, and inorganic gases. Sew. Ind. Wastes **22**:1432-1458.

Douglas, J. S. 1952. The story of the oceans. Dodd, Mead & Co., New York.

Droop, M. R. 1957. Auxotrophy and organic compounds in the nutrition of marine phytoplankton. J. Gen. Microbiol. **16**:286-293.

Drummond, J. C., and E. R. Gunther. 1934. Observations on fatty constituents of marine plankton. III. Vitamin A and D content of oils. J. Exp. Biol. **11**:203-209.

Dugan, J. 1956. Man under the sea. Harper & Row, Publishers, New York.

Dunbar, M. J. 1940. On the size, distribution, and breeding cycles of four marine planktonic animals from the arctic. J. Anim. Ecol. **9**:215-226.

Dunbar, M. J. 1942. Marine macroplankton from the Canadian eastern Arctic. II. Medusae, Siphonophora, Ctenophora, Pteropoda, and Chaetognatha. Can. J. Res. **20**[D] (3):71-77.

Dunbar, M. J. 1946. On *Themisto libellula* in Baffin Island coastal waters. J. Fish Res. Bd. Canada **6**:419-434.

Edmondson, C. H. 1933. Reef and shore fauna of Hawaii. Bernice P. Bishop Museum, Special publication 22. 295 p.

Edmondson, C. H. 1944. Incidence of fouling in Pearl Harbor. (Occasional papers.) Bernice P. Bishop Mus. **18**:1-34.

Edmondson, C. H. 1949. Seashore treasures. Pacific Books, Publishers, Palo Alto, Calif. 144 p.

Edmondson, C. H., and W. M. Ingram. 1939. Fouling organisms in Hawaii. Bernice P. Bishop Mus. 14:251-300.

Ehrlich, P. 1968. The population bomb. Ballantine Books, Inc., New York.

Eibl-Eibesfeldt, I. 1966. Land of a thousand atolls (Maldive and Nicobar islands). The World Publishing Co., Cleveland.

Ekman, S. 1940. Begründung einer statistischen Methode in der regionalen Tiergeographie. Nova Acta Reg. Soc. Sci. Upsaliensis (ser. 4) 12:2.

Ekman, S. 1953. Zoogeography of the seas. Sidgwick & Jackson, Ltd., London.

Ellis, D. V. 1959. The benthos of soft sea-bottom in arctic North America. Nature 184:79-80.

Elton, C. 1946. The ecology of animals. 2nd ed. John Wiley & Sons, Inc., New York.

Elton, C. 1948. The ecology of invasions by animals and plants. John Wiley & Sons, Inc., New York.

Emery, K. O. 1956. Deep-standing internal waves in California basins. Limnol. Oceanogr. 1:35-41.

Emery, K. O., and C. O'D. Iselin. 1967. Human food from ocean and land. Science 157:1279-1281.

Endean, R. 1957. The biogeography of Queensland's shallow-water echinoderm fauna (excluding Crinoidea), with a rearrangement of the faunistic provinces of tropical Australia. Australian J. Mar. Freshwater Res. 8 (3):233-273.

Engel, L., and editors of *Life*. 1961, 1963. The sea. (Life Nature Library.) Time, Inc., New York.

Enright, J. T., and W. M. Hammer. 1967. Vertical diurnal migration and endogenous rhythmicity. Science 157:937-941.

Epstein, S., R. Buchsbaum, H. A. Lowenstein, and H. C. Urey. 1951. Carbonate-water isotopic temperature scale. Bull. Geol. Soc. Amer. 62:417-426.

Epstein, S., R. Buchsbaum, H. A. Lowenstein, and H. C. Urey. 1953. Revised carbonate-water isotopic temperature scale. Bull. Geol. Soc. Amer. 64:1315-1326.

Eriksen, A., and D. T. Lawrence. 1940. Occurrence and cause of pollution in Gray's Harbor. Bull. no. 2. Washington State Pollution Commission.

Eshleman, R. von. 1969. The atmospheres of Mars and Venus. Sci. Amer. 220 (3):78-88.

Esterly, C. O. 1911. Diurnal migrations of *Calanus finmarchicus* in the San Diego region during 1909. Int. Rev. Hydrobiol. Hydrogr. 4:140-151.

Esterly, C. O. 1919. Reaction of various plankton animals with reference to their diurnal migrations. Univ. California Pub. Zool. 19:1-83.

Evans, I. O. [ed.]. 1962. Sea and seashore. Frederick Warne & Co., Inc., New York.

Evenau, M., L. T. Evans, and A. P. Hughes. 1965. Photoenvironment. *In* E. J. Bowen [ed.] Recent progress in photobiology. Academic Press, Inc., New York.

Ewing, M., and W. L. Donn. 1956. The theory of ice ages. Science 123:1061.

Ewing, M., and W. L. Donn. 1958. The theory of ice ages. II. The theory that certain local terrestrial conditions caused Pleistocene glaciation is discussed further. Science 127:1159.

Eyden, D. 1923. Specific gravity as a factor in the vertical distribution of plankton. Proc. Cambridge Phil. Soc. Biol. Sci. 1:49-55.

Fager, E. W. 1957. Determination and analysis of recurrent groups. Ecology 38:586-595.

Fairbridge, R. 1966. Encyclopedia of oceanography. Reinhold Publishing Corp., New York.

Farran, G. P. 1947. Vertical distribution of plankton, *Sagitta, Calanus,* and *Metridia,* off the south coast of Ireland. Proc. Roy. Irish Acad. [B] 51:121-136.

Feder, H. M. 1966. Cleaning symbiosis in the marine environment. *In* S. M. Henry [ed.] Symbiosis. Vol. I. Academic Press, Inc., New York.

Fischer-Piette, E. 1931. Sur la pénétration des diverses espèces marines sessiles dans les estuaires et sa limitation par l'eau douce. Ann. Inst. Oceanogr. Monaco 10:213-243.

Fish, C. J. 1925. Seasonal distribution of the plankton of the Woods Hole region. Bull. Bur. Fish. (Washington) 41:91-179.

Fish, C. J. 1935. Marine biology and paleoecology. J. Paleontol. 9:92-100.

Fish, C. J. 1954. Preliminary observations on biology of boreo-arctic and subtropical oceanic zooplankton populations. Symposium on Marine and Fresh-Water Plankton in the Indo-Pacific. Bangkok 3-9.

Fish, C. J., and M. W. Johnson. 1937. The biology of the zooplankton population in the Bay of Fundy and Gulf of Maine, with special reference to production and distribution. J. Biol. Bd. Canada 3:189-321.

Fitch, J. E. 1953. Common marine bivalves of California. California Department of Fish and Game. Fisheries bull. no. 90. 102 p.

Flattely, F. W., and C. L. Walton. 1922. The biology of the seashore. New York, The Macmillan Co.

Fleming, R. H. 1957. General features of the oceans. *In* J. W. Hedgpeth [ed.] Treatise on marine ecology and paleoecology. Vol. I. Ecology. Geological Society of America, New York.

Flora, C. J., and E. Fairbanks. 1962. The sound and the sea. Pioneer Printing Co., Bellingham, Wash.

Forbes, E., and R. Goodwin-Qustin. 1959. Natural history of the European seas. Jon Van Voorst, London.

Ford, E. 1923. Animal communities of the level

sea bottom adjacent to Plymouth. J. Mar. Biol. Ass. U. K. **13**:164-224.

Forester, G. R. 1953. A new dredge for collecting burrowing animals. J. Mar. Biol. Ass. U. K. **32**:193-198.

Forester, G. R. 1954. Preliminary note on a survey of Stoke-Point rocks, with self-contained diving apparatus. J. Mar. Biol. Ass. U. K. **33**:341-344.

Fowell, R. R. 1939. Communities harbored by oar weeds (*Laminaria*). Proc. Swansea Sci. Field Natur. Soc. **2**:92-103.

Fox, D. L. 1947. Carotenoid and indolic biochromes of animals. Ann. Rev. Biochem. **16**:443-470.

Fox, D. L. 1950. Comparative metabolism of organic detritus by inshore animals. Ecology **31**: 100-108.

Fox, D. L. 1953. Animal biochromes and structural colors. John Wiley & Sons, Inc., New York.

Fox, D. L. 1957. Particulate organic detritus. *In* J. W. Hedgpeth [ed.] Treatise on marine ecology and paleoecology. Vol. I. Ecology. Geological Society of America, New York.

Fox, D. L., S. C. Crane, and B. H. McConnaughey. 1949. A biochemical study of the marine annelid worm, *Thoracophelia micronata:* its food, biochromes, and carotenoid metabolism. J. Mar. Res. **8**:567-585.

Fox, D. L., J. D. Isaacs, and I. F. Corcora. 1952. Marine leptopel: its recovery, measurement, and distribution. J. Mar. Res. **11**:29-46.

Fox, D. L., C. H. Oppenheimer, and J. S. Kittridge. 1953. Microfiltration in oceanographic research. II. Retention of colloidal micelles by adsorption filters and by filter-feeding invertebrates; proportions of dispersed organic and dispersed inorganic matter to organic solutes. J. Mar. Res. **12**:233-243.

Fox, H. M. 1939. The activity and metabolism of poikilothermal animals in different latitudes. Proc. Zool. Soc. London [A] **109**:141-156.

Foxton, P. 1956. Standing crop of zooplankton in the Southern Ocean. 'Discovery' Rep. **28**:193-235.

Fraenkel, G. S., and D. L. Gunn. 1940. The orientation of animals. Clarendon Press, Oxford.

Franc, A. 1951. Le zooplancton de la région de Dinard–Saint-Malo. Bull. Lab. Marit. Dinard **34**:25-40.

Fraser, C. M. 1942*a*. The collecting of marine zoological material in British Columbia waters. Can. Field Natur. **55**:1.

Fraser, C. M. 1942*b*. Marine zoology in the northeast Pacific. Trans. Roy. Soc. Canada Ser. 3, Vol. 36, pp. 1-18.

Fraser, J. H. 1947. Fiches d'identification du zooplancton. Nos. 9, 10. Thaliacea I, II. J. Cons. Perm. Int. Explor. Mer.

Fraser, J. H. 1948. Plankton in Scottish waters. Ann. Biol. (Copenhagen) **3**:43.

Fraser, J. H. 1949. Plankton of the Faeroe-Shet-land Channel and the Faeroes. June, Aug., 1947. Ann. Biol. (Copenhagen) **4**:27-28.

Fraser, J. H. 1949-1955. Plankton investigations from the Scottish research vessel. Ann. Biol. (Copenhagen) **4**:66-67; **6**:91-99; **7**:25; **7**:76-77; **8**:32-35; **8**:104-105; **9**:32-33; **10**:31.

Fraser, J. H. 1952. The Chaetognatha and other zooplankton of the Scottish area and their value as biological indicators of hydrographic conditions. Mar. Res. Scot. **2**:1-52.

Fraser, J. H. 1954. Zooplankton collections made by Scottish research vessel during 1953. Ann. Biol. (Copenhagen) **10**:99-101.

Fraser, J. H. 1955. The plankton of the waters approaching the British Isles in 1953. Mar. Res. Scot. No. 1.

Fraser, J. H. 1956, 1959. Scottish plankton investigations. Ann. Biol. (Copenhagen) **11**:26-27, 52-54; **14**:28-30, 66.

Fraser, J. H. 1962. Nature adrift: the story of plankton. Dufour Editions, Inc., Chester Springs, Pa.

Fraser, J. H., and A. Seville. 1949*a*. Macroplankton in the Faeroe Channel, 1948. Ann. Biol. (Copenhagen) **5**:61-64.

Fraser, J. H., and A. Seville. 1949*b*. Plankton distribution in Scottish and adjacent waters in 1948. Ann. Biol. (Copenhagen) **5**:61-64.

Frey, H., and S. Frey. 1961. One hundred and thirty feet down: handbook for hydronauts. Harcourt, Brace & World, Inc., New York.

Frost, N. 1937. Further plankton investigations. Report of the Fisheries Research Institute of Newfoundland, 1936-1937.

Gaardner, T., and H. H. Gran. 1927. Investigations of the production of plankton in the Oslo Fjord. Rapp. P.-v. Réun. Cons. Perm. Int. Explor. Mer **42**:1-48.

Gaardner, T., and R. Sparck. 1931. Biochemical and biological investigations of the west Norwegian oyster pools. Rapp. P.-v. Réun. Cons. Perm. Int. Explor. Mer **75**:47-58.

Gardiner, A. C. 1937. Phosphate production by planktonic animals. J. Cons. Perm. Int. Explor. Mer **12**:144-146.

Gauld, D. T. 1950. A fish cultivation experiment in an arm of a sea-lock. III. The plankton of Kyle Scottish. Proc. Roy. Soc. Edinb. [B] **64**:36-64.

George, P. C. 1953. The marine plankton of the coastal waters of Calicut, with observations on hydrological conditions. J. Zool. Soc. India **5**:76-107.

Gerard, R. D., and W. J. Worzel. 1967. Condensation of atmospheric moisture from tropical maritime air masses as a freshwater resource. Science **157**:1300-1302.

Gilbert, K., and F. McNeill. 1959. The Great Barrier Reef and adjacent isles. Coral Press Pty., Ltd., Paddington, Sidney, Australia.

Gillbricht, M. 1952a. Die Planktonverteilung in der Irminger See im Juni 1955. Ber. Deut. Komm. Meeresforsch. 15:260-275.

Gillbricht, M. 1952b. Untersuchungen zur Produktionsbiologie des planktons in der Kieler Bucht. Kieler Meeresforsch. 8:173-191.

Gilmartin, M. 1958. Some observation on the lagoon plankton of Eniwetok Atoll. Pacific Sci. 12:313-316.

Gislén, T. 1930. Epibioses of the Gullmar Fjord. II. K. svenska Vetensk-Akad. Stockholm 4:1-376.

Glover, R. S. 1955. Science and the herring industry. Advances Sci. 44:426-434.

Glover, R. S. 1961. Biogeographical boundaries: the shapes of distributions. In M. Sears [ed.] Oceanography. Publication no. 67. American Association for the Advancement of Science, Washington, D. C.

Glover, R. S. 1967. The continuous plankton recorder survey of the North Atlantic. In N. B. Marshall [ed.] Aspects of marine zoology. Academic Press, Inc., New York.

Gohar, H. A. S. 1934. Partnership between fish and anemone. Nature 134:291.

Goldberg, E. D. 1957. Biogeochemistry of trace metals. In J. W. Hedgpeth [ed.] Treatise on marine ecology and paleoecology. Vol. I. Ecology. Geological Society of America, New York.

Goldberg, E. D., M. Baker, and D. L. Fox. 1952. Microfiltration in oceanographic research. I. Marine sampling with the molecular filter. J. Mar. Res. 11:194-204.

Goldman, C. R. 1962. A method of studying nutrient-limiting factors in situ in water columns isolated by polyethylene film. Limnol. Oceanogr. 7:99-101.

Goodwin, R. H. 1961. Connecticut's coastal marshes, a vanishing resource. Bulletin no. 2. Connecticut Arboretum, New London. 36 p.

Gordienke, P. A. 1961. The Arctic Ocean. Sci. Amer. 204 (5):88.

Gotto, R. V. 1951. Some plankton records from Strangford Lough, County Down. Irish Natur. J. Biol. 10:162-164.

Gourley, D. R. H. 1952. Failure of P^{32} to exchange with organic phosphorous compounds. Nature (London) 169:192.

Graham, H. W. 1941. Plankton production in relation to character of water in the open Pacific. J. Mar. Res. 4:189-197.

Graham, H. W. 1943. Chlorophyll content of marine plankton. J. Mar. Res. 5:153-160.

Graham, H. W., and H. Gay. 1945. Season of attachment and growth of sedentary marine organisms at Oakland, Calif. Ecology 26:375-386.

Graham, M. 1956. Sea fisheries: their investigation and use in the United Kingdom. Edward Arnold Publishers, Ltd., London.

Grainger, E. H. 1959. The annual oceanographic cycle at Igloolik in the Canadian Arctic. I. The zooplankton and physical and chemical observations. J. Fish. Res. Bd. Canada 16:453-501.

Gran, H. H. 1912. Pelagic plant life. In J. Murray and J. Hjort [ed.] Depths of the ocean. The Macmillan Co., New York.

Gran, H. H. 1927. The production of plankton in the coastal waters off Bergen, Mar.-Apr., 1922. Rep. Norweg. Fish. Mar. Res. Invest. 3 (8):1-74.

Gran, H. H. 1929a. Quantitative investigations during expeditions with the "Michael Sars" (July-Sept., 1924). Rapp. P.-v. Réun. Cons. Perm. Int. Explor. Mer 56 (5):1-50.

Gran, H. H. 1929b. Investigation of the production of plankton outside the Romsdalsfjord 1926-1927. Rapp. P.-v. Réun. Cons. Perm. Int. Explor. Mer 56 (6):1-112.

Gran, H. H. 1931. On the conditions for the production of plankton in the sea. Rapp. P.-v. Réun. Cons. Perm. Int. Explor. Mer 75:37-46.

Gran, H. H. 1932. Phytoplankton, methods and problems. J. Cons. Perm. Int. Explor. Mer 7:343-358.

Gran, H. H., and T. Braarud. 1935. Quantitative study of phytoplankton in the Bay of Fundy and Gulf of Maine (including observations on hydrography, chemistry, and turbidity). J. Biol. Bd. Canada 1 (5):279-467.

Green, J. 1968. The biology of estuarine animals. In R. P. Bales [ed.] Biology Series. University of Washington Press, Seattle.

Grier, M. C. 1941. Oceanography of the North Pacific Ocean, Bering Sea, and Bering Strait [a bibliography]. Library Series. Vol. II. University of Washington Press, Seattle. 319 p.

Grøntved, J. 1952. Investigations on the phytoplankton in the southern North Sea, May, 1947. Medd. Komm. Danm. Fisk. Havund. (Ser. Plankton) 5 (5):1-49.

Grøntved, J. 1957. Sampler for underwater macrovegetation. J. Cons. Perm. Int. Explor. Mer 22:293-297.

Grøntved, J. 1958. Underwater macrovegetation. J. Cons. Perm. Int. Explor. Mer 24:32-42.

Gross, F. 1937. Notes on the culture of some marine plankton organisms. J. Mar. Biol. Ass. U. K. 21:753-768.

Gross, F., and E. Zeuthen, 1948. The buoyancy of plankton diatoms: a problem of cell physiology. Proc. Roy. Soc. [ser. B] 135:382-389.

Guberlet, M. L. 1936. Animals of the seashore. Metropolitan Press, Portland, Ore.

Guberlet, M. L. 1956. Seaweeds at ebb tide. University of Washington Press, Seattle.

Guilecher, A. 1958. Coastal and submarine morphology. John Wiley & Sons, Inc., New York.

Gullion, E. A. [ed.]. 1968. Uses of the seas. The American Assembly, Columbia University. Prentice-Hall, Inc., Englewood Cliffs, N. J.

Gunter, G. 1957. Temperature. *In* J. W. Hedgpeth [ed.] Treatise on marine ecology and paleoecology. Vol. I. Ecology. Geological Society of America, New York.

Gunter, G., and W. E. Gunter. 1947. Mass mortality of marine animals on the lower west coast of Florida, Nov., 1946, to Jan., 1947. Science **105**:256-257.

Gunter, G., and W. E. Gunter. 1956. Some relations of faunal distributions to salinity in estuarine waters. Ecology **37**:616-619.

Gunter, G., and W. E. Gunter. 1958. A study of an estuarine area with water-level control in the Louisiana marsh. Proc. Louisiana Acad. Sci. **21**: 5-34.

Gunter, G., R. H. Williams, C. C. Davis, and F. G. W. Smith. 1948. Catastrophic mass mortality of marine animals and coincident phytoplankton bloom on the west coast of Florida, Nov., 1946, to Aug., 1947. Ecol. Monogr. **18**:310-324.

Günther, K., and K. Deckert. 1950. Wunderwelt der tief See. F. A. Herbig Verlagsbuchhandlung (walterkahnert), Berlin, Grunewald. [Trans. by E. W. Dickes. 1956. Creatures of the deep. George Allen & Unwin, Ltd., London.]

Gunther, E. R. 1934. Fatty constituents of marine plankton. J. Exp. Biol. **11**:173-197.

Gunther, E. R. 1936. A report on oceanographical investigations in the Peru Coastal Current. 'Discovery' Rep. **13**:109-276.

Gurney, R. 1939. Bibliography of the larvae of decapod Crustacea. The Royal Society, London. i-viii, 123 p. (Available as reprints from Stechert-Hafner, 1960.)

Gurney, R. 1942. Larvae of decapod Crustacea. Ray Society, London. i-viii, 306 p. (Available as reprints from Stechert-Hafner, 1960.)

Halstead, B. W. 1959. Dangerous marine animals. Cornell Maritime Press, Inc., Cambridge, Md.

Hamilton, W. J. 1967. Analysis of bird navigation experiments. *In* K. E. F. Watt [ed.] Systems analysis in ecology. Academic Press, Inc., New York.

Hansen, K. V. 1951. On the diurnal migration of zooplankton in relation to the discontinuity layer. J. Cons. Perm. Int. Explor. Mer **17**:231-241.

Hansen, K. V. 1959. Investigations of the primary production and the quantitative distribution of the macroplankton in the North Atlantic. International Oceanographic Congress Preprints. American Association for the Advancement of Science, Washington, D. C.

Hardy, A. C. 1926. A new method of plankton research. Nature **118**:630.

Hardy, A. C. 1936*a*. The continuous plankton recorder. 'Discovery' Rep. **11**:457-510.

Hardy, A. C. 1936*b*. The arctic plankton collected by the "Nautilus" Expedition, 1931. I. General account. J. Linn. Soc. London **39**:391-403.

Hardy, A. C. 1939. Ecological investigation with the continuous plankton recorder. Hull Bull. Mar. Ecol. **1**:1-57.

Hardy, A. C. 1941. Plankton as a source of food. Nature **147**:695-696.

Hardy, A. C. 1947. Experiments on the vertical migration of plankton animals. J. Mar. Biol. Ass. U. K. **26**:467-526.

Hardy, A. C. 1953. Essays in marine biology. Oliver & Boyd, London. 144 p.

Hardy, A. C. 1955. A further example of the patchiness of plankton distribution. (Papers in Marine Biology and Oceanography.) Deep-Sea Res. (suppl.) **3**:7-11.

Hardy, A. C. 1956, 1960. The open sea: its natural history. Part 1. The world of plankton. Part 2. Fish and fisheries. Houghton Mifflin Co., Boston.

Hardy, A. C. 1958. Toward prediction in the sea. *In* A. A. Buzzati-Traverso [ed.] Perspectives in marine biology. University of California Press, Richmond.

Hardy, A. C., and R. Bainbridge. 1954. Experimental observations on the vertical migrations of plankton animals. J. Mar. Biol. Ass. U. K. **33**: 409-448.

Hardy, A. C., and E. R. Gunther. 1935. The plankton of the South Georgia whaling grounds and adjacent waters, 1926-1927. 'Discovery' Rep. **11**:1-456.

Hardy, A. C., and W. N. Paton. 1947. Experiments on vertical migration of plankton animals. J. Mar. Biol. Ass. U. K. **26**:467-526.

Harris, E. 1957. Radiophosphorus metabolism in zooplankton and microorganisms. Can. J. Zool. **35**:769-782.

Harris, E. 1959. The nitrogen cycle in Long Island Sound. Bull. Bingham Oceanogr. Coll. **17**:31-65.

Harris, J. E. 1953. Physical factors involved in the vertical migration of plankton. Quart. J. Microbiol. Sci. **94**:537-550.

Harris, J. E., and U. K. Wolfe. 1955. A laboratory study of vertical migration. Proc. Roy. Soc. (Biol.) **144**:329-354.

Harrison, R. J., and J. E. King. 1965. Marine mammals. Hutchinson University Library, London.

Hart, T. J. 1934. On the phytoplankton of the southwest Atlantic and the Bellingshausen Sea, 1929-1931. 'Discovery' Rep. **8**:1-268.

Hart, T. J. 1942. Phytoplankton periodicity in the

Antarctic surface waters. 'Discovery' Rep. **21**:263-348.

Hart, T. J. 1953. Plankton of the Benguela Current. Nature (London) **171**:631-634.

Hart, T. J., and R. I. Currie. 1960. The Benguela Current. 'Discovery' Rep. **31**:123-298.

Hartman, W. D., and T. F. Goreau. 1970. Jamaican coralline sponges: their morphology, ecology, and fossil relatives. Symp. Zool. Soc. London **25**:205-243. [Proposes new class of sponges, the Sclerospongiae.]

Harvey, H. W. 1926. Nitrate in the sea. J. Mar. Biol. Ass. U. K. **14**:71-88.

Harvey, H. W. 1928. Biology, chemistry, and physics of seawater. Cambridge University Press, New York. 194 p.

Harvey, H. W. 1934a. Annual variation of planktonic vegetation. J. Mar. Biol. Ass. U. K. **19**:775-792.

Harvey, H. W. 1934b. Measurement of phytoplankton population. J. Mar. Biol. Ass. U. K. **19**:761-773.

Harvey, H. W. 1942. Production of life in the sea. Biol. Rev. **17**:221-246.

Harvey, H. W. 1945. Recent advances in the chemistry and biology of seawater. Cambridge University Press, New York.

Harvey, H. W. 1947. Manganese and growth of phytoplankton. J. Mar. Biol. Ass. U. K. **26**:562-579.

Harvey, H. W. 1949. On manganese in sea and fresh waters. J. Mar. Biol. Ass. U. K. **28**:155-164.

Harvey, H. W. 1950. Production of living matter in the sea. J. Mar. Biol. Ass. U. K. **29**:97-136.

Harvey, H. W. 1957, 1960. The chemistry and fertility of seawater. Cambridge University Press, New York.

Harvey, H. W., L. H. N. Cooper, M. V. Lebour, and F. S. Russell. 1935. Plankton production and its control. J. Mar. Biol. Ass. U. K. **20**:407-441.

Hasle, G. R. 1959. A quantitative study of phytoplankton from the equatorial Pacific. Deep-Sea Res. **6**:38-59.

Haynes, B. C. 1947. Techniques of observing the weather. John Wiley & Sons, Inc., New York. 272 p.

Hazen, W. E. [ed.]. 1964. Readings in population and community ecology. W. B. Saunders Co., Philadelphia. 388 p.

Hedgpeth, J. W. 1951. Classification of estuarine and brackish waters and the hydrographic climate. In H. S. Ladd and others [ed.] Report of the Committee on a Treatise on Marine Ecology and Paleoecology, 1950-1951. Geological Society of America, New York.

Hedgpeth, J. W. [ed.]. 1957. Treatise on marine ecology and paleoecology. Vol. I. Geological Society of America, New York.

Heinrich, A. K. 1962. Life histories of plankton animals and seasonal cycles of plankton communities in the oceans. J. Cons. Perm. Int. Explor. Mer **27**:15-24.

Henderson, G. T. D. 1954. The young of fish and fish eggs. Hull Bull. Mar. Ecol. **3**:215-252.

Hendler, R. W. 1959. Self-absorption correction for carbon-14. Science **130**:772-777.

Hensen, V. 1911. Das Leben im Ozean nach Zahlungen seiner Bewohner. Erg. Plankton-Expdn. der Humboldt-stiftung. Lipsius & Tischer, Leipzig. 406 p.

Hentschel, E. 1936. Allgemeine Biologie des Südatlantischen Ozeans. Wiss. Erg. Deut. Atlantischen Exped. auf Meleor, 1925-1927. Vol. XI, 344 p.

Hentschel, E., and H. Wattenberg. 1930. Plankton und Phosphat in der Oberfläshenschicht des Südatlantischen Ozeans. Ann. Hydrogr. (Berlin) **58**:273-277.

Herald, E. S. 1961. Living fishes of the world. Doubleday & Co., Inc., Garden City, N. Y.

Herring, P. J. 1967. Pigments of plankton at the sea surface. In N. B. Marshall [ed.] Aspects of marine zoology. Academic Press, Inc., New York.

Hersey, J. B. [ed.]. 1967. Deep-sea photography. The Johns Hopkins Press, Baltimore.

Hesse, R., W. C. Allee, and K. Schmidt. 1951. Ecological animal geography. Part 2, Marine. 2nd ed. John Wiley & Sons, Inc., New York.

Hewson, E. W., and R. W. Longley. 1944. Meteorology: theoretical and applied. John Wiley & Sons, Inc., New York. 468 p.

Heyderdahl, T. 1950. Kon Tiki. Rand McNally & Co., Chicago. 304 p.

Hickman, C. P. 1967. Biology of the invertebrates. The C. V. Mosby Co., St. Louis.

Hida, T. A. 1957. Chaetognaths and pteropods as biological indicators in North Pacific waters. Special Scientific Report in Fisheries, no. 215. Fish and Wildlife Service, U. S. Department of the Interior, Washington, D. C.

Hida, T. A., and J. E. King. 1955. Vertical distribution of zooplankton in the central equatorial Pacific (July-Aug., 1952). Special Scientific Report on Fisheries, no. 144. Fish and Wildlife Service, U. S. Department of the Interior, Washington, D. C. 22 p.

Hill, M. N. [ed.]. 1962-1963. The sea. Vol. I. Physical oceanography. Vol. II. Composition of seawater and descriptive oceanography. Vol. III. The earth beneath the sea: history. John Wiley & Sons, Inc., New York.

Hock, C. W. 1941. Marine chitin-decomposing bacteria. J. Mar. Res. **4**:99-106.

Hoffman, C. 1956. Untersuchungen über die Remineralization des Phosphors im Plankton. Kieler Meeresforsch. **12**:25-36.

Holmboe, J., G. E. Forsythe, and W. S. Gustin.

1945. Dynamic meteorology. John Wiley & Sons, Inc., New York.

Holme, N. A. 1949a. Fauna of sand and mud banks near mouth of Exe estuary. J. Mar. Biol. Ass. U. K. **28**:189-237.

Holme, N. A. 1949b. A new bottom sampler. J. Mar. Biol. Ass. U. K. **28**:323-332.

Holme, N. A. 1950. Bottom fauna of the Great West Bay. J. Mar. Biol. Ass. U. K. **29**:163-183.

Holmes, R. W. 1958. Surface chlorophyll "A," surface primary production and zooplankton volumes in the eastern Pacific Ocean. Rapp. P.-v. Réun. Cons. Perm. Int. Explor. Mer **144**:109-116.

Holmes, R. W. 1960. Physical, chemical, and biological observations in the eastern tropical Pacific Ocean Scottish Expedition (April-June, 1958). Special Scientific Report on Fisheries, no. 345. Fish and Wildlife Service, U. S. Department of the Interior, Washington, D. C.

Hoyle, G., and T. Smyth. 1963. Giant muscle fibers in a barnacle *Balanus nubilus* Darwin. Science **139**:49-50.

Huberty, M. R., and W. L. Flock. 1959. Natural resources. McGraw-Hill Book Co., New York.

Hulbert, E. M., J. H. Ryther, and R. R. C. Guillard. 1960. Phytoplankton of the Sargasso Sea, off Bermuda. J. Cons. Perm. Int. Explor. Mer **25**:115-128.

Hull, C. H. J. 1960. Oxygen balance in an estuary. J. San. Eng. Div., Proc. Amer. Soc. Civ. Eng. **86** (SA6):105-120.

Hutchinson, G. E. 1953. The concept of pattern in ecology. Proc. Acad. Natur. Sci. (Philadelphia) **105**:1-12.

Hutner, S. H., and J. J. A. McLaughlin. 1958. Poisonous tides. Sci. Amer. **199** (2):92-98.

Hyman, L. H. 1940. The invertebrates. Vol. I. Protozoa through Ctenophora. McGraw-Hill Book Co., New York.

Hyman, L. H. 1951a. The invertebrates. Vol. II. Platyhelminthes and Rhynchocoela. McGraw-Hill Book Co., New York.

Hyman, L. H. 1951b. The invertebrates. Vol. III. Acanthocephala, Aschelminthes, Entoprocta. McGraw-Hill Book Co., New York.

Hyman, L. H. 1955. The invertebrates. Vol. IV. Echinodermata. McGraw-Hill Book Co., New York.

Hyman, L. H. 1959. The invertebrates. Vol. V. Smaller coelomate phyla. McGraw-Hill Book Co., New York.

Hynes, H. B. N. 1960. Biology of polluted waters. University Press, Liverpool.

Ikard, F. N. 1967. Oily wastes at sea. Science **157**:625.

Ingle, R. M., and D. P. de Sylva. 1955. The red tide. Educ. Ser. Mar. Lab. Univ. Miami **1**:30.

Inglis, C. C., and F. J. I. Kestner. 1958. Long-term effects of training walls, reclamation, and dredging on estuaries. Min. Proc. Instr. Civ. Eng. **9**:193-216.

Iselin, C. O. D. 1939. Some physical factors which may influence the productivity of New England's coastal waters. J. Mar. Res. **2**:75-85.

Ivanoff, A. 1953. Au sujet de la diffusion de la lumière par l'eau de mer. Ann. Geophys. **9**:26-27.

Ivanoff, A. 1957. Contributions à l'étude des propriétés optiques de l'eau de mer en Bretagne et en Corse, et à la théorie de la polarisation sous-marine. Ann. Geophys. **13**:22-53.

Ivanoff, A. 1958a. Au sujet de l'utilisation d'un diagramme P-β pour caractériser les masses d'eau océanique. C. R. Acad. Sci. (Paris) **246**:2636-2639.

Ivanoff, A. 1958b. Essai d'hydrologie optique entre Nice et la Corse. C. R. Acad. Sci. (Paris) **246**:3492-3496.

Ivanoff, A. 1959a. Introduction à une étude des propriétés diffusants des eaux de la baie de Naples. Pub. Staz. Zool. (Napoli) **31**:33-43.

Ivanoff, A. 1959b. Optical method of investigation of the oceans: The P-β diagram. J. Opt. Soc. Amer. **49**:103-104.

Ivanoff, A. 1961. A new water sampler and a new scattering polarizing meter for optical investigations of the oceans. International Oceanographic Congress. Preprints. American Association for the Advancement of Science, Washington, D. C.

Ivanoff, A., N. Jerlov, and T. H. Waterman. 1961. Comparative study of irradiance, beam transmittance, and scattering in the sea near Bermuda. Limnol. Oceanogr. **6** (2):129-148.

Ivanoff, A., and T. H. Waterman. 1958a. Elliptical polarization of submarine illumination. J. Mar. Res. **16**:255-282.

Ivanoff, A., and T. H. Waterman. 1958b. Factors, mainly depth and wavelength, affecting the degree of underwater light polarization. J. Mar. Res. **16**:283-307.

Iversen, E. S., 1968. Farming the edge of the sea. Fishing News (Books) Ltd., London.

Jacobs, M. B., and M. E. Ewing. 1969. Suspended particulate matter: concentration in the major oceans. Science **163**:380-383.

Japanese Hydrographic Office. 1956. Plankton observation. Equapac. Expedition. Unpublished report.

Jaschnow, W. A. 1939a. Plankton productivity in the southwestern part of the Barents Sea. Trans. Inst. Mar. Fish. Oceanogr. U.S.S.R. **4**:201-224.

Jaschnow, W. A. 1939b. Reproduction and seasonal variations in distribution of different stages of *Calanus finmarchicus* of the Barents Sea. Trans. Inst. Mar. Fish. Oceanogr. U.S.S.R. **4**:225-244.

Jaschnow, W. A. 1940. On the plankton productivity

in the northern seas of U.S.S.R. Société des Naturalistes de Moscou, Moscow. 84 p.

Jerlov, N. G. 1944. A standard transparency meter. Svenska Hydrogr.-Biol. Komm. Skriften **19**:1-6.

Jerlov, N. G. 1951. Optical studies of ocean waters. (Reports of the Swedish Deep-Sea Expedition.) Phys. Chem. **3**:1-59.

Jerlov, N. G. 1953. Particle distribution in the ocean. (Reports of the Swedish Deep-Sea Expedition.) Phys. Chem. **3**:73-97.

Jerlov, N. G. 1955a. Factors influencing the transparency of the Baltic waters. Medd. Oceanogr. Inst. Göteborg **25**:1-19.

Jerlov, N. G. 1955b. Particulate matter in the sea as determined by means of the Tyndall meter. Tellus **7**:218-225.

Jerlov, N. G. 1958. Adriatic thalassographic cruise, 1955. II. Distribution of suspended material in the Adriatic Sea. Arch. Oceanogr. Limnol. **11**:227-250.

Jerlov, N. G. 1959. Maxima in the vertical distribution of particles in the sea. Deep-Sea Res. **5**:173-184.

Jerlov, N. G., and F. Koczy. 1951. Photographic measurements of daylight in deep water. (Reports of the Swedish Deep-Sea Expedition.) Phys. Chem. **3**:1-59, 61-69.

Jerlov, N. G., and J. Piccard. 1959. Bathyscaphe measurements of daylight penetration into the Mediterranean. Deep-Sea Res. **5**:201-204.

Jespersen, P. 1923. Dr. Thorild Woulff's plankton collections in the waters west of Greenland. Metazoa. Medd. Grønland **64**:103-160.

Jespersen, P. 1924. On the quantity of macroplankton in the Mediterranean and Atlantic. Int. Rev. Hydrobiol. Hydrog. **12**:102-115.

Jespersen, P. 1935. Quantitative investigations in the distribution of macroplankton in different oceanic regions. Dana. Rep. no. 7. 44 p.

Jespersen, P. 1940. Investigations on the quantity and distribution of zooplankton in Icelandic waters. Medd. Komm. Danm. Fisk. Havund. (Ser. Plankton) **3** (5):1-77.

Jespersen, P. 1944. Investigations on the food of herring and the macroplankton in the waters around the Faeroes. Medd. Komm. Danm. Fisk. Havund. (Ser. Plankton) **3** (7):1-44.

Jespersen, P., and M. F. S. Russell. 1939-1952. Fiches d'identification du zooplancton. J. Cons. Perm. Int. Explor. Mer, no. 1-49.

Jitts, H. R. 1957. The C^{14} method for measuring CO_2 uptake in marine productivity studies. C.S.I.R.O. Aust. Div. Fish. Oceanogr. Rep. **8**:1-12.

Jitts, H. R., and B. D. Scott. 1961. Determination of zero-thickness activity in Geiger counting of C^{14} solutions used in marine productivity studies. Limnol. Oceanogr. **6**:116-123.

Johns, W. 1968. Estuaries: America's most vulnerable frontiers. American Wildlife Federation. 16 p.

Johnson, M. W. 1939. The correlation of water movements and the dispersal of pelagic larval stages of certain littoral animals, especially the sand crab, *Emerita*. J. Mar. Res. **2**:236-245.

Johnson, M. W. 1942. Notes on zooplankton. Records of observations. Scripps Institution of Oceanography. Vol. I. Oceanographic observations on E. W. Scripps cruises of 1938. University of California Press, Berkeley.

Johnson, M. W. 1943. Underwater sounds of biological origin. Division of Water Research, report no. U28. University of California Press, Berkeley. 20 p.

Johnson, M. W. 1948. Sound as a tool in marine ecology. J. Mar. Res. **7**:433-458.

Johnson, M. W. 1949. Zooplankton as an index of water between Bikini Lagoon and the open sea. Trans. Amer. Geophys. Union **30**:238-244.

Johnson, M. W. 1954. Plankton of the northern Marshall Islands. Geological survey. Professional paper no. 260-F. Part 2. F. 301-314. Washington, D. C.

Johnson, M. W. 1958. Observations on inshore plankton collected during summer, 1957, at Point Barrow, Alaska. J. Mar. Res. (Thompson Anniversary vol.) **17**:272-281.

Johnson, M. W., F. A. Everest, and R. W. Young. 1947. Role of snapping shrimp (*Crangon* and *Synalpheus*) in production of underwater noise in the sea. Biol. Bull. Mar. Biol. Lab. Woods Hole **93**:122-138.

Johnson, M. W., and R. C. Miller. 1935. Seasonal settlement of shipworms, barnacles, and other wharf-piling organisms at Friday Harbor, Washington. Univ. Washington Pub. Oceanogr. **2**:1-18.

Johnson, T. W. 1968. Saprobic marine fungi. *In* G. C. Ainsworth and A. S. Sussman [ed.] The fungi. Vol. 3. Academic Press, Inc., New York.

Johnson, T. W., and F. K. Sparrow. 1962. Fungi in oceans and estuaries. Hafner Publishing Co., Inc., New York.

Johnston, R. 1955. Biologically active compounds in the sea. J. Mar. Biol. Ass. U. K. **34**:185-195.

Johnston, R. 1959. Antimetabolites and marine algae. International Oceanographic Congress Preprints. American Association for the Advancement of Science, Washington, D. C.

Johnstone, J. 1908. Conditions of life in the sea. Cambridge University Press, New York. 332 p.

Johnstone, J., A. Scott, and H. C. Chadwick. 1924. The marine plankton, with special reference to investigations made at Port Erin, Isle of Man; a handbook for students and amateur workers. University Press of Liverpool, Ltd. Hodder & Stoughton, Ltd., London.

Jones, G. E. 1959. Biologically active organic substances in seawater. International Oceanographic Congress Preprints. American Association for the Advancement of Science, Washington, D. C.

Jones, N. S. 1950. Marine bottom communities. Biol. Rev. 25:283-313.

Jones, N. S. 1956. Fauna and biomass of muddy sand deposit off Port Erin. J. Anim. Ecol. 25:217-252.

Jordon, D. S. 1905. A guide to the study of fishes. 2 vols. Henry Holt, New York.

Jørgensen, C. B. 1966. Biology of suspension feeding. Pergamon Press, Inc., New York.

Juday, C. 1943. The utilization of aquatic resources. Science 97:456-458.

June, F. C., and J. L. Chamberlin. 1959. Role of the estuary in the life history and biology of Atlantic menhaden. Proceedings of the Gulf and Carribbean Fisheries Institute, 11th annual session, 1958.

Kalber, F. A. 1959a. Hypothesis on role of tide marshes in estuarine productivity. Estuarine Bull. 4:3, 14-15.

Kalber, F. A. 1959b. Where does the shoreline begin? Delaware Conserv. 3 (3):4-6.

Kalber, F. A. 1960. Pilot study of nutrients in Delaware River estuary. Project F-13-R-3. Ref. 60-7. Federal Aid to Fish and Wildlife Recovery. University of Delaware Marine Laboratory.

Kalle, K. 1953. Der einfluss des englischen Küstenwassers auf den Chemismus des Wasserkörper in der südlichen Nordsee. Ber. Deut. Komm. Meeresforsch. 13:130-135.

Kalle, K. 1957. Chemische Untersuchungen in der Irmiger See im Juni 1955. Ber. Deut. Komm. Meeresforsch. 14:313-328.

Kampa, E. M., and B. P. Boden. 1956. Light generation in an ionic scattering layer. Deep-Sea Res. 4:73-92.

Kandler, R. I. 1950. Jahreszeitliches Vorkommen und unperiodisches Auftreten von Fischbrut, Medusen und Dekapod-Larven im Fehmernbelt in den Jahren 1934-1943. Ber. Deut. Komm. Meeresforsch. 12:47-85.

Kelley, D. W. 1966. Ecological studies of the Sacramento–San Joaquin estuary. California Fish and Game, fisheries bull. no. 133.

Ketchum, B. H. 1947. Biochemical relations between marine organisms and their environment. Ecol. Monogr. 17:309-315.

Ketchum, B. H. 1957. The effects of the ecological system on the transport of elements in the sea. Pub. no. 551 (pp. 52-59). National Academy of Sciences–National Research Council.

Ketchum, B. H., J. C. Ayers, and R. F. Vaccaro. 1952. Processes contributing to the decrease of coliform bacteria in a tidal estuary. Ecology 33:247-258.

Ketchum, B. H., C. L. Carey, and M. Briggs. 1949. Preliminary studies on the viability and dispersal of coliform bacteria in the sea: limnological aspects of water supply and waste disposal. American Association for the Advancement of Science, Washington, D. C.

Ketchum, B. R., N. Corwin, and D. J. Keen. 1955. The significance of organic phosphorus determinations in ocean waters. Deep-Sea Res. 2:172-181.

Ketchum, B. H., and D. J. Keen. 1948. Unusual phosphorus concentrations in the Florida "red tide" seawater. J. Mar. Res. 7:17-21.

Ketchum, B. H., J. H. Ryther, C. S. Yentsch, and N. Corwin. 1958. Productivity in relation to nutrients. Rapp. P.-v. Réun. Cons. Perm. Int. Explor. Mer 144:132-140.

Ketchum, B. H., and D. H. Shonting. 1958. Optical studies of the particulate matter in the sea. Ref. no. 58-15. Woods Hole Oceanographic Institute.

Keys, A., E. H. Christensen, and A. Krogh. 1935. The organic metabolism of seawater, with special reference to the ultimate food cycle in the sea. J. Mar. Biol. Ass. U. K. 20:181-196.

Kielhorn, W. V. 1952. The biology of surface zone zooplankton of a boreo-arctic ocean area. J. Fish. Res. Bd. Canada 9:223-264.

Kilby, J. D. 1955. Fishes of two Gulf coastal marsh areas of Florida. Rep. U. S. Fish. Comm. 27:241-267.

Kincaid, T. 1942. Biotic and economic relations of plankton. California Fish Game 28:210-215.

King, J. E., T. S. Austin, and M. S. Doty. 1957. Preliminary report on Expedition East Tropic. Special Scientific Report on Fisheries, no. 201. Fish and Wildlife Service, U. S. Department of the Interior. 155 p.

King, J. E., and J. Demond. 1953. Zooplankton abundance in the Central Pacific. U. S. Fish Wild. Serv. Fish. Bull. 54:111-114.

King, J. E., and T. S. Hida. 1957. Zooplankton abundance in Hawaiian waters, 1953-1954. Special Scientific Report on Fisheries, no. 221. Fish and Wildlife Service, U. S. Department of the Interior.

Kitching, J., T. Macan, and H. Gilson. 1934. Studies in sublittoral ecology. I. A submarine gully in Wemburg, South Devon. J. Mar. Biol. Ass. U. K. 19 (2):667-705.

Knight-Jones, E. W., and S. Z. Quasim. 1955. Responses of some marine plankton animals to changes in hydrostatic pressure. Nature (London) 175:941.

Knox, G. A. 1963. The biogeography and intertidal ecology of the Australian coasts. Oceanogr. Mar. Biol. Annu. Rev. 1:341-404.

Knudsen, M. 1927. Bottom sampler for hard bottom. Medd. Komm. Danm. Fisk. Havund. (Ser. Plankton) 3:3-4.

Knudsen, N. 1901. Hydrographic tables. 2nd ed. G. E. C. Gad, Copenhagen. 63 p.

Knudsen, V. O., R. S. Alford, and J. W. Emling. 1948. Underwater ambient noise. J. Mar. Res. (Sverdrup 60th Anniversary vol.) 7:410-429.

Kofoid, C. A., and A. S. Campbell. 1929. A conspectus of the marine and fresh water ciliata belonging to the suborder Tintinnoinea, with descriptions of new species principally from the Agassiz Expedition to the Eastern Tropical Pacific. Univ. California Pub. Zool. 34:1-403.

Kofoid, C. A., and R. C. Miller. 1927. Marine borers and their relation to marine construction on the Pacific coast. The San Francisco Bay Marine Piling Committee, San Francisco.

Kohlmeyer, J., and E. Kohlmeyer. 1964. Synoptic plates of higher marine fungi. 2nd ed. Weinheim J. Cramer, New York.

Kohn, A. J., and P. Helfrich. 1957. Primary organic productivity of a Hawaiian coral reef. Limnol. Oceanogr. 2:241-251.

Komai, T. 1951. The nematocysts in the ctenophore *Euchlora rubra*. Amer. Nat. 85:73.

Kon, S. K. 1958. Some thoughts on biochemical perspectives in marine biology. *In* A. A. Buzzati-Traverso [ed.] Perspectives in marine biology. University of California Press, Richmond.

Kort, V. G. 1962. The Antarctic Ocean. Sci. Amer. 207 (3):113.

Kow, Tham Ah. 1953a. Plankton calendar of Singapore Strait, with suggestions for a simplified methodology. Section II. Tudo- Pacific Fisheries Council. Proceedings of the 4th meeting, Quezon City, The Philippines.

Kow, Tham Ah. 1953b. A preliminary study of the physical, chemical, and biological characteristics of Singapore Strait. Colonial Office. Fish. Pub. 1 (4):1-65.

Kreps, E., and N. Verjbinskaya. 1930. Seasonal changes in the phosphate and nitrate content and in hydrogen ion concentration in the Barents Sea. J. Cons. Perm. Int. Explor. Mer 5:327-346.

Krey, J. 1958a. Chemical determinations of net plankton, with special reference to equivalent albumin content. J. Mar. Res. (Thompson Anniversary vol.) 17:312-324.

Krey, J. 1958b. Chemical methods of estimating the standing crop of phytoplankton. Rapp. P.-v. Réun. Cons. Perm. Int. Explor. Mer 144:20-27.

Krey, J. 1961. Der Detritus im Meere. J. Cons. Perm. Int. Explor. Mer 26:263-280.

Kriss, A. E. 1959. Microbiology and the chief problems in the Black Sea. Deep-Sea Res. 5:193-200.

Kriss, A. E., I. E. Mishustina, N. Mitskevich, and S. S. Abyzov. 1961. Microorganisms as hydrological indicators in seas and oceans. IV. Deep-Sea Res. 7:225-236.

Kriss, A. E., I. E. Mishustina, N. Mitskevich, and

E. V. Zemtsova. 1964. Microbial population of oceans and seas. 1st Eng. transl. by K. Syers; edited by G. E. Fogg. 1967. Edward Arnold (Publishers) Ltd., London.

Kriss, A. E., and E. A. Rukina. 1952. Biomass of microorganisms and their rates of reproduction in oceanic depths. Zh. Obshch. Biol. 12:349-362.

Krizenecky, J. 1925. Untersuchungen über die Assimilations Fähigheit der Wassertiere für im Wasser gelöste Nahrstoff. Biol. Gen. 1:79-149.

Krogh, A. 1931. Dissolved substances as food of aquatic organisms. Rapp. P.-v. Réun. Cons. Perm. Int. Explor. Mer 75:7-26. (Also, Biol. Rev. 6:412-442.)

Krogh, A. 1934a. Conditions of life in the ocean. Ecol. Monogr. 4:421-429.

Krogh, A. 1934b. Life at great depths in the ocean. Ecol. Monogr. 4:430-439.

Krogh, A. 1939. Osmotic regulations in aquatic animals. Cambridge University Press, New York. 242 p.

Krogh, A., and R. Sparck. 1936. A new bottom sampler for investigation of the micro-fauna of the sea bottom. Kongl. Danske. Vidensk. Selsk. Biol. Medd. 13:1-12.

Krumbein, W. C. 1936. Application of logarithmic moments to size-frequency distributions of sediments. J. Sed. Petrology 6:35-47.

Kuenen, P. H. 1941. Geochemical calculations concerning the total mass of sediments in the earth. Amer. J. Sci. 39:161-190.

Kuenen, P. H. 1950. Marine geology. John Wiley & Sons, Inc., New York.

Kunne, C. 1935. Die Verbreitung der grosseren Planktontiere (ausser Fischbrut). *In* W. Mielck and C. Kunne. 1935. Helgoland Wiss. Meeresunters. 19 (7):62-118.

Kunne, C. 1950. Die Nahrung der Meerstiere. II. Das Plankton. *In* Handbuch der Seefischerei Nordeuropas. Verlag Stuttgart.

Kusmorskaja, A. 1940. Changements saisonnières du plancton de la Mer d'Okhotsk. Bull. MOIP, Biol. 49:3-4. (F.s.)

Kuznetsov, S. 1955. Use of radioactive isotopes for photosynthesis and chemosynthesis study in bodies of water. Papers presented by U.S.S.R. at International Conference on Peaceful Uses of Atomic Energy.

Lack, D. 1945. The ecology of closely related species, with special reference to the cormorant (*Phalacrocorax carbo*) and shag (*P. aristotelis*). J. Anim. Ecol. 14 (1):12-16.

Ladd, H. S., and J. I. Tracy. 1949. The problem of coral reefs. Sci. Monthly 69:297-305.

Laevastu, P. 1959. A review of marine pollution and the work of the F.A.O., Biology Branch, in this subject. Food and Agricultural Organization. Rome, Mimei. 58 p.

Lane, F. W. 1957. Kingdom of the octopus; the life history of the Cephalopoda. Jarrolds Publishers (London) Ltd., London.

Lang, K. 1948. Contribution to the ecology of *Priapulus caudatus* Lam. Ark. Zool. **41** (5):1-12.

Langmuir, I. 1938. Surface motion of water induced by wind. Science **87**:119-123.

Lanyon, W. E., and W. N. Tavolga. 1960. Animal sounds and communications. Hafner Publishing Co., Inc., New York.

Larson, R. L., and F. N. Spiess. 1969. East Pacific Rise: a near-bottom geophysical profile. Science **163**:68-70.

Lauff, G. H. [ed.]. 1967. Estuaries. (Symposium, Mar.-Apr., 1964.) American Association for the Advancement of Science, Washington, D. C. 733 p.

Lawson, G. W. 1954. Rocky shore zonation on the Gold Coast. J. Ecol. **44** (1):153-170.

Lawson, G. W. 1955. Rocky shore zonation in the British Cameroons. J. West Africa Sci. Ass. **1** (2): 78-88.

Lawson, G. W. 1957a. Seasonal variation of intertidal zonation on the coast of Ghana in relation to tidal factors. J. Ecol. **45**:831-860.

Lawson, G. W. 1957b. Some features of intertidal ecology of Sierra Leone. J. West Africa Sci. Ass. **3** (2):166-174.

Leavitt, B. B. 1935. Quantitative study of vertical distribution of larger zooplankton in deep water. Biol. Bull. Mar. Biol. Lab. Woods Hole **68**:115-130.

Leavitt, B. B. 1938. Quantitative vertical distribution of macroplankton in the Atlantic Ocean Basin. Biol. Bull. Mar. Biol. Lab. Woods Hole **74**:376-394.

Le Boeuf, B. J., and R. S. Peterson. 1969. Social status and mating activity in elephant seals. Science **163**:91-93.

Lebour, M. V. 1917. Microplankton of Plymouth Sound from the region beyond the breakwater. J. Mar. Biol. Ass. U. K. **11**:133-182.

Lebour, M. V. 1922, 1923. The food of plankton organisms. J. Mar. Biol. Ass. U. K. **12**:644-677 (part I); **13**:70-92 (part II).

Lee, M. O. [ed.]. 1964-1965. Biology of the Antarctic seas. Llano, G. A. [ed.]. Vol. 2. American Geophysical Union, National Academy of Sciences–National Research Council. Pub. no. 1190, 1297.

Legare, J. E. H. 1957. Qualitative and quantitative distribution of plankton in the Strait of Georgia in relation to certain oceanographic factors. J. Fish. Res. Bd. Canada **14**:551-552.

Legaud, M. 1957. Orsom III. Résultats biologiques de l'expédition Equapac. Institute Français d'Océanie. Rapp. Sci. no. 1 (processed report).

Leighton, D., and R. A. Boolootian. 1963. Diet and growth in the black abalone, *Haliotis cracherodii*. Ecology **44**:227-238.

Lewin, R. A. 1954. A marine *Stichococcus* sp. which requires vitamin B_{12} (cobalamin). J. Gen. Microbiol. **10** (1):92-96.

Lewis, J. R. 1964. The ecology of rocky shores. English Universities Press, Ltd., London.

Light, S. F., R. I. Smith, F. A. Pitelka, D. P. Abbott, and F. M. Weesner. 1954. Intertidal invertebrates of the central California coast. (S. F. Light's "Laboratory and Field Text in Invertebrate Zoology." Revised by Smith and others.) University of California Press, Richmond. 446 p.

Lillick, L. C. 1938. Preliminary report on the phytoplankton of the Gulf of Maine. Amer. Midl. Natur. **20**:624-640.

Lillick, L. C. 1940. Phytoplankton and planktonic protozoa of the offshore waters of the Gulf of Maine. II. Qualitative composition of the planktonic flora. Trans. Amer. Phil. Soc. **31**:193-237.

Limbaugh, C. 1961. Cleaning symbiosis. [Prepared by Howard Feder from field notes of the late Conrad Limbaugh.] Sci. Amer. **205** (2):42-49.

Lindberg, G. U. 1959. A list of the fauna of sea waters of south Sakhalin and south Kurile Islands: investigation of Far Eastern seas [in Russian]. C. R. Acad. Sci. (U.S.S.R.) **6**:173-256.

Lipman, C. B. 1926. The concentration of seawater as affecting its bacterial population. J. Bacteriol. **12**:311-313.

Loftas, T. 1969. The last resource—man's exploitation of the sea. H. Hamilton, London.

Lohmann, H. 1911. Über das Nannoplankton und die Zentrifugierung. Int. Rev. Hydrobiol. Hydrogr. **4**:1-38.

Lohmann, H. 1920. Die Bevölkerung des Ozeans mit Plankton nach den Ergebnissen der Zentrifugenfänge wahrend der Ausreise der "Deutschland," 1911. Zugleich ein Beitrag zur Biologie des Atlantischen Ozeans. Arch. Biontol. **4** (3):1-617.

Longhurst, A. R. 1959. Benthos densities off tropical West Africa. J. Cons. Perm. Int. Explor. Mer **25**:21-28.

Loosanoff, V. L., and H. C. Davis. 1951. Delayed spawning of lamellibranchs by low temperature. J. Mar Res. **10**:197-202.

Loosanoff, V. L., H. B. Engle, and C. A. Nomejka. 1955. Differences in intensity of settling of oysters and starfish. Biol. Bull. **109**:75-81.

Loye, D. P., and D. A. Proudfoot. 1946. Underwater noise due to marine life. J. Acoust. Soc. Amer. **18** (2):446-449.

Lucas, C. E. 1936. On certain interactions between phytoplankton and zooplankton under experimental conditions. J. Cons. Perm. Int. Explor. Mer **11**:343-361.

Lucas, C. E. 1938. Some aspects of integration in

plankton communities. J. Cons. Perm. Int. Explor. Mer **13**:309-321.

Lucas, C. E. 1947. Ecological effects of external metabolites. Biol. Rev. **22**:270-295.

Lucas, C. E. 1949. External metabolites and ecological adaptation. Symp. Soc. Exp. Biol. **111**:336-356.

Lucas, C. E. 1955. External metabolites in the sea. (Papers in Marine Biology and Oceanography.) Deep-Sea Res. (supp.) **3**:139-148.

Lucas, C. E. 1956. Plankton and basic production. *In* M. Graham [ed.] Sea fisheries: their investigation in the United Kingdom. Edward Arnold Publishers, Ltd., London.

Lucas, C. E. 1961. Interrelationships between acquatic organisms mediated by external metabolites. *In* M. Sears [ed.] Oceanography. Pub. no. 67. American Association for the Advancement of Science, Washington D. C.

Luck, J. M. 1957. Man against his environment: the next 100 years. Science **126**:903.

Lund, E. J. 1936. Some facts relating to the occurrence of dead and dying fish on the Texas coast during June, July, and August, 1935. Annual Reports, Texas Game, Fish, and Oyster Commission, 1934-1935, pp. 47-50.

MacArthur, R. H., and J. H. Connell. 1966. The biology of populations. John Wiley & Sons, Inc., New York. 226 p.

MacGinitie, G. E. 1939. Littoral marine communities. Amer. Midl. Natur. **21**:28-55.

MacGinitie, G. E., and N. MacGinitie. 1949. Natural history of marine animals. McGraw-Hill Book Co., New York.

Mackintosh, N. A. 1934. Distribution of macroplankton in the Atlantic sector of the Antarctic. 'Discovery' Rep. **9**:67-158.

Mackintosh, N. A. 1937. Seasonal circulation of Antarctic macroplankton. 'Discovery' Rep. **16**:367-412.

Mankowski, W. 1948. Macroplankton investigations in the Gulf of Gdańsk in June and July, 1946. Ann. Biol. (Copenhagen) **3**:110.

Mare, M. F. 1940. Plankton production of Plymouth and the mouth of the English Channel in 1935. J. Mar. Biol. Ass. U. K. **24**:461-482.

Mare, M. F. 1942. A marine benthic community, with special reference to the microorganisms. J. Mar. Biol. Ass. U. K. **25**:517-554.

Margalef, R. 1958. Temporal succession and spatial heterogeneity in phytoplankton. *In* A. A. Buzzati-Traverso [ed.] Perspectives in marine biology. University of California Press, Richmond.

Margalef, R. 1961. Communication of structure in planktonic populations. Limnol. Oceanogr. **6**:124-128.

Marshall, E. K., and H. W. Smith. 1930. The glo-

merular development of the vertebrate kidney in relation to habitat. Biol. Bull. **59**:125-153.

Marshall, N. B. 1954. Aspects of deep-sea biology. Hutchinson's Scientific & Technical Publications, London.

Marshall, N. B. [ed.]. 1967. Aspects of marine zoology. Symposium no. 19 of the Zoological Society of London. Academic Press, Inc., New York.

Marshall, P. T. 1958. Primary production in the Arctic. J. Cons. Perm. Int. Explor. Mer **23**:173-177.

Marshall, S. M. 1925. A survey of Clyde plankton. Proc. Roy. Soc. Edinb. [B.] **45**:117-141.

Marshall, S. M. 1933. The production of microplankton in the Great Barrier Reef region. Great Barrier Reef Expedition. Sci. Rep. Brit. Mus. (N.H.) **2**:111-158.

Marshall, S. M. 1948. Further experiments on the fertilization of a sea loch (Loch Craiglin). J. Mar. Biol. Ass. U. K. **27**:360-379.

Marshall, S. M., and A. P. Orr. 1927. Relation of plankton to some chemical and physical factors in the Clyde Sea area. J. Mar. Biol. Ass. U. K. **14**:837-868.

Marshall, S. M., and A. P. Orr. [ed.]. 1953. Essays in marine biology. Oliver & Boyd, London.

Martin, D. F. 1968. Marine chemistry. Analysis of seawater, Marcel Dekker, Inc., New York.

Matthews, G. V. T. 1955. Bird navigation. Cambridge University Press, New York.

Mayor, A. G. 1924. Structure and ecology of Samoan reefs. Carnegie Institute papers. Dep. Mar. Biol. **19**:1-25.

McAllister, C. D., T. R. Parsons, and J. D. H. Strickland. 1960. Primary productivity and fertility at Station "P" in the northeastern Pacific Ocean. J. Cons. Perm. Int. Explor. Mer **25**:240-259.

McClane, A. J. [ed.]. 1965. McClane's Standard fishing encyclopedia and international angling guide. Holt, Rinehart & Winston, Inc., New York.

McClendon, J. F. 1917. Standardization of a new colorimetric method for determination of CO_2 tension and CO_2 and O_2 of seawater. J. Biol. Chem. **30**:256-288.

McConnaughey, B. H. 1949. Mesozoa of the family Dicyemidae from California. Univ. California Pub. Zool. **55**:1-34.

McConnaughey, B. H. 1951. The life cycle of the dicyemid mesozoa. Univ. California Pub. Zool. **55**:295-336.

McConnaughey, B. H. 1964. The determination and analysis of plankton communities. *In* G. Rahardjo [ed.] Penelitian Laut di Indonesia. (Special number.) 40 p.

McConnaughey, B. H. 1966. The mesozoa. *In* E. L. Dougherty and others [ed.] The lower invertebrates. University of California Press, Richmond.

McConnaughey, B. H. 1968. The Mesozoa. *In* M. Florkin and B. T. Sheer [ed.] Chemical zoology. Vol. II. Academic Press, Inc., New York.

McConnaughey, B. H., and D. L. Fox. 1949. The anatomy and biology of the marine polychaete *Thoracophelia mucronata* (Treadwell) Ophelcidae. Univ. California Pub. Zool. 47:319-340.

McConnaughey, B. H., and E. I. McConnaughey. 1954. Strange life of the dicyemid mesozoans. Sci. Monthly 79:277-284.

McEwen, G. F. 1916. Summary and interpretation of the hydrographic observations made by the Scripps Institution for Biological Research of the University of California, 1908-1915. Univ. California Pub. Zool. 15 (3):1-38.

McMurrich, J. P. 1916. Notes on plankton of the British Columbia coast. Trans. Roy. Soc. Canada (ser. 3) 10:75-85.

McMurrich, J. P. 1917. Winter plankton in the neighborhood of St. Andrens, 1914-1915. Contributions to Canadian biology, 1915-1916. Department of Naval Service, Ottawa.

Meglitsch, P. A. 1967. Invertebrate zoology. Oxford University Press, New York.

Menard, H. W. 1961. The East Pacific Rise. Sci. Amer. 205 (6):52.

Menard, H. W. 1964. Marine geology of the Pacific. McGraw-Hill Book Co., New York.

Menard, H. W. 1967. Sea-floor spreading, topography, and the second layer. Science 157:923-924.

Menon, K. S. 1931. Preliminary account of the Madras plankton. Rep. Indian Mus. 33:489-516.

Menon, K. S. 1945. Observations on the seasonal distribution of the plankton of the Trivandrum coast. Proc. Indian Acad. Sci. 22B (2):31-62.

Menzel, D. W., and J. H. Ryther. 1960. Annual cycle of primary production in the Sargasso Sea off Bermuda. Deep-Sea Res. 6:351-367.

Menzel, D. W., and J. H. Ryther. 1961a. Nutrients limiting the production of phytoplankton in the Sargasso Sea, with special reference to iron. Deep-Sea Res. 7:276-281.

Menzel, D. W., and J. H. Ryther. 1961b. Zooplankton in the Sargasso Sea off Bermuda and its relation to organic production. J. Cons. Perm. Int. Explor. Mer 26:250-259.

Mesolella, K. J. 1967. Zonation of uplifted Pleistocene coral reefs on Barbados, West Indies. Science 156:638-640.

Metcalf, M. M. 1930. Salinity and size. Science 72:526-527.

Meyers, S. P., and E. S. Reynolds. 1957. Incidence of marine fungi in relation to wood borer attack. Science 126:969.

Miles, J. W., and G. W. Pearce. 1957. Rapid method for measurement of rate of sorption of DDT by mud surfaces. Science 126:169.

Miller, R. C., W. D. Ramage, and E. L. Lazier. 1928. Physical and chemical conditions in San Francisco Bay, especially in relation to the tides. Univ. California Pub. Zool. 31:201-267.

Miller, S. L. 1953. A production of amino acids under possible primitive earth conditions. Science 117:528-529.

Miller, S. L., and H. C. Urey. 1958. Organic compound formation on the primitive earth. *In* E. Hutchings [ed.] Frontiers in science. Basic Books, Inc., Publisher, New York.

Millott, N. [ed.]. 1967. Echinoderm biology. (Symposium of the Zoological Society of London, no. 20.) Academic Press, Inc., New York.

Milne, A. 1940. The ecology of the Tamar estuary. J. Mar. Biol. Ass. U. K. 24:69-87.

Miner, R. W. 1950. Field book of seashore life. G. P. Putnam's Sons, New York.

Miyajima, M. 1934. La question de "l'eau rouge." Un péril pour les huitres perlières. Bull. Soc. Gén. Agricul. Pêche 41:97-110.

Miyake, Y., and K. Saruhasi. 1956. On the vertical distribution of dissolved O_2 in the ocean. Deep-Sea Res. 3:242-247.

Moberg, E. G. 1926. Chemical composition of marine plankton. Proceedings of the 3rd Pan-Pacific Scientific Congress, Tokyo.

Moestafa, S. H., and B. H. McConnaughey. 1966. *Catostylus ouwensi* (Rhizostomeae, Catostylidae), a new jellyfish from Irian (New Guinea), and *Ouwensia catostyli* n. gen., n. sp., parasitic in *C. ouwensi*. Treubia 27:1-9.

Moore, H. B. 1930, 1931. Muds of the Clyde Sea area. J. Mar. Biol. Ass. U. K. 16:597-607; 17:325-358.

Moore, H. B. 1937. Marine fauna of the Isle of Man. Proc. Liverpool Biol. Soc. 50:38-57.

Moore, H. B. 1949. The zooplankton of the upper waters of the Bermuda area of the North Atlantic. Bull. Bingham Oceanogr. Coll. 12 (2):1-97.

Moore, H. B. 1958. Marine ecology. John Wiley & Sons, Inc., New York.

Moore, H. B., and E. G. Corwin. 1956. Effects of temperature illumination and pressure on vertical distribution of zooplankton. Bull. Mar. Sci. Gulf Carib. 6:273-287.

Moore, H. F. 1899. An inquiry into the feasibility of introducing useful marine animals into the waters of Great Salt Lake. U. S. Fish. Comm. Rep., 1899.

Morgan, B. 1956. Diving with safety, U. S. Divers, Los Angeles.

Morris, R. W., and L. R. Kittleman. 1967. Piezoelectric property of otoliths. Science 158:368-370.

Mullin, J. P., and J. B. Riley. 1955. Spectrophotometric determination of NO_3 in natural waters, with particular reference to seawater. Anal. Chim. Acta 12:464-480.

Murphy, G. I. 1961. Oceanography and variations

in the Pacific sardine population. California Coop. Fish. Invest. Rep. 8:55-64.

Murphy, J., and J. P. Riley. 1958. Single-solution method for determination of soluble phosphate in seawater. J. Mar. Biol. Ass. U. K. 37:9-14.

Murphy, R. C. 1938. Birds of the high seas: albatrosses, petrels, man-o'-war birds, and tropic birds. (Color paintings: Wings over the bounding main, by Allen Brooks.) Nat. Geog. Mag. 74:226-251.

Murray, J. 1895. Summary of the scientific results obtained at the sounding, dredging, and trawling stations of H.M.S. Challenger. 'Challenger' Rep. Part I: pp. xxxiii-lii, 1-176; xix, 797-1608.

Murray, J., and J. Hjort. 1912. The depths of the ocean. Hist. Natur. Class, vol. 37. Reprint, 1964. Hafner Publishing Co., Inc., New York.

Muscatine, L. 1967. Glycerol excretion by symbiotic algae from corals and *tridacna* and its control by the host. Science 156:516-519.

Myers, J. J., C. H. Holm, and R. F. McAllister [ed.]. 1969. Handbook of ocean and underwater engineering. McGraw-Hill Book Co., New York.

National Academy of Sciences–National Research Council. 1957. Effects of atomic radiation on oceanography and fisheries. Pub. no. 651. Washington, D. C. 137 p.

National Geographic Society. 1965. Wondrous world of fishes. The Society, Washington, D. C.

Needham, J. 1930. On the penetration of marine organisms into fresh water. Biol. Zentralbl. 50: 504-509.

Nelson, P. R., and W. T. Edmondson. 1955. Limnological effects of fertilizing Bear Lake, Alaska. Fisheries Bull. no. 102. Fish and Wildlife Service, U. S. Dept. of the Interior, Washington, D. C.

Neushul, M., W. D. Clarke, and D. W. Brown. 1967. Subtidal plant and animal communities of the southern California islands. *In* N. Philbrick [ed.] Biology of California islands. Santa Barbara Botanical Garden, Santa Barbara.

Newcombe, C. L. 1935. Certain environmental factors of a sand beach in the St. Andrew's region, New Brunswick, with a preliminary designation of the intertidal communities. J. Ecol. 23:334-355.

Newcombe, H. B. 1957. Magnitude of biological hazard from strontium-90. Science 126:549.

Newell, I. M. 1948. Marine molluscan provinces of western North America: A critique and a new analysis. Proc. Amer. Phil. Soc. 92:155-166.

Newman, W. A. 1970. Acanthaster: a disaster? (with reply by R. H. Chesher). Science 167:1274-1275.

Nicholls, G. D., H. Curl, and V. T. Bowen. 1959. Spectrographic analysis of marine plankton. Limnol. Oceanogr. 4:472-478.

Nicol, J. A. C. 1967. The biology of marine animals. 2nd ed. John Wiley & Sons, Inc., New York.

Nielsen, S. E. 1937. On the relation between the quantities of phytoplankton and zooplankton in the sea. J. Cons. Perm. Int. Explor. Mer 19:309-328.

Nikolsky, G. V. 1963. Ecology of fishes. (Transl. by L. Birkett), Academic Press, Inc., New York.

Norman, J. R. 1963. A history of fishes. 2nd ed. Hill & Wang, Inc., New York.

Norman, J. R., and F. C. Fraser. 1949. Field book of giant fishes. G. P. Putnam's Sons, New York.

Norris, K. S. [ed.]. 1966. Whales, dolphins, and porpoises. Proceedings of the 1st International Symposium on Cetacean Research (Aug., 1963), Washington, D. C. University of California Press, Richmond. 785 p.

Odum, E. P. 1961. Role of tidal marshes in estuarine production. N. Y. State Conserv. 15 (6):12-15.

Odum, H. T. 1953. Factors controlling marine invasion into Florida fresh waters. Bull. Mar. Sci. Gulf Carib. 3:134-156.

Odum, H. T. 1956. Efficiencies, size of organisms, and community structure. Ecology 37:593-597.

Odum, H. T., and E. P. Odum. 1955. Trophic structure and productivity of a windward coral reef community on Eniwetok Atoll. Ecol. Monogr. 25: 291-320.

Oliver, W. R. S. 1923. Marine littoral plant and animal communities in New Zealand. Trans. New Zealand Inst. 54:496-545.

Olson, T. A., and F. J. Burgess [ed.]. 1967. Pollution and marine ecology. John Wiley & Sons, Inc., New York.

Oparin, A. I. 1957. The origin of life on the earth. 3rd ed. (Transl. from Russian by A. Synge.) Oliver & Boyd, Edinburgh.

Oparin, A. I. 1961. Life: its nature, origin, and development. (Transl. from Russian by A. Synge.) Academic Press, Inc., New York.

Oppenheimer, C. H. 1955. Effect of marine bacteria on development and hatching of pelagic fish eggs, and the control of such bacteria by antibiotics. Copeia No. 1, pp. 43-49.

Oppenheimer, C. H. [ed.]. 1963. Symposium on marine microbiology. Charles C Thomas, Publisher, Springfield. Ill.

Orr, A. P., and F. W. Moorhouse. 1933. Physical and chemical conditions in mangrove swamps. Great Barrier Reef Expedition. Sci. Rep. Brit. Mus. (N. H.) 2:102-110.

Orton, H. H. 1920. Sea temperature, breeding, and distribution in marine animals. J. Mar. Biol. Ass. U. K. 12:339-366.

Orton, J. H. 1923. Some experiments on rate of growth in a polar region (Spitzbergen) and in England. Nature (London) 3:146-148.

Ostenfeld, C. H. [ed.]. 1916. International plankton catalog J. Cons. Perm. Int. Explor. Mer 170:1-87.

Oster, R. H., and G. L. Clarke. 1935. Penetration of red, green, and violet components of daylight into Atlantic waters. J. Opt. Soc. Amer. 25:84-91.

Ostroumoff, A. 1896. Résultats scientifiques de l'expédition d'Almariai. Bull. Acad. Sci. (St. Petersburg ser.) 5:4-5, 389-408.

Ostwald, W. 1902. Zur Theorie des Planktons. Biol. Zentralbl. 22:596-605, 609-638.

Paddock, W., and P. Paddock. 1967. America's decision: Who will survive? Little, Brown & Co., Boston.

Parker, R. H. 1955. Changes in invertebrate fauna attributable to salinity changes in bays of central Texas. J. Paleontol. 29:193-211.

Parker, R. H. 1956. Macro-invertebrate communities as indicators of sedimentary environments in east Mississippi Delta regions. Bull. Amer. Ass. Petrol. Geol. 40:295-376.

Patil, A. M. 1951. Study of the marine fauna of the Karwar coast and neighboring islands. J. Bombay Natur. Hist. Soc. 50:128-139.

Patten, B. C. 1959. An introduction to the cybernetics of the ecosystem: the trophic-dynamic aspect. Ecology 40:221-231.

Pearse, A. S. 1936. Migrations of animals from sea to land. Duke University Press, Durham, N. C.

Pearse, A. S. 1950. The emigrations of animals from the sea. Sherwood Press, Dryden, N. Y.

Pearse, A. S., and G. Gunter. 1957. Salinity. In J. W. Hedgpeth [ed.] Treatise on marine ecology and paleoecology. Geological Society of America, New York.

Pearse, A. S., H. J. Humm, and G. W. Wharton. 1942. Ecology of sand beaches at Beaufort, N. C. Ecol. Monogr. 12:135-190.

Pennak, R. W. 1943. An effective method of diagramming diurnal movements of zooplankton organisms. Ecology 24:405-407.

Pequegnat, W. C. 1958. Whales, plankton, and man. Sci. Amer. 198:84-90.

Perrier, R. 1936. La faune de la France en tableaux synoptiques illustrés. Vol. 1. Delgrave, Paris.

Petersen, C. G. 1913. The animal communities of the sea bottom and their importance for marine zoogeography. Rep. Dan. Biol. Sta. 21:1-44.

Petersen, C. G. 1915. Investigations concerning the valuation of the sea. Rep. Dan. Biol. Sta. 23:27-29.

Petersen, C. G. 1918. The sea bottom and its production of fish food. I. Apparatus. Rep. Dan. Biol. Sta. 25:1-62.

Petersen, C. G. 1924. Necessity for quantitative methods in investigation of animal life on the sea bottom. Proc. Zool. Soc. London 94:687-694.

Petersen, C. G., and P. Boysen-Jensen. 1911. Valuation of the sea. I. Animal life of the sea bottom, its food and quantity. Rep. Dan. Biol. Sta. 20:1-79.

Pettersson, H. 1954. The ocean floor. Yale University Press, New Haven, Conn.

Pfeffer, G. 1889. Zur Fauna von Süd Georgien. Mitt. Naturforsch. Mus. Hamburg 6.

Pfleger, F. B. 1960. Ecology and distribution of recent Foraminifera. Johns Hopkins Press, Baltimore.

Phifer, L. D. 1934. Phytoplankton of Eastsound, Washington, Feb.-Nov., 1932. Univ. Washington Publ. Oceanogr. 1:97-110.

Philbrick, N. [ed.]. 1967. Proceedings of the symposium on the biology of the California Islands. Santa Barbara Botanical Garden, Santa Barbara.

Phillips, A. G. 1917, 1922. Analytical search for metals in Tortuga's marine organisms. Papers of the Department of Marine Biology, nos. 11, 18. Carnegie Institute, Washington, D. C.

Piccard, A. 1956. Earth, sky, and sea. Oxford University Press, New York.

Pickford, G. E., and B. H. McConnaughey. 1949. The Octopus bimaculatus problem: a study in sibling species. Bull. Bingham Oceanogr. Coll. 12 (4):1-66.

Pittendrigh, C. S. 1958. Perspectives in the study of biological clocks. In A. A. Buzzati-Traverso [ed.] Perspectives in marine biology. University of California Press, Richmond.

Platt, R. B., and J. F. Griffiths. 1964. Environmental measurement and interpretation. Reinhold Publishing Corp., New York.

Plunkett, M. A., and N. W. Rakestraw. 1955. Dissolved organic matter in the sea. (Papers in Marine Biology and Oceanography.) Deep-Sea Res. (supp.) 3:12-14.

Polikarpov, G. G. 1964. Radioecology of aquatic organisms. Scripta Bechnica. Transl. by Schultz and Klement, 1966. Reinhold Publishing Corp., New York.

Pomeroy, L. R. 1960. Residence time of dissolved phosphate in natural waters. Science 131:1731-1732.

Pomeroy, L. R., and F. M. Bush. 1959. Regeneration of phosphate by marine animals. International Oceanographic Congress Preprints. American Association for the Advancement of Science, Washington, D. C.

Ponomareva, L. A. 1957. Zooplankton of the West Kara Sea and Baidaratskaya Bay. Trans. Inst. Oceanol. Acad. Sci. U.S.S.R. 20:228-245.

Poore, M. E. C. 1962. Successive approximation in descriptive ecology. In J. B. Cragg [ed.] Advances in ecological research. Vol. I. Academic Press, Inc., New York.

Powell, W. M., and G. L. Clarke, 1936. The reflection and absorption of daylight at the surface of the ocean. J. Opt. Soc. Amer. 26:111-120.

Powers, P. A. 1932. Cyclotrichium meunieri, sp.

nov. (Protozoa, Ciliata): cause of red water in the Gulf of Maine. Biol. Bull. **63**:74-80.

Prasad, R. R. 1954. Observations on the distribution and fluctuation of planktonic larvae off Mandapan. Symposium on Marine and Freshwater Plankton in the Indo-Pacific, Bangkok, 1954.

Pratt, D. M. 1949. Experiments in the fertilization of a saltwater pond. J. Mar. Res. **8**:36-59.

Pratt, D. M. 1950. Experimental study of the phosphorus cycle in fertilized salt water. J. Mar. Res. **9**:29-54.

Proudman, J. 1953. Dynamic oceanography. John Wiley & Sons, Inc., New York.

Putter, A. 1909. Die Ernährung der Wassertiere und der Stoffhaushalt der Gewässer. Gustav Fischer, Jena, E. Germany.

Quastler, H. 1959. Information theory of biological integration. Amer. Natural. **93**:245-254.

Rae, K. M. 1950. The continuous plankton recorder survey; the plankton around the north of the British Isles in 1950. Ann. Biol. **7**:72-76.

Rae, K. M. 1958. Parameters of the marine environment. *In* A. A. Buzzati-Traverso [ed.] Perspectives in marine biology. University of California Press, Richmond.

Rae, K. M., and Rees, C. B. 1947. The copepoda in the North Sea. Hull Bull. Mar. Ecol. **1**:171-238.

Rafter, T. A., and G. J. Fergusson. 1957. "Atombomb effect"—recent increase of carbon-14 content of the atmosphere and biosphere. Science **126**:557.

Rakestraw, N. W. 1936. The occurrence and significance of nitrate in the sea. Biol. Bull. Mar. Biol. Lab. Woods Hole **71**:133-167.

Rakestraw, N. W., and D. E. Carritt. 1948. Some seasonal chemical changes in the open ocean. J. Mar. Res. **7**:362-369.

Ray, C., and E. Ciampi. 1956. The underwater guide to marine life. A. S. Barnes & Co. Inc., Cranbury, N. J.

Ray, D. L. [ed.]. 1958. Marine boring and fouling organisms (1st Friday Harbor symposium, 1957). University of Washington Press, Seattle.

Raymont, J. E. G. 1955. Fauna of an intertidal mud flat. (Papers in Marine Biology and Oceanography.) Deep-Sea Res. (supp.) **3**:178-203.

Raymont, J. E. G. 1963. Plankton and productivity in the oceans. A Pergamon Press Book, The Macmillan Co., New York.

Raymont, J. E. G. 1966. The production of marine plankton. *In* J. B. Cragg [ed.] Advances in ecology research. Vol. 3. Academic Press, Inc., New York.

Raymont, J. E. G., and B. G. A. Carne. 1959. Zooplankton of South Hampton water. International Oceanographic Congress Preprints. American Association for the Advancement of Science, Washington, D. C.

Raymont, J. E. G., and R. J. Conover. 1961. Further investigations on the carbohydrate content of marine zooplankton. Limnol. Oceanogr. **6**:154-164.

Raymont, J. E. G., and S. Krishnaswamy. 1960. Carbohydrates in some marine planktonic animals. J. Mar. Biol. Ass. U. K. **39**:239-248.

Raymont, J. E. G., and R. S. Miller. 1962. Production of marine zooplankton with fertilization in an enclosed body of sea water. Int. Rev. Hydrobiol. Hydrogr. **47**:169-209.

Redfield, A. C. 1934. On the proportions of organic derivatives in seawater and their relation to the composition of the plankton. (James Johnstone memorial volume.) Liverpool University Press, Liverpool.

Redfield, A. C. 1939. The history of a population of *Limacina retroversa* during its drift across the Gulf of Maine. Biol. Bull. Mar. Biol. Lab. Woods Hole **76**:26-47.

Redfield, A. C. 1941. The effect of circulation of water on the distribution of the calanoid community in the Gulf of Maine. Biol. Bull. Mar. Biol. Lab. Woods Hole **80**:86-110.

Redfield, A. C. 1958. The biological control of chemical factors in the environment. Amer. Sci. **46**:205-221.

Redfield, A. C., and A. B. Keys. 1938. The distribution of ammonia in the waters of the Gulf of Maine. Biol. Bull. Mar. Biol. Lab. Woods Hole **74**:83-92.

Redfield, A. C., H. P. Smith, and B. H. Ketchum. 1937. The cycle of organic phosphorus in the Gulf of Maine. Biol. Bull. Mar. Biol. Lab. Woods Hole **73**:421-443.

Rees, C. B. 1939. Plankton of the upper reaches of the Bristol Channel. J. Mar. Biol. Ass. U. K. **23**:397-415.

Rees, C. B. 1940. Preliminary study of the ecology of a mud flat. J. Mar. Biol. Ass. U. K. **24**:185-199.

Rees, W. J. [ed.]. 1966. The cnidaria and their evolution. (Symposia of the Zoological Society of London, no. 16.) Academic Press, Inc., New York.

Reid, G. K. 1961. Ecology of inland waters and estuaries. Reinhold Publishing Corp., New York.

Remane, A. 1959. Die interstitielle Fauna des Meeressands. Proceedings of the 15th International Congress of Zoology, London.

Renouf, L. P. W. 1939. Faunistic notes from the south coast of County Cork, Eire. Ann. Mag. Natur. Hist. (ser. 11) **4**:520-525.

Reuszer, H. W. 1933. Marine bacteria and their role in the cycle of life in the sea. III. Distribution of bacteria in the ocean waters and muds about Cape Cod. Biol. Bull. Mar. Biol. Lab. Woods Hole **65**:480-497.

Rice, T. R. 1954. Biotic influences affecting population growth of planktonic algae. Fisheries bull. no. 54. Fish and Wildlife Service, U. S. Department of the Interior.

Richards, F. A. 1957a. Oxygen in the ocean. In J. W. Hedgpeth [ed.] Treatise on marine ecology and paleoecology. Vol. I. Ecology. Geological Society of America, New York.

Richards, F. A. 1957b. Current aspects of chemical oceanography. In L. H. Ahrens, F. Press, K. Rankama, and S. K. Runcorn [ed.] Progress in physics and chemistry of the earth. Vol. 2, pp. 77-128. Pergamon Press, Inc., New York.

Richards, F. A., and B. B. Benson. 1961. Nitrogen/argon and nitrogen isotope ratios in two anaerobic environments: the Cariaco Trench in the Caribbean Sea, and Dramsfjord, Norway. Deep-Sea Res. 7:254-264.

Richards, F. A., and T. G. Thompson. 1952. Estimation and characterization of plankton populations by pigment analysis. II. A spectrophotometric method for the estimation of plankton pigments. J. Mar. Res. 11:156-172.

Richards, F. A., and R. F. Vaccaro. 1956. The Cariaco Trench, an anaerobic basin in the Caribbean Sea. Deep-Sea Res. 3:214-228.

Ricketts, E., and J. Calvin. 1968. Between Pacific tides. 4th ed. Revised by J. Hedgpeth. Stanford University Press, Stanford, Calif.

Rider, B. F., and M. G. Mellon. 1946. Colorimetric determination of nitrates. Ind. Eng. Chem. Anal. Edit. 18:96-99.

Riedl, R. J. 1953. Quantitativ okölogische Methoden mariner Turbellarienforschung. Österr. Zool. Zeitschr. 4:108-145.

Riedl, R. J. 1954. Unterwasserforschung im Mittlemeer. Nchurw. Rundsch. 2:65-71.

Riedl, R. J. M. 1969. Gnathostomulida from America. Science 163:445-452.

Riley, G. A. 1937. Significance of the Mississippi River drainage for biological conditions in the northern Gulf of Mexico. J. Mar. Res. 1:64-74.

Riley, G. A. 1937-1938. Plankton studies. I. A preliminary investigation of the plankton of the Tortugas region. J. Mar. Res. 1:335-352.

Riley, G. A. 1939. Plankton studies. II. The western North Atlantic, May-June, 1939. J. Mar. Res. 2:145-162.

Riley, G. A. 1941. Plankton studies. V. Regional summary. J. Mar. Res. 4:162-171.

Riley, G. A. 1942. Relation of vertical turbulence and spring diatom flowerings. J. Mar. Res. 5:67-87.

Riley, G. A. 1946. Factors controlling phytoplankton population on Georges Bank. J. Mar. Res. 6:54-73.

Riley, G. A. 1947a. Seasonal fluctuations of the phytoplankton populations in New England coastal waters. J. Mar. Res. 6:114-125.

Riley, G. A. 1947b. Theoretical analysis of zooplankton population of Georges Bank. J. Mar. Res. 6:104-113.

Riley, G. A. 1955. Review of the oceanography of Long Island Sound. (Papers in Marine Biology and Oceanography.) Deep-Sea Res. (supp.) 3:224-238.

Riley, G. A. 1957. Phytoplankton of the north central Sargasso Sea. Limnol. Oceanogr. 2:252-270.

Riley, G. A. [ed.]. 1963. Marine biology. Vol. I. First International Interdisciplinary Conference on Marine Biology. Princeton, N. J., Oct. 29 to Nov. 1, 1961. Hafner Publishing Co., Inc., New York.

Riley, G. A., and R. Arx. 1949. Theoretical analysis of seasonal changes in the phytoplankton of Pusan Harbor, Korea. J. Mar. Res. 8:60-72.

Riley, G. A., and D. F. Bumpers. 1946. Phytoplankton-zooplankton relationship on Georges Bank. J. Mar. Res. 6:33-47.

Riley, G. A., and S. A. M. Conover. 1956. Oceanography of Long Island Sound, 1952-1954. III. Chemical oceanography. Bull. Bingham Oceanogr. Coll. 15:47-61.

Riley, G. A., A. Stommel, and D. F. Bumpers. 1949. Quantitative ecology of the plankton of the western North Atlantic. Bull. Bingham Oceanogr. Coll. 12:1-169.

Risebrough, R. W., R. J. Huggett, J. J. Griffin, and E. D. Goldberg. 1968. Pesticides: translantic movements in the Northeast Trades. Science 159:1233-1235.

Ritter, W. E., E. L. Michael, and G. P. McEwen. 1915. Hydrographic, plankton and dredging records of the Scripps Institution for Biological Research of the University of California, 1901-1912. Univ. California Pub. Zool. 15:1-206.

Roaf, H. 1910. Contributions to the physiology of marine invertebrates. J. Physiol. (London) 39:438-452.

Robertson, R. J., and H. E. Wirth. 1934a. Free ammonia, albuminoid nitrogen, and organic nitrogen in waters of the Pacific Ocean off the coasts of Washington and Vancouver Island. J. Cons. Perm. Int. Explor. Mer 9:187-195.

Robertson, R. J., and H. E. Wirth. 1934b. Report on the free ammonia, albuminoid nitrogen, and organic nitrogen in the waters of the Puget Sound area, during the summers of 1931 and 1932. J. Cons. Perm. Int. Explor. Mer 9:15-27.

Rona, E., and C. E. Miliani. 1969. Absolute dating of Caribbean cores P6304-8 and P6304-9. Science 163:66-67.

Rose, M. 1925. Contributions à l'étude de la biologie du plancton. Arch. Zool. Exp. Gen. **64**:387-542.

Ross, D. M., and L. Sutton. 1967. Swimming sea anemones of Puget Sound: swimming of *Actinostola* new species in response to *Stomphia coccinea*. Science **155**:1419-1421.

Roughly, T. C. 1936. Wonders of the Great Barrier Reef. Angus & Robertson, Ltd., London.

Rounsefell, G. A., and W. H. Everhart. 1953. Fishery science: its methods and application. John Wiley & Sons, Inc., New York.

Rubinoff, I. 1968. Central American sea level canal: possible biological effects. Science **161**:857-861.

Runnstrom, S. 1932. Eine Uebersicht über das Zooplankton des Herda und Hjeltefjordes. Bergens Mus. Aarb., 1931, no. 7, pp. 147-179.

Russell, F. S. 1925-1934. Vertical distribution of marine macroplankton. I-XII. J. Mar. Biol. Ass. U. K. **13 to 19.**

Russell, F. S. 1933. Seasonal distribution of macroplankton as shown by catches in 2-meter ring trawl in offshore waters off Plymouth. J. Mar. Biol. Ass. U. K. **19**:17-82.

Russell, F. S. 1934. The zooplankton. II, III, IV. British Museum Great Barrier Reef Expedition. Sci. Rep. Brit. Mus. (N. H.) **2**:176-201, 159-176, 203-276.

Russell, F. S. 1935. On the value of certain plankton animals as indicators of water movements in the English Channel and North Sea. J. Mar. Biol. Ass. U. K. **20**:309-332.

Russell, F. S. 1937. Seasonal abundance of pelagic young of teleostean fishes in the Plymouth area. IV. The year 1936, with notes on the conditions, shown by the occurrence of plankton indicators. J. Mar. Biol. Ass. U. K. **21**:679-686.

Russell, F. S. 1939. Hydrographical and biological conditions in the North Sea as indicated by plankton organisms. J. Cons. Perm. Int. Explor. Mer **14**:171-192.

Russell, F. S., and C. M. Younge. 1928. The seas: our knowledge of life in the sea and how it is gained. Frederick Warne & Co., Inc., New York.

Russell, R. C. M., and D. H. McMillan. 1952. Waves and tides. Hutchison's Scientific & Technical Publications, London.

Rustad, D. 1952. Zoological notes from the biology station, University of Bergen. Arb. 1951. Naturvid. Rekke. 1952, 1-2.

Rustad, E. 1946. Experiments on photosynthesis and respiration at different depths in the Oslo Fjord. Nytt. Mag. Naturvid. **85**:223-229.

Rutten, M. G. 1962. The geological aspects of the origin of life on earth. Elsevier Publishing Co., Amsterdam and New York.

Ruud, J. 1926. Quantitative investigations of plankton at Lofoten, March-Apr., 1922-1924. Rep. Norweg. Fish. Mar. Res. Invest. 3 (7):1-30.

Ryther, J. H. 1954. Ecology of phytoplankton blooms in Moriches Bay and Great South Bay, Long Island, N. Y. Biol. Bull. Mar. Biol. Lab. Woods Hole **106**:198-209.

Ryther, J. H. 1956. Photosynthesis in the ocean as a function of light intensity. Limnol. Oceanogr. **1**:61-70.

Ryther, J. H. 1959. Potential productivity of the sea. Science **130**:602-608.

Ryther, J. H. 1960. Seasonal and geographic range of primary production in the western Sargasso Sea. Deep-Sea Res. **6**:235-238.

Ryther, J. H. 1963. Geographic variations in productivity in the sea. *In* M. N. Hill [ed.] The sea; ideas and observations on progress in the study of the seas. Vol. 2. John Wiley & Sons, Inc., New York.

Ryther, J. H. 1969. Photosynthesis and fish production in the sea. Science **166**:72-76. (An important article, which gives further confirmation to the pessimistic outlook regarding further expansion of marine fisheries. "At present rate of expansion" they cannot last over a decade. Exploitation already is close to the maximum possible for sustained yield and is exceeding it in several fisheries.)

Ryther, J. H., and R. F. Vaccaro. 1954. Comparison of the oxygen and C^{14} methods of measuring marine photosynthesis. J. Cons. Perm. Int. Explor. Mer **20**:25-34.

Ryther, J. H., and C. S. Yentsch. 1957. Estimation of phytoplankton production in the ocean from chlorophyll and light data. Limnol. Oceanogr. **2**:281-286.

Ryther, J. H., and C. S. Yentsch. 1958. Primary production of continental shelf waters off New York. Limnol. Oceanogr. **3**:327-335.

Sanders, H. L. 1956. Oceanography of Long Island Sound, 1952-1954. X. Biology of marine bottom communities. Bull. Bingham Oceanogr. Coll. **15**:345-414.

Sanders, H. L. 1958. Benthic studies in Buzzard's Bay. I. Animal-sediment relationships. Limnol. Oceanogr. **3**:245-258.

Sanders, H. L. 1960. Benthic studies in Bozzard's Bay. II. The meiofauna. Limnol. Oceanogr. **5**:121-137.

Sanders, H. L. 1960. Benthic studies in Buzzard's Bay. III. The structure of the soft-bottom community. Limnol. Oceanogr. **5**:138-153.

Sanders, H. L., and R. R. Hessler. 1969. Ecology of the deep-sea benthos. Science **163**:1419-1424.

Sargent, M. C., and T. S. Austin. 1954. Biologic economy of coral reefs. Bikini and nearby atolls, Marshall Islands. Prof. Pap. U. S. Geol. Surv. 260-E, pp. 293-300.

Savage, R. E. 1931. The relation between feeding herring off the coast of England and plankton in surrounding waters. Fish. Invest. London (ser. 2) 12 (3):1-88.

Savilov, A. I. 1957. Biological aspect of bottom fauna groupings of the north Okhotsk Sea. Trudȳ Inst. Okeanol. 20:88-170.

Scagel, R. F. 1957. Annotated list of marine algae of British Columbia and northern Washington, with keys to genera. Canadian Department of Northern Affairs and Natural Resources, Ottawa.

Schenck, H. G., and A. M. Keen. 1936. Marine molluscan provinces of western North America. Proc. Amer. Phil. Soc. 76:921-938.

Schlieper, C. 1935. Neuere Ergebnisse und Probleme aus dem Gebiet des Osmoregulation wasserlebender Tiere. Biol. Rev. 10:334-360.

Schlieper, C. 1968. Methoden der meeresbiologischen Forschung. Gustav Fischer, Jena.

Schmalhausen, I. I. 1968. The origin of terrestrial vertebrates. (Translated from Russian by L. Kelso.) Academic Press, Inc., New York.

Schultz, L. P., and E. M. Stern. 1948. The ways of fishes. D. Van Nostrand Co., Inc., New York.

Sculthorpe, C. D. 1967. The biology of aquatic vascular plants. St. Martin's Press, Inc., New York.

Sears, M. 1941. Notes on the phytoplankton on Georges Bank in 1940. J. Mar. Res. 4:247-257.

Sears, M. 1954. Notes on the Peruvian Coastal Current. I. An introduction to the ecology of Pisco Bay. Deep-Sea Res. 1:141-169.

Sears, M. [ed.]. 1961. Oceanography. Symposium no. 67. American Association for the Advancement of Science, Washington, D. C.

Segerstråle, S. G. 1953. Increase in salinity of the inner Baltic and its influence on the fauna. Soc. Scint. Fennica 13 (15):1-7.

Segerstråle, S. G. 1957. The Baltic Sea. In J. W. Hedgpeth [ed.] Treatise on marine ecology and paleoecology. Vol. I. Ecology. Geological Society of America, New York.

Segerstråle, S. G. 1959. Brackish-water classification, a historical survey. Arch. Oceanogr. Limnol. (supp.) 11:7-13.

Seiwell, G. E. 1935a. Annual organic production and nutrient phosphorus requirement in the tropical western North Atlantic. J. Cons. Perm. Int. Explor. Mer 10:20-32.

Seiwell, G. E. 1935b. Note on iron analyses of Atlantic coastal waters. Ecology 16:663-664.

Seiwell, G. E. 1939. Daily temperature variations in the western North Atlantic. J. Cons. Perm. Int. Explor. Mer 14:357-369.

Seiwell, H. R. 1942. Analysis of vertical oscillations in the southern North Atlantic. Trans. Amer. Phil. Soc. 85:136-158.

Sellius, G. 1733. Historia naturalis Teredinis seu xylophagi marine Tubuloconchoidis speciatum Belgin. Trajecti ad Rhenum.

Seshaiya, R. V. 1959. Estuarine hydrology and biology. Curr. Sci. 28:54-56.

Shapeero, W. L. 1961. Phylogeny of Priapulida. Science 133:879-880.

Shapeero, W. L. 1962. Distribution of Priapulus caudatus Lam. on the Pacific coast of North America. Amer. Midl. Nat. 68 (1):237-241.

Sharp, C. A. 1962. Polynesian navigation. In J. Golson [ed.] Polynesian navigation. Polynesian Society Memoir no. 34. Polynesian Society, Inc., Wellington, N. Z. [Critique of Sharp's theory.]

Sharp, C. A. 1963. Ancient voyagers in Polynesia. University of California Press, Berkeley.

Shelford, V. E. 1930. Geographic extent and succession in Pacific North American intertidal (Balanus) communities. Pub. Puget Sound Biol. Sta. 7:216-223.

Shelford, V. E., and E. D. Towler. 1925. Animal communities of the San Juan Channel and adjacent areas. Pub. Puget Sound Biol. Sta. 5:33-73.

Shelford, V. E., A. O. Weese, L. A. Rice, D. I. Rasmussen, and A. MacLean. 1935. Some marine biotic communities of the Pacific coast of North America. I. General survey of the communities. Ecol. Monogr. 5:251-332.

Shepard, F. 1948. Submarine geology. Harper & Brothers, New York.

Shepard, F. 1959. The earth beneath the sea. The Johns Hopkins Press, Baltimore.

Shepard, F., and O. Emery. 1946. Submarine photography of the California coast. J. Geol. 54:306-321.

Shimada, B. M. 1958. Diurnal fluctuations in photosynthetic rate and chlorophyll "A" content of phytoplankton from eastern Pacific waters. Limnol. Oceanogr. 3:336-339.

Shirshov, P. P. 1937. Season changes of the phytoplankton of the polar seas in connection with the ice regime. Trans. Arctic Inst. Leningrad 82:47-113.

Shirshov, P. P. 1938. Oceanological observations. Dokl. Akad. Nauk. S.S.S.R. 19:569-580.

Shropshire, R. G. 1944. Plankton harvesting. J. Mar. Res. 5:185-188.

Sieburth, J. McN., and P. R. Burkholder. 1959. Antibiotic activity of antarctic phytoplankton. International Oceanographic Congress Preprints. American Association for the Advancement of Science, Washington, D. C.

Sillén, L. G. 1967. The ocean as a chemical system. Science 156:1189-1197.

Simons, E. G. 1957. An ecological survey of the Upper Laguna Madre. Publ. Inst. Mar. Sci. (Port Aransas, Texas) 4 (2):156-200.

Sindermann, C. J. 1957. Diseases of fishes of the Western Atlantic. V. Parasites as indicators of herring movements. Res. Bull. 27. Marine Department of Sea and Shore Fisheries. 30 p.

Skopintsev, B. 1960. Organic matter in natural water (aquatic humus). Trudȳ Gos. Okeanogr. 17 (29).

Skorikow, A. S. 1905. Beobachtungen über das Plankton der Newa. Biol. Zentralbl. 25:5-19.

Sleggs, G. F. 1927. Marine phytoplankton in the region of La Jolla, Calif., during the summer of 1924. Bull. Scripps Inst. Oceanogr. (Tech. Ser.) 1:93-117.

Smayda, T. J. 1958. Biogeographical studies of marine phytoplankton. Oikos 9:158-191.

Smith, F. G. W., R. H. Williams, and C. C. Davis. 1950. An ecological survey of the tropical waters adjacent to Miami. Ecology 31:119-146.

Smith, G. M. 1944. Marine algae of the Monterey Peninsula. Stanford University Press, Stanford, Calif.

Smith, J. E. [ed.]. 1968. "Torrey Canyon" pollution and marine life. A report of the Plymouth Laboratory. Cambridge University Press, New York.

Smith, M. 1926. Monograph of the sea snakes. Reprint, 1964. Hafner Publishing Co., Inc., New York.

Sorokin, Yu. 1955. Determination of chemosynthesis value in the water of the Rybinskoye Reservoir by the use of C^{14}. Dokl. Akad. Nauk. S.S.S.R. 105:1343.

Sorokin, Yu. 1957. Determination of chemosynthesis efficiency in methane and hydrogen oxidation in bodies of water. Mikrobiologiia 26 (13).

Sorokin, Yu. 1958. Study of chemosynthesis in mud deposits by the use of C^{14}. Mikrobiologiia 27 (206).

Soto, A. R., and W. B. Deichmann. 1967. Major metabolism and acute toxicity of aldrin, dieldrin, and endrin. Envir. Res. 1 (4).

Sousa, E. S. 1 va, E. de Sousa, and Santos-Pinto, J. dos. 1949. O plancton da Baia de S. Martiaho do Porto. Bull. Soc. Portug. Sci. Natur. (ser. 2) 2:203-241.

Southward, A. J. 1961, 1962. The distribution of some plankton animals in the English Channel and western approaches. J. Mar. Biol. Ass. U. K. 41:17-35; 42:275-375.

Spencer, C. P. 1954. Studies on the culture of a marine diatom. J. Mar. Biol. Ass. U. K. 33:265-290.

Spencer, C. P. 1956. Bacterial oxidation of ammonia in the sea. J. Mar. Biol. Ass. U. K. 35:621-630.

Spooner, G. M. 1933. Observations on the reactions of marine plankton to light. J. Mar. Biol. Ass. U. K. 19:385-438.

Spooner, G. M., and H. B. Moore. 1940. Ecology of the Tamar estuary. VI. Macrofauna of intertidal muds. J. Mar. Biol. Ass. U. K. 24:283-330.

Sproston, N. G. 1949. Preliminary survey of plankton of the Chu-san region, with a review of relevant literature. Sinensia 20:58-161.

Stanbury, F. A. 1931. Effect of light of different intensities, reduced selectivity, and nonselectivity on rate of growth of *Nitzschia closterium*. J. Mar. Biol. Ass. U. K. 17:633-653.

Starr, T. J., M. E. Jones, and D. Martinez. 1957. Production of vitamin B_{12}-active substances by marine bacteria. Limnol. Oceanogr. 2:114-119.

Steele, J. H. 1956. Plant production on the Fladen Ground. J. Mar. Biol. Ass. U. K. 35:1-33.

Steele, J. H. 1958. Production studies in the northern North Sea. Rapp. P.-v. Réun. Cons. Perm. Int. Explor. Mer 144:79-84.

Steele, J. H. 1959. Quantitative ecology of marine phytoplankton. Biol. Rev. 34:129-158.

Steele, J. H., and C. S. Yentsch. 1960. The vertical distribution of chlorophyll. J. Mar. Biol. Ass. U. K. 39:217-226.

Steeman, E. N. 1962. Inactivation of the photochemical mechanism in photosynthesis as a means to protect cells against high light intensities. Physiol. Plantarum 15 (1):161-171.

Steemann-Nielsen, E. 1935. The production of phytoplankton at the Faeroe Isles, Iceland, E. Greenland, and in the waters around. Medd. Komm. Danm. Fisk. Havund. (Ser. Plankton) 3 (1):1-93.

Steemann-Nielsen, E. 1937a. The annual amount of organic matter produced by the phytoplankton in the sound off Helsingør. Medd. Komm. Danm. Fisk. Havund. (Ser. Plankton) 3 (3):1-37.

Steemann-Nielsen, E. 1937b. On the relation between the quantities of phytoplankton and zooplankton in the sea. J. Cons. Perm. Int. Explor. Mer 12:147-153.

Steemann-Nielsen, E. 1940. Die Produktionsbedingungen des Phytoplanktons in Übergangsgebiet zwischen der Nord- und Ostsee. Medd. Komm. Danm. Fisk. Havund. (Ser. Plankton) 3 (4):1-55.

Steemann-Nielsen, E. 1951. The marine vegetation of the Isefjord: a study of ecology and production. Medd. Komm. Danm. Fisk. Havund. (Ser. Plankton) 5 (4):1-114.

Steemann-Nielsen, E. 1952. Use of radio-active carbon (C^{14}) for measuring organic production in the sea. J. Cons. Perm. Int. Explor. Mer 18:117-140.

Steemann-Nielsen, E. 1954. On organic production in the oceans. J. Cons. Perm. Int. Explor. Mer 19:309-328.

Steemann-Nielsen, E. 1958a. The balance between phytoplankton and zooplankton in the sea. J. Cons. Perm. Int. Explor. Mer 23:178-198.

Steemann-Nielsen, E. 1958b. Experimental methods for measuring organic production in the sea. Rapp. P.-v. Réun. Cons. Perm. Int. Explor. Mer 144:38-46.

Steemann-Nielsen, E. 1958c. A survey of recent Danish measurements of the organic productivity in the sea. Rapp. P.-v. Réun. Cons. Perm. Int. Explor. Mer 144:92-95.

Steemann-Nielsen, E., and V. K. Hansen. 1959. Light adaptation in marine phytoplankton populations and its interrelation with temperature. Physiol. Plantarum 12:353-370.

Steemann-Nielsen, E., and E. A. Jensen. 1957. Primary oceanic production: the autotrophic production of organic matter in the oceans. 'Galathea' Rep. 1:49-136.

Steinbeck, J., and E. F. Ricketts. 1941. Sea of Cortez. The Viking Press, Inc., New York.

Steiner, M. 1938. Zur Kenntnis der Phosphat-Kreislaufes in Seen. Naturwissenschaften 26:723-724.

Stephen, A. C. 1930. Studies on the Scottish marine fauna: additional observations on the fauna of the sandy and muddy areas of the tidal zone. Trans. Roy. Soc. Edinb. [B.] 56:521-535.

Stephen, A. C. 1931, 1932. Notes on the biology of some lamellibranchs. J. Mar. Biol. Ass. U. K. 17:277-300; 18:51-68.

Stephen, A. C. 1938. Production of large broods in certain marine lamellibranchs, with possible relation to weather conditions. J. Anim. Ecol. 7:130-143.

Stephenson, T. A. 1947. The constitution of the intertidal fauna and flora of South Africa. Ann. Natal Mus. 10 (3):261-358, 11 (2):207-324.

Stephenson, T. A., and A. B. Stephenson. 1949. Universal features of zonation between tide marks on rocky coasts. J. Ecol. 37 (2):289-305.

Stephenson, T. A., and A. B. Stephenson. 1950. Life between tidemarks in North America: Florida Keys. Ecology 38:354-402.

Stephenson, T. A., and A. B. Stephenson. 1952. Life between tidemarks in North America: North Florida and the Carolinas. Ecology 40:1-49.

Stephenson, T. A., and A. B. Stephenson. 1954. Life between tidemarks in North America: Nova Scotia and Prince Edward Island. Ecology 42:14-70.

Stephenson, T. A., and A. B. Stephenson. 1961. Life between tidemarks in North America: Vancouver Island. Ecology 49:1-29, 227-243.

Stephenson, W. 1949. Certain effects of agitation upon the release of phosphate from mud. J. Mar. Biol. Ass. U. K. 28:371-380.

Stepheus, G. C., and R. A. Schiuske. 1957. Uptake of amino acids from sea water by ciliary-mucoid feeding animals. Biol. Bull. Mar. Biol. Lab. Woods Hole 111:356-357. (See also Ibid. 115:341-342 and Limnol. Oceanogr. 6:175-181.)

Stommel, H. 1958. The Gulf Stream. University of California Press, Richmond. 202 p.

Sverdrup, H. U. 1953. On conditions for the vernal blooming of phytoplankton. J. Cons. Perm. Int. Explor. Mer 18:287-295.

Sverdrup, H. U., M. W. Johnson, and R. H. Fleming. 1946. The oceans: their physics, chemistry, and general biology. Prentice-Hall, Inc., Englewood Cliffs, N. J. 1087 p.

Swallow, J. C. 1957. Some further deep-current measurements using naturally buoyant floats. Deep-Sea Res. 4:93-104.

Swallow, J. C., and L. V. Worthington. 1961. Observations of a deep countercurrent in the western North Atlantic. Deep-Sea Res. 8:1-19.

Talling, J. F. 1961. Photosynthesis under natural conditions, Ann. Rev. Pl. Physiol. 12:133-154.

Tavolga, W. N. [ed.]. 1964. Marine bioacoustics. The Macmillan Co., New York.

Taylor, W. R. 1957. Marine algae of the northeastern coast of North America. University of Michigan Press, Ann Arbor.

Taylor, W. R. 1962. Marine algae from the tropical Atlantic Ocean. V. Algae from the Lesser Antilles. Contr. U. S. Natl. Herbarium 36 (2):43-62.

Teal, J. M. 1962. Energy flow in the salt marsh ecosystem of Georgia. Ecology 43:614-624.

Thienemann, A. 1920. Die Grundlagen der Biocoenotik. In Festschrift für Zschokke. Kober, Basel.

Thienemann, A. 1926. Der Nahrungskreislauf im Wasser. Verh. d. Zool. Ges. Leipzig (Zool. Auz.) 31:29-79.

Thomas, W. H. 1959. Evaluation of phytoplankton production measurements. Bull. Scripps Inst. Oceanogr. (Tech. Ser.) 52 (22):8-11.

Thompson, E. F., and H. C. Gilson. 1937. Chemical and physical investigations. (John Murray Expedition, 1933-1934.) Sci. Rep. Brit. Mus. (N. H.) 2:15-81.

Thompson, T. G., and R. W. Bremer. 1935. Occurrence of iron in the water of the northeast Pacific. J. Cons. Perm. Int. Explor. Mer 10:39-47.

Thomson, S. H. 1961. What is happening to our estuaries? Transactions of 26th American Wildlife and Natural Resources Conference, Washington, D. C.

Thorson, G. 1936. Larval development, growth, and metabolism of arctic marine invertebrates. Medd. Grønland 100:1-155.

Thorson, G. 1946. Reproduction and larval development of Danish marine invertebrates. Medd. Komm. Danm. Fisk. Havund. (Ser. Plankton) 4:1-523.

Thorson, G. 1950. Reproductive and larval ecology of marine bottom invertebrates. Biol. Rev. 25:1-45.

Thorson, G. 1952. Zur jetzigen Lage der marinen Bodentier—Ökologie. Verh. Deut. Zool. Ges. Wilhelmshaven, 1951, pp. 270-327.

Thorson, G. 1955. Modern aspects of marine level-bottom animal communities. J. Mar. Res. 14:387-397.

Thorson, G. 1958. Parallel level-bottom communities: their temperature adaptation and food animals. *In* A. A. Buzzati-Traverso [ed.] Perspectives in marine biology. University of California Press, Richmond.

Tinbergen, N. 1953. The herring gull's world: a study of the social behavior of birds. William Collins & Sons Co., Ltd., London. 255 p. [Rev. ed. 1961. Basic Books, Inc., Publishers, New York.]

Towler, E. D. 1930. Analysis of intertidal communities of the San Juan Archipelago. Pub. Puget Sound Biol. Sta. 7:223-240.

Tully, J. P., and A. J. Dodimead. 1957. Canadian oceanographic research in the North Pacific Ocean. Pacific Oceanographic Group, Nanaimo, B. C. Fisheries Research Board of Canada.

Turekian, K. K. 1968. The oceans. Prentice-Hall, Inc., Englewood Cliffs, N. J.

Turner, R. D. 1966. A survey and illustrated catalog of the Teredinidae. Museum of Comparative Zoology of Harvard University, Cambridge, Mass.

UNESCO. 1968. Monographs on oceanographic methodolgy. Zooplankton sampling UNESCO Publications Center, New York.

Urey, H. C., A. Lowenstein, S. Epstein, and C. R. McKinney. 1951. Oxygen isotope method—temperature of calcareous structures. Bull. Geol. Soc. Amer. **62**:399.

Ussing, H. H. 1938. The biology of some important plankton animals in the fjords of East Greenland. Medd. Grønland **100**:101-108.

Utinomi, H. 1956. Colored illustrations of seashore animals of Japan. Hoikush. Osaka, Japan. 167 p.

Vaccaro, R. F., and J. H. Ryther. 1960. Marine phytoplankton and the distribution of nitrate in the sea. J. Cons. Perm. Int. Explor. Mer **25**:260-271.

Välikangas, I. 1933. Über die Biologie der Ostsee als Brackwassergebiet. Verh. int. Verein. theor. angew. Limnol. **6**:1.

Vallentyne, J. R. 1957. The molecular nature of organic matter in lakes and oceans, with lesser reference to sewage and terrestrial soils. J. Fish. Res. Bd. Canada **14**:33-82.

Vaughan, R. W. 1919. Corals and the formation of coral reefs. Smithsonian report, 1917. Smithsonian Institution, Washington, D. C.

Vaughan, T. W. 1930. The oceanographical point of view. Contrib. Mar. Biol. Stanford University Press, pp. 40-56

Veevers, H. G. 1951. Photography of the sea floor. J. Mar. Biol. Ass. U. K. **30**:215-221.

Veevers, H. G. 1952. A photographic survey of certain areas of sea floor near Plymouth. J. Mar. Biol. Ass. U. K. **31**:215-221.

Verjbinskaya, N. 1932. Observations on the nitrate

changes in the Barents Sea. J. Cons. Perm. Int. Explor. Mer **7**:47-52.

Verrill, A. E. 1880, 1882, 1885. Remarkable marine fauna occupying the outer banks. Amer. J. Sci. (ser. 3) **20, 23, 29.**

Vinogradov, A. P. 1938. Chemical composition of marine plankton. Trudy Inst. Okeanol. **2**:97-112.

Vinogradov, A. P. 1953. The elementary chemical composition of marine organisms. Memoir 2. Sears Foundation for Marine Research, New Haven, Conn. 647 p.

Vinogradova, N. G. 1959. The zoogeographical distribution of the deep-water bottom fauna in the abyssal zone of the ocean. Deep-Sea Res. **5**:205-208.

Vishniac, W. 1968. Autotrophy; energy availability in the sea and scope of activity. *In* C. H. Oppenheimer [ed.] Marine biology. Vol. IV. (Proceedings of Fourth International Conference on Marine Biology, 1966.) New York Academy of Sciences, New York.

Visscher, J. P. 1927. Nature and extent of fouling of ships' bottoms. Bull. Bur. Fish. (Washington) **43**:193-252.

Waksman, S. A. 1933. Distribution of organic matter in the sea bottom and chemical nature and origin of marine humus. Soil Sci. **36**:125-147.

Waksman, S. A., C. L. Carey, and H. W. Reuszer. 1933. Marine bacteria and their role in the cycle of life in the sea. I. Decomposition of marine plant and animal residues by bacteria. Biol. Bull. Mar. Biol. Lab. Woods Hole **65**:57-79.

Waksman, S. A., and M. Hotchkiss. 1937. Viability of bacteria in seawater. J. Bacteriol. **33**:389-400.

Waksman, S. A., and C. E. Renn. 1936. Decomposition of organic matter in seawater. III. Factors influencing the rate of decomposition. Biol. Bull. Mar. Biol. Lab. Woods Hole **70**:472-483.

Waksman, S. A., H. W. Reuszer, and C. L. Carey. 1935. Decomposition of organic matter in seawater by bacteria. J. Bacteriol. **29**:531-543.

Waksman, S. A., and U. Vartiovaara. 1938. Adsorption of bacteria by marine bottom. Biol. Bull. Mar. Biol. Lab. Woods Hole **74**:56-63.

Walford, L. A. 1958. Living resources of the sea: opportunities for research and expansion. The Ronald Press Co., New York. 321 p.

Waller, H. O., and W. Polski. 1959. Planktonic Foraminifera of the Asiatic shelf. Contr. Cushman Found. **10**:123-126.

Wangersky, P. J. 1952. Isolation of ascorbic acid and rhamnosides from seawater. Science **115**:685.

Wangersky, P. J. 1959. Dissolved carbohydrates in Long Island Sound, 1956-1958. Bull. Bingham Oceanogr. Coll. **17**:87-94.

Waterman, T. H. 1954. Polarization patterns in submarine illumination. Science **120**:927-932.

Waterman, T. H. 1960. Interaction of polarized light and turbidity in the orientation of *Daphnia* and *Mysidium*. Z. Vergl. Physiol. **43**:149-172.

Waterman, T. H., R. F. Nennemacher, F. A. Chace, and G. L. Clarke. 1939. Diurnal vertical migrations of deep-water plankton. Biol. Bull. Mar. Biol. Lab. Woods Hole **76**:256-279.

Watson, A. T. 1928. Observations on the habits and life history of *Pectinaria (Lagis) koreni* Mgr. Proc. Trans. Liverpool Biol. Soc. **42**:25-60. [Edited and with introduction by P. Fauvel.]

Watson, G. E. 1966. Seabirds of the tropical Atlantic Ocean. Smithsonian Identification Manuals Series. Smithsonian Institution, Washington, D. C.

Watt, K. E. [ed.]. 1966, 1967. Systems analysis in ecology. Academic Press, Inc., New York.

Wegener, A. 1912. Die Entstehung der Kontinente. Peterm. Mitt. 185-195; 253-256; 305-309. Shorter version in Geol. Rundsch. 3 (4):276-292.

Wegener, A. 1924. The origin of continents and oceans. Transl. from 3rd German ed. by J. G. A. Skerl. Methuen & Co., Ltd., London.

Wegener, A. 1966. The origin of continents and oceans. Transl. from 4th German ed. by J. Biram. Dover Publications, Inc., New York.

Wentworth, C. K. 1926. Methods of mechanical analysis of sediments. Univ. Iowa Stud. Natur. Hist. 11 (11):3-52.

Wetmore, A. 1936. Birds of the northern seas. Natn. Geog. Mag. **69** (1):95-122.

Weyl, P. K. 1968. Precambrian environment and the development of life. Science **161**:158-160.

Wiborg, K. F. 1954. Investigations on zooplankton in coastal and offshore waters of western and northwestern Norway. Rep. Norweg. Fish Mar. Res. Invest. **11**:1-246.

Wiborg, K. F. 1955. Zooplankton in relation to hydrography in the Norwegian Sea. Rep. Norweg. Fish Mar. Res. Invest. 11 (4):1-66.

Wiborg, K. F. 1960. Investigations on pelagic fry of cod and haddock in coastal and offshore areas of northern Norway in July-August, 1957. Rep. Norweg. Fish Mar. Res. Invest. 12 (8):1-18.

Wickstead, J. H. 1958. A survey of the larger zooplankton of Singapore Strait. J. Cons. Perm. Int. Explor. Mer **23**:341-353.

Wickstead, J. H. 1959. A predatory copepod. J. Anim. Ecol. **28**:69-72.

Wieser, W. 1952. Investigations on the microfauna of seaweeds inhabiting rocky coast. J. Mar. Biol. Ass. U. K. **31**:145-174.

Wieser, W. 1953. Die Beziehung zwischen Mundhöhlengestalt, Ernaehrungsgeweise und Vorkommen bei freilebenden marinen Nematoden. Ark. Zool. 2 (4):439-484.

Wieser, W. 1954. Boitrage zur Kenntnis der Nematoden submariner Höhlen. Ergebnisse der Österreichischen Tyrrhenia Expedition, 1952. Teil II. Österr. Zool. Zeitschr. 5 (½):172-230.

Wieser, W. 1957. Free-living marine nematodes. *In* Reports Lunds University Chile Expedition, 1948-1949. Vol. 4. Ecology. Acta Univ. Lund. Andra Afd. N.S. **52** (13):1-115.

Wieser, W. 1959. Free-living nematodes and other small invertebrates of the Puget Sound beaches. Univ. Washington Publ. Zool. **19**:1-179.

Wilbur, K. M., and C. M. Yonge. 1964, 1966. Physiology of Mollusca. 2 vols. Academic Press, Inc., New York.

Wiles, W. W. 1960. Use of pore concentration in the tests of planktonic Foraminifera for correlation of quaternary deep sea sediments. Bull. Geol. Soc. Amer. **70**:1699.

Williston, S. W. 1914. Water reptiles of the past and present. University of Chicago Press, Chicago. 251 p.

Wilson, D. P. 1951*a*. Larval metamorphosis and the substratum. Ann. Biol. **27**:491-501.

Wilson, D. P. 1951*b*. Life of the shore and shallow sea. Nicholson & Watson, London.

Wilson, D. P. 1958. Problems in larval ecology related to localized distribution of bottom animals. *In* A. A. Buzzati-Traverso [ed.] Perspectives in marine biology. University of California Press, Richmond.

Wilson, D. P., and F. J. Armstrong. 1958. Biological differences between seawaters: experiments in 1954 and 1955. J. Mar. Biol. Ass. U. K. **37**:331-348.

Wimpenny, R. S. 1936. The distribution, breeding, and feeding of some important plankton organisms of the southwest North Sea in 1934. I. *Calanus finmarchicus* (Gunn.), *Sagitta setroa* (J. Müller), and *Sagitta elegans* (Verrill). Fish. Invest. London (ser. 2) **15** (3):1-53.

Wimpenny, R. S. 1938. Diurnal variation in the feeding and breeding of zooplankton related to the numerical balance of the zoophytoplankton community. J. Cons. Perm. Int. Explor. Mer **13**:323-336.

Wimpenny, R. S. 1946. The size of diatoms. II. Further observations on *Rhizolenia styliformis* (Brightwell). J. Mar. Biol. Ass. U. K. **26**:271-284.

Wimpenny, R. S. 1958. Carbon production in the sea at the Smith's Knoll Light Vessel. Rapp. P.-v. Réun. Cons. Perm. Int. Explor. Mer **144**:70-72.

Wimpenny, R. S. 1966. The plankton of the sea. American Elsevier Publishing Co., Inc., New York.

Wolff, T. 1960. The hadal community: an introduction. Deep-Sea Res. **6**:95-124.

Wood, E. J. F. 1950. Investigations of underwater fouling. I. Role of bacteria in early stages of fouling. Aust. J. Mar. Freshw. Res. **1**:85-91.

Wood, E. J. F. 1953. Heterotrophic bacteria in marine environments of eastern Australia. Aust. J. Mar. Freshw. Res. **4**:140-200.

Wood, E. J. F. 1963. Relative importance of groups of algae and protozoa in marine environments of the southwest Pacific and east Indian oceans. *In* C. H. Oppenheimer [ed.] Symposium on marine microbiology. Charles C Thomas, Publisher, Springfield, Ill.

Wood, E. J. F., and J. Ferguson. 1965. Marine microbial ecology. Reinhold Publishing Corp., New York. 243 p.

Woodmansee, R. A. 1958. Seasonal distribution of zooplankton off Chicken Key in Biscayne Bay, Florida. Ecology **39**:247-262.

Woodwell, G. M., C. F. Wurster, and P. A. Isaacson. 1967. DDT residues in an East Coast estuary: a case of biological concentration of a persistent insecticide. Science **156**:821-824.

Worthington, L. V. 1954. Preliminary note on the time scale in North Atlantic circulation. Deep-Sea Res. **1**:244-251.

Wust, G. 1930. Meridional Schichtung und Tiefenzirculation in den west Halften der drei Ozeane. J. Cons. Int. Perm. Explor. Mer **5**:7-21.

Wust, G. 1954. Die zonal Verteilung von Salzgehalt, Niedershlag, Verdunstung, Temperatur, und Dichte an der Oberflache der Ozeane. Kieler Meeresforsch. **10**:137-161.

Wyrtki, K. 1961. The thermohaline circulation in relation to the general circulation in the oceans. Deep-Sea Res. **8**:39-64.

Yankwich, P. E., T. H. Norris, and J. L. Juston. 1947. Correcting for the absorption of weak beta particles in thick samples. Ind. Eng. Chem. Anal. **1950**:439-441.

Yentsch, C. S., and J. H. Ryther. 1957. Short-term variations in phytoplankton chlorophyll and their significance. Limnol. Oceanogr. **26**:140-142.

Yentsch, C. S., and J. H. Ryther. 1959. Relative significance of the net phytoplankton and nannoplankton in the waters of Vineyard Sound. J. Cons. Int. Perm. Explor. Mer **24**:231-238.

Yentsch, C. S., and R. F. Scagel. 1958. Diurnal study of phytoplankton pigments: an *in situ* study in Eastsound, Washington. J. Mar. Res. (Thompson Anniversary vol.) **17**:567-583.

Yonge, C. M. 1928. Feeding mechanisms in the invertebrates. Biol. Rev. **3**:21-76.

Yonge, C. M. 1958. Ecology and physiology of reef-building corals. *In* A. A. Buzzati-Traverso [ed.] Perspectives in marine biology. University of California Press, Richmond.

Zalokkar, M. 1942. Les associations sous-marines de la côte adriatique au-dessous de Velebit. Bull. Soc. Bot. Genève **33**:172-195.

Zenkevitch, L. A. 1947, 1951. The fauna and biological productivity of the sea. Vols. I and II. Moscow. 506 and 588 p.

Zenkevitch, L. A. 1961. Certain quantitative characteristics of the pelagic and bottom life of the ocean. *In* M. Sears [ed.] Oceanography. Pub. no. 67. American Association for the Advancement of Science, Washington, D. C.

Zenkevitch, L. A. 1963. Biology of the seas of the U.S.S.R. Transl. by S. Botcharskaya. John Wiley & Sons, Inc., New York.

Zenkevitch, L. A., and J. A. Birstein. 1956. Studies of deep-water fauna and related problems. Deep-Sea Res. **4**:54-64.

Zenkevitch, L. A., V. Brodsky, and M. Idelson. 1928. Materials for the study of the productivity of the sea bottom in the White, Barents, and Kara seas. J. Cons. Perm. Int. Explor. Mer **3**:372-379.

ZoBell, C. E. 1941. Studies on marine bacteria. I. Cultural requirements of the heterotrophic aerobes. J. Mar. Res. **4**:42-75.

ZoBell, C. E. 1946. Marine microbiology. Chronica Botanica Press, Waltham, Mass.

ZoBell, C. E. 1949. Influence of hydrostatic pressure on growth and viability of terrestrial and marine bacteria. J. Bacteriol. **57**:179-189.

ZoBell, C. E. 1957. Marine bacteria. *In* J. W. Hedgpeth [ed.] Treatise on marine ecology and paleoecology. Vol. I. Ecology. Geological Society of America, New York.

ZoBell, C. E., and E. C. Allen. 1935. Significance of marine bacteria in fouling of submerged surfaces. J. Bacteriol. **29**:239-251.

Zweig, G. [ed.]. 1963-1967. Analytical methods for pesticides, plant growth regulators, and food additives. Vol. 1 (1963). Principles, methods, and general applications. Vol. 2 (1964). Insecticides. Vol. 3 (1964). Fungicides, nematocides, soil fumigants, rodenticides, and food additives. Vol. 4 (1964). Herbicides. Vol. 5 (1967). Additional principles, applications, and analyses for pesticides. Academic Press, Inc., New York.

Annotations to the bibliography

The following are groupings of publications representative of certain types of publication or of particular areas of marine biology. The student should realize that the bibliography given in this book is by no means a complete list of relevant publications. It is merely a sampling. Most of the references cited will furnish leads to numerous other important and interesting contributions.

PUBLICATIONS FOR THE GENERAL READER AS WELL AS FOR MARINE BIOLOGISTS

Bates, 1960; Berrill, 1951, 1959; Beebe, 1928, 1935; Carrington, 1957, 1960; Carson, 1951, 1955; Cousteau and Dumas, 1953; Cowen, 1960; Douglas, 1952; Dugan, 1956; Loftas, 1969; Piccard, 1956; Steinbeck and Ricketts, 1941; Walford, 1958.

The following are good examples of highly illustrated, colorful books for young and old natural history lovers: Berrill, 1966; Engel and editors of *Life*, 1961; *Life* volumes: *The World We Live In* and *The Sea;* Amos, 1966.

Scientific American has devoted special issues and numerous articles to oceanic and marine biological studies. Good examples of regional studies are the May, 1961 (arctic), and September, 1962 (antarctic), issues. Ten of these articles were reprinted in book form as: *The Ocean,* a *Scientific American* book, published by the W. H. Freeman Co., San Francisco, in 1969. The *National Geographic Magazine* contains many fine regional studies on a less technical level. Indexes to these periodicals will give leads to many rewarding, beautifully illustrated articles.

SOME BOOKS HELPFUL ALONG THE SEASHORE: PHOTOGRAPHS, DRAWINGS, AND DISCUSSION OF ORGANISMS COMMONLY SEEN AND THEIR NATURAL HISTORY AND ECOLOGY

Carson, 1955 (eastern coast, U. S.); Ricketts and Calvin, 1968 (western coast, U. S.); Light and others, 1954, revised by Smith and others (Keys, West Coast animals); Guberlet, 1936, 1956 (West Coast animals and seaweeds); MacGinitie and MacGinitie, 1949 (natural history, marine animals); Flattely and Walton, 1922 (biology of the seashore); Flora and Fairbanks, 1962 (natural history, Puget Sound region); Wilson, 1951b (life of shore and shallow waters); Smith, 1944 (marine algae, West Coast); Dawson, 1956 (marine algae); Lewis, 1964 (ecology of rocky shores); Iverson, 1968 (farming the edge of the sea).

SOME BOOKS ADDRESSED PRIMARILY TO MARINE BIOLOGISTS AND OCEANOGRAPHERS

Russell and Younge, 1928; Marshall, 1954, 1967; Hardy, 1956; Murray and Hjort, 1912; Stommel, 1958; Friedrich, 1969. The foregoing works are of a general nature and contain much of interest to laymen as well as to professional marine biologists. The following are intended primarily for professional marine biologists and oceanographers. Sverdrup, Johnson, and Fleming, 1946; Hedgpeth [ed.],

1957 (2-volume treatise); Buzzati-Traverso [ed.], 1958; Fairbridge, 1966; Nitikin [ed.], 1959; Barnes, 1959; Raymont, 1963; Wimpenny, 1966; Oppenheimer [ed.], 1963; Davis, 1955; Kriss and others, 1964; Moore, 1958; Nicol, 1959; Green, 1968.

SOME RECURRENT PUBLICATIONS, JOURNALS, ANNUAL REVIEWS, ETC. OF INTEREST TO MARINE BIOLOGISTS

Advances in Marine Biology (F. S. Russell [ed.], (annual; Academic Press, Inc.)

Biological Bulletin of the Marine Biology Laboratory, Woods Hole (monthly)

Bulletin of the Bingham Oceanographic Collection (Yale) (has been discontinued)

Bulletin of Environmental Contamination and Toxicology

Current Contents in Marine Sciences (FAO monthly —a bibliographic aid)

Deep-Sea Research (Pergamon Press). (The 1955 supplement to Vol. 3, Papers in Marine Biology and Oceanography, dedicated to H. B. Bigelow, one of the leaders in development of oceanographic studies in America, contains numerous interesting contributions by various scientists and a complete bibliography of the scientific publications of H. B. Bigelow.)

Ecology (American Ecological Society)

Environmental Research (annual; Academic Press, Inc.)

International Journal of Oceanology and Limnology (Omnipress)

International Marine Science Quarterly (UNESCO)

Internationale Revue der gesampten Hydrobiologie

Journal du Conseil (Copenhagen)

Journal of Experimental Marine Biology and Ecology

Journal of the Marine Biological Association, United Kingdom

Journal of Marine Research (Sears Foundation)

Limnology and Oceanography (American Society of Limnology and Oceanography)

Marine Biology (international journal begun 1967-1968)

Oceanic Citation Journal (Oceanic Research Institute; contains abstracts as well as citations)

Oceanography and Marine Biology (annual review; Allen & Unwin, London)

ECOLOGY

General and marine: Allee and others, 1949; Clements and Shelford, 1927; Hedgpeth [ed.], 1957 (2-volume treatise on marine ecology and paleoecology); Hesse and others, 1951 (ecological animal geography); Moore, 1958; Odum, 1953, 1956.

Analytical methods in ecology: Fager, 1957 (analysis of recurrent groups); Gislin, 1930 (culmination of Peterson school of particulate approach to community analysis in ecology); Lucas, 1938 (integration in plankton communities); Margalef, 1958 (information theory in ecological analysis); McConnaughey, 1964 (analysis of field data to determine communities, revision and extension of Fager's method); Newell, 1948 (analysis of linearly disposed communities or provinces of organisms, revision and extension from methods proposed by Schenck and Keen); Patten, 1959 (cybernetics of ecosystem); Petersen, 1918 (statistical analysis of soft bottom benthic communities); Poore, 1962 (successive approximation in descriptive ecology); Schenck and Keen, 1936 (analysis of linearly disposed biologic provinces); Wieser, 1958 (particulate and comparative approaches compared); Williams, 1965 (and numerous other articles by Williams, and by Williams and Lambert in British Journal of Ecology (multivariate analysis and information theory in ecological analysis).

Population ecology: Boughey, 1968; Hazen [ed.], 1964; MacArthur and Connell, 1966.

Zoogeography and regional studies: Ekman, 1953 (best general summary of zoogeography of world's oceans); Knox, 1963 (good example of regional biographical study—intertidal biogeography of Australian–New Zealand region).

The great multivolume reports of various scientific expeditions are also mines of information. Among the greatest of these, the *'Challenger' Reports*, some fifty volumes, resulting from the first world-wide oceanographic expedition, opened up a new era in marine biology and oceanography and remain as a classic with which all oceanographers and marine biologists should be familiar. A few of the other notable expedition reports, regional in scope, are those from the Siboga expedition (Dutch East Indies), the Great Barrier Reef expedition (northeast Australia), the "Galathea" expeditions, the John Murray expedition, the *'Discovery' Reports*, and the Prince Albert of Monaco expeditions, the John Hancock expeditions, and the "Albatross" expeditions.

INVERTEBRATE ZOOLOGY

The literature in this area is too vast to even begin to summarize here. Some good recent American books are the five-volume treatise of Hyman (1940-1959), and the recent texts of Barnes (1968), Hickman (1967), and Meglitsch (1967). Ricketts and Calvin (as revised by Hedgpeth), 1968, contains a useful annotated bibliography arranged by groups.

PLANKTON

Some books devoted to plankton: Davis, 1955 (keys and illustrations for identification); Fraser, 1962; Hardy, 1956; Johnstone and others, 1924; Ostenfeld, 1916 (international catalog); Raymont, 1963

(probably the best general book on plankton and marine production).

Some shorter works of a general nature relating to plankton: Bernard, 1955; Bogorov, 1934, 1941, 1956, 1958; Buljan, 1959 (influence of forest on coastal plankton); Cushing, 1953, 1954, 1959a; Hardy, 1956, 1960; Harvey, 1942, 1947, 1950; Harvey and others, 1935; Kincaid, 1942; Lebour, 1922, 1923; Lucas, 1956; Nielsen, 1937; Raymont, 1966; Redfield, 1934; Ryther, 1954, 1956; Steele, 1959.

Some examples of special studies of plankton organisms from particular points of view: Bentley, 1960 (influence of plant hormones on); Eyden, 1923 (relation to specific gravity); Gunther, 1934 (fatty constituents); Spooner, 1933 (reactions to light).

Some general plankton studies from various areas: Apstein, 1907, 1910 (Ceylon); Bigelow, 1914, 1917, 1922, 1926a (Atlantic coast off northern U. S.); Bogorov, 1939 (Arctic, in relation to ice forecasting); Brandt and Apstein, 1901-1912 (North Atlantic); Brunelli, 1941 (Laguna Veneta, Italy); Bsharah, 1957 (Florida Current); Burckhardt, 1899 (White Sea); Candeias, 1930 (Portugal); Cleve, 1901 (Indian Ocean and Malay Archipelago); Coleman and Segrove, 1955 (New England waters); Cowles, 1930 (Chesapeake Bay, offshore); Delsman, 1939 (Java Sea); Fish, 1925 (Woods Hole, seasonal); Fraser, 1948, 1949, 1954, 1955, 1956 (British Isles region); Frost, 1937 (Newfoundland); Gauld, 1950 (Scottish sea lakes); George, 1953 (India, coastal); Gilmartin, 1958 (Eniwetok lagoon, Central Pacific); Gotto, 1951 (Ireland); Gran, 1927 (Norway); Gunther, 1936 (Peru Current); Hardy, 1936b (Arctic), Hart, 1953 (Benguela Current); Hentschel and Wattenberg, 1930 (South Atlantic); Holmes, 1958 (eastern Pacific); Japanese Hydrographic Office, 1956 (equatorial Pacific); Jespersen, 1923, 1924, 1935, 1944 (macroplankton, North Atlantic and Mediterranean); Kow, 1953 (Singapore Strait); Jaschnow, 1939, 1940 (Arctic and Barents Sea); Kusmorskaja, 1940 (Sea of Okhotsk); Wood, 1963 (southwest Pacific and east Indian Ocean); Yamazi, 1958 (Japan).

PHYTOPLANKTON

Biology: Bainbridge, 1957 (size, shape, density of phytoplankton clouds); Cushing, 1958 (carbon content); Doty and Ogun, 1957 (daily periodicity in production); Droop, 1957 (nutrition of); Gran, 1912, 1927, 1929, 1931 (quantitative studies and conditions for production); Harvey, 1934, 1947 (measurement of, and relation to manganese); Margalef, 1958 (succession and heterogeneity); Rice, 1954 (biotic influences on growth of); Rustad, 1946 (photosynthesis and light at various depths); Smayda, 1958 (biogeography of); Sieburth and Burkholder, 1959 (antibiotic activity of); Sverdrup,

1953 (conditions for vernal bloom); Vaccaro and Ryther, 1960 (relation to NO_3).

Some regional or area studies of phytoplankton: Allen, W. E., 1919-1924 (Southern California); Bigelow, 1922, and Bigelow and others, 1940 (Gulf of Maine); Bogorov and Beklemishev, 1955 (western North Pacific); Riley, 1942, 1946, 1947 (New England region), 1957 (Sargasso Sea); Riley and Conover, 1956 (Long Island Sound); Corlett, 1953 (oceanic, Atlantic); Gran and Braarud, 1935 (Bay of Fundy, Gulf of Maine); Grøntved, 1952 (North Sea); Hart, 1934 (Bellingshausen Sea and southwestern Atlantic); Hasle, 1959 (equatorial Pacific); Hulbert and others, 1960 (Sargasso Sea); Lillick, 1938, 1940 (Gulf of Maine and environs); Phifer, 1934 (Puget Sound); Ryther, 1954 (Long Island Sound); Sears, 1941 (Georges Bank, Gulf of Maine); Shirshov, 1937, 1938 (polar seas, relation to ice regime); Sleggs, 1927.

PRIMARY PRODUCTION AND PRODUCTIVITY OF MARINE WATERS

General works on production: Raymont, 1963; Ryther, 1963 (geographical variations in productivity), 1954 (ecology of blooms), 1956 (photosynthesis as a function of light intensity), 1959 (potential productivity), 1969 (marine production and world fisheries).

Methods of studying production: Braarud, 1961 (cultivation methods); Brand, 1934 (determination of carbon and nitrogen in small samples); Creitz and Richards, 1955 (estimation by pigment analysis); Doty, 1956, 1959 (C^{14} method); Gran, 1927, 1929, 1931 (oxygen method); Gross, 1937 (culture); Harvey, 1934, 1942, 1950 (measurement of production and review summary of production); Harvey and others, 1935 (production and its control); Krey, 1958, 1961 (chemical methods); Richards and Thompson, 1952 (pigment analysis); Ryther and Yentsch, 1957 (chlorophyll and light data); Steemann-Nielsen, 1952, 1954, 1958, and Steemann-Nielsen and Jensen, 1957 (experimental methods, especially C^{14}).

Some studies of production in various regions and areas: Curl, 1960 (northern coastal South America); Currie, 1953 (Benguela Current); Gaardner and Gran, 1927 (Norway); Gaardner and Sparck, 1931 (Norway); Gillbricht, 1952b (Kieler Bucht); Graham, 1941, 1943 (open Pacific); Gran, 1927-1931 (North Atlantic and off Norway); Hansen, 1959 (North Atlantic); Holmes, 1958 (eastern Pacific); Jaschnow, 1939, 1940 (Arctic Ocean and Barents Sea); Marshall, P. T., 1958 (Arctic); McAllister and others, 1960 (northeastern Pacific); Riley, 1946 (Georges Bank), 1955 (Long Island Sound), 1957 (Sargasso Sea); Riley and Conover, 1956 (Long Island Sound); Riley and others, 1949 (North Atlantic); Rose, 1925 (Bay of Algiers);

Ryther, 1963 (Sargasso Sea); Steele, 1958 (North Sea); Steemann-Nielsen, 1935, 1937, 1940, 1951 (eastern North Atlantic); Wimpenny, 1958 (Smith's Knoll Light Vessel).

ZOOPLANKTON

Identification: Jespersen and Russell, 1939-1952 (identification).

Diurnal vertical migration and vertical distribution—relation to the deep-scattering layer: Boden and Kampa, 1967 (relation to light); Bogorov, 1946 (vertical migration in polar seas); Clarke, G. L., 1936 (relation to light); Cushing, 1951, 1955 (review and experiments); David, 1961 (relation to speciation); Enright and Hammer, 1967 (massive experiments involving several species); Esterly, 1919 (experiments on reactions); Hansen, 1951 (relation to the discontinuity layer); Hardy, 1947, and Hardy and Bainbridge, 1954 (experiments on rates of movement, etc., and use of plankton wheel); Hida and King, 1955 (vertical distribution in equatorial Pacific); Leavitt, 1935, 1938 (movements of macroplankton); Moore and Corwin, 1956 (temperature and light); Russell, 1925-1934 (vertical distribution); Russell, 1935 (behavior in relation to light, value as indicators of water masses in English Channel, general review); Waterman and others, 1939 (deep-water plankton).

Zooplankton as indicators of hydrographic conditions and biogeographic boundaries: Bary, 1959. (See also section on diurnal vertical migration and vertical distribution, above, and section on general studies in particular regions or areas, below.)

Other aspects of zooplankton behavior: Bainbridge, 1949 (movements in a diatom gradient); Bossanyi, 1958 (demersal natant zooplankton); David, 1967 (neuston); Denton and Shaw, 1962 (buoyancy of gelatinous zooplankton).

Some general studies in particular regions or areas: Bernard, 1955b (Bay of Algiers, seasonal); Bernstein, 1934 (Kara Sea); Bigelow, 1938 (Bermuda, medusae); Bigelow and Sears, 1939 (New England, volumes on zooplankton); Bogorov and Vinogradov, 1955 (northwestern Pacific); Clarke, G. L., 1940 (comparative richness, coastal and offshore Atlantic); Deevey, 1948, 1952, 1956 (Block Island Sound, Long Island Sound); Delsman, 1939 (Java Sea); Fraser, 1947, 1952, 1955 (off the British Isles); Jespersen, 1923, 1924, 1935, 1940, 1944 (macroplankton of North Atlantic, Mediterranean, quantity in various oceans, etc.); Johnson, 1954, 1958 (Marshall Islands, Pacific, and Point Barrow, Alaska); Mackintosh, 1934, 1937 (Antarctic, relation of circulation and depth changes to life cycle and maintenance of populations in the region); Menzel and Ryther, 1961 (Sargasso Sea); Moore, H. B., 1949 (Bermuda); Riley, 1947 (Georges Bank, Gulf of Maine); Runnstrom, 1932 (Norway, fjords); Rus-

sell, 1934 (Great Barrier Reef, Australia); Savage, 1931 (off England); Shropshire, 1944 (plankton harvesting).

Relations between zooplankton and phytoplankton: Bainbridge, 1953; Beklemishev, 1957; Clarke, G. L., 1939; Nielsen, 1937; Steemann-Nielsen, 1937.

Continuous plankton recorder studies: Glover, 1967; Hardy, 1926, 1936a, 1939, 1958; Henderson, 1954; Rae and Rees, 1947; Rees, 1954.

Nannoplankton: Aurich, 1949; Lohmann, 1911; Yentsch and Ryther, 1959.

SHORELINE BIOLOGY

Estuaries, tidal flats, river mouths, etc.: Two general treatments of estuarine biology: Green, 1968 (biology of estuarine animals); Lauff [ed.], 1967 (symposium on estuaries).

Other references: Alexander and others, 1932 (salinity foreshore waters); Amos, 1952, 1954, 1955, 1956 (Delaware estuary); Barlow, 1955 (processes determining distribution of zooplankton in estuaries); Beanland, 1940 (Dover estuary); Cooper and Milne, 1938-1939 (ecology of the Tamar estuary); Day, 1951, and Day and others, 1952, 1953 (ecology of South African estuaries); Fischer-Piette, 1931 (penetration of sessile marine animals into estuaries); Goodwin, 1961 (vanishing Connecticut estuaries); Gunter and Gunter, 1956, 1958 (Louisiana estuaries); Hedgpeth, 1951 (classification of brackish and estuarine environments); Holme, 1949a, 1950 (Exe estuary and Great West Bay, Great Britain); Hull, 1960 (oxygen balance in estuaries); Johns, 1968 (vulnerability of estuaries); June and Chamberlin, 1959 (estuaries and menhaden); Kalber, 1959a, 1960 (estuaries and tide marshes); Kelley, 1966 (ecology of San Joaquin and Sacramento estuaries); Milne, 1940 (ecology of Tamar estuary); Odum, E. P., 1961 (tide marshes and estuarine production); Pratt, 1949, 1950 (fertilization of saltwater pond); Raymont, 1955 (ecology of tidal mud flat); Rees, 1940 (ecology of tidal mud flat); Reid, 1961 (inland waters and estuaries); Segerstråle, 1953, 1957, 1959 (Baltic region, classification of brackish waters); Seshaiya, 1959 (estuarine hydrology); Shelford, 1930 (Pacific Northwest); Shelford and Towler, 1925 (animal communities); Shelford and others, 1935 (marine communities); Spooner and Moore, 1940 (ecology of Tamar estuary); Teal, 1962 (energy flow in a salt marsh).

Penetration of marine animals into estuaries and fresh waters: Annandale, 1922 (into the Ganges River); Fischer-Piette, 1931; Needham, 1930; Odum, 1953; Pearse, 1936, 1950 (immigrations from sea to land); Schmalhausen, 1968 (origin land vertebrates).

Open coast and continental shelf: For general treatments, especially of the intertidal, see Ricketts and Calvin, 1968; Stephenson and Stephenson, 1949 (universal features of intertidal life), and 1950, 1952, 1954, 1961 (accounts of the intertidal in representative regions of North America); Wilson, 1951*b* (life of shore and shallow sea).

Other references: Blegvad, 1914, 1916, 1925, 1929 (Danish waters and mortality due to ice); Brotskaja and Zenkevich, 1939 (Barents Sea); Bruce, 1928 (physical factors, sandy beach); Buchanan, 1968 (shelf benthos off African Gold Coast); Coe and Allen, 1937 (growth on test blocks, Southern California); Cole, 1956 (commercial shellfish); Coleman, 1933 (intertidal zonation); Drach, 1948 (rocky shore communities); Ellis, 1960 (infaunal benthos, North American Arctic); Ford, 1923 (off Plymouth); Fowell, 1939 (*Laminaria* communities); Fox, 1950 (organic detritus and inshore animals); Gilsen, 1930 (Gullmar Fjord); Grøntved, 1957, 1958 (seaweeds); Jones, 1950, 1956 (bottom communities); Kitching and others, 1934 (sublittoral ecology); Knox, 1963 (intertidal, Australia—New Zealand); Lawson, 1954-1957 (zonation, rocky shores, West Africa); Longhurst, 1959 (benthos densities of tropical West Africa); Loosanoff and Davis, 1951 (ecology, growth, and settlement of larval bivalves); MacGinitie, 1935 (littoral communities); Mare, 1942 (microorganisms of benthic community); Moore, 1930, 1931 (mud of Clyde Sea); Neushul and others, 1967 (subtidal, California islands); Newcombe, 1935 (sand beach ecology); Parker, 1955, 1956 (ecology of Texas coast and Mississippi River delta); Pearse and others, 1942 (ecology of eastern coast sand beach); Petersen, 1913, 1915, 1918, 1924 (Danish benthic communities); Petersen and Boysen-Jensen, 1911 (evaluation of benthic life); Remane, 1959 (interstitial life in sand); Sanders, 1956, 1958, 1960 (benthos, eastern coast, U. S.); Savilov, 1957 (Sea of Okhotsk); Stephen, 1930 (sandy mud ecology); Stephenson, 1947 (intertidal, South Africa); Thorson, 1936, 1946, 1950, 1952, 1955, 1958 (ecology of benthos, especially with respect to reproduction and larval life and settlement, and comparison of equivalent benthic communities of different latitudes and depths); Towler, 1930 (intertidal, Puget Sound); Wieser, 1952 (microfauna of seaweeds); Wilson, 1951*a*, 1958 (larval ecology and settlement); Wilson and Armstrong, 1958 (biological differences between different seawater masses); Yonge, 1928 (feeding mechanisms).

Boring and fouling organisms: The best general summary of work with these groups is contained in Ray [ed.], 1958. A few other works dealing with these problems or the organisms involved are Coe and Allen, 1937 (growth on test blocks); Doochin and Smith, 1951 (fouling in relation to currents); Edmondson, 1944, and Edmondson and Ingram 1939 (fouling organisms in Hawaii); Graham and Gay, 1945 (at Oakland, Calif.); Johnson and Miller, 1935 (shipworms at Friday Harbor); Kofoid and

Miller, 1927 (shipworms in San Francisco Bay); Turner, 1966 (survey of shipworms); Visscher, 1927 (fouling of ship bottoms); ZoBell and Allen, 1935 (role of bacteria in initiating fouling).

CORALS AND MANGROVE

Corals: One of the most extensive studies of coral reefs was made by an expedition sponsored by the British Museum of Natural History. Their reports are embodied in a series of volumes: *Scientific Reports of the Great Barrier Reef Expedition, 1928-1929, British Museum of Natural History.* One of the earliest and greatest scientific studies of coral reefs is that of Darwin, 1897. This was recently reissued as a paperback by the University of California. A more recent book is that of Gilbert and McNeill, 1959.

Some other references on corals: Bakus, 1964, 1966, 1969 (effects of fish-grazing on organisms of coral reefs); Daly, 1910 (Pleistocene glaciation and coral reefs); Kohn and Helfrich, 1957 (productivity of coral reef); Ladd and Tracy, 1949 (general discussion of coral reefs); Mayor, 1924 (structure and ecology of Samoan reefs); Mesolella, 1967 (Pleistocene reefs on Barbados); Roughly, 1936 (Great Barrier Reef); Sargent and Austin, 1954 (biological production and economy of coral reef); Yonge, 1958 (ecology and physiology of reef corals).

Mangrove: Davis, 1940 (ecology and geological role of mangroves in Florida); Orr and Moorhouse, 1933 (physical and chemical conditions in mangrove swamps).

DEEP-SEA BIOLOGY

Bruun, 1957 (general discussion); Bruun and others, 1956 ("Galathea" expedition); Sanders and Hessler, 1969 (ecology of deep-sea benthos); Vinogradova, 1959 (abyssal benthos); Waksman, 1933 (chemical and organic matter of sediments, marine humus); Zenkevitch and Birstein, 1956 (deep-water fauna).

OTHER REFERENCES CONCERNING MARINE ORGANISMS

Mass mortalities and red tides: Brongersma-Sanders, 1957 (review article on mass mortalities, all causes); Allen, W. E., 1946 (at La Jolla, Calif., 1945); Blegvad, 1929 (intertidal, due to ice); Collier, 1958 (biochemical aspects of red tides); Galtsoff (red tides); Gunter and Gunter, 1947, and Gunter and others, 1948 (red tide, mass mortality, Florida); Ingle and de Sylva, 1955 (red tide); Ketchum and Keen, 1948 (phosphorus concentrations in Florida red tide); Lund, 1936 (red tides along the Texas coast); McLaughlan, 1958 (poison tides); Powers, 1932 (red tide, Gulf of Maine, caused by a ciliate).

Underwater sounds of biological origin: Burkenroad, 1931; Dobrin, 1947; Knudsen and others, 1948; Johnson, 1943, 1948; Johnson, Everest, and Young, 1947; Loye and Proudfoot, 1946; Tavolga [ed.], 1964.

Biological clocks: Cold Spring Harbor Symposium, 1960; Brahmachary, 1967; Brown, 1958 (Brown and co-workers have published numerous other articles on this subject, not listed in this bibliography); Pittendrigh, 1958.

Behavior of ecological significance: Allee, 1931 (animal aggregations); Baer, 1951 (ecology of animal parasites); Dales, 1966 (symbiosis in marine organisms); Feder, 1966 (cleaning symbiosis); Fraenkel and Gunn, 1940, (orientation); Thorson, 1946 (reproductive and larval behavior). A new journal, begun in 1968, is entitled *Communications in Behavioral Biology* (S. A. Weinstein [ed.] Academic Press, Inc., New York).

MARINE MICROBIOLOGY

Burkholder, 1959, and Burkholder and Burkholder, 1956 (vitamins and bacteria); Carey, 1938 (nitrifying bacteria); Hock, 1941 (chitin-digesting bacteria); Johnson, 1968 (saprobic marine fungi); Johnson and Sparrow, 1962 (fungi in oceans and estuaries); Ketchum and others, 1949, 1952 (coliform bacteria in seawater); Krogh and Sparck, 1936 (bottom-sampler for microbiology); Kriss, 1959, Kriss and Rukina, 1952, and Kriss and others, 1961, 1964 (1967) (oceanic microbiology); Kuznetsov, 1955 (use of C^{14} to study microbial chemosynthesis; Kriss says his method is invalid); Lipman, 1926 (effect of dilution of seawater on marine bacteria); Meyers and Reynolds, 1957 (relation of marine fungi to wood borer attack); Oppenheimer, 1955, 1958, 1963 (marine microbiology); Reuszer, 1933 (bacteria in the cycle of the sea); Skopintsev, 1950 (marine humus); Sorokin, 1955, 1957, 1958 (C^{14} measurements of chemosynthesis; subject to same criticism as Kuznetsov, above); Spencer, 1956 (bacterial oxidation of ammonia in the sea); Starr and others, 1957 (production of vitamin B_{12} by marine bacteria); Waksman, 1933, Waksman and Hotchkiss, 1937 (role of bacteria in the cycle of the sea, and viability of bacteria in seawater); Wood and Ferguson, 1965 (marine microbial ecology); ZoBell, 1941, 1946, 1949, 1957 (marine bacteria).

PHYSICAL AND CHEMICAL OCEANOGRAPHY, AND METEOROLOGY IMPORTANT FOR MARINE BIOLOGY

Some books of importance: Sverdrup and others, 1946 *(The Oceans);* Sears [ed.], 1961 *(Oceanography);* Fairbridge, 1966 *(Encyclopedia of Oceanography);* Myers and others [ed.], 1969 *(Handbook of Ocean and Underwater Engineering);* Hill [ed.], 1962-1963 *(The Sea,* 3 vols.); Hewson and others, 1944 *(Meteorology);* Proudman, 1953 *(Dynamic Oceanography);* Holmboe and others, 1945 *(Dynamic Meteorology).*

The annual review volumes, *Oceanography and Marine Biology,* edited by Barnes, also contain many valuable contributions. A good presentation of physical and biological oceanography is also to be found in *Oceanography,* 1966, Publication no. 1492 of the National Academy of Sciences–National Research Council, issued in 1967, and in a projection of governmental goals and possible programs in this field in *Marine Science Affairs—a Year of Transition,* issued by the Government Printing Office the same year, and prepared by the National Council on Marine Resources and Engineering Development.

Some individual research reports: Atkins, 1926 (phosphate and phytoplankton), 1953 (copper content of seawater); Atkins and Poole, 1940, 1958 (angular distribution of light in the sea); Braarud and Klem, 1931 (hydrographical and chemical coastal Scandinavian waters); Callaway, 1957 (oceanographical and meteorological observations, Pacific); Clarke, G. L., 1936, 1939, 1941 (penetration and absorption of light in seawater); Currie, 1953 (upwelling, Benguela Current); Dakin and Colefax, 1935 (oceanographical observations off Australia); Deacon, 1933, 1937 (hydrology of South Atlantic and Antarctic oceans); Emery, 1956 (internal waves); Evenau and others, 1965 (photoenvironment); Grainger, 1959 (annual oceanographical cycle, Canadian Arctic); Grier, 1941 (oceanography, North Pacific and Bering Sea); Gunter, 1957 (temperature); Gunther, 1936 (Peru Current); Iselin, 1939 (physical factors and productivity); Ivanoff, 1953, 1957, 1958, 1959, 1961 (light); Jacobs and Ewing, 1969 (suspended particulate matter in major oceans); Jerlov, 1944, 1951, 1953, 1955, 1958, 1959 (light); Kalle, 1939 (color of the sea); Kampa and Boden, 1956 (light generation in ionic scattering layer); Ketchum and Shonting, 1958 (optical studies of particulate matter in sea); Knudsen, 1901 (hydrographic tables); McEwen, 1916 (hydrographical observations at Scripps Institution); Miller and others, 1928 (physicochemical conditions in San Francisco Bay in relation to tides); Oster and Clarke, 1935 (light); Polikarpov, 1964 (important book—radioecology of aquatic organisms); Riley and Conover, 1956 (Long Island Sound); Russell and McMillan, 1952 (waves and tides); Seiwell, 1942 (vertical oscillations); Thompson and Gilson, 1937 (chemicophysical observations, Murray expedition); Waterman, 1954, 1960 (polarized light); Wust, 1954 (zonal divisions of upper ocean waters with respect to various factors).

Some studies of currents and ocean circulation: Boden and Kampa, 1953 (cascading); Cooper and Vaux, 1949, Cooper, 1961 (cascading); Cromwell, 1953 (circulation in meridional plane in equatorial

Pacific); Hart, 1953, Hart and Currie, 1960 (Benguela Current); Sears, 1954 (Peruvian Current); Stommel, 1958 (Gulf Stream); Swallow, 1957, Swallow and Worthington, 1961 (deep currents); Worthington, 1954 (time scale in Atlantic circulation); Wust, 1930 (deep circulation and its divisions); Wyrtki, 1961 (thermohaline circulation).

Temperature and salinity: Boyko [ed.], 1965 (book, salinity and aridity); Fox, H. M., 1939 (poikilotherms in different latitudes); Metcalf, 1930 (salinity and size); Pearse and Gunter, 1957 (salinity, general review).

Theory of ice ages: Ewing and Donn, 1956; Donn and Ewing, 1966.

CHEMISTRY OF SEAWATER AND OF MARINE ORGANISMS

Chemistry of seawater: Armstrong, 1954 (phosphorus and silicon), 1957 (iron), 1968 (inorganic suspended matter); Atkins, 1922 (influence of pH on algal cells), 1953 (seasonal variation in copper); Bernard, 1939, 1958 (fertility of Mediterranean water); Brandhorst, 1958 (nitrification and dentrification, tropical Pacific); Brouardel and Fage, 1953 (oxygen next to the sediment); Bruce and Hood, 1969 (phosphate in Texas bays); Chow and Thompson, 1952, 1954 (copper); Clowes, 1938 (phosphate and silicate, Southern Ocean), 1950 (hydrology, South African waters); Cooper, 1933, 1934, 1935, 1938, 1948, 1951, 1952, 1956, 1961 (chemistry and circulation of seawaters and relation to fisheries, etc.); Goldberg, 1957 (trace metals); Gourley, 1952 (phosphorus); Harris, E., 1959 (nitrogen cycle); Harvey, 1926, 1928, 1934, 1942, 1945, 1947, 1949, 1957 (important series of articles and a short book on physics and chemistry of seawater); Hoffman, 1956, 1960 (remineralization of phosphorus); Kalle, 1953, 1957 (chemistry of shallow seas); Ketchum and others, 1955, 1958 (phosphorus); Kreps and Verjbinskaya, 1930 (nutrients, Barents Sea); Martin, 1968 (marine chemistry); McClendon, 1917 (carbon dioxide and oxygen); Miyake and Saruhasi, 1956 (vertical distribution of oxygen); Mullin and Riley, 1955 (nitrate); Murphy and Riley, 1958 (phosphate); Pomeroy, 1960, (dissolved phosphate—time in water); Pomeroy and Bush, 1959 (phosphate regeneration); Rakestraw, 1936 (nitrate); Rakestraw and Carritt, 1948 (chemistry of ocean, seasonal changes); Redfield, 1958 (biological control of chemical environment); Redfield and Keys, 1938 (ammonia); Redfield and others, 1937 (organic phosphorus); Richards, 1957 (2 articles, general, and oxygen); Richards and Benson, 1961, and Richards and Vaccaro, 1956 (anaerobic environments); Rider and Mellon, 1946 (nitrate); Riley and Conover, 1956 (chemistry, Long Island Sound); Robertson and Wirth, 1934b (chemistry, Puget Sound); Seiwell, 1935 (2 articles, phos-

phorus in tropical western Atlantic and iron in Atlantic coastal waters); Spencer, 1956 (oxidation of ammonia); Steiner, 1938 (phosphorus cycle); Stephenson, W., 1949 (phosphate, release from mud); Thompson and Bremer, 1935 (iron); Verjbinskaya, 1932 (nitrate).

Organic matter and biologically active compounds in the sea: Collier, 1953, 1958 (organic compounds in the water, their effects, and biochemical aspects of red tides); Cowey, 1956 (vitamin B_{12} in coastal waters); Daisley and Fisher, 1958 (vertical distribution, vitamin B_{12}); Fox, 1957, and Fox and others, 1952, 1953 (marine leptopel and colloidal micelles); Harvey (several articles—see section on chemistry of seawater, above); Johnston, 1955, 1959 (biologically active compounds and antimetabolites); Jones, G. E., 1959 (biologically active compounds); Ketchum, 1947, 1957 (biochemical relations between organisms and ecosystem, and transport of elements in sea); Keys and others, 1935 (organic metabolism of seawater, ultimate food cycle); Krizenecky, 1925 (dissolved nutrients as possible source of animal food); Krogh, 1931 (same); Putter, 1909 (same); Sillen, 1967 (ocean as a chemical system); Brand and Rakestraw, 1937-1947 (organic nitrogenous matter in the sea); Vallentyne, 1957 (molecular nature of organic matter in water).

Biochemistry of marine organisms: Belser, 1959 (bioassay of marine organisms); Harris, 1957 (radiophosphorus in metabolism zooplankton and microorganisms); Herring, 1967 (pigments, surface plankton); Ketchum and others, 1955, 1958 (productivity in relation to nutrients); Kon, 1958 (biochemical perspectives); Lewin, 1954 (*Stichococcus* and vitamin B_{12}); Lucas, 1947, 1949, 1955, 1961 (external metabolites); Phillips, 1917, 1922 (metals in marine organisms); Raymont and Conover, 1961 (carbohydrates in marine organisms); Vinogradov, 1938, 1953 (chemical composition, marine organisms); Wangersky, 1952, 1959 (ascorbic acid, rhamnosides, and dissolved carbohydrates); Wilson, 1951, Wilson and Armstrong, 1958 (biological differences between seawaters).

SUBMARINE GEOLOGY

Books: Guilecher, 1958 (coastal and submarine morphology); Hill, 1963 (Vol. III. The sea); Kuenen, 1950 (marine geology); Menard, 1964 (marine geology. Pacific); Pettersson, 1954 (ocean floor); Shepard, 1948, 1959 (submarine geology).

Shorter works: Andel, 1968 (deep-sea drilling); Bascom, 1961 (deep-sea drilling); Boden and Kampa, 1953 (geological effects, cascading and water movements); Kuenen, 1941 (calculations, total mass of sediments in the earth); Menard, 1953, 1955, 1961, 1967 (sea floor spreading, fractures, etc.); Rona and Miliani, 1969 (dating Caribbean cores); Shepard and Emery, 1946 (submarine photography,

California coast); Larson and Spiess, 1969 (East Pacific Rise); Veevers, 1951, 1952 (photography, sea floor).

ORIGIN AND EARLY EVOLUTION OF LIFE

Bernal, 1967; Calvin, 1961 (chemical evolution); Miller, 1953 (production, organic chemicals in artificial primitive anaerobic atmosphere); Miller and Urey, 1958 (origin of organic compounds in the primitive earth); Oparin, 1957, 1961 (origin of life); Rutten, 1962 (geological aspects of origin of life); Weyl, 1968 (role of thermocline in Precambrian evolution).

METHODS

Books: Barnes, 1959a (apparatus and methods, oceanography, 2 vols.); Costello and others, (handling marine eggs and embryos); Frey and Frey, 1961 (handbook for hydronauts); Meyers and others, 1969 (handbook, ocean and underwater engineering); Schlieper, 1968 (methods of marine biological research); Watt [ed.], 1966, 1967 (systems analysis, electronic and computer methods); Zweig [ed.], 1963-1967 (analytical methods for pesticides, 5 vols.).
See also recent annual instrumentation issues of *Science.*
Other references: Ahlstrom, 1952 (instrumentation); Allen and Nelson, 1910 (culture of plankton organisms); Belser, 1959 (bioassay, organic substances); Braarud, 1961 (use of culture methods for analysis of environmental influences); Brand, 1934, Brand and others, 1937-1947 (chemical methods and bioassay methods); Creitz and Richards, 1955 (estimation of plankton by pigment analysis and millipore filtration); Forester, 1953 (dredge for burrowing animals); Goldberg and others, 1952 (microfiltration); Goldman, 1962 (method of studying nutrient-limiting factors in the water column); Grøntved, 1957 (macrovegetation sampler); Hardenberg (plankton sampling different depths); Hardy, 1936a (plankton indicator); Holme, 1949b (bottom-sampler); Knudsen, 1927 (bottom-sampler for hard bottoms); Kow, 1953a (making plankton calendar); Morgan, 1956 (diving safely); Mullin and Riley, 1955 (nitrate, spectrophotometric determination); Murphy and Riley, 1958 (phosphate determination); Petersen, 1918 (methods and analysis for benthic communities on soft substrate); Richards and Thompson, 1952 (spectrophotometric method, plankton analysis by pigments); Rider and Mellon, 1946 (nitrate); Riedl, 1953 (methods for study of meiobenthos).
Methods using radioactive isotopes: Calvin and others, 1949 (C^{14} methods); Devereux, 1967 (oxygen isotope ratios for temperature determination at time of deposition in calcareous structures); Doty,

1956, 1959 (C^{14} method); Epstein and others, 1951, 1953 (oxygen isotope ratio method); Jitts, 1957, and Jitts and Scott, 1961 (C^{14} method); Steemann-Nielsen, 1952, 1958 (C^{14} method); Yankwich and others, 1947 (correction for C^{14} method).

MARINE BIOLOGICAL RESOURCES AND THEIR USE

Books: Huberty and Flock, 1959 (natural resources —good chapter on marine resources); Graham, 1956 (sea fisheries); Loftas, 1969 (exploitation of marine resources); Walford, 1958 (living resources from the sea).
Some shorter references: Carter, 1968 (scramble for oil leases); Emery and Iselin, 1967 (food from oceans and land); Gerard and Lamar, 1967 (fresh water from sea wind); Glover, 1955 (herring); Hardy, 1936b, 1941 herring and plankton, plankton for food); Juday, 1943 (utilization of fisheries); Pequegnat, 1958 (whales and krill); Raymont and Miller, 1962 (production in enclosed fertilized marine water body); Shropshire, 1944 (plankton harvesting).

IMPACT AND PLIGHT OF MAN

Books: Carson, R., 1962 (*Silent Spring,* effects of pesticides and other environmental pollution); Castro, 1952 (*Geography of Hunger);* Commoner, 1966 (*Science and Survival);* Ehrlich, 1968 (*The Population Bomb);* Hynes, 1960 (*Biology of Polluted Waters);* Laevastu, 1959 (*A Review of Marine Pollution);* Olson and Burgess [ed.], 1967 (*Pollution and Marine Ecology);* Paddock and Paddock, 1967 (coming famine and our choices); Polikarpov, 1964 (1966) (radioactive pollution); Smith, J. E. [ed.], 1968 ("Torrey Canyon" disaster). In addition to the above books, the report of the National Academy of Sciences–National Research Council, 1957 (effects of atomic radiation on oceanography and fisheries), should be read.
Shorter articles: Carter, 1968 (oil leases on continental shelf), 1970 (estuary in crisis); Chesher, 1969 (destruction of Pacific corals by sea star); Chichester, 1965 (pesticides); Davis, 1948 (industrial copper); Eriksen and Lawrence, 1940 (pollution in Gray's Harbor, Wash.); Goodwin, 1961 (disappearing estuaries); Ikard, 1967 (oil pollution); Inglis and Kestner, 1958 (alteration of estuaries); Johns, 1968 (vulnerability of estuaries); Luck, 1957 (man against his environment); Miles and Pearce, 1967 (pesticides); Newcombe, 1957 (biological hazard of strontium-90); Rafter and Fergusson, 1957 (increase of atmospheric C^{14} because of testing, etc.); Risebrough and others, 1968 (pesticides); Rubinoff, 1968 (sea level canal); Soto and Deichmann, 1967 (pesticides); Thompson, S. H., 1961 (estuaries); Woodwell, Wurster, and Isaacson, 1967 (insecticides).

Index

---

*Boldface numbers refer to pages with illustrations.

415